TEACHING AND LEARNING ABOUT SCIENCE

Teaching and Learning about Science

Language, Theories, Methods, History, Traditions and Values

By

Derek Hodson
OISE, University of Toronto, Canada

SENSE PUBLISHERS
ROTTERDAM / BOSTON / TAIPEI

A C.I.P. record for this book is available from the Library of Congress.

ISBN 978-94-6091-051-7 (paperback)
ISBN 978-94-6091-052-4 (hardback)
ISBN 978-94-6091-053-1 (e-book)

Published by: Sense Publishers,
P.O. Box 21858, 3001 AW Rotterdam, The Netherlands
http://www.sensepublishers.com

Printed on acid-free paper

To Susie – for making it all worthwhile

CONTENTS

PREFACE

In a previous book, *Towards Scientific Literacy: A Teachers' Guide to the History, Philosophy and Sociology of Science* (Hodson, 2008), I presented a critical reading of the vast and complex literature encompassing the history of science, philosophy of science and sociology of science (HPS). The prime purpose of that book was to identify some key ideas in HPS for inclusion in the school science curriculum, in line with the prominence given to HPS in recent international debate in science education and the numerous influential reports on science education that identify the centrality of HPS to scientific literacy. This book discusses how that particular selection of HPS ideas can be assembled into a coherent curriculum and presented to students in ways that are meaningful, motivating and successful.

I am categorically not arguing for a formal course in history, philosophy and sociology of science. In its undiluted form, HPS is too demanding for most school age students. More importantly, it is too dry and dusty to be of sufficient interest to more than a handful of students. Instead, I take what might be described as an anthropological approach in which scientists are studied as a group of people with considerable social, economic and political importance in our society, a group that has its own distinctive language, body of knowledge, investigative methods, history, traditions, norms and values. This approach is chosen both for the theoretical overview it provides and for its motivational value, especially for students from sociocultural groups currently under-served by science education and under-represented in science.

Over the past two decades, findings from science education research, international debate on the guiding purposes of science education and the nature of scientific and technological literacy, official and semi-official reports on science education (including recommendations from prestigious organizations such as AAAS and UNESCO) and concerns expressed by scientists, environmentalists and engineers about current science education provision and the continuing low levels of scientific attainment among the general population, have led to some radical re-thinking of the nature of the science curriculum. There has been a marked shift of rhetorical emphasis in the direction of: (i) considerations of the nature of science, (ii) model-based reasoning, (iii) inquiry-based learning, (iv) scientific argumentation, and (v) use of language- rich learning experiences (reading, writing, talking) to enhance concept acquisition and development. Two years ago, a major report on future directions for science education published by the US-based

National Research Council (Duschl, et al., 2007) stated: "students who understand science: (a) know, use and interpret scientific explanations of the natural world; (b) generate and evaluate scientific evidence and explanations; (c) understand the nature and development of scientific knowledge; (d) participate productively in scientific practices and discourse" (p. 334). These findings, arguments and pronouncements seem to point very clearly in the direction of regarding science education as a study of *scientific practice*, rather than the traditional emphases on acquisition of content and understanding of "the scientific method" (in its narrow sense). This book is an attempt to present a comprehensive, research-based account of how such a vision could be translated into curriculum practice. It is written as a contribution to critical consideration of the curriculum image of science, scientists and scientific practice, and the values that underpin it, seeking change wherever deemed necessary, and thereby exerting pressure for change on science itself.

Chapter 1 identifies the key role of HPS in achieving critical scientific literacy. Chapter 2 surveys the research literature concerning students' understanding of HPS issues. Chapter 3 surveys the parallel research literature focused on teachers' understanding of HPS issues and the ways in which that understanding influences their curriculum decision-making and, in turn, impacts students' HPS understanding. Chapter 4 introduces the anthropological approach to teaching and leaning *about science* – in particular, King and Brownell's (1966) argument that the academic disciplines (including science) can be considered in terms of eight primary characteristics: a *community*, an *expression of human imagination*, a *domain*, a *tradition*, a *syntactical structure* (a distinctive mode of inquiry and cluster of methods for generating and validating new knowledge), a *substantive structure* (a complex framework of concepts, propositions, laws, models and theories), a *specialized language*, and a *valuative and affective stance*. Each of these characteristics is considered in Chapters 4 to 10, though discussion of some characteristics extends over more than one chapter. For example, history of science (science as a tradition) is prominent in Chapters 4, 5, 6, 7, 8 and 10. In addition to introducing the anthropological approach, Chapter 4 addresses the notion of science as a community of practice, addresses some important demarcation issues, and considers error, bias and misconduct in science. Chapter 5 presents further thoughts on demarcation in relation to pseudosciences, traditional knowledge, technology, and what Layton et al. (1993) refer to as "practical knowledge for action". Chapter 6 focuses on the substantive structure of science; Chapter 7 concerns the syntactical structure of science. Chapter 8 addresses some key issues relating to the distinctive language of science; Chapter 9 focuses on the use of language-rich curriculum experiences (talking, reading, writing and listening) in assisting students to acquire and develop conceptual understanding and enhance their understanding of HPS issues. Chapter 10 considers science as a historical tradition, the distinctive values of science, and the driving forces of contemporary scientific practice. The chapter ends by posing a question about whether these underlying values can and should be changed.

The book can be read in a number of ways. Some readers might prefer to read Chapters 2 and 3 *after* Chapters 4 to 10, on the grounds that interpretation of research findings relating to students' and teachers' HPS understanding is easier in the light of prior discussion of the eight disciplinary characteristics. My own view is that Chapters 2 and 3 provide important pointers to the kind of approach that teachers can and should be using in teaching about science, scientists and scientific practice, and so they are better located early in the book. Chapters 6, 7, 8 and 10 can be read in any order. The nature of scientific knowledge, the investigative methods of science and the language of science are closely inter-twined and discussion of any one aspect is considerably enhanced by discussion of the others. It is also enhanced by an appreciation of the history, traditions, norms and underlying values of science. Numerous cross-references are included to assist readers make these connections. Because the theoretical concerns and pedagogical strategies discussed in Chapter 9 have direct relevance to matters addressed in Chapters 6, 7, 8 and 10, some readers might prefer to read this chapter immediately after Chapter 6 or Chapter 7, or even leave it until they have read Chapter 10.

The book represents my thoughts about science and science education following a career extending over more than forty years. My thinking has been informed by experiences as a researcher in synthetic organic chemistry, a science teacher in four schools, and a university-based science educator in five universities, extending over four countries. It has been enriched by countless discussions with students, colleagues and friends. My intention has been to provide a resource to stimulate further debate about the many ways in which considerations in HPS impact science education. More provocatively, how such debate might change the science education we provide and the science in which we engage.

Derek Hodson
Auckland
May, 2009

ACKNOWLEDGEMENTS

This book could not have been written without the unwavering love, inspiration, encouragement and support of my wife, Sue Hodson. To you, Susie, I extend my heartfelt thanks. I also thank the numerous colleagues and students at the Ontario Institute for Studies in Education, University of Hong Kong and University of Auckland for enriching my thinking and stimulating my imagination through many discussions of science education issues extending over many years.

I also extend my thanks to Professor Richard Duschl, the Open University, Taylor and Francis and John Wiley & Sons Inc. for permission to reproduce figures and tables originally published elsewhere.

- Figure 4.1 is reproduced with permission of the Open University (Milton Keynes, United Kingdom). Source: Hodson, D. (1993) Teaching and learning about science: Considerations in the philosophy and sociology of science. In D. Edwards, E. Scanlon & D. West (Eds.), *Teaching, Learning and Assessment in Science Education* (pp. 5–32). London: Paul Chapman/Open University (Figure 1, p. 7). An earlier version of the diagram appeared in: Hodson, D. (1990). Making the implicit explicit: A curriculum planning model for enhancing children's understanding of science. In D.E. Herget (Ed.), *More History and Philosophy of Science in Science Teaching* (pp. 292–310). Tallahassee, FL: Florida State University Press.
- Figure 6.1 is reproduced with permission of John Wiley & Sons Inc. Source: Driver, R., Newton, P. & Osborne, J. (2000). Establishing the norms of scientific argumentation in classrooms. *Science Education, 84*(3), 287–312 (Figure 1, p. 296) Copyright 2000. These authors had adapted a figure previously published in: R.N. Giere (1997). *Understanding Scientific Reasoning.* 3rd edition. Fort Worth, TX: Holt, Rinehart & Winston.
- Figure 6.2 is reproduced with permission from Dr Richard Duschl, Waterbury Chair Professor, College of Education, Penn State University. Source: Duschl, R.A. (1990). *Restructuring Science Education: The Importance of Theories and their Development.* New York: Teachers College Press (p. 87). The author had adapted a figure previously published in: L. Laudan (1984). *Science and Values: The Aims of Science and Their Role in Scientific Debate.* Berkeley, CA: University of California Press (p. 63).
- Figure 8.1 is reproduced with permission from Taylor and Francis (http://www.informaworld.com) Source: Simon, S., Erduran, S. & Osborne, J. (2006). Learning to teach argumentation: Research and development in the science classroom. *International Journal of Science Education, 28*(2–3), 235–260 (Figure 1, p. 240).

CHAPTER 1
SCIENTIFIC LITERACY AND THE KEY ROLE OF HPS

Some 20+ years ago, Ezra Shahn (1988) declared that scientific illiteracy is a serious and persistent problem.

> At one level it affects nations; because large parts of their populations are not adequately prepared, they cannot train enough technically proficient people to satisfy their economic and defense needs. More basically it affects people; those who are science illiterate are often deprived of the ability to understand the increasingly technological world, to make informed decisions regarding their health and their environment, to choose careers in remunerative technological fields and, in many ways, to think clearly. (p. 42)

This passage raises important questions about the kind of scientific and technological knowledge we should include in the curriculum and about the levels of attainment in acquisition of that knowledge we should be seeking. It highlights the key distinction between an education that prepares students for a career as a professional scientist, engineer or technician and an education that focuses on wider citizenship goals. It also prompts questions such as "Who benefits from increased levels of scientific literacy?" and "What is science education for?" Inevitably, different answers to these questions are given by different stakeholders in the science education enterprise: students, teachers, parents, scientists, politicians, employer groups, groups of "concerned citizens", science education researchers, science teacher educators, science curriculum developers, and all the other groups and institutions with an interest in science and technology education. What emerges by way of science curriculum at any particular time is the outcome of argument, negotiation and compromise among these groups, some of whom wield more power than others. For convenience, and following the suggestion of Thomas and Durant (1987), arguments for raising general levels of scientific and technological literacy, sometimes expressed in terms of enhancing public understanding of science and technology, can be categorized into perceived benefits to science, benefits to individuals and benefits to society as a whole.

Benefits to science are seen largely in terms of increased numbers of recruits to science-based professions (including medicine and engineering), greater support for scientific, technological and medical research, and more realistic public expectations of science. Little by way of elaboration needs to be said about the first argument, save that increased recruitment might also result in increased diversity within the community of scientists. Increased numbers of women and members of

1

ethnic minority groups, currently under-represented in science and technology in many countries, would do much to enrich the science-based professions and might serve to re-direct and reorient priorities for research and development – a matter that will be addressed, albeit briefly, later in the book. With regard to the other perceived benefits for science, it could be argued that confidence in scientists, continuing public support for science, and the high levels of public funding science currently enjoys, depend on citizens having some general understanding of what scientists do and how they do it. More significantly, perhaps, support depends on whether the public *values* what scientists do. Research by Miller (2004) suggests that the overwhelming majority of citizens (at least in the United States) are supportive of current levels of funding for scientific research and consider that this research has contributed substantially to raising both the standard and the quality of life. For many years there appeared to be no cause for the scientific establishment to be concerned about lack of support for research in fields such as materials science, biological sciences and medicine; whatever public doubts existed tended to focus on moral-ethical issues raised by developments in such fields as stem cell research, genetic manipulation and xenotransplantation, and occasional concern about the thoroughness and dependability of drug testing protocols. As Giddens (1991) observes, "lay attitudes towards science, technology and other esoteric forms of expertise … tend to express the same mixed attitudes of reverence and reserve, approval and disquiet, enthusiasm and apathy, which philosophers and social analysts (themselves experts of sorts) express in their writings" (p. 7). However, it is probably true to say that there has been a significant decline in public confidence in science and scientists in recent years as a consequence of the BSE episode (the so-called "mad cow disease") in the United Kingdom and concerns about bird flu, SARS, West Nile Virus and other transmissible diseases – fuelled by the wild swings between panic and complacency in the news media. Skepticism is now rife regarding the bland assurances about health risks associated with nuclear power stations, overhead power lines and mobile phones; there is unease about the emergence of so-called "superbugs" in hospitals, anxiety about the environmental impact of genetically engineered crops and, until the last year or so, serious concern about the failure of scientists to reach consensus about the causes, significance and proper response to global warming and climate change. It is worth noting, with regard to such matters, that public perception of science and scientists is impacted asymmetrically: episodes that worry the public and threaten trust in science always seem more vivid, compelling and memorable than events likely to generate approval and support. Publicity about the harmful side effects of insecticide use, for example, is likely to be more influential on public opinion than reports about the development of a new high yield food crop. Among some sections of the public there is mounting concern about the increasing domination of scientific and technological research by commercial, governmental and military interests, the increasing vulnerability of science and scientists to the pressures of capitalism and politics, and the increased secrecy and vested interest that result. As Baskaran and Boden (2006) comment, Western science seems to have moved from being "an activity committed to the production of public goods for public benefit to the producer of private goods primarily for commercial (profitable) exploitation" (p. 43). The close link between science and commerce in the field of genetic engineering has been a particular trigger for deepening mistrust of scientists. Indeed, Ho (1997) claims that "practically all established molecular geneticists have some direct or indirect connection with industry, which will set limits on what the scientists can and will do research on ... compromising their integrity as independent scientists" (p. 155). In consequence, as Barad (2000) notes, "the public senses that scientists are not owning up to their biases, commitments, assumptions, and presuppositions, or to base human weaknesses such as the drive for wealth, fame, tenure, or other forms of power" (p. 229). In its third report, the (UK) House of Lords Select Committee on Science and Technology commented on what it perceives as a "crisis of trust":

> Society's relationship with science is in a critical phase ... On the one hand, there has never been a time when the issues involving science were more exciting, the public more interested, or the opportunities more apparent. On the other hand, public confidence in scientific advice to Government has been rocked by a series of events, culminating in the BSE fiasco,[1] and many people are deeply uneasy about the huge opportunities presented by areas of science including biotechnology and information technology, which seem to be advancing far ahead of their awareness and assent. In turn, public unease, mistrust and occasional outright hostility are breeding a climate of deep anxiety among scientists themselves. (Select Committee, 2000, p. 11)

More than 45 years ago, Schwab (1962) advocated a shift of emphasis for school science away from the learning of scientific knowledge (the products of science) towards an understanding of the processes of scientific inquiry on the grounds that it can ensure "a public which is aware of the conditions and character of scientific enquiry, which understands the anxieties and disappointments that attend it, and which is, therefore, prepared to give science the continuing support which it requires" (p. 38). More significantly, a focus on history, philosophy and sociology of science (HPS) in the school science curriculum can play a key role in bringing about a more *critical* understanding of scientists and scientific practice. A related argument advanced by Shamos (1993) is that enhanced scientific literacy is a defence against what he sees as the anti-science and neo-Luddite movements that are (in his words) "threatening to undermine science". The school science curriculum, he argues, "should be the forum for debunking the attempts of such fringe elements to distort the public mind, first by exposing their tactics, and then by stressing over and over again the central role in science of objective, reproducible evidence" (p. 71).

Arguments that scientific and technological literacy brings benefits to *individuals* come in a variety of forms. It is commonly argued, for example, that scientifically and technologically literate individuals have access to a wide range of employment opportunities and are well-positioned to respond positively and competently to the introduction of new technologies in the workplace. In recent years, this has been especially true in industries that make extensive use of information and communications technology. In addition, it is argued that those who are scientifically literate are better able to cope with the demands of everyday life in an increasingly technology-dominated society, although even casual observation of technological innovation shows that advances are generally in the direction of increased user-friendliness, so that *less* expertise is needed to cope with a new technology than was needed for the old. Competency is built into the technology! Atkin and Helms (1992) express this view particularly well.

> Ordinary, intelligent human beings can get along perfectly well without a knowledge of Newton's Laws, an understanding of the atomic nature of matter, or even how everyday consumer devices (like TV sets or automobiles) operate. They do it all the time. That is not to say that people are not benefited by such information, of course. But to claim a major place for science education in the schools on the basis of its essentiality in personal and social functioning – as by likening its centrality to reading or calculating – is an exaggeration that misleads the teacher, the student, and the public. (p. 4)

A stronger case is that scientifically literate individuals are better positioned to evaluate and respond appropriately to the supposed "scientific evidence" used by advertizing agencies and politicians, and better equipped to make important decisions that affect their health, security and economic well-being. The point at issue here is that particular scientific knowledge, skills and attitudes are essential for everyday life in a complex, rapidly changing and science/technology-dominated society. What that knowledge comprises will be discussed later in the chapter.

Some have argued for the intellectual, aesthetic and moral-ethical benefits conferred on individuals by scientific literacy. For example, C.P. Snow (1962) referred to science as "the most beautiful and wonderful collective work of the mind of man", thus making it as crucial to contemporary culture as literature, music and fine art. Richard Dawkins (1998) expresses similar views: "the feeling of awed wonder that science can give us is one of the highest experiences of which the human psyche is capable. It is a deep aesthetic passion to rank with the finest that music and poetry can deliver" (p. x). Schibeci and Lee (2003) argue for the deep personal benefits of studying and engaging in science, which they characterize as an "intellectually enabling and ennobling enterprise" (p. 188). Others have claimed that appreciation of the ethical standards and code of responsible behaviour within the scientific community will lead to more ethical behaviour in the wider community – that is, the pursuit of scientific truth regardless of personal interests, ambitions and prejudice (part of the traditional image of the objective and dispassionate scientist) makes science a powerful carrier of moral values and ethical principles: "Science is in many respects the systematic application of some highly regarded human values – integrity, diligence, fairness, curiosity, openness to new ideas, skepticism, and imagination" (AAAS, 1989, p. 201). In other words, scientific literacy does not just result in more skilled and more knowledgeable people, it results in *wiser* people – that is, people well-equipped to make morally and ethically superior decisions.

Arguments that increased scientific literacy brings benefits to society as a whole include the familiar and increasingly pervasive economic argument and the claim that it would enhance democracy and promote more responsible citizenship. The first argument sees scientific literacy as a form of human capital that sustains the economic well-being of a country. Put simply, continued economic development brought about by enhanced competitiveness in international markets depends on science-based research and development, technological innovation and a steady supply of well-trained scientists, engineers and technicians. The Government of Canada has long promoted this view:

> Our future prosperity will depend on our ability to respond creatively to the opportunities and challenges posed by rapid change in fields such as information technologies, new materials, biotechnologies and telecommunications ... To meet the challenges of a technologically driven economy, we must not only upgrade the skills of our work force, we must also foster a lifelong learning culture to encourage the continuous learning needed in an environment of constant change. (Government of Canada, 1991, pp. 12 & 14)

As I will argue in Chapter 8 with respect to the distinctive language of science, the discourse of any group is suffused with assumptions, beliefs, attitudes and values. Thus, teaching newcomers to use the discourse of the social group is a crucial element in the business of enculturating them into a community of practice, helping to create a particular view of the world and to foster particular attitudes and habits. Skilful and continued deployment of a particular discourse creates a particular social reality. Indeed, rhetoric becomes reality and those who think differently and have different values are regarded as deviant or aberrant. Thus, Lankshear et al. (1996) argue that the pressures exerted by business and industry on schools to provide more "job ready" people can be seen as part of an overt sociotechnical engineering practice in which new capitalism is creating "new kinds of people by changing not just their overt viewpoints but their actual practices" (p. 22). In short, the business community is re-engineering people in its own image! The impact on Ontario schools has been considerable. It is now rare to find teachers who are prepared to question Ministry of Education directives, much less to design and implement curriculum experiences that challenge these new societal assumptions or address controversial issues. Instead, most teachers comply with all policy guidelines, ensure coverage of the prescribed curriculum content as efficiently as possible, avoid anything controversial, implement standardized assessment schemes to measure designated learning outcomes, and fill in the myriad boxes in the ever-expanding catalogue of official report cards (Alsop, 2009).

At the opposite end of the political spectrum is the argument that increased scientific literacy will promote more *democratic* decision-making (by encouraging everyone to exercise their democratic rights) and more *effective* decision-making (by encouraging people to exercise their democratic rights more wisely and responsibly).[2] In the words of Chen and Novick (1984), enhanced scientific literacy is a means "to avert the situation where social values, individual involvement, responsibility, community participation and the very heart of democratic decision making will be dominated and practiced by a small elite" (p. 425). Democracy is strengthened when *all* citizens are equipped to confront and evaluate socio-scientific issues (SSI) knowledgeably and rationally, rather than (or as well as) emotionally,[3] and to make informed decisions on matters of personal and public concern. As Dawkins (1998) remarks: "lawyers would make better lawyers, judges better judges, parliamentarians better parliamentarians and citizens better citizens if they knew more science and, more to the point, if they reasoned more like scientists" (p. 113). It is also the case, as Sagan (1995) notes, that some scientists would be better scientists and some engineers would be better engineers if they were more cognizant of the social, cultural, ethical, environmental and economic consequences of what they do. In Hodson (2008) I promoted the view that scientific literacy is the driving force for sociopolitical action – an argument that will be explored at length in a later book. Roth and Barton (2004) make essentially the same point: "critical scientific literacy is inextricably linked with social and political literacy in the service of social responsibility" (p. 10). In common with Roth and Lee (2002, 2004), they recognize that significant impact on SSI decision-making is more likely through collective action than individual efforts, thus shifting the ultimate focus of education for scientific literacy towards effective public practice, summed up by the increasingly popular notion of *enhanced public engagement with science*.

One further argument for seeking enhanced scientific literacy is that it might also be the most effective way to address: (i) the naïve trust that many people have in whatever information they obtain from a few minutes searching the Internet, (ii) the suspicion with which many people regard any argument that deploys statistics, on the grounds that "statistics can prove anything", (iii) the increasing acceptance of New Age beliefs, such as iridology, reflexology and the healing properties of crystals, and (iv) the continuing fascination that many people have with astrology and the paranormal.

DEFINING SCIENTIFIC LITERACY

Given all this rhetoric and debate about scientific literacy, and the large number of official and semi-official documents promoting it (AAAS, 1993, UNESCO, 1993; National Research Council, 1996; Council of Ministers of Education, 1997; Millar & Osborne, 1998; Organization for Economic Cooperation and Development, 1999; Goodrum et al., 2000; Department of Education (RSA), 2002), one might expect: (i) clear consensus regarding its definition and (ii) the availability of unambiguous guidelines concerning a curriculum capable of achieving it. Such is not the case, as reviews by Gräber and Bolte (1997), Laugksch (2000), De Boer (2001) and Roberts (2007) make abundantly clear.[4]

It would be a very odd state of affairs for someone to claim scientific literacy and to admit ignorance of the major theoretical frameworks of biology, chemistry, physics and the earth sciences. Scientific literacy clearly has a content component, though the precise nature of that content is a matter of debate – a debate that is outside the scope of this book. What *is* of concern in this book are those elements of the history of science, philosophy of science and sociology of science that constitute a satisfactory understanding of the nature of science (NOS), long regarded as a major component of scientific literacy and an important learning objective of science curricula.[5] Indeed, the promotion of NOS in official curriculum documents has become so prominent that Dagher and BouJaoude (2005) have stated: "improving students' and teachers' understanding of the nature of science has shifted from a *desirable* goal, to being a *central* one for achieving scientific literacy" (p. 378, emphasis added). In making the case for NOS knowledge in the curriculum, Driver et al. (1996) contend that in addition to its intrinsic value, NOS understanding enhances learning of science content, generates interest in science and develops students' ability to make informed decisions on socioscientific issues based on careful consideration of evidence. There is also an argument, advanced by Erduran et al. (2007) that NOS knowledge (and the wider notion of HPS understanding) is of immense value to *teachers*, making them more reflective and more resourceful.

In many ways, the long-standing confusion over terms such as "literacy", "illiteracy" and "literate", where some writers see literacy as mere functional competence while others see it as a sensitive awareness of the complexities of language, is mirrored in the use of the term "scientific literacy". Some see "being scientifically literate" as the capacity to read, with reasonable understanding, lay articles about scientific and technological matters published in newspapers and magazines, or posted on the Internet. For example, Miller (2000, 2004) defines it in terms of reading and making sense of the Tuesday science section of *The New York Times*. Others regard scientific literacy as possession of the knowledge, skills and attitudes deemed necessary for a career as a professional scientist or engineer. We need to ask whether scientific literacy is more akin to what a "literate" or well-educated person would know and be able to do, or more akin to a basic or functional literacy – that is, being able to read at an acceptable level of comprehension?[6] Some years ago, Atkin and Helms (1992) asked two key questions about scientific literacy. First, does a person need to know science in the same sense that they need to know their mother tongue? Second, is the ability to use scientific knowledge in the way one uses language essential for adequate functioning and responsible citizenship? To both questions, their answer was "No". An alternative question is: "Does one need to be literate in order to achieve scientific literacy?" Now the answer is clearly "Yes", regardless of whether the argument for scientific literacy focuses on the preparation of future scientists or the education of responsible citizens.[7] Engagement in science is not possible without a reasonable level of literacy. As Anderson (1999) states: "reading and writing are the mechanisms through which scientists accomplish [their] task. Scientists create, share, and negotiate the meanings of inscriptions – notes, reports, tables, graphs, drawings, diagrams" (p. 973). Scientific knowledge cannot be articulated and communicated except through text, and its associated symbols, diagrams, graphs and equations. Moreover, the specialized language of science makes it possible for scientists to construct an alternative interpretation and explanation of events and phenomena to that provided by ordinary, everyday language. Indeed, it could be said that learning the language of science is synonymous with (or certainly coincident with) learning science, and that *doing* science in any meaningful sense requires a reasonable facility with the language.[8] It is scientific language that shapes our ideas, provides the means for constructing scientific understanding and explanations, enables us to communicate the purposes, procedures, findings, conclusions and implications of our inquiries, and allows us to relate our work to existing knowledge and understanding.

> Without text, the social practices that make science possible could not be engaged: (a) the recording and presentation and re-presentation of data; (b) the encoding and preservation of accepted science for other scientists; (c) the peer reviewing of ideas by scientists anywhere in the world; (d) the critical re-examination of ideas once published; (e) the future connecting of ideas that were developed previously; (f) the communication of scientific ideas between those who have never met, even between those who did not live contemporaneously; (g) the encoding of variant positions; and (h) the focusing of concerted attention on a fixed set of ideas for the purpose of interpretation, prediction, explanation, or test. The practices centrally involve texts, through their creation in writing and their interpretation, analysis, and critique through reading. (Norris & Phillips, 2008, p. 256)

Kintgen (1988 – cited by Osborne, 2002, p. 213) identifies four stages of literacy: first, the *signature* stage, that is, the ability to read and write one's own name; second, the *recitation* stage, where an individual is able to read all (or most) of the words in a passage but has little understanding of the overall meaning or its significance and implications; third, the *comprehension* stage, that is, the ability to make sense of unfamiliar material at the literal level; fourth, the *analytical* stage, comprising the ability to analyze, interpret, critique and evaluate text, including media reports. It is this fourth level of literacy that we should be seeking, a stage of development that necessitates substantial conceptual understanding as well as linguistic capability. Proficient and critical reading of science text, whether first order or second order literature, involves more than just recognizing all the words and being able to locate specific information; it also involves the ability to (i) determine when something is an inference, a hypothesis, a conclusion or an assumption, (ii) distinguish between an explanation and the evidence for it, and (iii) recognize when the author is asserting a claim to "scientific truth", expressing doubt or engaging in speculation. Without this level of interpretation the reader will fail to grasp the essential scientific meaning. Put simply, learning to think and reason scientifically requires a measure of familiarity and facility with the forms and conventions of the language of science (Norris & Phillips, 2003). It is not solely a matter of recognizing the words, it is also an ability to comprehend, evaluate and construct arguments that link evidence to ideas and theories, and use that understanding in new situations. Thus, teaching about the language of science, and its use in scientific argumentation, should be a key element in science education at all levels. It is the strength and plausibility of the argument and its supporting evidence that scientists consider when judging any claim to knowledge. All aspects of the "case" need to be appraised: Was the method well-chosen and well-executed? How trustworthy is the data? Does the conclusion follow from the data and its interpretation? Are there alternative interpretations and conclusions? Does the argument take account of existing knowledge, and does it build on it or refute it? Does the conclusion constitute new knowledge? How plausible is the overall case? And so on. The construction and appraisal of argument is a crucial dimension of scientific practice. Consequently, understanding the nature of scientific arguments and being able to construct and evaluate them is a crucial element of scientific literacy. Thus, we need to provide frequent and rich opportunities for students to explore and use the language of science: to read and write science, discuss the meaning of scientific text, note how ideas are supported by evidence, construct plausible arguments and evaluate arguments constructed by others. Because most people obtain the bulk of their knowledge of contemporary science and technology from television, newspapers, magazines and the Internet, the capacity for active critical engagement with scientific text is a crucial element of scientific literacy.[9] Indeed, it could be claimed that it is the *most important* element.

> To be fully scientifically literate, students need to be able to distinguish among good science, bad science and non-science, make critical judgements about what to believe, and use scientific information and knowledge to inform decision making at the personal, employment and community level. In other words, they need to be *critical consumers* of science. This entails recognizing that scientific text is a cultural artifact, and so may carry implicit messages relating to interests, values, power, class, gender, ethnicity and sexual orientation. (Hodson, 2008, p. 3)

LITERACY OR LITERACIES?

What else can be regarded as essential components of scientific literacy? Understanding the complex relationships among science, technology, society and environment? Knowing about the historical development of the "big ideas" of science and the sociocultural and economic circumstances that led to the development of key technologies? Knowing something about the complex organizational structure of scientific practice and its systems of control and ownership? Being aware of contemporary applications of science? Having the ability to use science in everyday problem solving? Holding a personal view on controversial issues that have a science and/or technology dimension? Possessing a basic understanding of global environmental issues?

We can begin to answer these questions by taking note of common practice in linguistics, and especially the teaching of English as a second language. The notion of a "general understanding" of English has been replaced by an array of functional understandings that are situation specific. Courses are routinely devised along the lines of English for foreign-trained nurses practising in the United Kingdom or English for Russian engineers working in the United States, on the grounds that literacy needs are highly context dependent. The same principle applies to scientific knowledge and its associated scientific literacy. For example, while biochemists are interested in the chemical reactions constituting digestion, nutritionists are more concerned with "calorie counts", and while an evolutionary biologist might choose to classify an animal as mammal, reptile or fish, an ecologist would prefer to employ categories such as lives on land and lives in water, or carnivore, insectivore and herbivore. How we choose to classify depends on (i) what we know (our conceptual understanding) and (ii) our purpose in classifying. Is the classification intended to inform the study of evolutionary processes? Is it part of preparation for ecological work? Or is it to help us find our way around the local zoo? Similarly, because their primary concerns are different, physicists and engineers confront the world with very different theoretical frameworks – for example, a commitment to the principle of conservation of energy versus daily concern about heat loss (and the means to minimize it) and deployment of a theoretical position that seems to envisage heat as a fluid that flows. Indeed, because the problems they face may be unique and ill-defined, engineers and other technologists often have to create purpose-made theoretical models by drawing on content from several scientific disciplines. It follows that there is value in thinking about different scientific literacies for different purposes and for different sociocultural contexts.[10]

Shen (1975) identifies three such categories: *practical* scientific literacy, *civic* scientific literacy and *cultural scientific* literacy. Practical scientific literacy is knowledge that can be used by individuals to cope with life's everyday problems relating to diet, health, consumer preferences, technological competence, and so on; civic scientific literacy comprises the knowledge, skills, attitudes and values necessary to play a full and active part in key decision-making in areas such as use of natural resources, energy policy, moral-ethical issues relating to medical and technological innovations, and environmental protection; cultural scientific literacy involves knowing something of the ideas and theories of science that constitute major cultural achievements and the sociocultural and intellectual environment in which they were produced. The term cultural scientific literacy is used to signal belief that answers to deeply-rooted questions such as the nature and origin of life, and of the cosmos, collectively constitute a cultural heritage and resource to which everyone should have access. Layton et al. (1993) have described this aspect of scientific literacy as "recognition and appreciation of "the cathedrals of science", science as a majestic achievement of the human intellect and spirit" (p. 15)

In the context of this book the notion of environmental literacy is particularly helpful. Although the term itself is not universally accepted, with some writers opting for "ecological literacy", "education for sustainability" or even "ecological citizenship" (Hart, 2007), there is some general agreement on its three major components: (i) knowledge and understanding of the environment and the impact of people on it (including content knowledge such as the hydrological cycle, food webs, mechanisms of climate change and ozone depletion); (ii) attitudes and values that reflect feelings of concern for the environment and foster a sensitive environmental ethic; (iii) a sense of responsibility to address issues and resolve environmental problems through participation and action, both as individuals and collectively. The importance of environmental or ecological literacy to life in the 21st century is graphically captured by David Orr's (1992) comments on the consequences of not giving it prominence in the curriculum.

> A generation of ecological yahoos without a clue why the color of the water in their rivers is related to their food supply, or why storms are becoming more severe as the planet warms. The same persons as adults will create businesses, vote, have families, and above all, consume. If they come to reflect on the discrepancy between the splendor of their private lives in a hotter, more toxic and violent world, as ecological illiterates they will have roughly the same success as one trying to balance a checkbook without knowing arithmetic. (p. 86)

Miller and Kimmel (2001) have defined *biomedical* literacy as the specialized knowledge needed by an individual facing a personal decision on gene therapy or seeking to understand and participate in debate on research involving embryonic stem cells. Although it is possible to identify other specific scientific literacies – for example, *industrial* scientific literacy, for management personnel and plant workers in specific manufacturing environments, *professional* scientific literacy, to serve the needs of those whose future work lies in science and science-related professions, *recreational* scientific literacy, focusing on the science underpinning gardening, scuba-diving, car maintenance, and so on – there is little to be gained by doing so here, save to argue for the adoption of the term *critical* scientific literacy, where "critical" is taken to mean rigorous, analytical, logical, thorough, open-minded, skeptical, careful and reflective. I use the term *critical scientific literacy* on the grounds that the most important function of scientific literacy is to confer a measure of intellectual independence and personal autonomy: first, an independence from authority; second, a disposition to test the plausibility and applicability of principles and ideas for oneself, whether by experience or by a critical evaluation of the testimony of others; third, an ability to form intentions and choose a course of action in accordance with a scale of values that is self-formulated. In other words, the fundamental purpose of scientific literacy (and technological literacy) is to help people *think for themselves* and *reach their own conclusions* about a range of issues that have a scientific and/or technological dimension. As Désautels et al. (2002) put it: "Instead of a liberal education, we seek a liberating education" (p. 266).

The science underpinning socioscientific issues is sometimes complex and uncertain. It is frequently outside the knowledge store of most citizens. Even among scientists, only those working at the research frontier are likely to have a full understanding of all the key elements. As Lewontin (2002) points out, this poses a problem for a democratic society.

> On the one hand, the behavior of the state is supposed to reflect the popular will, as determined either by a direct appeal to the opinion of the people or

11

through the intermediary of their elected representatives. On the other hand, the esoteric knowledge and understanding required to make rational decisions in which science and technology are critical factors lie in the possession of a small expert elite. Even within the ranks of "scientists" only a tiny subset have the necessary expertise to make an informed decision about a particular issue. (p. 28)

Although we are increasingly dependent on experts, it is undesirable to cede all deliberation and all policy decisions to a particular small group of experts. Every citizen needs sufficient understanding *about* the relevant science (if not understanding *of* the science) to play a part. Every citizen needs to develop what Lorraine Code (1987) calls "a policy of circumspection" and what McPeck (1981) calls "reflective skepticism" – that is, the disposition to question and to seek the opinions of others on all the science that underpins the issues they confront in everyday life. Code (1987) notes that "one of the most important and difficult steps in learning who can be trusted is realizing that authority cannot create truth" (p. 248). Balance is the key: not blind acceptance of the views espoused by those who are seen, or see themselves, as experts; not cynicism and distrust of all experts. Guy Claxton (1997) captures the essence of this position particularly well: "[students] need to be able to see through the claims of Science to truth, universality, and trustworthiness, while at the same time not jumping out of the frying-pan of awe and gullibility, in the face of Science's smugness and superiority, into the fire of an equally dangerous and simplistic cynicism, or into the arms of the pseudo-certainties of the New Age" (p. 84). Balance is encapsulated in the notion of intellectual independence. As Munby (1980) notes: "One can be said to be intellectually independent when one has all the resources necessary for judging the truth of a knowledge claim independently of other people" (p. 15).

As I will argue in a subsequent book, a measure of *politicization* of students can be achieved by assisting them to recognize that: (i) the benefits of technological innovations can sometimes be accompanied by unwelcome and unexpected side effects, including social dislocation and adverse environmental impact; (ii) the benefits and hazards are not always distributed equably within and between societies, creating major sociopolitical concerns; and (iii) new scientific developments and new technologies frequently create complex moral and ethical dilemmas that challenge currently accepted values and beliefs. Confronting these realities, formulating their own position and planning appropriate action will require students to invest their scientific and technological literacy with a sharp critical edge. Part of that critical edge entails understanding both the power and limitation of scientific discourse and argumentation, and "being able to discriminate between Science and Scientism – the illicit attempt to give warrant and status to one's claims by presenting them as if they were Scientifically proven or justified; and developing the disposition to do so in the course of daily life" (Claxton, 1997, p. 83 – capitals in original).

SCIENCE AND TECHNOLOGY

As evident in the foregoing, any discussion of scientific literacy raises questions about technology, the relationship between science and technology, and the meaning of technological literacy. While science can be regarded as a search for explanations of phenomena and events in the natural world, technology is the means by which humans modify nature to meet their needs and wants, and better serve their interests. While it is easy to think of technology in terms of artifacts (computers, aeroplanes, microwave ovens, pesticides and fertilizers, birth control pills, water treatment plants, power stations, and the like), it is important to remember that it also includes the knowledge, skills and infrastructure necessary for the design, manufacture, operation and maintenance of those artifacts. Although there can be some important differences between them in terms of purposes, concepts, procedures and criteria for judging acceptability of solutions, science and technology are closely related. Clearly, scientific understanding of the natural world is the basis for much of contemporary technological development. The design of computer chips, for instance, depends on detailed understanding of the electrical properties of silicon and other materials, and the design of a drug to fight a particular disease is made possible by knowledge of how the structures of proteins and other biological molecules determine their interactions. Conversely, technology is essential to much contemporary scientific research. Current rates of progress would be impossible without microscopes, telescopes, infra red spectrometers, particle accelerators and DNA sequencers, while running simulations to explore the complexities of the models that meteorologists build to study global climate change is just one example of how contemporary science requires elaborate computer technology. In fields such as high energy physics and nanotechnology it is very difficult, if not impossible, to disentangle science and technology. Indeed, it often makes more sense to speak of *technoscience*. Johnson (1989) sums up the relationship between science and technology as follows:

> Technology is the application of knowledge, tools, and skills to solve practical problems and extend human capabilities. Technology is best described as process, but it is commonly known by its products and their effects on society. It is enhanced by the discoveries of science and shaped by the designs of engineering. It is conceived by inventors and planners, raised to fruition by the work of entrepreneurs, and implemented and used by society ... Technology's role is doing, making and implementing things. The principles of science, whether discovered or not, underlie technology. The results and actions of technology are subject to the laws of nature, even though technology has often preceded or even spawned the discovery of the science on which it is based. (p. 1 – cited by Lewis & Gagel, 1992, p. 127)

Arnold Pacey's classic work, *The Culture of Technology* (published in 1983), defines technological products and practices in terms of a *technical* aspect (knowledge, skills, techniques, tools, machines, resources, materials and people), an *organizational* aspect (including economic and industrial activity, professional

13

activity, users, consumers and trade unions) and a *cultural* aspect (goals, values, beliefs, aspirations, ethical codes, creative endeavour, etc.). In similar vein, Carl Mitcham (1994) conceptualizes technology in terms of four aspects: (i) as objects, artifacts and products; (ii) as a distinctive form of knowledge, separate from science; (iii) as a cluster of processes (designing, constructing or manufacturing, evaluating, systematizing, etc.); and (iv) as volition (the notion that technology is part of our human will and, therefore, an intrinsic part of our culture). Both writers note that technology reflects our needs, interests, values and aspirations, thus providing a valuable counter to the somewhat bleak views of technological determinism portrayed in many science fiction stories and movies – the idea that current technology determines future technology and that human beings must adapt to its dictates (see Winner, 1977, for an extended discussion). Remarks such as "You can't stop progress", "It's inevitable" and "That's what we will have to get used to" are commonplace and reveal a strong sense of individual and collective disempowerment and a feeling that technological development and change are in the hands of others, if not of technology itself. An essential step in pursuit of critical scientific and technological literacy is the application of a social and political critique capable of challenging the notion of technological determinism. We *can* control technology and its social and environmental impact, if we have the will and the political literacy to do so. More significantly, we can *control the controllers* and redirect technology in such a way that adverse environmental impact is reduced (if not entirely eliminated) and issues of freedom, equality and justice are kept in the forefront of discussion in the establishment of policy. These matters are outside the scope of this book but will be dealt with at length in a book currently in preparation.

In arguing the case for the cultivation of technological thinking to be an integral part of every student's education, the Association for Science Education (1988) identified four inter-related strands: *technological literacy* – familiarity with the content and methods of a range of technologies; *technological awareness* – recognition of the personal, moral, social, ethical, economic and environmental implications of technological developments; *technological capability* – the ability to tackle a technological problem, both independently and in cooperation with others; *information technology* – competence and confidence in the technological handling of information. In a similar attempt to delineate the field of technology education, Layton (1993) employs a complex classification system based on six "functional competencies":

— Technological awareness or *receiver competence*: the ability to recognize technology in use and acknowledge its possibilities.
— Technological application or *user competence*: the ability to use technology for specific purposes.
— Technological capability or *maker competence*: the ability to design and make artifacts.
— Technological impact assessment or *monitoring competence*: the ability to assess the personal and social implications of a technological development.

- Technological consciousness or *paradigmatic competence*: an acceptance of, and an ability to work within, a "mental set" that defines what constitutes a problem, circumscribes what counts as a solution and prescribes the criteria in terms of which all technological activity is to be evaluated.
- Technological evaluation or *critic competence*: the ability to judge the worth of a technological development in the light of personal values and to step outside the "mental set" to evaluate what it is doing to us.

A similar approach has been adopted by Gräber et al. (2002), culminating in a 7-component competency-based model of scientific literacy comprising subject competence, epistemological competence, learning competence (using different learning strategies to build personal scientific knowledge), social competence (ability to work in a team on matters relating to science and technology), procedural competence, communicative competence and ethical competence. My own inclination is simply to extend the three major elements of science education listed in endnote 8 to include technology, together with the dimensions of politicization and preparation for sociopolitical action.

- *Learning science and technology* – acquiring and developing conceptual and theoretical knowledge in science and technology, and gaining familiarity with a range of technologies.
- *Learning about science and technology* – developing an understanding of the nature, methods and language of science and technology; appreciation of the history and development of science and technology; awareness of the complex interactions among science, technology, society and environment; and sensitivity to the personal, social, economic, environmental and moral-ethical implications of particular technologies.
- *Doing science and technology* – engaging in and developing expertise in scientific inquiry and problem-solving; developing confidence and competence in tackling a wide range of "real world" technological tasks and problems.
- *Engaging in sociopolitical action* – acquiring (through guided participation) the capacity and commitment to take appropriate, responsible and effective action on science/technology-related matters of social, economic, environmental and moral-ethical concern, and the willingness to undertake roles and responsibilities in shaping public policy related to scientific and technological developments at local, regional, national and/or global levels.

Sometimes there is value in teachers emphasizing the differences between science and technology, sometimes it is more important and more interesting to direct attention to the similarities. Sometimes it is important for students to think in a purely scientific way, sometimes it is crucial that they learn to think in a technological way (e.g., like an architect, doctor or engineer) – a frame of mind predicated on a willingness to draw on knowledge from a range of disciplines in order to address complex real world issues and problems. Some issues of demarcation between science and technology are considered in Chapter 5.

Any discussion of technological literacy inevitably raises important issues relating to computer technology. The World Wide Web, computer-aided design, word processing, data processing and electronic transfer of information have become the engines of economic growth and have fundamentally changed the ways we learn, communicate and do business. *Computer literacy* extends well beyond the acquisition of basic computer skills, to encompass: (i) consideration of the socioeconomic impact of computer technology and the globalization it has accelerated; (ii) the legal issues and moral-ethical dilemmas associated with open access to information, censorship and data protection; (iii) the capacity to evaluate information for accuracy, relevance and appropriateness; and (iv) the ability to detect implied meaning, bias and vested interest. In these latter dimensions there is significant overlap with *media literacy* – an area of concern that is well outside the scope of this book. It should also be noted that scientific and technological literacy necessarily includes some basic understanding of mathematics, such as familiarity with simple algebraic equations and the capacity to interpret graphical and statistical data. Duschl et al. (2007) note that encouraging students to express their ideas in mathematical form (especially graphical representation) helps them to clarify and develop their scientific understanding, and can sometimes lead to them "noticing new patterns or relationships that otherwise would not be grasped" (p. 153). The notion of scientific and technological literacy I wish to develop also includes some historical understanding of the dependence of science on mathematics. For example, Kepler could not have derived his laws of planetary motion without knowledge of conic sections accumulated by Greek mathematicians some 1800 years earlier, Hilbert's theory of integral equations was essential for the development of quantum mechanics, and Riemann's differential geometry was integral to Einstein's theory of relativity. It is also reasonable to conclude that the extraordinary growth in scientific knowledge since the 17th century is attributable, in large part, to developments in mathematics - and, in particular, to the invention of differential and integral calculus.

TOWARDS A CURRICULUM

The foregoing discussion leads me to conclude that *critical* scientific and technological literacy comprises a number of basic components.

- A general understanding of some of the fundamental concepts, ideas, principles and theories of science, and the ability to use them appropriately and effectively.
- Some knowledge of the ways in which scientific knowledge is generated, validated and disseminated.
- Familiarity with the form, structure and rhetorical purposes of scientific language.
- The capacity to read and interpret scientific data and, at a general level, to evaluate their validity and reliability.
- The ability to evaluate a scientific argument or claim to knowledge, and to present one's own ideas clearly, concisely and appropriately.

- General interest in science, together with a willingness and capacity to update and acquire new scientific knowledge.
- Understanding of basic technological concepts, principles of design and criteria of evaluation.
- Possession of a range of basic hands-on technical/technological skills, the capacity and confidence to tackle technological problems, and the willingness to develop new knowledge and skills.
- Some awareness of the sociocultural and cognitive circumstances surrounding the history and development of some of the "big ideas" of science, the origin and development of important technologies, and the development of mathematics.
- An appreciation of the complexity of inter-relationships among science, technology, society and environment.
- A commitment to critical understanding of contemporary socioscientific issues at the local, regional, national and global levels, including their historical roots and underlying values, together with a willingness to take appropriate and responsible action, and encourage others to do so.
- The capacity and willingness to address moral-ethical issues associated with scientific research and the deployment of scientific knowledge and technological innovations.

It is immediately apparent that issues in the history, philosophy and sociology of science impact on all these components, either directly or indirectly. This is true even of the first item, if one takes the view that understanding and using an item of scientific knowledge entails knowledge of its role and status as well as its strengths, weaknesses and relationships to other knowledge items. In other words, as argued in Hodson (2008), scientific literacy thus constructed is located just as much in learning *about* science and in *doing* science as it is in learning science. No science curriculum can equip citizens with thorough first-hand knowledge of all the science underlying every important issue. Moreover, much of the scientific knowledge learned in school, especially in the rapidly expanding fields of the biological sciences, will be out- of-date within a few years of leaving school and so of little value in addressing socioscientific issues. However, science education can enable students to understand the significance of knowledge presented by others, and it can enable them to evaluate the validity and reliability of that knowledge and to understand why scientists often disagree among themselves on such major matters as climate change (and its causes) without taking it as evidence of bias or incompetence. What is too often unrecognized by science teachers, science textbooks and curricula, and by the wider public, is that dispute is one of the key driving forces of science. Hence controversial issues – no matter whether the controversy focuses on conceptual issues, procedural issues or moral-ethical issues – should be an essential component of a curriculum aimed at critical scientific literacy. Indeed, I would argue that teachers have a duty to assist students in confronting and resolving such matters.

> Education should not attempt to shelter children from even the harsher contro-
> versies of adult life, but should prepare them to deal with such controversies
> knowledgeably, sensibly, tolerantly and morally. (Qualifications & Curriculum
> Authority, 1998, p. 56)

Of course, students also need to know that scientists can and do make mistakes, and that bias and incompetence do sometimes occur. In other words, they need to have a clear understanding of what counts as *good* science (i.e., a well-designed inquiry and a well-argued conclusion) and be able to identify the nature of any inadequacies, errors and biases. In common with the case for intellectual inde-pendence made earlier in the chapter, Fourez (1997) argues that a key element of scientific literacy is knowing when scientists and other "experts" can be trusted and when their motives and/or methods should be called into question. His point is that while there is often a need to access and utilize expert opinion we do not need to do so uncritically. We can evaluate the quality of data and argument for ourselves, we can look at the extent of agreement among experts and the focus of any disagreements, and we can look at the "track record" of all those who pro-fess expertise. In similar vein, Norris (1995) makes the point that "nonscientists' belief or disbelief in scientific propositions is not based on direct evidence for or against those propositions but, rather, on reasons for believing or disbelieving the scientists who assert them" (p. 206). Fourez (1997) also argues that scientifically literate individuals know when to leave "black boxes" closed (i.e., making use of a concept, idea or tool in an instrumental way, without necessarily striving to understand it) and when it is important to open it (i.e., to strive for detailed understanding). Consequently, he says, the science curriculum should teach "the right way to use black boxes" (p. 913) as well as the right way to use standard, well-established scientific knowledge, theories, models, metaphors and inter- or multi-disciplinary models of various kinds.

What I am advocating for inclusion in the school science curriculum, as I argue at length in Hodson (2008), are those elements of the history, philosophy and sociology of science that will enable all students to leave school with robust knowledge about the nature of scientific inquiry and theory building, an under-standing of the role and status of scientific knowledge, an ability to understand and use the language of science appropriately and effectively, the capacity to analyze, synthesize and evaluate knowledge claims, some insight into the socio-cultural, economic and political factors that impact the priorities and conduct of science, a developing capacity to deal with the moral-ethical issues that attend some scientific and technological developments, and some experience of con-ducting authentic scientific investigations for themselves and by themselves.[11] While we cannot provide all the science knowledge that our students will need in the future (indeed, we do not know precisely what knowledge they will need), and while much of the science they will need to know has yet to be discovered, we *do* know what knowledge, skills and attitudes will be essential to appraising scientific reports, evaluating scientific arguments and forming a personal opinion about the science and technology dimensions of real world issues. This realization constitutes a powerful argument for a major shift of emphasis from learning science to learning *about* science. An emphasis on *what* we know (the current emphasis) rather than how we know (part of the approach being advocated) too often leaves students only able to justify their beliefs by reference to the authority of the teacher or textbook. As Östman (1998) points out, the constant focus on a "correct explanations" approach encourages and reinforces the "companion meaning" that the products of the scientific enterprise are "something that everyone will agree upon, if they just use their senses to smell, taste, listen to, and look at nature" (p. 57). In other words, science is seen as a simple and straightforward route to the truth about the universe; scientific knowledge is seen as authoritative, and students are steered towards conformity with received "official" views rather than towards intellectual independence. In contrast, as Lemke (1998) notes, science is "a very human activity, full of biases and accidents, driven by egos and budgets, competitive and sometimes venal, locked into the agendas of larger institutions and social movements: a wonderful and terrible human comedy, like every other part of life" (p. 2). Science is much more than its products: *what* science says about the world is only part of the story. A much bigger and more important part concerns the ways in which scientists generate and validate that knowledge, establish research priorities and use knowledge to address real world problems.

> However much we may teach them about electrical circuits, redox reactions, or genetic recombination, or even about controlled experimentation and graphical analysis of quantitative covariation, how much better able does this really make them to decide when they trust expert opinion and when they should be skeptical of it? If we teach more rigorously about acids and bases, but do not tell students anything about the historical origins of these concepts or the economic impact of the technologies based on them, is the scientific literacy we are producing really going to be useful to our students as citizens? (Lemke, 2001, pp. 299 & 300)

If students acquire good learning habits and positive attitudes towards science during the school years it will be relatively easy for them to acquire additional scientific knowledge later on, as and when the need arises, provided that they have also acquired the language skills to access and evaluate relevant information from diverse sources. Of course, to be scientifically literate in the sense I am arguing for in this chapter, students will not only need the language skills to access knowledge from various sources, but also the capacity to participate comfortably in public debate about SSI and express their knowledge, views, opinions and values in a form appropriate to their purpose and the audience being addressed.

The robust familiarity with key issues in the history, philosophy and sociology of science essential to critical scientific literacy requires lengthy and close contact with someone already familiar with them – that is, with a teacher or scientist who can provide appropriate guidance, support, experience and criticism. Formal education is crucial in laying the appropriate foundation. At present, in many schools, students often acquire NOS knowledge *by accident*, picking up ideas more by what is absent than from any deliberate intention on the part of the teacher

(Munby & Roberts, 1998). Hodson (2008) identifies and clarifies a number of major items for inclusion in the curriculum from the vast literatures of history of science, philosophy of science and sociology of science. This book is concerned with ways of "translating" these items into a robust and coherent curriculum.

Three points about curriculum construction should be made at this stage. First, the goal of improving NOS understanding is often hampered by stereotyped images of science and scientists consciously or unconsciously built into school science curricula (Hodson, 1998b) and perpetuated by science textbooks (McComas, 1998). As a first priority, therefore, teachers need to present students with a more authentic science education – that is, an approach to learning science that has elements in common with the practices of the scientific community. Gilbert (2004) describes authentic science education in terms of four major characteristics.

> First, it would more faithfully represent the processes by which science is conducted and its results are socially accepted: it should be more historically and philosophically valid. Second, it would reflect the core element of creativity that has made science one of the major cultural achievements of humanity in recent centuries. Third, it would provide a minimalist network of ideas with which to provide satisfactory explanations of phenomena in the world-as-experienced. Lastly, it would be capable of underpinning those technological solutions to human problems that are the basis of prosperous economies, social well-being, and the health of individuals (p. 116).

Second, the formulation of a prescribed list of NOS items or a set of "required beliefs" about the nature of science is both educationally undesirable and inappropriate to the goal of *critical* scientific literacy. The intent of the learning *about* science component of the curriculum is not inculcation of particular views, and we should no more tolerate this position than we should approve the relativist position in which any view is regarded as equally acceptable to any other. Rather, the intention is to confront students with a range of alternatives, engage them in critical debate, with appropriate stimulus and guidance, and assist them in formulating their own views – views for which they can argue and for which they can provide a robust and coherent "warrant for belief".

Third, the drive for authenticity should be tempered by considerations of appropriateness. Not everything in the literature of history, philosophy and sociology of science is appropriate for young minds – see Brush (1974), Matthews (1998), Allchin (2004a) and Bell (2004) for a discussion of this proposition, and Davson-Galle (2008) for a vigorous argument *against* NOS as an element of a compulsory science curriculum. In my view, we should select NOS items for the curriculum in relation to other educational goals – for example, motivating students and assisting them in developing positive attitudes towards science, paying close attention to the cognitive goals and emotional demands of specific learning contexts, creating opportunities for students to experience *doing* science for themselves, enabling students to address complex socioscientific issues with critical understanding, attending to concerns of multiculturalism and antiracism, and so on.

The degree of sophistication of the NOS items we include should be appropriate to the stage of cognitive and emotional development of the students and should be compatible with other long and short-term educational goals. There are numerous goals for science education, and education in general, that can, will and *should* impact on decisions about the NOS content of lessons. Our concern is not just good philosophy of science, good sociology of science or good history of science, not just authenticity and preparation for sociopolitical action ... but the educational needs and interests of the students – *all* students. Selection of NOS items should consider the *changing* needs and interests of students at different stages of their science education, as well as taking cognizance of the views of "experts" and promoting the wider goals of (i) authentic representation of science and (ii) politicization of students. (Hodson, 2008, p. 182)

Attention in this book is largely concentrated on curriculum structures and teaching and learning methods capable of ensuring that all students are properly equipped to: (i) confront and solve everyday problems that have a science and/or technology dimension; (ii) comprehend the significance of technological developments, understand the science that underpins them and evaluate their value and appropriateness; and (iii) address socioscientific and technological issues in local, regional, national and global contexts, formulate a personal position on the issues and take appropriate action.

NOTES

[1] The report notes that "in the run-up to the BSE crisis in 1996, uncertainties were insufficiently acknowledged, with advisers and ministers representing the lack of any "sound scientific" evidence for a risk as evidence that it could not occur. It was thus common for politicians to deliver unequivocal assurances of safety" (Select Committee, 2000, p. 1).

[2] Irwin and Wynne (1996) note that this argument first appeared in the 1940s in a document published by the Association of Scientific Workers, under the title *Science and the Nation*. It also featured prominently in The Royal Society's influential report *The Public Understanding of Science* (1985).

[3] The role of emotions in decision-making on controversial issues will be examined in a subsequent book.

[4] Roberts (2007) reviews a number of publications that address the meaning of scientific literacy (SL). He categorizes them in terms of five basic approaches or "conceptual methodologies": first, the historical approach, tracing the development of notions of SL since the late 1950s; second, approaches that try to identify "types" or "levels" of SL; third, discussion of SL focused on the notion of literacy itself; fourth, discussion of SL in terms of the needs and interests of practising scientists; fifth, an approach that identifies situations or contexts in which SL is presumed to be necessary or valuable.

[5] In recent publications, Lederman (2006, 2007) seeks to restrict use of the term *nature of science* to the characteristics of scientific knowledge (i.e., to epistemological considerations). This is an odd notion, given that much of our scientific knowledge and, therefore, consideration of its status, validity and reliability, is intimately bound up with the design, conduct and reporting of scientific investigations. Moreover, teaching activities focused on NOS often include empirical investigations. Thus, as Ryder (2009) points out, the conduct of scientific inquiry and NOS are related both conceptually and pedagogically. In common with definitions adopted by Osborne et al. (2003), Clough (2006), Clough and Olson (2008), Hodson (2008) and Wong and Hodson (2009), my conception of NOS encompasses the characteristics of scientific inquiry, the role and status of the scientific knowledge it generates, how scientists work as a social group, and how science impacts and is impacted by the social context in which it is located.

[6] Osborne (2002) resists the notion that scientific literacy is a quality that an individual possesses or does *not* possess. Rather, he says, "scientific literacy exists on a continuum between being totally illiterate (and totally dependent on others) to acknowledged expertise (and minimal intellectual dependence). Knowing and understanding both some of the content and the appropriate use of language of science is an essential component on the path towards such scientific literacy" (p. 214).

[7] I emphatically reject the suggestion by Klopfer (1969) that this distinction should be reflected in a differentiated school science curriculum: "One curricular stream ... designed for students planning to enter careers as scientists, physicians, and engineers ... the other ... designed for students who will become the nonscientist citizenry ... housewives, service workers, salesmen, etc. ... Differentiation of students (should) begin at about age fourteen when they choose the high school they will attend" (p. 203). My own view is more closely aligned with Jerry Wellington's (2001) argument that every student's science education should comprise components that reflect the three major justifications for particular science content: (i) the intrinsic value of science education; (ii) the citizenship argument; (iii) utilitarian arguments. In an attempt at compromise, the authors of *Beyond 2000: Science Education for the Future* (Millar & Osborne, 1998) state that science education between the ages of 5 and 16 (the years of compulsory schooling in the UK) should comprise a course to enhance general scientific literacy, with more specialized science education delayed until later years: "the structure of the science curriculum needs to differentiate more explicitly between those elements designed to enhance 'scientific literacy', and those designed as the early stages of a specialist training in science, so that the requirement for the latter does not come to distort the former" (p. 10). Donnelly (2005) is vociferous in his condemnation of this proposal and the potentially adverse impact it would have on the science education of the majority of students by greatly reducing the theoretical content of the curriculum.

[8] In a number of publications (Hodson, 1992a, 1994, 1998a) I have argued that science education is best regarded as comprising three major elements: learning science (an emphasis on content), learning *about* science (concern with HPS issues and science-technology-society-environment (STSE) dimensions); *doing* science (first hand experience of scientific inquiry). In recent years (Hodson, 2003), I have added a fourth component: *engaging in sociopolitical action.*

[9] A survey in the United Kingdom showed that most school-age students spend more time accessing science and science-related materials on the Internet than they do in science lessons (Department for Education & Skills, 2002 – cited by Wellington & Britto, 2004, p. 208). In general, the popular media report what we might call "frontier science" or "science-in-the-making" (Latour, 1987) and, necessarily, citizens depend on journalists to keep them informed of advances in science and technology; even professional scientists are unable to keep up with the primary literature in *all* fields and so must depend on others to keep them abreast of change and development.

[10] Science education for (responsible) citizenship, with its emphasis on confronting complex SSI at local, regional, national and global levels, throws up the same kind of issue. Different notions of good or responsible citizenship are evident in different societal contexts: "The concept of good citizenship in an Islamic theocracy is not the same as in the USA, Australia or other industrialized democracies. Likewise, acting as a good citizen with respect to a global science-related issue may conflict with actions taken in response to more local, regional or national concerns" (Jenkins, 2006, p. 207). Murcia (2009) represents scientific literacy by means of a rope metaphor: interwoven threads of conceptual knowledge, NOS knowledge and understanding of the social, political and economic context in which science is located. Individuals acquire understanding and expertise in each of these threads, but the particular balance between the threads is determined by that person's interests, dispositions and aspirations.

[11] The authors of *Taking Science to School*, a recent report from the National Research Council (Duschl et al., 2007), argue along similar though narrower lines when they state that students in school who are *proficient in science*: "(1) know, use, and interpret scientific explanations of the natural world; (2) generate and evaluate scientific evidence and explanations; (3) understand the nature and development of scientific knowledge; and (4) participate productively in scientific practices and discourse" (p. 36).

CHAPTER 2
RESEARCH ON STUDENTS' VIEWS OF NOS

In planning how best the goal of critical scientific literacy and its attendant understanding of NOS knowledge can be attained we should take steps to ascertain the position from which we are likely to "begin the journey". Some 40 years ago, David Ausubel (1968) remarked: "If I had to reduce all of educational psychology to just a single principle, I would say this: Find out what the learner already knows and teach him accordingly" (p. 337). This applies just as much to teaching NOS knowledge as to any other aspect of science education. Just as students already have some *prior* understanding of many of the entities, phenomena and events in the natural world they encounter in science lessons, so they have prior conceptions of science and scientists. These views, which will inevitably impact the ways in which they respond to classroom events, particularly those relating to the collection and interpretation of data, may be different from the views that form the intended learning *about* science components of the curriculum, and will certainly be different from the ideas about science implied by the basic components of *critical* scientific literacy articulated in Chapter 1.

Students will, of course, use their existing NOS views as a lens through which to make sense of events in the classroom, with considerable potential for misunderstanding if the teacher plans her/his lessons in accordance with one view of science and students interpret the activity in accordance with a different one. Over the past 25 years or so, extensive research into students' alternative frameworks of understanding in science has alerted teachers to the potential learning difficulties likely to arise from a mismatch between a student's conceptual framework and that assumed by the teacher in designing curriculum experiences. Significant mismatch between a student's understanding of the role, status and origin of scientific knowledge and the views implicit in the teaching and learning activities deployed, or made explicit in NOS teaching, create problems that are equally serious and potentially damaging. Furthermore, students' NOS views may be particularly robust and resistant to change because they are constantly reinforced by powerful out-of-school experiences. Books, newspapers, television programmes, movies, advertisements, museum exhibits and Internet Websites are all likely to carry messages about science and images of scientists, and it is inevitable that students will be impacted by them.

If science teachers are to intervene effectively and shift students' understanding *about* science in the desired direction, they need reliable information about

23

students' likely understanding of NOS issues and, when possible, how their views develop over time and in response to particular kinds of interventions. If we can identify the kind of experiences that stimulate change and development we are in a much stronger position to design effective curricula for achieving critical scientific literacy. Teacher educators and curriculum developers also need to know about teachers' NOS views and how they influence curriculum decision-making and, thereby, impact on students' views. This has been a fruitful area of research in recent years, though it is fair to say that much more remains to be done. My intention here is not to provide an exhaustive review of the literature; that task has already been admirably carried out by Lederman (1992, 2007) and Abd-El-Khalick and Lederman (2000). Rather, my purpose is to engage in critical reflection on this vast literature with a view to identifying some findings that are particularly relevant to the themes developed and elaborated throughout this book, findings that may provide some guidance on curriculum structure, design of teaching and learning activities, development of curriculum materials, appropriate NOS-oriented programmes for preservice teacher education and priorities for continuing teacher professional development.

RESEARCH METHODS

Methods for ascertaining students' and/or teachers' views about science and scientists include questionnaires and surveys, interviews, small group discussions, writing tasks and classroom observations (particularly in the context of hands-on activities). As discussed below, each has its strengths and weaknesses. Necessarily, researchers who use questionnaire methods decide what should count as legitimate research data *before* the data collection process begins; those who use classroom observation (and, to a lesser extent, those who use interview methods) are able to make such decisions *after* data collection. They also have the luxury of embracing multiple perspectives.

Among the best-known NOS-oriented questionnaires are the *Test on Understanding Science* (TOUS) (Cooley & Klopfer, 1961), the *Nature of Science Scale* (NOSS) (Kimball, 1967), the *Nature of Science Test* (NOST) (Billeh & Hasan, 1975) and the *Nature of Scientific Knowledge Scale* (NSKS) (Rubba, 1976; Rubba & Anderson, 1978), together with a modified version (M-NSKS) developed by Meichtry (1992). Instruments dealing with the processes of science, such as the *Science Process Inventory* (SPI) (Welch, 1969a), the *Wisconsin Inventory of Science Processes* (WISP) (Welch, 1969b) and the *Test of Integrated Process Skills* (TIPS) (Dillashaw & Okey, 1980; Burns et al., 1985) could also be regarded as providing information on understanding some key aspects of the nature of science. While questionnaires are quick and easy to administer, they can be overly restrictive, incapable of accommodating subtle shades of meaning and susceptible to misinterpretation. Only very rarely do they afford respondents the opportunity to explain why they have made a particular response to a questionnaire item. It may well be that the same response from two respondents arises from quite different understanding and reasoning, while similar reasoning by two respondents may result in different responses.

As pointed out in Hodson (2008, chapter 2), most of the early instruments were constructed in accordance with a particular philosophical position and are predicated on the assumption that all scientists behave in the same way. Hence teacher and/or student responses that do not correspond to the model of science assumed in the test are judged to be "incorrect", "inadequate" or "naïve". Lucas (1975), Koulaidis and Ogborn (1995), Alters (1997) and Lederman et al. (2002) provide an extended discussion of this issue. It is also the case that many of these instruments pre-date significant work in the philosophy and sociology of science, and so are of severely limited value in further studies. Like research in science itself, research in science education is a product of its time and place, and there is a constant need to update its methods. Reviews by Lederman (1992, 2007) and Lederman et al. (1998, 2000) describe several additional NOS instruments that take into account the work of more recent and even contemporary scholars in philosophy and sociology of science. Most notable among these newer instruments are *Conceptions of Scientific Theories Test* (COST) (Cotham & Smith, 1981), *Views on Science-Technology-Society* (VOSTS) (Aikenhead et al., 1989), the *Nature of Science Survey* (Lederman & O'Malley, 1990), and the *Views of Nature of Science Questionnaire* (VNOS) (Lederman et al., 2002) and its several subsequent modifications (see Lederman, 2007).

Sometimes the complexity and subtlety of NOS issues makes it difficult to find appropriate language for framing questions. Opinions in philosophy of science are rarely of the kind that can be readily expressed in short statements with which all individuals can confidently and unambiguously agree or disagree. If it is difficult for the researcher to find the right words, how much more difficult is it for the respondent to capture the meaning they seek to convey? In consequence, it cannot be assumed that the question and/or the answer will be understood in exactly the way it was intended, especially with younger students and those with poor language skills. Far from solving these problems, multiple-choice items can sometimes exacerbate them by leaving no scope for expressing doubt or subtle shades of difference in meaning. One response is to administer the questionnaire orally or to adopt an interview-based approach.

Despite some interesting recent developments in questionnaire design, one problem remains unsolved: students do not always interpret questionnaire items in the way the designers intended. The designers of VOSTS attempted to circumvent this problem by using multiple choice items derived from student writing and interviews to provide a number of different "position statements" (sometimes up to 10 positions per item), including "I don't understand" and "I don't know enough about this subject to make a choice" (Aikenhead et al., 1987; Aikenhead & Ryan, 1992). It is the avoidance of the forced choice and the wide range of aspects covered (definitions, influence of society on science/technology, influence of science/technology on society, influence of school science on society, characteristics of scientists, social construction of scientific knowledge, social construction of technology, and nature of scientific knowledge) that give the instrument such enormous research potential. Nevertheless, as Abd-El-Khalick and BouJaoude (1997) and Botton and Brown (1998) point out, VOSTS was conceived and written within a North American sociocultural context and, in consequence, may have limited validity in non-Western contexts. In response to concerns like these, Tsai and Liu (2005) have developed a survey instrument that they claim is more sensitive to sociocultural influences on science and students' views of science. It focuses on five characteristics of scientific knowledge and its development: (i) the role of social negotiations within the scientific community; (ii) the invented and creative nature of science; (iii) the theory-laden nature of scientific investigation; (iv) the cultural influences on science; and (v) the changing and tentative nature of scientific knowledge. Rooted in similar concerns about the socioculturally-determined dimensions of NOS understanding is the *Thinking about Science* instrument designed by Cobern and Loving (2002) as both a pedagogical tool (for preservice teacher education programmes) and a research tool for assessing views of science in relation to economics, the environment, religion, aesthetics, race and gender.

Lederman and O'Malley (1990) utilized some of the design characteristics of VOSTS to develop an instrument comprising just seven fairly open-ended items (e.g., "Is there a difference between a scientific theory and a scientific law? Give an example to illustrate your answer"), to be used in conjunction with follow-up interviews. Not all students were prompted to focus their answers on the tentativeness of science, as the authors intended, and although the interview stage was able to go some way towards re-focusing attention, these findings serve to reiterate the difficulty of attempting to interpret students' understanding from their written responses to researcher-generated questions.[1] Because of differences in underlying philosophical position, and the likely low correlation between scores obtained using different instruments, questionnaire data obtained from different student or teacher populations should only be used for comparative purposes if the same instrument is used throughout, and even then the specific limitations and biases of the instrument should be clearly acknowledged.

Frustrated by these problems, some researchers and teachers incline to the view that more useful information can be obtained, especially from younger students, by relying solely on open-ended methods such as the Draw-a-Scientist Test (DAST) (Chambers, 1983). In his initial study, Chambers used this test with 4807 primary (elementary) school children in Australia, Canada and the United States. He identified seven common features in their drawings, in addition to the almost universal representation of the scientist as a man: laboratory overall; spectacles (glasses); facial hair; "symbols of research" (specialized instruments and equipment); "symbols of knowledge" (books, filing cabinets, etc); technological products (rockets, medicines, machines); and captions such as "Eureka" (with its attendant lighted bulb), $E=mc^2$ and think bubbles saying "I've got it" or "A-ah! So that's how it is". In the 25 years since Chambers' original work, students' drawings have changed very little (Fort & Varney, 1989; Symington & Spurling, 1990; Mason et al., 1991; Jackson, 1992; Newton & Newton, 1992, 1998; Rosenthal, 1993; Matthews, 1994; Huber & Burton, 1995; She, 1995, 1998;

Rahm & Charbonneau, 1997; Barman, 1997, 1999; Finson, 2002; Fung, 2002),[2] with research indicating that the stereotype emerges round about grade 2 and is well-established and held by the majority of students by grade 5.

When Losh et al. (2008) invited 70 students in grade 1, 58 students in grade 2 and 78 in grade 3 to draw a teacher and a veterinarian ("animal doctor" was the term used) as well as a scientist, 70% of the teachers were recognizably female, 53% of animal doctors were recognizably female and 60% of the scientists were recognizable male. Ambiguous gender figures were fairly common among grade 1 boys (28% for scientists, 31% for teachers and 48% for veterinarians), leading the researchers to speculate that other researchers had provided clear instructions on gender assignment. Fung (2002) also notes a high incidence (22%) of gender ambiguity among the drawings of 675 Hong Kong students in grades 2, 4, 6, 8 and 10. Despite their more regular exposure to teachers than to the other occupations, students in the Losh et al. (2008) study drew fewer details for teachers, providing more beakers, lab coats or animals for scientists or veterinarians than they drew chalkboards, books or pencils for teachers. Scientists smiled less and were seen as less attractive overall than teachers and veterinarians.

However, there is a strong possibility that researchers can be seriously mis-led by the drawings children produce. As Newton and Newton (1998) point out, "their drawings reflect their stage of development and some attributes may have no particular significance for a child but may be given undue significance by an adult interpreting them" (p. 1138). Even though young children invariably draw scientists as bald men with smiling faces, regardless of the specific context in which the scientist is placed (except in the study described above), it would be unwise to assume that children view scientists as especially likely to be bald and contented. As Claxton (1990) reminds us, children compartmentalize their knowledge and so may hold several images of the scientist – for example, the everyday comic book or cartoon version, the "official", approved curriculum version for use in school, and their personal (and perhaps private) view. It is not always clear which version DAST is accessing, nor how seriously the student took the drawing task. Simply asking students to "draw a scientist" might send them a message that a "typical scientist" exists (Boylan et al., 1992). There is also the possibility, especially with students in upper secondary school or university, that the response is intended to make a sociopolitical point – for example, that there are too few women or members of ethnic minority groups engaged in sci-ence. Scherz and Oren (2006) argue that asking students to draw the scientist's workplace can be helpful, while Rennie and Jarvis (1995a) suggest that students should be encouraged to annotate their drawings in order to clarify meaning and intention. Further insight into students' views can be gained by talking to them about their drawings and the thinking behind them, asking them if they know anyone who uses science in their work (and what this entails), or presenting them with writing tasks based on scientists and scientific discovery. Miller (1992, 1993) advocates the following approach: "Please tell me, in your own words, what does it mean to study something scientifically?" When given the opportunity to discuss their drawings and stories with the teacher, even very young children will provide detailed explanations (Sumrall, 1995; Sharkawy, 2006; Tucker-Raymond et al., 2007). Interestingly, discussion seems to take a different course when this task is set in science lessons than it does in other areas of the curriculum. It is also increasingly evident that young children's responses to open-ended writing tasks involving science, scientists and engineers are not stable and consistent: accounts and stories produced in science lessons are very different from those produced in language arts lessons (Hodson, 1993a). Students may even provide significantly different oral and written responses to nature of science questions (Roth & Roychoudhury, 1994).

While less restrictive, instruments designed for more flexible and open-ended responses, such as the *Images of Science Probe* (Driver et al., 1996), small group discussion and situated-inquiry interviews (Welzel & Roth, 1998; Ryder et al., 1999), sometimes pose major problems of interpretation for the researcher. So, too, do observation studies, unless supported by an interview-based follow-up capable of exploring the impact of context on student understanding. While interviews hold out the possibility of accessing underlying beliefs, their effectiveness can be severely compromised by the asymmetric power relationship between interviewer and interviewee, regardless of whether the interviewer is the teacher or an independent researcher. In an interview situation, some students may be shy or reluctant to talk; they may feel anxious or afraid; they may respond in ways that they perceive to be acceptable to the interviewer, or expected by them. Observation via audio or video recording of group-based tasks involving reading, writing and talking, practical work, role play, debating and drama constitute a less threatening situation for students, though even here there can be problems. Any classroom activity can be impacted by complex and sometimes unpredictable social factors. Indeed, it would be surprising if activities taking place in the very public venues of classrooms were *not* influenced by the social goals of students: making friends, feeling comfortable, impressing others, "attending to their image", and so on. Students carrying out activities in school science classrooms are not "free agents", no more than scientists *doing* science in their laboratories are free agents. Both are constrained by and driven by social pressures and social aspirations. Both sets of activities take place in spaces where image, reputation, power and prestige play a very significant role, spaces where social skills are crucial and awareness of social norms is crucial. As in science itself, authority and prestige can play an important role in the classroom, especially in group-work situations, with the views of the most insistent or the most prestigious group members being likely to prevail. All who have had dealings with adolescents (especially their own children) are uncomfortably aware of the power of peer group influences and the drive towards conformity with the group's norms, rules, language, dress and behaviour, and the stress induced by failure to gain admission to the valued group. The point is that the ways in which a group engages in classroom activities depends on the nature of the social interactions within the group to an extent at least equal to its dependence on the students' cognitive understanding of what is required of them. These complicating factors can mask or distort the NOS understanding we hope to infer from conversations and actions. In short, all approaches to ascertaining NOS views carry a risk that the characterization or description of science ascribed to the research subject is, in some measure, an artifact of the research method.

The context in which an interview question, questionnaire item or assessment task is set and, indeed, whether there is a specific context at all, can have major impact on an individual's response. Decontextualized questions (such as "What is your view of a scientific theory?" or "What is an experiment?") can seem infuriatingly vague to students and can be met with seeming incomprehension. Use of such questions can pose major problems of interpretation for the researcher. Conversely, context-embedded questions have domain specific knowledge requirements that may sometimes preclude students from formulating a response that properly reflects their NOS views. Moreover, respondents may feel constrained by restriction of the question to one context and unable to communicate what they know about the many significant differences in the ways that scientists in different fields conduct investigations. Familiarity with the context, understanding of the underlying science concepts, interest in the situation, and opportunity to utilize knowledge about other situations, are all crucial to ensuring that we access the students' authentic NOS understanding. Put simply, questions set in one context may trigger different responses from essentially the same questions set in a different context. To shed more light on this issue, Leach et al. (2000) conducted a study of 731 science students at upper secondary and undergraduate level in five European countries. Participating students were asked to respond to two groups of questions: first, some generalized, context-free items such as "Rate your agreement/disagreement with the two statements 'one data set always leads to one conclusion' versus 'different conclusions can legitimately be drawn from the same data' on a scale of 1–5"; second, a series of questions embedded in a specific subject context. Data analysis revealed three distinctive forms of reasoning: (i) *data focused reasoning* – scientific knowledge is a description of the material world, with differences of interpretation capable of being resolved by collecting further data; (ii) *radical relativist reasoning* – data cannot be used to determine whether a view is correct or incorrect, individuals are free to interpret observational and experimental data as they see fit; (iii) *theory and data related reasoning* – what scientists believe, what they do during investigations and what data are collected are closely inter-related, each being capable of influencing the others. The authors conclude that there is no evidence to indicate that a student's responses to generalized questions can be used to predict the way they are likely to respond to questions in a specific context. Nor is there any evidence to suggest that students hold unique epistemological positions that they use across a wide range of contexts.[3] Consequently, if questions and tasks are to be set in specific contexts there are important decisions to be made between school science contexts and real world contexts. This is especially important in research that addresses NOS views in the context of socioscientific issues (Sadler & Zeidler, 2004) and scientific controversies (Smith & Wenk, 2006).

It would be surprising if students did not have different views about the way science is conducted in school and the way science is conducted in specialist research establishments. Hogan (2000) refers to these different views as students' *proximal* knowledge of NOS (personal understanding and beliefs about their own science learning and the scientific knowledge they encounter and develop in science lessons) and *distal* knowledge of NOS (views they hold about the products, practices, codes of behaviour, standards and modes of communication of professional scientists). Sandoval (2005) draws a similar distinction between students' *practical* and *formal* epistemologies. Contextualized questions that ask students to reflect on their own laboratory experiences are likely to elicit the former, questions of a more general, de-contextualized nature ("What is science?" or "How do scientists validate knowledge claims?") are likely to elicit the latter. The problem for the researcher is to gauge the extent to which these differences exist and how they are accessed by different research probes. The problem for the teacher is to close the gap or, more in line with the arguments running through this book, to ensure that students are aware of the crucial distinctions as well as the similarities between science in school and science in the world outside school.

It almost goes without saying that there is significant potential for mismatch between what individuals state about their NOS understanding and what they do in terms of acting on that understanding, and this applies just as much to teachers as it does to students. Thus, another crucial consideration is whether we seek to ascertain *espoused* views or views *implicit in actions*. The former would probably be best served by questionnaires, writing tasks and interviews; the latter requires inferences to be drawn from observed behaviours and actions. It may also be the case that students hold significantly different views of science as they perceive it to be and science as they believe it *should* be – a distinction that Rowell and Cawthron (1982) were able to accommodate in their approach.

If we solve all these problems, we are still confronted with decisions about how to interpret the data. Should we adopt a nomothetic approach that focuses on the extent to which the students' or teachers' views match a pre-specified "ideal" or approved view? Any attempt to distinguish "adequate" NOS views from "inadequate" views involves judgements about the rival merits of inductivism and falsificationism, Kuhnian views versus Popperian views, realism versus instrumentalism, and so on. None of these judgements is easy to make, as discussion in Hodson (2008) has shown. Does it make more sense, then, to opt for an ideographic approach? Should we be satisfied to describe the views expressed by students and seek to understand them "on their own terms"? A major complicating factor is that neither students nor teachers will necessarily have coherent and consistent views across the range of issues embedded in the notion of NOS. Rather, their views may show the influence of several different, and possibly mutually incompatible philosophical positions. Older students, with more sophisticated NOS understanding, will have recognized that inquiry methods vary between science disciplines and that the nature of knowledge statements varies substantially with content, context and purpose. Few research instruments are sensitive to such matters. By assigning total scores rather than generating a profile of views, the research conflates valuable data that could inform the design of curriculum interventions. Rather than assigning individuals to one of several predetermined philosophical positions, it might make more sense to refer to their *Personal Framework of NOS Understanding*, and seek to highlight its interesting and significant features, an undertaking that could be facilitated by the use of repertory grids (as in the study by Shapiro, 1996).[4] Although Abd-El-Khalick (2004) found substantial inconsistencies in the NOS views of students at the undergraduate and graduate levels, he acknowledges that inconsistencies from the researchers' point of view may still be seen by the students as "a collection of ideas that make sense within a set of varied and personalized images of science" (p. 418).

A recent study by Ibrahim et al. (2009) does seek to consolidate data from a purpose-built questionnaire into NOS profiles. The questionnaire, *Views about Scientific Measurement* (VASM), which comprises six items addressing aspects of NOS and eight items dealing with scientific measurement, uses a common context (in earth sciences) and allows space for students to elaborate on their response or compose an alternative. The data, obtained from 179 science undergraduates, were found to cluster into four partially overlapping profiles: *modellers*, *experimenters*, *examiners* and *discoverers*. For modellers, theories are simple ways of explaining the often complex behaviour of nature; they are constructed by scientists and tested, validated and revised through experimentation. Creativity plays an important role in constructing hypotheses and theories, and in experimentation. When there are discrepancies between theoretical and experimental results, both theory and the experimental data need to be scrutinized. Experimenters also believe that scientists should use experimental evidence to test hypotheses and theories, but should do so in accordance with a strict scientific method. In situations of conflict, data have precedence over theories. Examiners regard the laws of nature as fixed and "out there" waiting to be discovered through observation, rather than constructed by scientists. Experimental work is essential; it is not informed by theory. Scientists may use both the scientific method and their imagination, but experimental data always have precedence over theories. Discoverers also believe that the laws of nature are out there waiting to be discovered through observation. Only experiments using the scientific method can be used to generate laws and theories. If experimental data conflict with a previously established theory, then both the theory and the data need to be checked. Interestingly, as a percentage of the total, the modeller profile was more common among students following a 4-year science foundation course than among physics majors.

Of course, different researchers make different decisions on these crucial questions. Sadly, however, not all researchers are appreciative of the consequences of the decisions they make for the validity, reliability and usefulness of the data they generate, though perhaps they are no more guilty in this respect than are researchers in other fields. Perhaps, as in many other situations, the most prudent approach is to use a range of methods and, of course, to subject all methods to close critical scrutiny, preferably by a team of evaluators with wide and varied experience. Despite the many caveats concerning the validity and reliability of research methods, it is incumbent on teachers, teacher educators and curriculum developers to pay careful attention to the findings of the rapidly growing number of studies that focus on students' and/or teachers' NOS views. For convenience, I have chosen to look separately (though only briefly) at students' views and teachers' views, with some "free floating" remarks concerning scientists' views.

STUDENTS' VIEWS OF SCIENCE AND SCIENTIFIC INQUIRY

It is always dangerous to generalize from research findings. Nevertheless, there seems to be a widespread view among students that science is an elitist activity, out of the reach of ordinary people, and that scientists are exceptionally smart and hard-working individuals, whose overriding goal is the betterment of life in general. Aikenhead et al. (1987) report that most students regard scientists as being motivated by curiosity (the need to "find out how things are") and the desire to "make the world a better place to live in", attributing these motives to scientists ahead of "earning a good salary" or "becoming famous". In terms of personal attributes, Ryan (1987) reports that scientists are generally seen by students as honest, trustworthy, objective, logical, methodical, analytical and open-minded.[5] It is fairly common for students to assess scientists very positively in terms of cognitive abilities, while being less complimentary with regard to their personality traits, social skills and moral-ethical standards (Ward, 1986; Andre et al., 1999; Song & Kim, 1999; Carlone, 2004). While they are respected for their expertise, scientists are often seen as drab, uninteresting, introverted, unemotional, insensitive, socially inept, "nerdy", work-obsessed individuals who are sometimes highly secretive[6] and occasionally sinister and dangerous. Scientists have few interests, and certainly not in music, movies or the arts in general. Most scientists do not have a "normal" family life: they may neglect their family or do not have a family because of their preoccupation with work. Indeed, they often go into the laboratory on their day off, working long hours on problems that have little relevance to people or social issues. Often, they are seen as careless of their appearance, generally untidy and disorganized; sometimes they even forget to eat because of their obsession with work. There is some evidence that stereotyped views relating to social issues are held more strongly by girls than by boys (Miller et al., 2006). Oddly, there is some evidence that the views of students whom Costa (1995) would characterize as "potential scientists" seem to hold particularly stereotyped views in this respect, views that are often surprisingly resistant to change. At the student teacher level, this situation seems to be reversed: those intending to teach science, presumably those with greater exposure to science and scientists, include fewer stereotypical features in their drawings of scientists than do those specializing in the liberal arts (Rosenthal, 1993). It almost goes without saying that an unfavourable image of scientists in respect of personal characteristics is bad for recruitment. As Sheffield (1997) comments, "Ask any teenager, or even any preteen, what she or he thinks that students gifted in mathematics and science look like, and it is likely that the answer will include an image that looks like the 'nerdy' scientist from *Back to the Future*: male, with glasses, a pocket protector, and a very strange hairdo ... It is nearly impossible to encourage students to do well in mathematics and science when they are faced with such ridiculous stereotypes everywhere they turn" (pp. 377 & 378, cited by Painter et al., 2006, p. 182). It is also bad for public confidence in science, and is likely to weaken political support for research and development.

> If, for whatever reason, people dislike the stereotypical scientists, they are less likely to support science. Why subsidize geeks to pursue their absurd and incomprehensible little projects? (Sagan, 1995, p. 363)

Reis and Galvão (2007) report on the ways in which the plots of science fiction stories written by students (in this case, by grade 11 students) can tell us a great deal about their perceptions of science and scientists. The authors note that some of the stories reveal the influence of the stereotypes and "catastrophic scenarios" depicted in movies, TV shows, books, comics and cartoons, including "the scientist as the hero and savior of society; the dangerous, ruthless scientist working on secret, controversial projects; and the scientist incapable of controlling the result of his work" (p. 1257). Follow-up interviews provide a valuable opportunity to challenge and possibly correct mistaken and stereotyped ideas about science, scientists and scientific practice.

There is some evidence (e.g., Barman, 1999) that students, at least in some parts of the world, are beginning to view scientists in a more "realistic way" – that is, as regular people rather than all-knowing geniuses, madmen, geeks or freaks. There is evidence that stereotypes weaken somewhat as students get older, possibly in response to the ways in which science is presented in school (She, 1998). Indeed, there is encouraging evidence that students' views can be favourably impacted by carefully designed interventions, especially those that focus on gender, ethnicity and physical disability (Flick, 1990; Bodzin & Gehringer, 2001; Bohrmann & Akerson, 2001; Sharkawy, 2006). However, Mbajiorgu and Iloputaife (2001) report that while an adapted VOSTS questionnaire and use of DAST indicated that the image of scientists held by preservice teachers had been favourably impacted by an NOS-oriented lecture programme, especially with respect to gender stereotyping, the changes seemed to be restricted to *espoused* views communicated during the research, and were unlikely to have significant influence on classroom or out-of-school behaviour. In one sense, this mirrors Hogan's (2000) observation of the disconnect or mismatch that often exists between students' understanding of the ways in which scientific knowledge is constructed and their understanding of how they themselves acquire new scientific knowledge.

It is probably fair to say that most students see science as being concerned with addressing natural phenomena through empirical investigations. Younger students see science as proceeding largely as an inductive process: making discoveries or "finding out things" through prolonged, careful and systematic observation. Older students often see science as a process of hypothesis testing. Generally, students regard observational evidence, especially that produced by experiment, as constituting the only legitimate support for a theory. There is little understanding that some theories (such as evolutionary theory or theories in plate tectonics, for example) are based on historical evidence (the fossil record), argument, and in recent years, perspectives developed through computer simulation. It is still common for students to believe that science is restricted to laboratory settings, and, as would be expected, "experiments" figure prominently in almost all students' descriptions of what scientists do. For many young children, an experiment is any practical activity, including measuring things or mixing things together in order to "see what happens", "make something happen" or "do something new" – a complex of views that probably reflects the tendency of teachers to refer to all hands-on activities as "experiments". As they gain more experience, students come to regard experiments as a reliable means of "finding out what actually happens", seeking a cause (of the effect they observe) or testing a prediction (usually to confirm it), provided that they are designed and conducted in accordance with the requirements of "the scientific method". It is commonly believed that careful adherence to the designated steps of this method is what gives scientists confidence in the knowledge the method generates.

The 12 year olds in the study conducted by Carey et al. (1989) are typical of many younger students in not regarding experiments as being guided by ideas, questions and assumptions. Indeed, the idea that experiments and other empirical investigations are conducted as part of the theory-building process in order to compile evidence for checking, testing and developing hypotheses and models, develops very slowly, and sometimes not at all (Driver et al., 1996). Because of the way practical work in school is organized, particularly with respect to writing a mandated and formulaic lab report, most students see experimental inquiry as a straightforward linear sequence of planning, conducting, interpreting and reporting. There is little appreciation of the reflexive nature of experimental design, little recognition that scientists frequently have to engage in revision and reorientation of the procedures in order to overcome initial shortcomings in design. In general, students are unaware that the scientific knowledge generated by experiments is provisional and subject to further critical scrutiny (Carey & Smith, 1993). It is particularly disturbing that all 11 science undergraduates in the study by Ryder et al. (1999) subscribed to the view that scientific knowledge can be proved beyond doubt by collecting appropriate empirical data. While this is a worrying enough view for future scientists to hold, it is considerably more worrying (in the context of this book's position on critical scientific literacy) when it is held by future teachers, physicians, journalists and politicians.

Using a questionnaire-based approach, Pomeroy (1993) found that scientists were more likely than science teachers to express traditional empiricist views about science. However, interviews with scientists often reveal a somewhat different picture. In the study conducted by Wong and Hodson (2009), all the scientists expressed the view that the specific approach to an inquiry is dependent on the nature of the problem, the knowledge we already possess, the personnel and technical facilities available, and so on. There is no one method! Although the scientists interviewed by Schwartz and Lederman (2006) noted that scientific investigation and theory building proceed in diverse ways, depending on the field, there was a clear indication by some (9 out of 24) that "claims made through experimental methods were more valid than claims made through non-experimental methods" (p. 13). Interestingly, one of the scientists in the Wong and Hodson (2009) study remarked that although controlled experiments constitute a particularly powerful means of establishing the validity of knowledge claims, no one experimental design can ensure perfectly valid and reliable data. Not only can small variations in method sometimes produce large differences in data, but seemingly identical procedures can produce different data because of differences in the quality of the instruments and the craft skills of the technicians. Discussion in both studies turned, as is usual in conversations with scientists about their methods, to the question of whether investigations need to be hypothesis-driven. In Schwartz and Lederman's (2006) study, chemists and life scientists said "Yes", while earth scientists and physicists said "No". Again, context seems to be the crucial consideration. One of the scientists interviewed by Wong and Hodson (2009) said that although investigations are not always hypothesis-driven, particularly in molecular biology (the scientist's own area of expertise), it is important to *pretend* that they are when submitting proposals for funding because funding agencies expect it and regard open-ended inquiries as little more than "fishing expeditions". A number of scientists reported that hypotheses are no longer necessary in many areas of study because recent advances in technology enable enormous amounts of data to be collected and analyzed in a matter of minutes. Interesting issues and problems for systematic study emerge by a process of "data mining". The research chemists interviewed by Samarapungavan et al. (2006) described their work in terms of trying to build molecules that theory suggests are possible, rather than in terms of testing a hypothesis or theory.[7] Failure to synthesize the molecule is not seen as falsification of the underlying theory so much as a stimulus to greater effort in finding a suitable route. While this study is important in highlighting how scientists in different disciplines use substantially different approaches to inquiry, it does seem to make the assumption that all research in chemistry is necessarily focused on synthesis. Although the authors report no data to that effect, it is reasonable to assume that sub-disciplines within chemistry will have their own distinctive and context-specific methods. Interestingly, Schwartz and Lederman (2008) report that differences in investigative approaches among the 24 scientists in their research sample are not illustrative of distinctive features of the broader sub-disciplines of chemistry, physics, biology and earth sciences; rather, they seem to arise from specific individual contexts and experiences.

When asked how they evaluate the quality of their work, the chemists interviewed by Samarapungavan et al. (2006) mentioned design criteria (such as elegance, efficiency and percentage yield), the judgement of others (via peer review, successful publication, level of research funding), the impact and fruitfulness of their work (both within the community and the wider society) and what the authors refer to as "cognitive factors": namely, "novelty, difficulty and completeness of the work" (p. 484). Students, however, rarely mentioned any criterion for judging the quality of research other than empirical adequacy. When Glasson and Bentley (2000) interviewed four scientists and two engineers, as part of a conference designed to acquaint science teachers with current research practice, all the scientists emphasized the importance of research design in obtaining valid and reliable data, though they noted that what constitutes good design is contingent on the specific circumstances of the inquiry. While both scientists and engineers made reference to connections between their research and societal issues (e.g., developing drugs to fight cancer, increasing agricultural production, etc.), the authors noted a tacit underlying assumption that science is free of bias and vested interest, and proceeds in a value-free and ethically neutral way.

SCIENTIFIC REASONING AND SCIENTIFIC KNOWLEDGE

There is evidence that many students perform less than satisfactorily in designing scientific investigations and interpreting data. Often they fail to generate or test hypotheses (on the occasions hypotheses are required), ignore critical variables, and interpret data in a way that supports prior beliefs. Kuhn et al. (1988) argue that scientifically literate people possess a set of "domain general" (i.e., generalizable and transferable) thinking skills that enable them to coordinate theory and evidence. Thus, they can consciously articulate the theories they hold, state what evidence supports or would refute them, and can justify why they hold a particular view, rather than some other view that might also explain the evidence. This set of abilities constitutes a major element in my notion of critical scientific literacy, as discussed in Chapter 1. Their research seems to show that these abilities develop slowly and steadily over time. Indeed, Kuhn (1997) states that until early adolescence, children tend not to think of theory and evidence as separate. If she is correct, the weaknesses in scientific reasoning listed above are a consequence of students' failure to recognize that their current beliefs can (and maybe should) be modified in the light of evidence. When people think only *with* their theories rather than *about* them, say Kuhn et al. (1988, 1995), they are unable to consider alternatives. Further, these authors say, it is only by considering alternatives, and seeking to identify "what is not", that one can begin to achieve confidence in one's knowledge of "what is".

Two issues are relevant to consideration of Kuhn's conclusions. First, the propositions with which the research subjects are presented seem to grossly misrepresent the nature of scientific theory; second, the activities in which Kuhn and her co-workers ask students to engage bear little resemblance to real scientific inquiry. For example, it is difficult to see how studying some fairly trivial data with a view to choosing the more likely of some possible causes, such as whether the smoothness of a tennis ball affects a player's service (or not) or whether apples or oranges are better at warding off a cold (Kuhn et al., 1988), can be regarded as an exploration of the applicability and robustness of a scientific theory. Similar doubts focus on the researchers' use of an exercise in manipulating variables that might affect the speed of a toy boat or toy car (Kuhn et al., 1995). Much more convincing is Samarapungavan's (1992) finding that the ability to use a theory and argue for its appropriateness (in terms of consistency with the evidence) is largely a matter of familiarity with the particular concepts involved. In other words, effective theory use constitutes a set of "domain specific" reasoning skills (see also Brown, 1990; Koslowski, 1996). When students have this familiarity and have interest in the situation or phenomenon under consideration, Samarapungavan (1992) argues, children as young as six are able to differentiate theories on grounds of logical and empirical consistency, generalizability, parsimony and scope. Similarly, Sodian et al. (1991) found that grade 1 and grade 2 students can distinguish belief from evidence, and can recognize disconfirming evidence when the situation does not make high-level knowledge demands or present a major conflict with students' prior beliefs. Schauble et al. (1991a, 1995) and Schauble (1996) have also pointed out the crucial role played by conceptual knowledge. Put simply, students can reason better when they have a rich conceptual background on which to draw and, thereby, are better positioned to investigate more thoroughly and systematically (see also, Dunbar, 1993). When students have only rudimentary conceptual understanding it is difficult for teachers and researchers to ascertain whether a poorly constructed explanation or argument is a consequence of lack of understanding of the content or lack of understanding of the nature of scientific explanations. It is also the case that scientific reasoning is not simple common sense; rather, it is part of a cultural tradition and its characteristics have, to an extent, to be learned (Brewer & Samarapungavan, 1991). Developing expertise in scientific reasoning is a process of "bootstrapping" between conceptual/theoretical knowledge and careful consideration of evidence: considerations of theory constrain data collection and interpretation, and the data constrain, refine and elaborate theory. Confirmation of the centrality of conceptual understanding can be found in Kuhn's (2007) observation that students who have successfully distinguished between causal and non-causal variables in one context cannot always utilize that understanding successfully in a different context. Further elaboration of the continuing dispute between advocates of domain-general and domain-specific arguments is outside the scope of this book. Kuhn and Pearsall (2000), Metz (2000, 2004), Zimmerman (2000), Lehrer and Schauble (2006) and Schauble (2008) provide much useful research data and background theorizing relevant to the dispute.

Driver et al. (1996) characterize students' reasoning about theory and evidence in terms of whether it is *phenomenon-based*, *relation-based* or *model-based*. In phenomenon-based reasoning, an approach used only by the youngest students in the research sample (12 year olds), a theory is a taken-for-granted statement of fact, and no distinction is made between observations and explanation. Explanations are simply a re-statement of the observations, and investigations of phenomena are largely random or *ad hoc*. In relation-based reasoning (the commonest form of reasoning among 12 and 16 year olds), a distinction *is* made between observation and explanation. Commonly, the explanation is a generalization from empirical data or a correlation between stated variables arrived at by inductive reasoning and expressed in the same kind of language categories as the observations, but with no reference to any underlying mechanism. In model-based reasoning (used by only a minority of students, even at age 16) there is a discontinuity between explanation and observations. Explanations involve posited entities, couched in different language from the observations (for example, the macro behaviour of substances explained in terms of atoms and molecules), and are clearly the outcome of creative endeavour by scientists. Observational evidence constitutes a reason for subscribing to a particular explanation or theory, and so the appearance of new evidence is an incentive to shift allegiance to a model that better deals with it. Although these categories form a clear hierarchy in terms of the complexity and sophistication of the reasoning, with model-based reasoning being the target learning,[8] it seems that students who have reached the level of model-based reasoning in some subject areas may still revert to a less sophisticated approach on occasions, especially when unfamiliar with the associated concepts. They may also be *disinclined* to use this "higher level" approach when the context is of little interest or importance to them – clearly, an important message for all teachers. It is alarming that when Windschitl (2004) invited 14 preservice secondary school teachers to conduct an empirical investigation of their choice and design, *none* employed model-based reasoning. Following a longitudinal study of 15 primary school students in Western Australia, Tytler and Peterson (2004) conclude that the relation-based reasoning category is insufficiently refined to capture the nuances of different student reasoning patterns. They propose a narrower definition of the category and the insertion of an additional category, which they call *concept-based* reasoning.

– *Relation-based reasoning* – explanations involve the identification of relations between observable or taken-for-granted entities, rather than a search for an underlying cause; explanatory approaches tend to be confirmatory and uncritical; they emerge from the data in an unproblematic way.
– *Concept-based reasoning* – explanations are cast in terms of conceptual entities; experimentation is guided by hypotheses; the role of disconfirming evidence is acknowledged as significant.

The authors point out that their definition of relation-based reasoning has much in common with the *engineering* model of experimentation proposed by Schauble et al. (1991b), where "the purpose is seen as the achievement of a successful outcome rather than identification of cause, and where there is a focus on confirming and manipulating variables relevant to an outcome rather than test for disconfirming evidence" (p. 113).

Using a purpose-made instrument called *Nature of Science Interview*, which includes questions about the goals of science, the nature of scientific questions, the role of experiment, the theory-dependence of experiment and the processes by which scientists change their ideas, Carey and Smith (1993) have also characterized students' epistemological understanding in terms of a 3-level hierarchy. At level 1, scientific knowledge is assumed to comprise a collection of true facts about the world, acquired in piecemeal and unproblematic fashion from observations and experiments. At this stage, no distinctions are made among scientists' ideas, investigative approaches and findings; students have no appreciation of the role of scientists' ideas in guiding their investigations, no notion that data from investigation provides evidence for ideas, and no appreciation of the uncertainty and tentative nature of scientific knowledge. At level 2, scientific knowledge comprises a collection of *tested* ideas about "how things work" and "why things happen"; scientists design experiments involving measurements and observations in order to test their ideas, which they abandon or revise if the test does not confirm them. Students at this stage have developed the notions of explanation and hypothesis testing (though hypotheses are seen as data-generated rather than theory-generated) and are probably slightly ahead of students characterized as being at the "relation-based" stage in the Driver et al. (1996) model. At level 3, explanations are recognized as based on theoretical entities that are not part of the simple observation categories. Scientific knowledge comprises well-tested and coherent explanatory frameworks that scientists use to make predictions, construct specific hypotheses, design inquiries and interpret the evidence generated. Theories are recognized as inherently conjectural and hypothesis-driven, experiments are seen as a means of developing them further. Smith et al. (2000) have refined the coding system by introducing two intermediate levels. At level 1.5, there are signs of an awareness of the role of ideas in scientists' work, although the nature of ideas is vague. At level 2.5, there is a deepening understanding of the conceptual/explanatory nature of hypotheses/theories, awareness that the same pattern of results can be interpreted in more than one way, and recognition that theories may be broader in scope than hypotheses and can affect hypothesis testing.

Carey and Smith (1993) found that Grade 7 students had average level scores of 1.0 across the various dimensions. In Honda's (1994) study of Grade 11 students, cited by Smith et al. (2000, p.358), scores averaged at 1.39, although the majority of students reached level 2 on at least one dimension. In a study of 35 college freshmen intending to major in science, Smith and Wenk (2006) found that, in terms of differentiating between theory and evidence, students had average scores between 1.38 and 1.67, with most of their responses at level 1 or level 2; 12 students had average scores between 1.92 and 2.23, with most of their responses at level 2 or level 2.5; 13 students had average scores between 1.71 and 1.87, and were more evenly split between level 1.5 and level 2. Again, no student scored at level 3. Research showed that there was a strong relationship between students' level of differentiation of scientists' ideas and evidence, their awareness of the uncertainty and tentativeness of scientific knowledge, and their willingness to engage with scientific controversies.

No student with an average … score less than 1.7 (predominantly level 1 and 1.5) … regarded scientific knowledge as fundamentally uncertain … whereas almost all those with average … scores greater than 1.9 (consistent level 2 responders) did so. Those with average … scores between 1.7 and 1.9 were more variable, although the majority still regarded scientific knowledge as at least partially certain. Overall, there was a significant association between being a consistent level 2 responder … and being aware of the uncertainty of scientific knowledge … Similarly, there were clear relations between how students reasoned about a specific controversy … and their differentiation of evidence/hypotheses/theories ... and their understanding of the uncertainty of scientific knowledge ... For example, all students who systematically engaged with the controversy in deeper ways (e.g., by showing awareness that the scientists may have made interpretive errors) had been aware of the uncertainty of scientific knowledge. (Smith et al., 2000, p. 768)

Somewhat surprisingly, Thoermer and Sodian (2002) found no measurable differences between the scores of undergraduates studying science at a German university and those of graduate students in the same institution, though graduate students in physics did score substantially higher than all other groups.

CHANGE AND DEVELOPMENT

It would be surprising if students' educational experiences did *not* impact their progress towards more sophisticated epistemological thinking. In an effort to test this supposition, Smith et al. (2000) studied two grade 6 classes with different educational histories. Both classes had been taught by one teacher only throughout grades 1 to 6. However, the teacher in one class had used a constructivist approach in which small groups of students investigated phenomena and developed personal models to explain them through experimentation and dialogue within and among groups, while the teacher of the comparison class had adopted a more traditional approach using lectures, readings, outdoor experiences and art work. The students in the constructivist class had developed more sophisticated views about the goals of science (with a shift away from "doing things" and "gathering information" towards "developing ideas and understanding") and about the nature and purpose of experiments (a shift from "trying things out" and "finding answers" to "developing and testing ideas"). Indeed, 83% of students in the constructivist class had an average score of at least level 2, and *all of them* had average scores greater than 1.5 (see above, for an interpretation of these figures). In other words, they exhibited a level of sophistication usually only shown by students four or five grades more advanced. Indeed, the percentage of these grade 6 students expressing consistent level 2 views was higher than among the grade 11 students in Honda's (1994) study. The students in the constructivist class were more aware of the tentative and evolving nature of scientific knowledge and of the complex nature of theory appraisal; they recognized scientists as reflective and creative individuals who often collaborate with others. A number of them noted that scientists' current ideas probably influence what they choose to investigate, bias the design and interpretation of the inquiry, and cause them to resist new ideas proferred by others. Tucker-Raymond et al. (2007) report that a 10-week long course for 36 students in grades 1 to 3 involving extensive hands-on exploration, together with reading, writing and class discussion about science, brought about some significant shifts in students' NOS understanding: what the researchers call the *engineering view* of science (scientists "make things") decreased sharply, while the incidence of the *"inductive-explorer view"* and the *"developing knowledge view"* (scientists "find out things") both increased sharply – a clear reflection of the curriculum experiences provided. Somewhat surprisingly, research by Liu and Tsai (2008) indicates that science majors among first-year undergraduates have less sophisticated beliefs regarding the theory-laden and culturally-dependent aspects of science than non-science majors. Disturbingly, science education majors (intending teachers) had the poorest scores overall. The authors speculate that science and science education majors have simply been exposed for much longer to a curriculum that presents scientific knowledge as objective and universal (see also Schommer & Walker, 1997; Paulsen & Wells, 1998; Hofer, 2000; Palmer & Marra, 2004; Smith & Wenk, 2006).

Hogan and Maglienti (2001) present an intriguing account of some key differences between scientists and grade 8 students in the ways they reason about data. The researchers provided each participant with a series of ten conclusions from data made by someone else (a fictitious biology teacher), nine of which had a weakness of some kind (such as not accounting for all of the data or using vague or unspecific language). In the context of a 1:1 interview they asked the participants to evaluate the validity of the ten conclusions. Scientists prioritized empirical consistency of evidence and conclusions, the need to consider all the evidence and the requirement that the argument be free of logical inconsistencies. Students put more emphasis on their own prior theoretical understanding and whether the conclusion matched their own inferences from the data. They tended to form their own conclusions from the data and then use it as a check on the conclusions they were asked to evaluate. Ford (1998) reports similar findings when scientists and non-scientists were asked about the confidence they had in the claims located in a series of three articles in a popular scientific magazine. Scientists considered how data were collected and analyzed; non-scientists used personal experience and conformity with their own views. Differences are also apparent in the way scientists and non-scientists ascribe cause and effect: non-scientists tend towards a few simplistic causal structures (best of all, one structure) into which all new data is assimilated; scientists are more willing to consider a range of options (Chi, 1992).

Other research has indicated that in situations where students collect their own data, and those data do not point clearly to a conclusion, students will often impose their own patterns on the data based on their prior expectations. Thus, Millar and Lubben (1996) report that in situations where the data are inconclusive, students draw conclusions in line with their predictions; when data are more clearly at variance with predictions, students respect the data if their prediction was merely a guess but distort data to meet the prediction if it was based on a well-understood theory about the phenomenon under investigation. In a related study with 11, 14 and 16 year olds, Lubben and Millar (1996) note that many of the younger students made observations and took measurements "with no apparent awareness of the uncertainty associated with the measurement process or of the need to be able to defend their data as reliable" (p. 956). Across the age groups there was evidence of a progression in understanding of both the *processes of sound measurement* (from denial of the need to repeat measurements, through a search for recurring results and deliberate manipulation of control variables to ensure a satisfactory spread of results, to estimation of the likely range of results and identification of errors) and the *evaluation of measurements* (from assessing a single value on the basis of what is likely or expected, through selection in accordance with the place of the measurement in a sequence and selection of a recurring value, to using a full set to calculate mean or median). Studies of students' responses to anomalous data conducted by Chinn and Brewer (1993, 1998), Chinn and Malhotra (2002a) and Lin (2007) reveal a marked reluctance to give it proper critical attention. Students may simply ignore it, reject it as irrelevant or outside the domain of the theory, cast doubt on its reliability and appropriateness, express concern about its proper interpretation, put it to one side for later consideration (which rarely occurs), or reinterpret it to make it consistent with their existing views. Only very rarely will they use it to reject or modify their existing theoretical views, and even then it is often no more than a peripheral change or minor modification. Toplis (2007) notes that the demands of school lab work, with its severe constraints on time, materials and apparatus, and the high priority placed on "correct answers" in conventional assessment schemes, militate strongly against students responding appropriately to anomalous data. Even if they recognize anomalies, and even if they might be willing to collect further data, they are rarely in a position to do so. Interestingly, Toth and Klahr (2001) found that students experienced fewer problems in reasoning successfully from anomalous empirical evidence when they had collected the data themselves than when data were provided by the teacher or taken from a textbook. It is important to recognize that within the scientific community the response to anomalous data is profoundly affected by the reputation of the scientists who generate it, whether the methods used to generate it are well-established, and whether the anomalous data can be replicated by other research groups.

Samarapungavan et al. (2006) report some interesting differences between students and scientists in their response to anomalous data. Scientists drew a distinction between routine methodological error due to calibration faults, impure reagents, lapses in bench technique, and the like, and "productive anomalies", that is, unexpected results that are a stimulus to further investigation and theorizing. Undergraduates saw anomalous data as a clear indication of their own technical inadequacies, while high school students interpreted it as "getting the wrong result". As all science teachers are aware, it is common for students at all levels of education to respond to this situation by "massaging the data" (changing their observations to match the expected conclusion), working backwards from the conclusion found in the textbook or on the Internet in order to generate "new data", which is suitably adjusted so as not to seem "too good" or to "fit too well". Alternatively, students simply obtain the "right results" from friends.[9] Dunbar (1995) makes some interesting observations concerning the way inconsistent evidence is treated by scientists: less experienced scientists are more willing to maintain the hypothesis and reject the data than are more experienced scientists. While more experienced scientists show much less "confirmation bias" than their less experienced colleagues, they often display a "falsification bias": discarding good data that might actually confirm their hypothesis. Dunbar attributes this bias to frequent experience of being proved wrong.

In common with the students in several other studies, most of the 16 year olds interviewed by Driver et al. (1996) interpreted disagreements between scientists as evidence of sloppy work or incomplete data. Although some students did entertain the notion that personal interests and biases might impact the views of scientists, few students recognized that data can be interpreted in different ways because different theories are brought to bear.

> Evidence seems to be regarded as "information" or "facts", which tell us "how things are", rather than as raw material for conjecture about "how things might be". (p. 132)

In my experience, only a minority of students (even at 16 or 18 years of age) appreciate that theories are, in general, empirically under-determined. Nor do students always appreciate that creativity and imagination play a key role in their construction. Rather, many see theories as emerging relatively easily and painlessly from observations of phenomena and events or via experiment. When there is controversy, as in the case of evolutionary theory, for example, students may take the position that some theories are proven (by observation and experiment) while others are "just theories" – i.e., little more than "an idea" or speculation about how things might be explained.

Using interview methods and writing tasks, Joan Solomon and her colleagues have researched students' understanding of the nature of theory and its relationship to experiments. Ideas about theories appear to fall into three major categories: (a) theory as a hunch or guess about what will happen, (b) theory as an explanation for why things happen (very close to the view we might wish students to hold), (c) theory as a set of facts. In these studies, students sometimes used the expression "proper theories" for those explanations that have been shown by tests and experiments to be "true" (Duveen et al., 1993; Solomon et al., 1994a, 1996). As I report in Hodson (2008), some students in my Toronto-based study took the view that theories are just guesses or speculations. Comments such as "that's just a theory ... it isn't true" and "It's just something that *you* think" were widespread, in some sense reflecting the everyday saying: "it may be OK in theory but it doesn't work in practice". Intriguingly, some students in the Solomon et al. (1994) study seemed to see experiment as a "last resort" when no-one knows what to do next, as revealed in comments such as "We'll just have to do an experiment". Among older students in my Toronto-based study there was sometimes a distinction drawn between "science as it ought to be" (the kind of science that is described in textbooks) and "science as it is" (what goes on in research labs), as in the study by Rowell and Cawthron (1982) noted earlier. Students were not expressing the view that scientists behave differently when self-interest, commercial interests or military interests intervene (although this awareness may be an important aspect of critical scientific literacy); rather, they were suggesting that the ideals of scientific inquiry are so demanding that they are beyond the reach of most ordinary students. Interviews with students undertaking research projects as part of their final year programme at the University of Leeds revealed that undergraduates resemble secondary school students in regarding scientific knowledge as "provable beyond doubt" on the basis of experimentally acquired data (Ryder et al., 1999). In most cases, scientific inquiry was seen in terms of individual scientists seeking reliable data on which to base their conclusions, usually via experiment. Encouragingly, their appreciation of the role of theory in directing the nature of scientific inquiry became markedly more sophisticated during the course of the project work, though they remained relatively unaware of the internal and external social dimensions of scientific practice.

On a related theme, it is common for students to hold the view that when theories have gained substantial observational confirmation they become laws (Ryan & Aikenhead, 1992; Meyling, 1997; Brickhouse et al., 2002; Sandoval & Morrison, 2003; Dagher et al., 2004). The distinction between theories and laws is a recurring theme in the research conducted by Norman Lederman and his colleagues. The "approved position" is that laws describe relations and reg-ularities among entities that can be readily confirmed by observations, while theories are complex explanatory systems (including the capacity to explain why particular laws hold) that are not susceptible to simple observational test and are, generally, empirically under-determined. While philosophers of science and some educational researchers may choose to draw such a distinction, most scientists do not. Most students also fail to recognize this distinction, and many refer to a hierarchical relationship in which theories become laws when there is sufficient supporting evidence. Moreover, many secondary school students know full well that Newton's laws of motion are not laws at all, in the sense that they do not adequately describe the motion of objects travelling at close to the speed of light. They know, also, that the so-called gas laws do not describe the behaviour of real gases, though they do describe the behaviour of "ideal gases". Several of the scientists interviewed by Wong and Hodson (2009) said that "law" is a confusing, unhelpful and outmoded term; it indicates a status that is inconsistent with the basic premiss that *all* scientific knowledge is subject to modification when there is appropriate evidence and convincing argument for change. In my view this is a timely reminder that in our search for an authentic view of science we should take note of what scientists tell us about their day-to-day practices. After all, does *authentic* science not mean the activities of real scientists rather than some philosophy-inspired idealization of practice?

IMPACT ON LEARNING

It is reasonable to speculate that students' epistemological beliefs will impact their view of learning and their approach to problem-solving. Put simply, if stu-dents believe that scientific knowledge arises directly and unproblematically from observation they are likely to approach science learning in terms of memorizing facts, and they may expect to be able to *prove* that a hypotheses is correct (or false) by means of *the* scientific method. If they think of science as a continuing process of concept development and theory building they will be more inclined to focus on conceptual understanding. That is, students' epistemological beliefs may shape their metalearning assumptions and thus influence their learning orientations with respect to adoption of a performance orientation or a learning orientation (Kruglanski, 1989; Lidar et al., 2006).

In Tsai's (1998) study, a Chinese version of Pomeroy's (1993) questionnaire was used to ascertain the epistemological beliefs of approximately 200 high ability grade 8 students in a high school near Taipei City. Follow-up interviews identified 20 students with clearly articulated and strongly held views for subsequent interview concerning their learning orientations. Students who regard scientific knowledge as constructed through the creative efforts of scientists tended to employ more active and more constructivist learning methods; those students who believe that scientific knowledge arises from careful and systematic observation tended to use more rote learning strategies (see also, Halloun & Hestenes, 1998; Windschitl & Andre, 1998; Cavallo et al., 2003). The former group recognized the value of science in everyday life and were motivated to learn science largely by personal interest and curiosity; the latter group were motivated mainly by performance in examinations. Tsai reports that when students were asked to describe an ideal environment for learning science, the constructivist-oriented students emphasized opportunities to discuss with others, solve real life problems and control their own learning;[10] the empiricist group said that they learn best, and minimize the anxiety associated with making mistakes, when teachers present scientific ideas clearly, facilitate the assembly of a good set of notes and help them to build up reliable store of "correct facts". In a follow-up study with 1176 grade 10 students, those with a constructivist orientation were very critical of conservative teaching methods currently being used in their schools; they expressed a strong preference for approaches that would enable them to interact, negotiate meaning and build consensus with others, have enough time to integrate their new knowledge with their prior knowledge and experiences, exercise more control over their learning activities, and think independently (Tsai, 2000). Importantly, in the context of this book, constructivist students said that they try to use science in real life situations, while empiricists stated that science is not related to real life and has no value or application outside the classroom (see also Songer & Linn, 1991). Further, empiricist students seemed to lack a metacognitive perspective when learning science, and so focused on learning outcomes rather than learning procedures, while constructivist students tended to prefer learning in depth rather than breadth. This is significant, given that Hazari (2006) shows how a deep and narrow science curriculum in school (rather than broad and shallow) is strongly correlated with high levels of attainment in physics at university level, especially for female students.

Lin et al. (2004) show that students are likely to perform better on concept-based problem-solving (as distinct from algorithm-based problems) when they believe that observations are theory-laden, theories are constructed, and there is no single scientific method (scientists use whatever method is appropriate to the circumstances). Students who hold more traditional views, especially that there is one fixed, all-purpose method of scientific inquiry, perform less well on concept-based problems. However, their tendency to memorize information enables them to perform much better on assessment exercises that do not require knowledge integration and application. From lengthy interviews with 25 undergraduates studying chemistry, Havdala and Ashkenazi (2007) identified three individuals with clearly articulated but significantly different views about science, which the researchers designated as *empiricist-oriented* (observational data has priority over theory, which is seen as subjective), *rationalist-oriented* (theories are a sound guide to the validity and reliability of observations) and *constructivist-oriented* (both theory and observations are subjective and tentative). Subsequent observation of their responses to various laboratory-based activities revealed some fairly predictable differences in the way they prepared for the lab session, their approach to writing the lab report, the relative priorities afforded to theory and observation, and their ability to coordinate theory and empirical evidence. It was shown that over-confidence in one type of knowledge (theory or empirical knowledge) led to over-simplification of the relationship between theory and evidence.

SCIENCE AS A SOCIAL ENTERPRISE

Moss et al. (2001) state that, in general, grade 11 and 12 students' understanding of the nature of scientific knowledge (for example, that it requires evidence, and is tentative and developmental) is more complete than their understanding of the scientific enterprise. If the term "scientific enterprise" is taken to include internal and external social factors that impact on the conduct of science, and is not restricted to the specific methods employed in particular scientific inquiries (or to what some teachers continue to refer to as "*the* scientific method"), then I would readily concur. Despite the efforts of the STS movement, many students continue to believe that science occurs in something of a sociocultural vacuum – a view held by both preservice and inservice science teachers in Tairab's (2001) study.

Although some of the students in the study by Solomon et al. (1994) appreciated the importance of both collaboration and competition in stimulating scientific development, students generally have only a vague and fragmentary understanding of social factors internal to the practice of science. External influences are easier to recognize. Among the students with whom I have worked in Toronto schools, it is fairly common to hear the view that research funding should only be provided when scientists can demonstrate that their research will improve the quality of our lives in some tangible way. Science *should*, students often say, serve the public good. And many believe that it probably does. Although most students recognize how science (and, more particularly, technology) impact on society and the environment, with many being able to provide appropriate examples, they often have only a rudimentary understanding – beyond the issue of funding – of the ways in which sociocultural circumstances determine the kind of science that we do and may even impact the ways in which it is done. Interestingly, I have found greater awareness of these influences among students in New Zealand, possibly as a consequence of repeated emphasis on Maori perspectives in the New Zealand curriculum (both in science and in other subjects). Nevertheless, Hipkins et al. (2005), writing from a New Zealand location though not necessarily from a New Zealand perspective, surmise that STS has served as a distraction from epistemological issues because once teachers have addressed the social and environmental impact of technology (simplistically regarded as applied science), many feel that they have "done" the learning *about* science component of the curriculum, and contentedly return to content-oriented teaching.

These observations serve to raise important questions about the extent to which social and cultural identity impact on a student's NOS views. Cobern (1993, 1995, 1996) describes how different sociocultural environments produce different *worldviews* – defined as composite sets of consciously held and unconsciously held beliefs and values about the nature of reality and the generation of knowledge about it. These worldviews predispose an individual's thoughts and emotions, and their everyday behaviours and actions, in particular ways. It would be surprising, then, if they did not impact on students' views about science. Thus, we might anticipate some major cross-cultural differences in NOS understanding. Even *within* a particular society, the specific knowledge, beliefs, language, behaviours, values and aspirations acquired through membership of family and friendship groups may predispose students to particular views about science and scientists. So, too, might access to TV documentaries about science, visits to museums, zoos and science centres, all of which are contingent on students' sociocultural location and socioeconomic status. For some students, religion may be one of those sociocuturally located factors that impact quite substantially on NOS views, especially in the context of evolutionary theory or the moral-ethical issues surrounding genetic engineering and stem cell research (Dagher & BouJaoude, 1997, 2005; Roth & Alexander, 1997; Rudolph & Stewart, 1998; Ayala, 2000; Sinatra et al., 2003; Abd-El-Khalick & Akerson, 2004). Indeed, in some circumstances and with some individuals, it may well act as a boundary condition for entertaining scientific explanations at all; in effect, acting as a kind of roadblock that prevents students from recognizing and attending critically to data and ideas that are seen as constituting a challenge to religious convictions (Samarapungavan, 1997). Bybee (2001), Griffith and Brem (2004), Anderson (2007), Hermann (2008) and Martin-Hanson (2008) discuss pedagogical approaches suitable for teaching evolution as a controversial topic to students with strong religious beliefs, largely from a US perspective.

To investigate these influences, Dagher and BouJaoude (2005) used openended interviews to ascertain the perceptions of the nature of evolutionary theory held by 15 students majoring in biology at a private university in Beirut. Nine of the 15 students focused on the nature of the evidence and concluded that evolutionary theory is "deficient in meeting a criterion of 'proof' or evidence that is trustworthy" (p. 383). The authors state that although students may have only a fragmented and largely implicit view of what constitutes a scientific theory, they readily articulate some characteristics they claim to be "essential" or "crucial" when presented with a claim that they wish to reject because it clashes with strongly held religious beliefs, worldviews, moral-ethical values or social expectations. Thus, evolutionary theory was considered to be deficient as a scientific theory because there is no direct observational evidence, it is not subject to experimentation, it was not derived in accordance with the steps of the scientific method, and is unable to make testable predictions – a set of specifications that the theory of plate tectonics, for example, is not required to meet. It seems that when students feel that religious or cultural beliefs and/or moral-ethical values are threatened they readily cite grounds for rejection of a theory that are accepted as perfectly legitimate in other circumstances. Sinatra et al. (2003) show that students who see knowledge as tentative and subject to change, and have a disposition towards open-mindedness, can come to an understanding of scientific explanations of controversial topics such as human evolution or the chemical origin of life, even if they do not necessarily accept them as "true explanations". In other words, although they do not believe it, they understand why it is worthy of belief.

Problems resulting from a clash of worldviews have also been noted by Ogawa (1995) and Kawasaki (1996) with respect to some Japanese students and by Jegede (1998) with respect to African students. Aikenhead (1997) has reported at length on similar kinds of issues arising with First Nations students in Canada. For example, he argues that the interests of First Nations people in survival, coexistence and celebration of mystery are not in sympathy with the drive of Western science to achieve mastery of nature through objective knowledge based on mechanistic explanations. Nor are the holistic perspectives of Aboriginal knowledge, with its "gentle, accommodating, intuitive, and spiritual wisdom", in sympathy with reductionist Western science and its "aggressive, manipulative, mechanistic, and analytical explanations" (p. 220). Crossing the border into the community of science education is inhibited not only by cognitive barriers but also by some of the distinctive values-laden features of science, features that are often distorted by school science curricula into the scientistic cocktail of naïve realism, blissful empiricism, credulous experimentation, excessive rationalism and blind idealism noted by Nadeau and Désautels (1984). Sutherland (2005) notes that among the Cree students she interviewed, those who were able to distinguish clearly between Indigenous knowledge and Western scientific knowledge, and who could articulate different criteria of validity and applicability, were usually the most successful in learning science in school. Interestingly, Aikenhead and Otsuji (2000) report that Japanese teachers tend to view science from an holistic perspective, while Canadian teachers generally adopt an analytical and reductionist view. However, both groups of teachers seemed unaware of the ways in which science reflects and projects a particular set of socioculturally located beliefs and values that can constitute formidable barriers for some students. This research and theorizing tells us very clearly that careful consideration of issues in the history, philosophy and sociology of science are key to ensuring access to science for all students, and should be paramount in all efforts to re-build the curriculum.

Surprisingly, the Draw-a-Scientist Test tends to produce fairly consistent findings across cultures (Chambers, 1983; She, 1995, 1998; Parsons, 1997; Finson,

2002; Fong, 2002), although Song and Kim (1999) suggest that Korean students produce "slightly less stereotypical" drawings, especially with respect to gender and age. Generally, they draw younger scientists than their Western student counterparts – drawings that probably reflect the reality of the Korean scientific community. Using DAST and follow-up semi-structured interviews, Parsons (1997) elicited the images of scientists held by an all-female group of 20 high ability African American students. Eleven respondents drew and described their scientist as a White male, four as a Black male, two as a Black female and one as a White female. The most striking difference in the descriptions was that between the White male scientist and the Black male scientist. Both were perceived as intelligent and hard-working, but whereas White scientists were described as exhibiting the usual negatively stereotyped characteristics (being aloof, odd, and probably friendless), Black male scientists were portrayed as kind, respectful, easygoing and responsive to the needs of others. The author concludes that this picture of the Black scientist is consistent with what she refers to as "the Black Cultural Ethos" – the way African American people are expected to be, by other members of the community. In a study of 358 students in grades 1–7 in South West Louisiana, Sumrall (1995) found that African Americans (especially girls) produced less stereotyped drawings than Euro Americans with respect to both gender and race. Interestingly, the drawings of African American boys showed an equal division by race of scientist but an 84% bias in favour of male scientists. Indeed, many researchers have pointed out that girls are generally less stereotyped in their views about science and scientists than are boys, except in relation to social aspects (see earlier discussion). However, Tsai and Liu (2005) note that female Taiwanese students are less receptive than male students to the idea that scientific knowledge is created and tentative, rather than discovered and certain.

In a study of the images of science and scientists held by preservice teachers in Israel, Rubin et al. (2003) found that the drawings of Hebrew-speaking students conformed to the usual stereotypes of bald men in lab coats and wearing glasses, but the drawings of Arabic-speaking students showed men with beards and moustaches wearing traditional tunics (not lab coats). When asked to name five scientists, the Hebrew speakers tended to nominate Einstein, Newton, Galileo, Darwin and Pasteur, while the Arabic speakers often included Ibn Sina, al-Razi, al-Ghazali and al-Khawarizmi. Interestingly, when I tried a similar exercise with 27 serving teachers enrolled on an MEd programme at the University of Hong Kong, there was not a single Chinese scientist in anyone's list, though my Canadian students always include the name of Frederick Banting.[11] If Tao (2003) is correct in reporting that Hong Kong students generally hold very traditional stereotyped views on NOS issues, my finding should not have been surprising, even though it was disappointing. Of course, it should also be noted that stereotyped views *about* science do not prevent Hong Kong students performing exceptionally well in international tests of science attainment.

Liu and Lederman (2002) report that Taiwanese students display more sophisticated understanding of NOS issues (as measured by a variant of the VNOS questionnaire) than is usual among American students.[12] Although it should be noted that the Taiwanese population tested were designated as "gifted students", the authors report that an earlier study (by Lin) of 1670 senior high school students with a "more normal distribution of abilities" gave similar results. In contrast, Kang et al. (2005) suggest that Korean students (at grades 6, 8 and 10) are much more likely than Western students to hold the naïve view that scientific knowledge is true (in an absolute sense) because if this were not the case, "we wouldn't have to learn it in school". Theories change because "the old ones were proved to be wrong" by means of experiments, largely as a result of using better equipment or new technologies. Following a questionnaire-based study with more than 2000 grade 10 students in Turkish high schools, using a shortened and modified form of VOSTS, Dogan and Abd-El-Khalick (2008) conclude that the majority of students, and their teachers, hold naïve views on many aspects of NOS. More encouragingly, students and teachers hold "meritorious" or "informed" views with respect to the theory-driven nature of scientific observation (64.6% and 53.5%, repectively) and the tentativeness of scientific knowledge (68.2% and 72.7%, respectively). The researchers note that students living in regions that are "more European-like in culture (i.e., Marmara and Aegean)", tend to have more informed NOS views than those living in areas that are "more Eastern-like in their cultural overtones (i.e., Southeast and East Anatolia)" (p. 1102). Griffiths and Barman (1995) interviewed 96 students, drawn equally from high schools in Canada (mean age = 17.6 years), the United States (mean age = 16.9) and Australia (mean age = 16.5), about NOS issues – particularly, the nature of scientific inquiry, the role of observation, the tentative nature of scientific knowledge, and the status of scientific theories. While the very small samples preclude any firm international comparisons, there are some interesting and noteworthy differences. The lockstep view of scientific method was cited by 75% of US students but not at all by Australians; 40% of Canadian students, but only 15% of Australians and 0% of Americans, said that science changes over time because of the influence of new ideas; the theory dependence of observation was acknowledged by many more Canadian students than by Australian or US students. Of particular relevance to notions of critical scientific literacy, with its emphasis on politicization of students and promotion of responsible citizenship, is Liu and Lederman's (2007) Taiwan-based exploration of the interactive relationship between an individual's *worldview* and their conceptions of the nature of science. Most notably, student teachers with "informed" NOS views (as measured by a variant of VNOS) tend towards a biocentric view of the environment, while those with "naïve" views were more inclined to an anthropocentric view.

All of these findings point to one simple conclusion: if we change the curriculum we can be sure that student understanding will change, though we may not be able to predict in exactly what ways or to what extent. Also, given the ubiquity of the Internet and other contemporary ICT media, we can no longer rely on the old adage that "students won't learn what they are not taught". In the contemporary world, people may learn as much (if not more) about science from these sources as they do from the formal science curriculum. This realization makes it imperative that science teachers address NOS issues carefully, systematically and rigorously, and seek to subject the ideas students acquire from informal sources to rigorous critical scrutiny. Of course, as the research makes clear (if we need research to point out such an obvious conclusion), different curriculum experiences and out-of-school experiences result in students holding different views about science and scientists. As Driver et al. (1996) comment, the tendency of many students in the United Kingdom to regard experiments as a comparison of outcomes of different initial conditions is undoubtedly a consequence of the emphasis on "fair tests" in the National Curriculum for England and Wales. Also, the weakness of many students in model-based reasoning may be a direct consequence of the lack of emphasis on these matters in the classroom and textbooks, rather than a generalized inability to reason this way (see discussion in Chapter 6). As always, research has to be interpreted with great caution. Questions that spring to mind concern: (i) the potential for changing students' views through judicious curriculum interventions, and (ii) the extent to which students' views are a consequence of teachers' views transmitted implicitly or explicitly through the curriculum. It is this second matter that I address in Chapter 3.

NOTES

[1] This instrument has subsequently been modified to produce the *Views of Nature of Science* questionnaire (Lederman et al., 2002), and further modified by Lederman and his co-workers (Lederman, 2007) and by Wong and Hodson (2009).

[2] Finson et al. (1995) have developed a checklist for identifying and quantifying the components of students' drawings for more efficient data analysis.

[3] 614 students used data-focused reasoning on at least one occasion; 303 used radical relativist reasoning on at least one occasion; 728 used theory and data related reasoning on at least one occasion; 609 used both data focused and theory and data related reasoning; 239 used all three forms of reasoning.

[4] Repertory grids enable researchers to ascertain links between different facets of an individual's knowledge and understanding (and between understanding and actions) in quantitative form (Fransella & Bannister, 1977). Using them over the lifetime of a research project enables a developmental record of students' (or teachers') views to be built up. Because repertory grids often produce surprising data and highlight inconsistencies in respondents' views, they provide a fruitful avenue for discussion and exploration of ideas. For these reasons, Pope and Denicolo (1993) urge researchers to use them as "a procedure that facilitates a conversation" (p. 530).

[5] Given the situation 20 years on, and the suspicion with which scientific claims are sometimes greeted, one wonders whether similar results would be obtained today with respect to trustworthiness.

[6] Chambers (1983) notes that some students' drawings of scientists included notices on doors or cabinets that said "Private", "Top Secret" and "Keep Out".

[7] This perfectly describes my own experience as a researcher in synthetic organic chemistry some 40+ years ago.

[8] Chapter 4 includes some discussion of how explicit and reflective instruction in modelling can assist students in progressing through this hierarchy.

[9] Some teachers choose to avoid this problematic situation, especially with experiments that are central to building particular conceptual understanding, by engaging in what Nott and Smith (1995) and Nott and Wellington (1996) refer to as *conjuring* – that is, ensuring "correct results" by sleight of hand or manipulation of apparatus and materials.

[10] In a subsequent publication, Tsai (2003) reports that having a constructivist orientation increases the likelihood of students negotiating the meaning and significance of experiments with their peers.

[11] Frederick Banting, a Canadian, shared the 1923 Nobel prize for medicine with John MacLeod for their discovery of insulin.

[12] They also held a strongly stereotypic image that science is a product of Western society. Few students could name a Chinese scientist.

CHAPTER 3
RESEARCH ON TEACHERS' VIEWS OF NOS

It is reasonable to suppose that students' views *about* science are the outcome of a complex interaction between *curriculum experiences* (what students encounter in school science lessons) and *informal learning experiences* (what they learn via the popular media (leisure reading, including science fiction, movies, TV and radio, magazines and newspapers, Internet sites, advertizing, and so on) and from visits to museums, zoos, aquaria, nature reserves, field centres, hospitals, research centres, and the like. This book is primarily concerned with the learning *about* science content of curriculum experiences: the kind of experiences we commonly provide; the kind of experiences we *should* provide, but don't; the kind of experiences we *do* provide, but shouldn't. Part of that curriculum provision includes the need for teachers to address out-of-school influences on students' views about science, and whenever necessary take steps to counter them. Images of science and scientists in the popular media not only impact students' NOS views but also influence the decisions they make with respect to socioscientific issues. It is vitally important that teachers are cognizant of the nature of these influences.

Curriculum experiences are of two kinds: those that we explicitly plan and those that we do not. There are many explicit messages about science in textbooks, especially in those early chapters that tell students what science is and what scientists do; there are lots of explicit references to the nature of science and the history of science in STS-oriented curriculum materials; on occasions, teachers take time and trouble to emphasize particular features of science and scientific inquiry during laboratory activities or in class discussions. Just as frequently, however, "messages" about the nature of science and scientific practice are *not* consciously planned by the teacher. Rather, they are implicit messages located in the language we use, the kind of teaching and learning activities we employ (especially in laboratory and fieldwork settings), the examples of science and scientists in the real world we employ, the illustrative and biographical material included in textbooks, the assessment methods we use, and so on. What is at issue here is a very powerful *hidden* or implicit curriculum. In the words of Cawthron and Rowell (1978):

> Any science curriculum and its translation into practice embody images of the nature of man as scientist, of scientific knowledge and of the relationships between them. These are communicated, albeit implicitly and incidentally, just as surely as the subject matter itself. (p. 31)

At first glance it seems self-evident that teachers' own views about the nature of science will influence substantial aspects of their professional practice, including decisions about the design of learning experiences. At the very least, we might expect there to be a direct correspondence between teachers' views about scientific inquiry and the extent to which it is theory-driven, for example, and the way they design laboratory-based activities. It might also be expected that a teacher's views about the role and status of scientific theory, and the way in which scientific knowledge is generated, scrutinized and validated, will be evident in the design features of learning activities, especially in light of Tsai's (2002) finding that there is often a very close alignment of teachers' beliefs about teaching science, beliefs about learning science and views about the nature of science. Using an interview approach, the beliefs and views of 37 science teachers were separately categorized as "traditional", "process" or "constructivist" with respect to these three areas of concern.

> The "traditional" category perceives teaching science as transferring knowledge from teacher to students, learning science as acquiring or "reproducing" knowledge from credible sources, and scientific knowledge as correct answers or established truth. The "process" category perceives teaching science and learning science as an activity focusing on the processes of science or problem-solving procedures, and scientific knowledge as facts being discovered through "the" scientific method or by following codified procedures. The "constructivist" category views teaching science as helping students construct knowledge, learning science as constructing personal understanding, and science as a way of knowing. (p. 773)

Twenty-one of the teachers had fully congruent beliefs across the three domains: 15 were designated "traditional", four were regarded as "process" and two were deemed to be "constructivist".[1] Regrettably, the study did not extend to observation of these teachers' classroom practice. Hashweh's (1996) study of 35 science teachers *did* extend to consideration of classroom practice, revealing that teachers who hold a constructivist orientation towards scientific knowledge are more likely to take account of students' prior understanding, have a richer repertoire of teaching and learning strategies and adopt more successful strategies for effecting conceptual change than teachers with a positivist orientation. In short, Hashweh comments, there is a positive relationship between a "knowledge constructivist" and a "learning constructivist" orientation (p. 49). Similarly, Kang and Wallace (2005) and Kang (2007) observe that teachers with "more sophisticated" NOS views are more likely to present science as tentative knowledge and to adopt a constructivist pedagogy.

Benson's (1986) finding that views about science held by students *within* a class are often remarkably similar, while the views of students in *different* classes can be substantially different, lends further support to the contention that students' views are strongly influenced by curriculum experiences and that these, in turn, are determined by the teacher's own views about science. Similarly, in a study of science undergraduates at the University of Papua New Guinea, Kichawen et al. (2004) found that as students progressed through their programmes their NOS views increasingly resembled those of their tutors, although no formal instruction in NOS was included in the programme. In addition, Duschl (1983) found that teachers' beliefs about science (which were largely of a logical positivist nature in the sample he studied) influenced their choice of curriculum materials; Lantz and Kass (1987) found that three high school chemistry teachers taught the same content in significantly different ways because of differences in their understanding of the nature of chemistry; Lederman (1986) concluded that students' conceptions of the nature of science are positively influenced by science teachers who model an inquiry or problem-solving approach. Further evidence of a close cause and effect relationship among teachers' views, curriculum experiences and students' views about science is provided by Wolfe's (1989) study of teachers' practices with gifted students and by Zeidler and Lederman's (1987) finding that teacher language (realist or instrumentalist) had a significant effect on students' views about the nature of science. Further, Brickhouse (1990) found that teachers who regard science as a body of stable, fixed knowledge tend to use worksheet-driven or textbook-oriented practical work focused on "getting the right answer", whereas teachers who hold the view that science changes because of new observations and new ways of interpreting observational evidence tend to encourage discussion of the significance of observations in the light of theoretical considerations. In an experimental study, Dibbs (1982) found that he could impact quite markedly on students' views about scientific inquiry by basing his teaching unambiguously on inductivist, verificationist or hypothetico-deductivist approaches to experimentation, while Carey et al. (1989) showed that a purpose-built NOS teaching unit can effect a shift in the understanding of grade 7 students on matters relating to the construction of scientific knowledge. In contrast, Finson et al. (2006) found that the Draw-a-Scientist Test (DAST) revealed no notable differences in grade 9 students' images of scientists between groups taught by teachers categorized as predominantly *inquiry/constructivist* (emphasis on students following their own interests and actively engaged in inquiry-oriented learning; teachers as facilitators), *conceptual/contructivist* (with emphasis on careful sequencing of activities to assist students in concept acquisition and development) and *explicit/didactic* (mainly teacher-centred methods).[2]

If this body of research makes it reasonable to suppose that teachers' views about science constitute a significant influence on both the explicit and implicit curriculum, then it is important for all teachers to reflect on their own views, hold them up to critical scrutiny and, when found wanting, seek ways to develop and extend them (see Duschl, 2001, for a similar argument). Matthews (1992) tells us that "a teacher's epistemology is informally formed; it is picked up in the process of textbook education described by Kuhn. It consists of current prejudices which have been little bothered by historical information, or by philosophical analysis" (p. 30). While this may be true of some teachers, it is certainly not true of all. My experience is that many teachers devote a considerable amount of thought to nature of science issues, especially when designing hands-on learning activities. Unfortunately, their thinking is often uninformed and ill-directed. What they need is critical support. However, Matthews' remark does serve to remind us that within a group of teachers there is likely to be a wide variety of NOS views, with varying degrees of coherence and consistency. These views may be tentative or strongly held; they may or may not have been subjected to close critical scrutiny by the teacher, or by others. It makes eminently good sense, then, for teacher educators and curriculum developers to seek information concerning the views *about* science held by teachers and student teachers, and to investigate how they can be modified by inservice and preservice interventions.

Almost any of the questionnaire instruments designed for use with students (see Chapter 2) can be used with teachers, usually with only slight modification. A particularly useful tool for use with teachers in both preservice and inservice programs is the *Nature of Science Profile* developed by Nott and Wellington (1993). Respondents are invited to agree or disagree with the statements on a −5 to 0 to +5 scale (where +5 indicates strong agreement, −5 indicates strong disagreement, and a 0 rating indicates ambivalence). The scores on the 24 items are summed to create a profile in terms of five bipolar constructs: relativism versus positivism; inductivism versus deductivism; instrumentalism versus realism; contextualism versus decontextualism; and a pedagogical emphasis on process versus content. The principal value of this instrument lies in its ease of administration, wide scope and non-judgmental nature – key factors when dealing with student teachers and their apprehensions about impending teaching practice placements. In similar vein, Chen (2006) has developed an instrument for use with preservice secondary school teachers that seeks to ascertain their views on seven aspects of NOS (the tentativeness of scientific knowledge; the nature of scientific observation; scientific method(s); the relationships among hypotheses, laws and theories; objectivity and subjectivity in science; the role of creativity and imagination in science; the validation of scientific knowledge), together with their attitudes towards teaching about these particular elements of NOS. Of particular value when student teachers have gained some classroom experience is Nott and Wellington's (1996, 1998, 2000) "Critical Incidents" approach. In group settings, or in one-on-one interviews, teachers are invited to respond to descriptions of classroom events, many related to hands-on work in the laboratory, by answering three questions: What *could* you do in these circumstances? What *would* you do? What *should* you do? Responses, and the discussion that ensues, may reveal a great deal about teachers' views of science and scientific inquiry and, more importantly perhaps, how this understanding is deployed in classroom decision-making. Similar approaches using video and multimedia materials have been used by Hewitt et al. (2003), Wong et al. (2006) and Yung et al. (2007). An interesting variation adopted by Murcia and Schibeci (1999) uses a science-oriented newspaper article as a stimulus to thinking about questionnaire items.

Some 45 years ago, Miller (1963) compared the scores of students and their biology teachers using Klopfer and Cooley's TOUS instrument (see Chapter 2), reaching the alarming conclusion that the teachers did not understand science as well as their students. Nor, he said, did they understand it well enough to teach effectively (cited by Lederman, 1992, p. 340). Using the *Wisconsin In-*

ventory of Science Processes, Carey and Strauss (1968, 1970) also concluded that, in general, both serving science teachers and student teachers have inadequate NOS understanding, and that understanding is unrelated to other academic variables. However, Carey and Strauss did note that the views of both teachers and students can be "improved" by appropriate NOS-oriented courses. Leaving aside the tricky question of what constitutes "improvement", this observation prompts speculation that the massive upsurge of interest in NOS during the past 15 years, and the prominence it now has in many teacher education programmes, will have "improved" the NOS understanding of the current teaching force well beyond the woeful level lamented by Miller. Sadly, research continues to suggest that, with some notable exceptions, teachers' views still fall short of what researchers consider to be "adequate", "acceptable" or "desirable" (Koulaidis & Ogborn, 1989; Aguirre et al., 1990; Gallagher, 1991; King, 1991; Lederman, 1992, 1999; Pomeroy, 1993; Abell & Smith, 1994; Abd-El-Khalick & BouJaoude, 1997; Palmquist & Finley, 1997; Abd-El-Khalick et al., 1998; Haidar, 1999; Nott & Wellington, 1998; Murcia & Schibeci, 1999; Abd-El-Khalick & Lederman, 2000a; Abell et al., 2001; Lederman et al., 2001; Tairab, 2001; Southerland et al., 2003).

In general, the perceived weaknesses concern ignorance of the theory-laden nature of observation and experiment, belief in a fixed algorithmic method of scientific inquiry, uncertainty about the status of scientific knowledge (particularly the distinctions among theories, laws and models), the tendency to overlook the sociocultural embeddedness of scientific practice and the role of creativity and imagination in science. For example, it seems that teachers are no more successful than students in stating the difference between laws and theories, with a substantial number subscribing to the hierarchical view so prominent among students (Abd-El-Khalick et al., 1998), while science teachers in Gallagher's (1991) study put considerable emphasis on science as a body of established knowledge and had scant understanding of how that knowledge was formulated or validated. Even with respect to experimental investigations, a prominent part of most teachers' science education and a major element in the science curriculum in many educational jurisdictions, I have found that teachers are sometimes unable to operationalize variables in a way that would support unambiguous measurements, while many report claims that do not reflect the data.

In their efforts to teach students about scientific inquiry, teachers tend to organize hands-on activities in accordance with one of two possible models: naïve inductivism and hypothetico-deductivism. In the first scenario, observational evidence gathered by students is subjected to teacher-led scrutiny in order to reach a spurious consensus that leads to the desired generalization and subsequent explanation (Harris & Taylor, 1983; Hodson, 1996). The school version of Popper's hypothetico-deductive approach tends to focus on stringent testing of a hypothesis by systematic manipulation of variables based on a "fair test" approach. The preservice high school biology teachers in Lawson's (2002) study had a reasonable understanding of, and ability to use, a hypothetico-deductive approach to scientific inquiry provided that the hypothesized cause of an effect was directly observable. When indirect observational evidence was necessary, performance levels dropped alarmingly: arguments were frequently incomplete or confused, they sometimes included predictions that did not follow from the hypothesis nor feature in the proposed test, and they often failed to consider alternative hypotheses. Another difficulty for some teachers centres on reconciling belief in the tentative nature of scientific knowledge with assertions that (i) some claims are more valid or credible than others, and (ii) some knowledge is sufficiently well-established to be taken-for-granted and used as the basis for further theory-building – a problem that was discussed briefly in Hodson (2008).

In contrast to the many "gloom and doom" findings regarding teachers' NOS views, Cobern and Loving (2002) conclude that the preservice elementary school teachers with whom they have worked had a "judicious view of science that is an appropriate foundation for their further development as teachers of science" (p. 1028). It is tempting to ask whether the authors were lucky in their recruitment, less demanding than other researchers, or had provided a course that shifted student teachers from the more usual unsatisfactory views to these more "judicious" ones. Palmquist and Finley (1997) also found fairly encouraging NOS views among preservice teachers (secondary school teachers on this occasion). Significantly, they report that teacher education courses addressing NOS issues in the context of lesson design have helped student teachers to clarify and articulate their views, resolve internal conflicts and adopt what the authors call a "more contemporary view" of scientific method. Of course, any research that uses classroom observation as a means of judging the extent to which NOS issues are satisfactorily addressed is impacted by the researcher's own NOS views, and whatever differences there might be between them and the teachers' NOS views, as illustrated in the study by Bianchini and Colburn (2000).

By means of a 16-item instrument with an elaborate scoring system, Koulaidis and Ogborn (1988, 1989) sought to identify the views of individuals in relation to what the authors consider to be "the five main trends in the philosophy of science": inductivism, hypothetico-deductivism, contextualism (rationalist version), contextualism (relativist version) and relativism.[3] Their findings reveal that many teachers hold views about science that incorporate a diversity of elements drawn from more than one philosophical position, depending on the matter under consideration. Focusing separately on methodology, criteria of demarcation, patterns of scientific change and status of scientific knowledge, Koulaidis and Ogborn (1995) found that "inductivists in methodology tend to be rationalist contextualists on demarcation and on patterns of change in science, but are relativists regarding the status of scientific knowledge ... Contextualists with regard to methodology tend to be more rationalist concerning demarcation and the status of scientific knowledge ... Eclectics are in most cases eclectics on all four themes" (p. 279). These findings suggest that teachers may deploy different elements of their *Personal Framework of NOS Understanding* in different circumstances – for example, designing laboratory investigations compared with dealing with a specific episode in the history of science.

TEACHERS' VIEWS, CURRICULUM EXPERIENCES AND STUDENTS'
UNDERSTANDING OF NOS

A study of four experienced Taiwanese science teachers, two described by Tsai (2007) as having a constructivist epistemology of science and two identified as having a positivist orientation, found that the latter teachers tended to be teacher-oriented in approach, with a focus on knowledge acquisition. They made extensive use of lectures, tutorial problem-solving sessions and in-class tests. In contrast, the constructivist-oriented teachers focused on student understanding and application of conceptual knowledge. They engaged students in more laboratory and small-group inquiry and were more likely to use debates to challenge and critique students' understanding. There was some evidence of a relationship between teachers' and students' views about the theory-laden nature of scientific investigations, the creative elements of science, the tentativeness of scientific knowledge, the role of social negotiation within the scientific community and the sociocultural impact of science, but the effect size was small. An earlier study by Hodson (1993a) used some of the items from the Koulaidis and Ogborn instrument (see above), together with selected items from VOSTS, to identify teachers who seemed to possess a coherent, clearly identifiable and consistent "philosophic stance". Five such teachers (two inductivists, two hypothetico-deductivists and one rational contextualist) took part in a research study involving close examination of curriculum plans, worksheets, student lab books and assessment/evaluation materials, reinforced by observation of 15 hours of laboratory-based lessons. In post-lesson interviews, teachers were asked about the purposes of the lesson they had designed, what features they regarded as significant, whether it had gone as planned, the extent to which (and the ways in which) they believed it had been successful, and so on. The major finding was that teachers do *not* plan laboratory-based activities consistently in accordance with the NOS views they profess to hold, concentrating instead on the immediate concerns of classroom management and the demands of concept acquisition and development. Just as many of the university students interviewed by Rowell and Cawthron (1982) perceived a difference between science *as it is* and science as it *should be*, so teachers in Hodson's study seemed to have an ideal model of science teaching and learning (what they perceive science education *ought* to be like) and a somewhat different model of what is possible in a school laboratory, given the constraints of insufficient time, inadequate resources, unavailability of technician support,[4] the pressure of overcrowded syllabuses and the demands of public examinations. Interestingly, teachers changed their philosophic stance in response to changes in subject matter and perceived ability of the class, being more inclined towards an inductivist approach with biology topics and with those students regarded as lower in ability. Although the teachers seemed to be fairly consistent in presenting biology as inductivist and chemistry and physics as hypothetico-deductivist, the difference in philosophic stance between science disciplines seemed to reflect learning opportunities for concept acquisition arising from the particular subject matter rather than belief in a demarcation in methodology between the biological and physical sciences. The tendency to use inductivist approaches with students designated as "less able" seemed to be prompted by the widely held view that inductivism is easier to understand than hypothetico-deductivism.

Lederman's (1999) study of five biology teachers, ranging in experience from 2 to 15 years, also showed that teachers' conceptions of the nature of science do not necessarily determine (or, sometimes, even influence) classroom practice. Of much greater significance in determining the structure and conduct of lessons are concerns about classroom management (particularly important, of course, for less experienced teachers) and instructional goals relating to concept acquisition and motivation of students. When Abd-El-Khalick et al. (1998) observed the classroom activities organized by 14 preservice teachers with robust NOS views and a clearly stated intention to afford a measure of prominence to NOS-oriented teaching, they found scant evidence of teachers putting their views into practice. Barriers to NOS teaching perceived by the student teachers, when challenged to explain the mismatch between rhetoric and practice, included alternative priorities, particularly classroom management issues and the importance of teaching content, their own poor understanding of NOS and ignorance of how to assess NOS knowledge, lack of resources, opposition from supervising teachers, students' lack of interest in NOS and, as always, lack of time.[5] It is interesting to speculate on the relative weightings of these explanations, which clearly vary from individual to individual. As with most issues in education, the specific educational context in which teachers find themselves is crucial to their decision-making, making research findings difficult to generalize – see also, Tobin and McRobbie (1997) and Gwimbi and Monk (2003).

Following a 16-month action research intervention with four elementary school teachers, Waters-Adams (2006) concludes that there is little direct link between teachers' espoused views of NOS and their classroom practice. Rather, teachers' decision-making about classroom strategies is the outcome of a complex mix of tacit views about science (akin to Hogan's (2000) notion of a *proximal view of NOS* – i.e., views and beliefs built up through personal experience of teaching science) and a cluster of views and beliefs about the aims of education, the way children learn, and how curriculum should be structured. Clearly, there is enormous potential for conflict among ideas, especially when situation-specific constraints relating to time, equipment and other curriculum demands are factored in. Barnett and Hodson (2001) use the term *pedagogical context knowledge* for the knowledge that teachers use to inform decision-making in a particular educational situation: a cluster of (a) academic and research knowledge (science content knowledge, NOS knowledge, knowledge of how and why students learn, etc.), (b) pedagogical content knowledge (see Shulman, 1986, 1987), (c) professional knowledge (knowledge acquired, much of it unconsciously, through active participation in the community of science teachers), and (d) classroom knowledge (knowledge of their particular class and students). What is especially encouraging about the Waters-Adams (2006) study is that the action research experience enabled the teachers to explore their pedagogical context knowledge and to resolve some of the tensions inherent in all classroom situations and develop new approaches to teaching NOS, thus reinforcing Nott and Wellington's (1996) argument that teachers' knowledge of the nature of science (and also how to teach it) may be "as much formed by their teaching of science as informing their teaching of science" (p. 284).

Mismatches between teacher rhetoric and classroom practice are common, and are certainly not restricted to NOS issues. Also, for all kinds of reasons, mismatches are more common and more extensive among novice teachers than among experienced teachers. It is also common for mismatches to exist between what and how a teacher believes she/he has taught and what a researcher observes. And, indeed, between what either of these concludes about the lesson and what a student concludes. Many teachers claim that they have addressed NOS issues implicitly and indirectly through messages about science, scientists and scientific practice embedded in curriculum materials, classroom language and learning experiences (especially hands-on activities), but sometimes these messages are so subtly embedded that they are "invisible" to the majority of students (and often to the researcher-observer, too). The ability to use indirect approaches successfully is largely the province of experienced teachers who have built up a repertoire of techniques, the confidence to deploy them and the intuition to know when and how to do so, over many years. It is not an approach to be recommended to beginning teachers. They need to use, and should be encouraged to adopt, a more structured and overt approach. It is not surprising, therefore, that student teachers who were encouraged by Bell et al. (2000) to separate the teaching of NOS content from the teaching of science content were more successful in translating their own NOS views into lesson activities. They also gave nature of science teaching a higher priority than did the student teachers in an earlier study by Abd-El-Khalick et al. (1998), who were not given such advice. In fact, most of them used approaches based on the methods used for teaching NOS content in the teacher education programme itself. Whether the school students they taught were successful in acquiring this NOS understanding is another matter, and is not known, since no-one among the eleven student teachers attempted to assess NOS learning. Firstly, because they did not know *how* to assess it or felt that they had spent insufficient time on NOS teaching to warrant assessment; secondly, because NOS was not included in the learning objectives prescribed by the Head of Science in the practicum school.

A further question for consideration is whether more experienced teachers should be encouraged to separate NOS teaching from science content teaching. My intuitive feeling is that NOS teaching is more successful and of more interest to students (and possibly to teachers, too) when it is embedded in a conceptually rich context. It seems self-evident that understanding of the nature of theory, the concept of evidence and the notion of a sound experiment, for example, would all be enhanced by the ready availability of familiar examples. It also seems self-evident that NOS ideas (especially those relating to model, theory, experiment, inquiry, and so on) will be nuanced differently in different subject domains. Put simply, the methods of astrophysics are somewhat different from those of molecular biology and synthetic organic chemistry, raising some very different

NOS concerns. However, research evidence addressing this question is both scant and inconclusive. For example, Khishfe and Lederman (2006) observed that parallel groups of grade 9 students showed roughly the same improvement in all five aspects of NOS understanding on which their performance was evaluated (the distinction between observation and inference, and awareness that scientific knowledge is tentative, empirically based, subjective and the product of human imagination and creativity) regardless of whether the NOS teaching was integrated into the environmental science content (a topic on global warming) or was taught in a non-integrated way – that is, as "separate yet dispersed and intermittent with the global warming content" (p. 400), with no intentional connections made by the teacher. Similarly, Khishfe and Lederman (2007) found that three groups of grade 9 and grade 10 students studying a chemistry, biology or environmental science unit showed "improved" NOS understanding following an explicit reflective integrated approach, but the improvements were no greater than for the three parallel groups following a non-integrated explicit reflective approach. The improvement in the environmental science group's NOS views were "slightly higher" after the integrated approach than following the non-integrated approach, which the authors attribute to the motivational aspects of a controversial topic (global warming). There is little doubt that students' understanding of the science content of a particular topic will influence their appreciation of the associated NOS dimensions, and vice versa. There is no doubt, either, that the sociocultural context in which a topic is embedded and the affective responses it triggers will also impact the NOS understanding students glean from the learning activity. What research does not tell us, clearly and unambiguously, is the nature and extent of those influences, though simple common sense suggests that teachers should not attempt to teach NOS when students are struggling to get to grips with complex and difficult content issues.

Clough (2006) makes the point that *both* contextualized and decontextualized approaches are necessary, neither is sufficient, and teachers should purposefully scaffold back and forth between them, drawing students' attention to newly acquired NOS knowledge and using specific content to help students reinforce and apply their developing NOS understanding. He argues that decontextualized activities using discrepant events, black boxes, card games focused on NOS concepts, "feely bags" and pictorial items involving illusions and gestalt switches, and the like (see Clough, 1997; Lederman & Abd-El-Khalick, 1998) have an important role in introducing fundamental NOS issues in "familiar concrete ways that are not obstructed by unfamiliar science content, or historical stories that they can easily misinterpret" (p. 484).[6] Although they are often seen by teachers as "add-ons" and as taking time from "proper" science teaching (i.e., content-oriented teaching), these activities play a key role in setting up an NOS agenda for subsequent consolidation, elaboration and application. Many students find these activities intriguing and highly motivating, so they are particularly effective in directing attention to the specific features of NOS issues and stimulating deep cognitive processing, a conclusion reinforced by Heap's (2006) findings with preservice elementary teachers. Of course, students will be well aware that these activities do not represent authentic science, so they are, in themselves, insufficient to bring about the totality of NOS understanding we seek. Students need to apply the ideas in context-specific situations via content teaching and laboratory activities, augmented by historical and contemporary case studies of science. Subsequent chapters will elaborate on these strategies.

Clearly, teachers will not incorporate NOS-oriented teaching into their curriculum unless they believe it to be important and feasible. Despite the considerable efforts of many science educators over the past 20 years to promote the centrality of NOS understanding to notions of scientific literacy, substantial numbers of teachers remain unconvinced. Clearly, much more needs to be done in this regard. Teachers also need support in translating their NOS understanding into coherent teaching strategies and into appropriate and effective classroom activities; they need experience of using these strategies and activities; and they need critical support in evaluating them in action. Here is the paradox of educational innovation: teachers will only be convinced to adopt an approach when there is strong evidence that it works, but evidence is only forthcoming when some teachers adopt the approach. Akerson and Volrich (2006) generated the kind of evidence that might persuade some teachers during their study of the second author's successful efforts to impact favourably on grade 1 students' NOS views through purpose-made interventions focusing on observation-inference, the role of imagination and creativity in science, and the tentative nature of scientific knowledge. Hipkins et al. (2005) suggest that arguments based on the needs of "the knowledge society" might have sufficient "street credibility" to encourage more teachers to "take the plunge" – presumably on the grounds that NOS understanding enhances critical thinking.[7] Beginning teachers, whose principal concerns inevitably focus on issues of classroom management, need additional support in the form of ready-to-use instructional practices and management techniques that will enable them to achieve a level of comfort that permits attention to be focused in the direction of learning *about* science. Motivated by such considerations, Lederman et al. (2001) examined the NOS understanding of 15 preservice teachers and their implementation (or not) of NOS-oriented teaching activities. Four principal determining factors were identified: knowledge of NOS, knowledge of subject matter, pedagogical knowledge, intention to teach NOS. The authors nominate the final factor as the most critical, noting that "participants, regardless of NOS views or science background, did not teach in accordance with their NOS views if they had not internalized the importance of teaching NOS. Those with strong intentions to address NOS explicitly were more successful. Those participants with strong intentions, well-developed NOS views, and extensive knowledge of science content were most successful in their instructions" (p. 135).

In recent years, attention has increasingly turned to finding more effective ways of preparing both preservice and inservice teachers for the complexities of NOS teaching. Focuses include enhancing teachers' own NOS understanding, identifying effective classroom strategies for teaching NOS, and developing resource materials. One such study of a group of science teachers in the United Kingdom (Bartholomew et al., 2004) identified five overlapping and interacting "critical dimensions" that characterize and determine a teacher's ability to be effective in teaching *about* science: (i) teachers' knowledge and understanding of NOS and their confidence in, or anxiety about, teaching NOS; (ii) teachers' conceptions of their role (ranging from dispenser of information and knowledge to facilitator of learning); (iii) teachers' use of discourse (closed and authoritative versus open and dialogic); (iv) teachers' conception of learning goals (and the extent to which they are limited to knowledge gains or include the development of reasoning skills); and (v) the nature of classroom activities (particularly the shift from activities that are teacher selected, contrived and inauthentic to activities that are both owned by the students and scientifically authentic). Assigning teachers to a particular position on these five dimensions is not easy. First, because much has to be inferred from their actions in the hurly-burly of classroom action. Second, because it is unlikely that teachers will occupy a fixed position on each of the five dimensions; their exact position on any particular observed occasion may be impacted by the subject matter under consideration and the particular group of students being taught. Despite these difficulties and uncertainties, the authors have furnished teacher educators with a very useful analytical and diagnostic tool.

It almost goes without saying that there is likely to be enormous variation among teachers in terms of dimension 1. It is likely that many teachers are unaware of recent scholarship in history, philosophy and sociology of science (HPS), particularly the so-called science studies literature; substantial numbers of teachers were not exposed to NOS issues during their own science education, either at school or at university. Like any subject area, HPS includes ideas that are notoriously difficult to teach, and especially so for novices. So it is not enough for teachers to know one or two basic ideas in NOS; they need to be immersed in the subject in order to see the links and find suitable examples for classroom use. Put simply, teachers cannot teach what they do not know. And they certainly cannot teach NOS well without some considerable depth of NOS knowledge. Depth of NOS knowledge enables teachers to readily access a range of related examples, suitable demonstrations and historical episodes. To teach well, teachers need to be able to talk comfortably about many aspects of NOS, contextualize their teaching with episodes from the history of science and design lesson activities that can make the target NOS elements accessible and understandable to students. It is self-evident that lack of NOS knowledge will lead some teachers to experience high levels of anxiety when confronted with a request or requirement to teach about the nature of science. Moreover, many teachers will be uncomfortable with the fluid and tentative nature of NOS knowledge, and the methods appropriate to teaching it, preferring what they see as the "certainty" (or, at least, the stability) of science content knowledge.

ACQUIRING AND DEVELOPING NATURE OF SCIENCE KNOWLEDGE

Widespread dissatisfaction concerning teachers' understanding of the nature of science has prompted the development of numerous intervention strategies for enhancing their NOS knowledge. In general, the research findings with respect to teachers and student teachers are replicated in studies focused on enhancing school and university students' NOS understanding.

Research tells us that courses emphasizing history of science can be particularly beneficial, especially (and perhaps only) when they focus clearly on NOS issues (Abd-El-Khalick & Lederman, 2000b). For example, Lin (1998) and Lin and Chen (2002) show that students' NOS views, especially concerning the role and status of scientific theory, can be favourably impacted by studying important episodes in the history of science, particularly how and why scientists conducted experiments and used debate and argument to promote and defend their ideas. Dagher et al. (2004) provide some evidence that explicit, historically-oriented teaching focused on the nature of scientific theory can assist students to make progress (though not as much progress as the course designers had hoped to achieve) within "four core conceptions of theory": (i) equivalent to hypotheses; (ii) an idea with evidence; (iii) an explanation (ranging from tentative to established truth); (iv) an explanation based on evidence. More substantial and sustained progress was reported in a parallel study focused on assisting students to be more demanding and critical of the evidence presented in support of knowledge claims (Brickhouse et al., 2000). With younger students, science stories can sometimes have an impact on NOS views (see Solomon et al., 1992; Tao, 2003), although not always in the desired direction. As Tao (2003) reports, students sometimes use carefully selected elements of the stories to reinforce their "inadequate" or stereotyped view of science, just as they sometimes adjust data from experiments to fit their preconceptions. Both observations point to the need for timely and judicious teacher intervention. Mamlok-Naaman et al. (2005) report that a history of science approach with non-science-oriented grade 10 students (in an Israeli high school) resulted in greater awareness of the world of the scientist. In addition, the course had substantial impact on students' attitudes towards science and their understanding of the relationships between science and technology. Shortage of space precludes further elaboration of research findings on the efficacy of historical studies in promoting NOS understanding, except to note that McComas (2008) identifies approximately 80 NOS-oriented historical vignettes in books on HPS written for general readers.

Abd-El-Khalick and Lederman (2000a) have assembled a detailed review of 17 interventions designed to enhance teachers' conceptions of the nature of science – eight described as implicit approaches, nine as explicit approaches. Of course, it is not always easy to assign an implicit/explicit label without seeing the intervention "in action". The distinction resides not so much in differences in the kind of activities used (hands-on inquiries, historical case studies, lectures and readings, for example) as in the "extent to which learners are provided (or helped to come to grips) with the conceptual tools, such as some key aspects of NOS, that would enable them to think about and reflect on the activities in which they are engaged" (Abd-El-Khalick & Lederman, 2000a, p. 690). In an explicit approach, NOS understanding is regarded as "content", to be approached carefully and systematically, as with any other lesson content. It should be noted that regarding NOS knowledge as content does not entail a didactic or teacher-centred approach or the imposition of a particular view through exercise of teacher authority, but it does entail rejection of the belief that NOS understanding will just develop in students as a consequence of engaging in other learning activities. NOS should not be regarded as an incidental by-product of an activity; rather, it should be seen as the specific goal – i.e., as a designated cognitive learning outcome. As argued in Hodson (1996), the conceptual knowledge of science is not located "out there", waiting to be discovered by the undirected activities of students (one of the notions underpinning discovery learning). It has to be taught. The same argument applies to NOS knowledge. Our current stock of NOS knowledge has been constructed over many years by historians of science, philosophers of science, sociologists of science, scientists and science educators. Neither students nor student teachers are likely to discover it for themselves as a by-product of some other teaching and learning activity; they need guidance and support, for which, teachers need to devise systematic and purposeful strategies. Moreover, as with other aspects of science learning, students need opportunities for reflection on their ideas as a stimulus to reconstruction. Deeply held misconceptions about NOS may have been built up over years of experience, both in school and out of school, and may be shared with peers and reinforced by media representations of science and by textbooks. Such views will be very strongly held and are unlikely to be seriously challenged or displaced by implicit NOS messages embedded in text or classroom interactions. More than a quarter century ago, Posner et al. (1982) argued that cognitive change is likely when learners are dissatisfied with their current beliefs and understanding and have ready access to a new or better idea. To be acceptable, the new idea must meet certain conditions.

- It must be *intelligible* (understandable) – that is, the learner must understand what it means and how it can and should be used.
- It must be *plausible* (reasonable) – that is, it should be consistent with or capable of being reconciled with other aspects of the learner's understanding.
- It must be *fruitful* (useful) – that is, it should have the capacity to provide something of value to the learner. For example, solving important problems, facilitating new learning, addressing concerns, making valid and reliable predictions, suggesting new explanatory possibilities or providing new insight. For teachers, fruitfulness might include usefulness in designing learning opportunities for students.

Taking this view at face value, Hewson and Thorley (1989) describe the conceptual change approach to teaching and learning science as a matter of changing the status of rival conceptions with respect to the three conditions of intelligibility, plausibility and fruitfulness. Put simply, the teacher's task is to lower the status of the students' existing idea and raise the status of the new one.[8] However, a host of writers (including West & Pines, 1983; Bloom, 1992a,b; Pintrich et al., 1993; Demastes et al., 1995; Hennessey, 2003; Sinatra & Pintrich, 2003; Alsop, 2005; Sinatra, 2005; Johnston et al., 2006) have pointed out the ways in which this rationalist view of learning fails to recognize the complexity, uncertainty and fragility of learning and its susceptibility to a whole array of personal and social influences. These influences will be just as important with respect to the acquisition and development of NOS knowledge as they are to any other area of learning. Any or all of the following could impact on NOS learning, sometimes favourably and sometimes not: previous experiences; emotions, feelings, values and aesthetics; personal goals and motivation levels; views of learning;[9] social norms and aspirations; general feelings of well-being and satisfaction.

Work by Southerland et al. (2006) with five teachers enrolled on a graduate course focused on NOS found that development of "sophisticated" NOS views (as measured by VNOS and interview) and a commitment to implementing NOS-oriented teaching were profoundly influenced by particular *learning dispositions* (especially the drive to acquire meaningful understanding via deep processing and adoption of learning goals in preference to performance goals, being comfortable with ambiguity and having a willingness to examine one's beliefs and entertain alternatives) and *beliefs about learning and learners* (especially a view of learning as an achievable construction of knowledge for which successful strategies can be found, and a disposition to regard one's students as perfectly capable of learning NOS knowledge). Suffice it to say at this point that these arguments and these research findings lend support to Abd-El-Khalick and Lederman's (2000a) conclusion that "an explicit approach was generally *more* 'effective' in fostering "appropriate" conceptions of NOS among prospective and practising teachers" (p. 692). Most effective of all are those courses that have a substantial reflective component. For example, Lucas and Roth (1996) report substantial gains in NOS understanding during a course incorporating readings on NOS, reflective essays and class discussions, and opportunities for self-directed laboratory experiences; Akerson et al. (2000) report substantial improvements in elementary student teachers' NOS views when the science methods course required reflection on NOS, both orally and in writing, following a series of readings, case studies, debates and other activities. Akerson and Volrich (2006) note substantial improvement in the NOS understanding of grade 1 students when the teacher made repeated and explicit reference to NOS issues (focused largely on scientific observation, the tentativeness of scientific knowledge and creativity in science), encouraged students to keep a journal related to any NOS issues arising, and finished each lesson with class discussion triggered by the question, "How is what we did like what scientists do?" Heap (2006) also points out the centrality of reflection (in her case, the use of reflective journals) in changing, developing and consolidating NOS understanding. She also notes that generic content-free activities such as the "tricky tracks" and the "aging President" activities devised by Lederman and Abd-El-Khalick (1998) were particularly effective in stimulating shifts in inservice elementary teachers' views about the theory-dependence of observation, the distinction between observation and inference, and the role of creativity in science – thus reinforcing Clough's (2006) argument that both context embedded and context-free approaches are necessary (see above). It is also the case, predictably, that longer courses are more effective than shorter ones. For example, Liu and Lederman (2002) report no significant changes in

Taiwanese students' NOS views (measured by a questionnaire based on VNOS) following a one-week NOS-oriented science camp in an American university. The authors note that the students already had views that were more sophisticated than is common among students, so perhaps a course lasting just one week was unlikely to bring about further measurable change. It almost goes without saying that there is often enormous value in adopting the usual constructivist strategies: (i) ascertaining students' existing NOS views prior to instruction, (ii) providing opportunities for students to articulate their views, criticize the views of others and deal with criticism of their own views, and (iii) encouraging students to reflect on their changing NOS views (and possible reasons for the changes). See Ryder and Leach (2008) for some confirmation of this recommendation, together with further reinforcement of the importance of making NOS learning goals explicit and linking key NOS ideas with specific science content. Chapter 7 includes a more extensive discussion of constructivist-oriented approaches to teaching about NOS issues.

In order to test the robustness of the conclusion regarding explicit and reflective NOS instruction, Abd-El-Khalick (2005) provided 56 preservice secondary school science teachers with a purpose-made set of course units designed to address nature of science issues in an explicit, reflective way, and in doing so to focus on the needs of science teachers. Compared with students following a parallel, conventional science methods course, the target group developed deeper and more coherent NOS knowledge, as measured by VNOS-C (Lederman et al., 2002). In their teaching placements, they planned and implemented more explicit learning activities focused on NOS, and they established much richer and more extensive classroom discourse on NOS. Moreover, they were able to discuss ways in which their teaching could better accommodate their newly acquired NOS understanding. In other words, they had become much more reflective about their own teaching. With regard to acquisition of NOS knowledge, these findings are mirrored in a similarly motivated study conducted with grade 6 students by Khishfe and Abd-El-Khalick (2002). The target group of 36 students were involved in an inquiry-oriented approach to the physical sciences reinforced by reflective discussions of the target NOS elements. The comparison group of 29 students experienced no reflective content. At the outset, both groups of students had naïve views in the selected aspects of NOS, as ascertained by a combination of open-ended questionnaire and interview. After approximately ten weeks, many of the students in the target group expressed what the authors call "more informed" NOS views; the views of students in the other group were unchanged. This study provides some evidence that interventions are more successful when NOS issues are addressed within an inquiry-based approach to concept acquisition and development, a conclusion that is reinforced by the research findings of Kishfe (2008) with grade 7 students and Gess-Newcome (2002), Akerson and Buzzelli (2007), Akerson and Hanuscin (2007) and Akerson et al. (2007, 2008) at both the preservice and inservice teacher education levels. It prompts speculation on whether this approach is superior to an explicit and reflective approach based on historical case studies or on contemporary socioscientific issues. Clearly there is scope for (and need for) research in this regard (see Khisfe & Lederman, 2006). My own view is that NOS should be regarded as "knowledge in action", to be learned, applied and refined in specific contexts of use, particularly in planning lessons and confronting socioscientific issues. Celik and Bayrakçeken (2006) report that they were able to shift the NOS views of preservice teachers quite substantially through a course that embedded NOS content in a broader STS programme and adopted an inquiry-based learning approach to NOS issues, SSI and aspects of science pedagogy.

In an intriguing study blending explicit NOS instruction, opportunities for reflection, and on-the-job experience, Hanuscin et al. (2006) followed the developing understanding of a group of Teaching Assistants (TAs) working on a physical sciences foundation course for preservice primary school teachers. Group meetings involving all the teaching staff, attendance at the course lectures, supervision of laboratory activities, tutorial work with the students and opportunities to examine students' NOS views all contributed to the substantial improvement observed in the NOS knowledge of all the TAs over the course of the programme. In particular, the on-the-job experiences and the requirement to reflect on how NOS issues were exemplified in their work with students enabled the TAs to clarify the meaning of some problematic NOS concepts, such as creativity, tentativeness and subjectivity in science, identify and resolve inconsistencies in their understanding, and establish links between and among NOS items – for example, between subjectivity and cultural embeddedness. Together with the Waters-Adams (2006) action research study described earlier, this research has much to tell us about the professional development of science teachers with respect to NOS understanding.

In a study conducted by Akerson et al. (2006), a group of 19 preservice elementary teachers achieved substantial improvements in their understanding of aspects of NOS embedded in an explicit reflective instructional programme focused on NOS. However, interview and questionnaire data collected five months afterwards showed that not all the teachers had retained their "improved" NOS understanding. Indeed, some had reverted to their original views. It seems that some participants had experienced difficulty in personalizing the new NOS knowledge, in making it part of their personal framework of understanding. The authors interpret this difficulty in terms of under-developed metacognition, observing that those participants with strong metacognitive skills were much more successful in retaining the new understanding. It was also evident that the teachers who had retained their improved NOS views were able to provide examples of NOS-oriented activities they had used in class. The authors conclude that preservice teachers should be taught metacognitive strategies in conjunction with NOS knowledge, and should be given more extensive opportunities to reflect on the significance of their newly acquired NOS knowledge for the school curriculum and to use that understanding in designing curriculum materials and learning activities. Teacher educators might also consider the use of video material illustrating exemplary NOS-oriented teaching, with associated opportunities for critical group-based discussion (Wong et al., 2006; Yung et al., 2007).

We can safely conclude that whatever applies to student teachers with respect to the need for reflection and metacognition will also apply to students in school. Clearly, students are unlikely to engage in reflection unless teachers identify the key NOS elements on which to reflect and create opportunities for them to engage in reflection. What is also clear is that students are unlikely to take HPS courses seriously unless they are assessed. And it is not just assessment *per se* that is the priority; it is assessment that requires students to address NOS ideas critically and in meaningful contexts. There is already a long and unfortunate history of inappropriately designed assessment schemes distorting student learning by encouraging simple regurgitation of facts. Such an approach to NOS assessment would be positively harmful and immeasurably worse than no assessment at all.

It is interesting that the gains in NOS understanding consequent on exposure to explicit, reflective instruction are less substantial in relation to the sociocultural dimensions of science than with other NOS elements. Akerson et al. (2000) speculate that this is because the subtleties of the subjective and sociocultural influences on scientific practice are impossible to capture in a short course (in this case, for preservice teachers). One or two brief examples will not achieve it; detailed and richly textured case studies (both contemporary and historical) may do so – further reinforcement of the point made several times already in this chapter that the oft-used decontextualized approaches to NOS teaching need to be augmented by content-rich approaches. Dass (2005) reaches essentially the same conclusion when accounting for why a semester-long undergraduate history of science course focused on the sociocultural and political context of major scientific advances achieved only "modest gains". I would argue that disappointing outcomes are also a consequence of uncertainty about intended learning outcomes in this particular NOS domain, the inadequacy of assessment procedures for capturing student understanding, and the pervasiveness and power of images of science and scientists acquired through informal learning channels. Regarding factor 1, we should note that "the social and cultural embeddedness of scientific knowledge", the expression used by Lederman et al. (2002), can be interpreted in a number of ways: (i) scientific knowledge impacts on society, in the sense that it changes how we think and the ways in which live; (ii) social and cultural forces determine the priorities for scientific investigation and its associated technological development; (iii) scientific knowledge is socially constructed and, therefore, reflects the experiences, interests and attitudes of those who produce it; (iv) there are distinctive "sciences", such as African science, First Nations science and Islamic science, each of which claims to provide a satisfactory explanation of phenomena and events in the natural world. If teachers are unclear about precise learning goals relating to the sociocultural dimensions of science, as many are likely to be, there is likely to be a lack of clarity in lesson design. If researchers are unclear about the precise goals of NOS-oriented experiences, the assessment instruments they deploy will inevitably lack validity and reliability. With regard to the third factor listed above (the power of informal learning), it is crucial that teachers *know* the kind of views about science and scientists that students encounter, and that they take steps, in class, to address them and critique them. On occasions, they should seek to displace them.

In contradiction of the generalization that only explicit approaches can be successful in shifting students' NOS views in the desired direction, Craven et al. (2002) show that preservice elementary teachers, whose lack of robust conceptual knowledge in science can constitute a problem with a content-rich approach, can make substantial progress when given opportunities to explore and express both their tacit and explicit NOS knowledge through individual and collaborative writing tasks focused on criteria for distinguishing science from pseudoscience. Most of the students participating in the study came to accept that scientific knowledge is tentative rather than fixed and that scientific method is not fixed and invariant, but varies (to an extent) in relation to the specific context of the inquiry. As their views became more sophisticated, student descriptions increasingly referred to notions of validity, reliability, evidence-based conclusions, bias and interpretation. These terms had been entirely absent from their remarks at the outset.

As a young teacher I was made uncomfortably aware of the powerful impact of implicit curriculum messages when students, often many years later, recalled things that I had said, views I had expressed and values I had displayed – few of which I had planned. Jackson (1968), Freire (1972), Snyder (1973) and Gatto (2002) have all written at length on the "hidden curriculum" and the need for teachers to be aware of how its implicit messages can unconsciously influence students' knowledge, attitudes and values. It is interesting to speculate on why an implicit approach to NOS is often (usually) unsuccessful and to ask whether the advocacy of explicit over implicit relates only to particular aspects of NOS. Perhaps the messages concerning NOS are often too complex, too fluid and too subtle for students to recognize; perhaps they are swamped by messages concerning concept acquisition and concept development, which also reflect student and teacher priorities relating to content coverage. Inevitably, curriculum items to which teachers draw attention, and include in tests and examinations, will be given a much higher valuation by students. From my experience in multicultural and antiracist science education, and in environmental education (especially in its politicized form of education *for* the environment), I have long since recognized that too explicit an approach can be counter-productive. Preaching and moralizing is anathema to most students. As noted in endnote 5 in Chapter 1, my notion of NOS is much broader than that of Lederman et al. (2002) and Lederman (2006, 2007); it includes aspects of NOS that are heavily value-laden, relate to gender and ethnic bias, address topics with a substantial moral-ethical dimension, and so on. Perhaps these matters should be approached implicitly, while epistemological issues are approached explicitly. This is a matter deserving of further research. My own views on how the necessary guidance, challenge and support might be provided, and how appropriate and timely opportunities for reflection and reconstruction might be organized, are discussed at length in later chapters.

TEACHING *ABOUT* SCIENCE

It should be noted, again, that several studies have shown that whilst teacher knowledge of NOS is a necessary requirement, it is not *sufficient* to guarantee effective NOS-oriented teaching, or even any teaching of these matters at all (Abd-El-Khalick & Lederman, 2000a). Indeed, it may not even be the most important factor. Much more important, it seems, is the *commitment* to teach NOS and the capacity to organize an appropriate classroom environment (Schwartz & Lederman, 2002; Akerson & Abd-El-Khalick, 2003). The number one priority, then, is to ensure that teachers and student teachers recognize the importance of NOS teaching and its centrality to the notion of critical scientific literacy. Preservice and inservice courses, curriculum documents, teacher workshops, science teacher journals, conferences and teacher networks are the most appropriate vehicles. After many years in the role of students, prospective science teachers embark on preservice programmes with a complex array of knowledge, beliefs, attitudes, values, aspirations and expectations concerning science, the teaching and learning of science, and the role of teachers. When the content of teacher education courses presents and encourages significantly different ways of thinking about science education, student teachers need time to address the mismatches, resolve conflicts, establish new priorities and build what Mellado (1998) calls a "practical scheme of action". This is no easy task.

> For many teachers, enculturated in the habitus of traditional science teaching, this would require a shift in the conception of their own role from dispenser of knowledge to facilitator of learning; a change in their classroom discourse to one which is more open and dialogic; a shift in their conception of the learning goals of science lessons to one which incorporates the development of reasoning and an understanding of the epistemic basis of belief in science as well as the acquisition of knowledge; and the development of activities that link content and process in tasks whose point and value is transparent to their students. (Bartholomew et al., 2004, p. 678)

Martin-Diaz (2006) shows that there is widespread support for inclusion of NOS in the curriculum (at least, in Spain), with teachers who majored in the humanities being the most enthusiastic and those who majored in chemistry being the least supportive. Support is stronger for aspects of NOS that could be categorized as philosophical and historical than for those relating to societal issues and concerns. Interestingly, experienced teachers are more supportive than inexperienced teachers. There were no gender-related differences.

A further complicating factor militating against effective NOS teaching is that many elementary school teachers do not have sufficient scientific knowledge to use as an effective carrier of NOS teaching, and are unable to provide students with timely and appropriate examples and counter examples. Deep and robust knowledge in science, including secure knowledge of the evidence for holding particular beliefs and of the history of development of key scientific ideas, gives teachers room for manoeuvre and the ability to recognize NOS learning opportunities in the material being taught. Teachers need the capacity to illustrate NOS points with science examples and to enrich the teaching of science concepts with NOS discussion. For example, a study of three physics teachers (of varied teaching experience) by Taylor and Dana (2003) showed that understanding of the importance of controlled experimentation, and ability to recognize an uncontrolled or poorly controlled experiment, do not necessarily translate into an ability to design controlled experiments for class use unless the teacher also has extensive relevant subject knowledge of the matter under consideration. And even those teachers who do have the requisite science background may still find it difficult because they were not taught the epistemological aspects of the science they have learned. Robust understanding generates confidence to engage in dialogue with students and to use more interactive teaching and learning methods. It is often the case that teachers use highly formal language and adopt didactic teaching and learning methods when they reach the limits of their expertise in a subject. Unfortunately, formal teacher-centred approaches are not well-suited to addressing the complex, value-laden and sometimes speculative and controversial nature of NOS knowledge. Teachers are better advised to adopt a less formal approach involving articulation by students, attempts at clarification, collaborative critique and efforts to reach a genuine consensus through debate and argument rather than a teacher-directed pseudoconsensus. But such an approach is necessarily rooted in a deep and robust understanding of the ideas under discussion.

Achieving the necessary depth of NOS knowledge is not easy; it will certainly not be achieved by the occasional lecture on NOS. Rather, it needs to be fully integrated within preservice programmes. As we would anticipate from research findings on situated cognition, science content and context are crucial factors affecting the learning of NOS and its subsequent use and application. Put simply, learning NOS and teaching NOS are substantially easier in familiar contexts. Abd-El-Khalick (2001) argues that NOS instruction for preservice elementary teachers is more effective when embedded in a science course than in a science teaching methods course. My view is that NOS should be taught in the context of the science that teachers will be required to teach in school, and alongside the associated pedagogical content knowledge. The most frequently voiced criticism of teacher education programmes is that they do not focus specifically enough and clearly enough on what is expected in the classroom. Teaching NOS in a decontextualized way would only exacerbate these feelings; so would teaching NOS in one science context and expecting teachers to apply it in a different one. Prospective teachers should be given lots of opportunities, both individually and in groups, to plan, teach and evaluate lessons designed to promote NOS understanding. Insisting that student teachers plan to *assess* student learning in NOS is an important part of the message that NOS learning should be considered a cognitive learning goal and not a mere attitudinal or motivational "add-on". Opportunities to put this learning into practice during teacher placement should be guaranteed. NOS teaching is too important to be left to the whim of supervisory teachers. Enfield et al. (2008) describe in some detail how teaching guidelines focused on scaffolding activities designed to foster an explicit approach to teaching NOS can shift

73

the classroom practice of elementary school teachers quite substantially towards systematic consideration of key NOS issues. Unfortunately, these researchers did not ascertain whether these changed classroom practices impacted favourably on students' views. Hapgood et al. (2004) designed similar scaffolding structures (which they called "notebook texts") to guide and support grade 2 students in key aspects of scientific inquiry. Chapter 7 includes further discussion of this approach and similar work by Sandoval and Reiser (2004).

During their early years in the classroom, teachers with little science background might be well advised to consider restricting NOS teaching to topic areas in which they have a reasonable level of conceptual understanding and confidence. While particular course content and a particular course structure or orientation can create significant opportunities for NOS teaching, both can also act as a constraint. Just as particular content may or may not support an STS approach, some science content lends itself well to NOS-oriented teaching, and some does not. Indeed, a case can be made for choosing some curriculum content specifically for its value as a carrier for NOS teaching. In the increasingly common situations where nature of science is a mandated part of the curriculum it is not sufficient for teachers to have the general intention to teach NOS, they need the intention to teach *particular* NOS knowledge. While it is only very rarely that teachers are presented with a requirement to teach scientific knowledge with which they profoundly disagree, the situation with respect to NOS is somewhat different. It is quite possible that a particular course may require (overtly or covertly) a particular NOS position at variance with the teacher's. In such a situation, personality variables relating to conformist versus non-conformist tendencies will become important. Of course, the great advantage of the shift to mandated NOS teaching is that it brings with it an assessment requirement, and all teachers (as well as students and parents) are well aware that the assessment regime determines the worth and status of particular knowledge. Regrettably, the reality of modern schooling is that without an assessment dimension, NOS will be widely regarded as a "soft option", a diversion not intended to be taken seriously.

Reporting on the work of Project ICAN (Inquiry, Context, and the Nature of Science),[10] Kim et al. (2005) observe that teachers who have acquired sufficient NOS knowledge to be confident enough to teach it, and teachers who recognize that an explicit approach is better than an implicit one, do not necessarily recognize that a student-centred approach is better than a didactic one. They frequently function at an intermediate stage in which NOS teaching is separated from other learning and presented as content to be learned, almost in rote fashion. What these teachers lack, of course, is robust pedagogical content knowledge in relation to NOS, that is, a repertoire of related examples, explanations, demonstrations and historical episodes that would enable them to translate their NOS ideas into a form that makes it accessible and interesting to students. Such PCK-NOS is knowledge commonly acquired through experience. It enables teachers to talk comfortably about NOS issues, lead discussions, respond quickly and appropriately to questions, clarify misconceptions, provide good examples, and so on. It involves a complex of NOS knowledge, appropriate science content knowledge (including warrant for belief and history of development) and general pedagogical knowledge regarding student-centred learning. It is knowledge that establishes connections between what students do in class and what scientists do in labs, and between key NOS ideas and the conceptual and methodological structure of science. It is knowledge that locates important items of scientific knowledge in the sociocultural circumstances surrounding both their creation and their contemporary deployment. As Abd-El-Khalick and Lederman (2000a) say, "it is not enough for teachers to "know" that scientific knowledge is socially and culturally embedded. They should be able to use examples and/or simplified case histories from scientific practice to substantiate this claim and make it accessible and understandable to students' (p. 693). PCK-NOS is still an under-researched area of scholarship.

Perhaps the intermediate stage described by Kim et al. (2005) is essential as well as inevitable. Perhaps it is easier for teachers to make the transition to an explicit and reflective approach once confidence has been established within the relative "safety" of an explicit didactic approach. It gives breathing space for novice teachers to build up their PCK for NOS teaching. As teachers gain experience of NOS teaching and subject their practice to reflective criticism, they will acquire increasing expertise. In a sense, and in common with all aspects of teaching, teachers "haul themselves up by their bootstraps": beliefs about NOS and beliefs about NOS teaching are reflected in classroom practice, and reflection on classroom practice informs and changes beliefs about NOS and NOS teaching. However, an intermediate stage comprising a didactic approach to NOS knowledge does set up the possibility of a disturbing mismatch between the views of knowledge being made explicit and the view of knowledge implicit in its method of delivery. The kind of NOS-oriented curriculum for which I argued in chapter 1 would not be well-served by a traditional teacher-centred approach, a dilemma that many of the preservice teachers in the study conducted by Yilmaz-Tuzun and Topcu (2008) seemed unable to resolve. Even the commonly used question-answer approach – teacher-initiated question; student(s) response; teacher evaluation – sets the supposed discussion within an authoritarian framework. It has been characterized by Lemke (1990) as monologue masquerading as dialogue in pursuit of class control (but see Chapter 9 for a discussion of how teachers can make such exchanges more productive of learning). The mismatch could be further exacerbated by adoption of traditional assessment methods that seek particular "correct" answers, often within a multiple-choice format. At best, traditional methods require short responses within a highly prescribed framework. When students are encouraged to hold the view that scientific knowledge is negotiated within the community, yet observe their teacher dispensing this view and testing the extent to which it has been learned, they may not just be confused, they may be resentful or even outraged. They may become disengaged or antagonistic (Yerrick et al., 1998). These and other pedagogical issues relevant to an explicit reflective approach to NOS learning will be discussed at length in later chapters.

NOTES

[1] Even among those with "mixed" beliefs, 14 out of 16 were consistent across two belief systems (nine for teaching and learning; three for learning and NOS; two for teaching and NOS).

[2] The Draw-a-Science-Teacher-Teaching checklist was used to characterize participating teachers' teaching styles (Thomas et al., 2001).

[3] The two varieties of contextualism – the relativist version and the rationalist version – refer to broad positions derived, respectively, from Thomas Kuhn's early writing (e.g., *The Structure of Scientific Revolutions*) and his later writing (e.g., *The Essential Tension*). The five categories are fairly strictly defined in terms of responses to questionnaire items, and any respondent who deviates substantially from a particular profile is categorized as "eclectic".

[4] This research was conducted in New Zealand, where teachers rarely have the support of a lab technician.

[5] The authors speculate that the teachers may have consciously misled them and had expressed an interest in NOS teaching because they perceived that the researchers would approve such a view. Meanwhile, their principal concerns lay elsewhere – in content-oriented teaching, developing students' social skills, and the like.

[6] With regard to historical science stories, Tao (2003) points out that students are inclined to attend to aspects of science stories that fit their prior ideas about NOS and ignore aspects that do not fit their expectations. They will often reinterpret stories to fit their expectations.

[7] The colloquialisms are mine, and their use should not be blamed on the authors.

[8] Clough (2006) presents a very clear and detailed description of conceptual change theory in relation to NOS teaching.

[9] For teachers, "views of learning" refers to both one's own learning and the learning of students.

[10] ICAN is an NSF-funded teacher professional development project based at the Illinois Institute of Technology. It is designed to enhance K-12 teachers' knowledge and pedagogical knowledge related to NOS and scientific inquiry.

CHAPTER 4
MAKING NOS TEACHING
EXPLICIT AND REFLECTIVE

Research reviewed in Chapters 2 and 3 makes it abundantly clear that NOS teaching is much more likely to be successful when it is explicit and reflective. Implicit approaches, in which students are expected to acquire NOS understanding by "reading between the lines" as they engage in classroom activities, particularly practical work, have been shown to be, at best, only moderately successful in achieving their goals. The key question is *how* to make NOS teaching explicit without resorting to a free-standing course in history, philosophy and sociology of science? Moss et al. (2001) urge teachers to develop their own models of the nature of science, suited to their particular educational situation, rather than to seek a ready-to-use model designed elsewhere. I have considerable sympathy with this view.[1] I have little sympathy for the notion of formal courses in HPS, although I share Taber's (2008) view that teachers do need to establish clear learning goals for the NOS elements they include in the curriculum. As argued in Chapter 1, I would resist the call for a narrowly prescribed list of NOS items. Rather, I believe that we should address NOS issues as the need arises, and we should seek to confront students with a range of alternative views, provide the necessary support and guidance for them to engage in critical debate and argument, and assist them in formulating views for which they can construct a robust, coherent and appropriately argued justification. What is clear from the research outlined in Chapter 2 is that students' NOS views are often muddled and disorganized, often with several mutually contradictory elements. My notion of critical scientific literacy entails students building a coherent and internally consistent framework of NOS understanding.

Some years ago, I suggested that teachers might find it interesting, helpful and amusing to borrow the familiar model of "rational curriculum planning" (Tyler, 1949) and adapt it to the task of describing the scientific enterprise in terms of purpose, knowledge store, methods of inquiry and procedures for evaluation (Hodson, 1990a). An account of scientific practice organized in this way provides teachers with a model for planning a science curriculum that teaches more directly *about* science. In addition, it provides teacher educators with a structure for planning preservice and inservice courses on curriculum matters related to HPS and provides students with a useful means of thinking about science, in particular about the relationships between the purposes of science, the procedures and the-

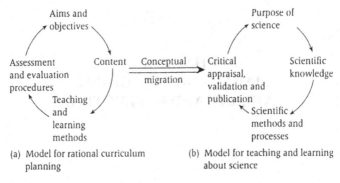

Figure 4.1. Adapting Tyler's rational curriculum planning model.

ories of science, and the mechanisms adopted by practitioners for evaluating the work of others. I noted at the time that the linear model for rational curriculum planning had long since been replaced, at least in theory, by a cyclic, reflexive and interactive model of curriculum planning (Figure 4.1a). By analogy, scientific practice should be regarded as cyclic and interactive. Although purposes, content, methods and assessment/evaluation can be considered separately, it is evident that they interact in complex and subtle ways, and students learning about science as well as teachers engaged in curriculum planning should be made aware of these interactions (Figure 4.1b).

Although the model did prove useful for the teachers with whom I have worked as a means of projecting some key NOS elements to the forefront during curriculum planning, different traditions of curriculum planning have intervened and few teachers are now familiar with the rational curriculum planning model. In searching the teacher education literature for a suitable alternative model with which today's teachers are likely to be familiar, I have been struck by the ubiquity of metaphors for teaching: teacher as broadcaster, teacher as gardener, teacher as tour guide, teacher as police officer, entertainer, director, *agent provocateur*, negotiator, circus ring master, ship's captain, chair person, scaffolder, manager, role model, and so on. One that suits my purpose is *teacher as anthropologist*[2] or "teacher as culture broker", as Aikenhead (2000, 2001a,b) calls it – an idea that was first proposed more than 25 years ago.

> Science and science education are cultural enterprises which form a part of the wider cultural matrix of society ... educational considerations concerning science must be made in the light of this wider perspective. (Maddock, 1981, p. 10)

Science can be taught in much the same way as an anthropologist teaches about another culture, that is, science can be regarded as a distinctive subculture involving a tribe of people with distinctive knowledge, language, rationality, customs, traditions, attitudes, values and norms. The concepts, procedures and language of science are recognized as cultural artifacts susceptible to systematic study; they are valid and robust within the cultural context in which they were developed and are currently deployed, but sometimes they have little relevance or meaning outside it; they are transmitted from generation to generation through community-approved educational practices.[3] And, of course, the community of scientists interacts with other communities in all kinds of ways. The nature and outcomes of these interactions, and the influence of vested interests, economic might and political power, are major focuses of attention in the kind of curriculum I propose.

Aikenhead's argument for deploying this particular metaphor is rooted in Giroux's (1992) notion of *border crossings*. Every individual has membership of several social groupings: family group, ethnic group, friendship group, employment group, possibly a religious group, sports group, leisure pursuits group, local community group or Internet chatroom and listserv group. Effective participation in these groups requires appropriate subcultural knowledge and skills, shared understanding, beliefs, language, code of behaviour, aspirations, values and expectations. As people move from one social context to another they are invariably required to change their way of speaking, acting and interacting with others, in order to be accepted. School is also a distinctive subculture with *its* own language, code of behaviour, values, goals and expectations. So, too, of course, is science and the school version of it. The greater the differences between a student's home and peer group subculture and the subcultures of school and school science, the more difficulties the student will encounter in "crossing the border", that is, in gaining access to school science and being successful there. It is fair to say that many teachers have seriously under-estimated the difficulties faced by some students. As Lemke (2001) comments, a student "spends most of every day, before and after science class, in other subject-area classes, in social interactions in school but outside the curriculum, and in life outside school. We have imagined that the few minutes of the science lesson somehow create an isolated and nearly autonomous learning universe, ignoring the sociocultural reality that students' beliefs, attitudes, values, and personal identities – all of which are critical to their achievement in science learning – are formed along trajectories that pass briefly through our classes" (p. 305). By shifting emphasis, the science teachers' job can be seen as helping students to gain an understanding of what, for many, are alien cultures (the subcultures of science, school and school science) and to assist them in moving freely and painlessly within and between these subcultures and the subcultures of home and community (Aikenhead, 1996, 1997, 2006; Cobern & Aikenhead, 1998; Aikenhead & Jegede, 1999; Hodson, 2001).

Unfortunately there are many students for whom the rules about the conduct of lessons, the conventions concerning who can speak and what can be spoken about (including what can be challenged), and the particular form of school talk and science talk, constitute a set of conventions and restrictions that are so formidable they are dissuaded even from seeking access to science education. It is all too much to cope with! Moreover, many students do not see themselves or their experiences, interests, aspirations, values and attitudes reflected in the science curriculum and are uncomfortable with the way science is presented, taught and assessed. It is little wonder they decide that science is not for them. What teachers need is a way of recognizing and dealing effectively with these barriers to access. What students need is a way of entering the subculture of science, using its knowledge and procedures to engage in interesting and important tasks, and leaving again with their sense of self intact. Better still, with their sense of self enriched by the experience. It seems almost superfluous to say that it is crucial that the science education we provide does not require students to give up or to compromise their cultural identity, aesthetic sensibilities or moral/ethical values. When presented with such a stark choice, many students do not choose science.

> When ... students' language and cultural experiences are in conflict with scientific practices, when they are forced to choose between the two worlds, or when they are told to ignore their cultural values ... [they] may avoid learning science. (Lee, 1997, p. 221)

The science teacher as anthropologist metaphor locates the purpose of science education in school in equipping students with the knowledge, self-knowledge and confidence to move freely between different worldviews, accepting each on its own terms and for its own purposes. As individuals participate in the activities of the science classroom, and strive to familiarize themselves with its language, norms, values and expectations, they begin to see themselves differently. The teacher's intention, of course, is that they begin to see themselves as (potential) scientists or, at least, as people who are scientifically literate in the senses discussed in Chapter 1. This sense of identity is not static; it is constantly being shaped and re-shaped by the day-to-day activities of the class and by the myriad influences outside school. Moreover, through interactions with other students, through discussion and argument, students also impact and change the classroom environment and stimulate other students to restructure their own sense of identity. Embedded in this particular interpretation of the science classroom is the teacher's obligation to scrutinize and challenge all ideas and values, and inculcate a willingness in students to do so. After all, this is where the critical element of critical scientific literacy is located. Ogawa (1995) develops broadly similar views in his notion of "multiscience teaching": helping students to move comfortably and effectively between and among personal science (including all forms of idiosyncratic beliefs and explanations for phenomena), Indigenous science (the communal beliefs of the specific cultural group to which one belongs) and Western modern science (as promoted through the curriculum). Aikenhead (1996) expresses the view I am advocating particularly well:

> Border crossings may be facilitated in classrooms by studying the subcultures of students' life-worlds and by contrasting them with a critical analysis of the sub-culture of science (its norms, values, beliefs, expectations, and conventional actions), consciously moving back and forth between life-worlds and the science-world, switching language conventions explicitly, switching conceptualizations explicitly, switching values explicitly, switching epistemologies explicitly. (p. 41)

The metaphor of science teacher as anthropologist focuses attention on the need to identify the barriers that currently deny access or make access to school science difficult for some students, and on the need to provide whatever support is necessary to help students overcome these barriers. It focuses attention on the necessity of making the distinctive features of the subculture of science explicit to students. It can be assumed that in teaching students about science we should endeavour to present an authentic view of science, a balanced view of NOS issues (see Hodson, 2008, chapters 9 and 10) and a critical view of science and scientific practice. Thus, the metaphor seems to be well-suited to my overarching goal of critical scientific literacy. Particularly helpful in its capacity to operationalize these ideas, despite being more than 40 years old, is the approach of King and Brownell (1966), who describe the disciplines of knowledge (including science) in terms of eight characteristics.

- As a *community* – a corps of competent people with a common intellectual commitment to building understanding.
- As a particular *expression of human imagination* – a distinctive way of making sense of phenomena and events, an idea that has much in common with Gardner's (1984) notion of multiple intelligences.
- As a *domain* – a particular sphere of concern, interest and activity.
- As a *tradition* – a history, comprising the activities, experiences, traditions and discourse of earlier practitioners.
- As a *syntactical structure* – a distinctive mode of inquiry and collection of methods for generating and validating new knowledge.
- As a *substantive structure* – a complex framework of concepts, propositions, laws, models and theories.
- As a *specialized language* – a form of "intellectual shorthand" for conveying meaning quickly, accurately and unambiguously, and a distinctive form and style of argument.
- As a *valuative* and *affective stance* – an array of fundamental beliefs about the nature of being (compare the notion of *worldview*), a complex of emotional and aesthetic sensibilities, and certain moral-ethical imperatives.

It is these characteristics that we need to make explicit in the curriculum by repeatedly and systematically drawing attention to them so that students become more aware of them, understand them and can deploy them to advantage. Studying these matters and making them explicit to students necessarily involves confronting and dispelling the many distortions and falsehoods about science that are commonly projected by school science curricula, school textbooks and the popular media.

While the metaphor of science teacher as anthropologist has significant educational potential, it has one serious shortcoming. Anthropologists do not and should not interfere in the cultures they study, nor should they become active participants. In contrast, science teachers aim to prepare students for participation in the subculture of science. Moreover, those of us with an interest in the politicization of students and promotion of active citizenship, and those of us who consider there to be serious problems with the way science is currently organized and managed, want students to become sufficiently knowledgeable, sufficiently committed and sufficiently politically astute to interfere and, where necessary and possible, *change* the way the community functions. We may also want our students to take their understanding of the sub-culture of science to other communities of practice and use it there in ways that are useful, and we may encourage them to bring ideas and values from other sub-cultures into science in order to enrich, develop and reorient it. Again, that is not something that anthropologists are supposed to do.

Both the metaphor of science teacher as anthropologist and the King and Brownell characterization of disciplines focus attention on demarcation criteria. What is science? Is it characterized and delineated mainly by its methods, its explanatory systems, its criteria of validity, its language or its underlying values and purposes? How does science differ from other forms of knowledge (or *disciplines* of knowledge as King and Brownell call them)? How much room for manoeuvre, change and adjustment within the eight dimensions is possible before we cross the border into another discipline, or create a new one? These are questions that teachers should ask themselves as they examine their own understanding of science, establish priorities for the curriculum, and design teaching and learning activities. They are questions that we should pose to students in our pursuit of critical scientific literacy. After all, distinguishing good science from bad science, detecting bias, distortion and fraud, differentiating among science, pseudoscience and non-science, and seeking ways to influence the priorities and practices of contemporary science, are among the key elements of critical scientific literacy.

DEMARCATION CRITERIA

Mahner and Bunge (1996) declare that "among philosophers of science there is no agreement as to how to characterize science, particularly, as how to demarcate it from nonscientific fields of inquiry" (p. 102). Essentially the same position is expressed by Laudan (1983): "it can be said fairly uncontroversially that there is no demarcation line between science and non-science, or between science and pseudo-science, which would win assent from a majority of philosophers" (p. 9). Nevertheless, questions of demarcation should be asked in school science, not least because the capacity to evaluate the claims of astrology, iridology and various kinds of "New Age science" is a crucial element of critical scientific literacy and one of the principal reasons for ensuring that students achieve it. If students cannot distinguish valid and reliable knowledge claims from invalid and unreliable ones, they are in no position to make the kind of decisions that I argued in Chapter 1 are central to the goal of critical scientific literacy. If science teachers cannot demarcate between science and non-science, they have no defence against those who would advocate the inclusion of creationism in the science curriculum, or its contemporary manifestation as "intelligent design".

Demarcation is not a trivial issue! The designation of a field of study as *scientific* is a much-sought-after status. The term carries with it a subtext of rationality, dependability, trustworthiness, rigour and reliability. Hence its frequent use in advertising. It is a status not readily conferred and is jealously guarded, especially by the scientific community, which has a vested interest in maintaining its own high status and would not countenance any weakening of that status by "suspect" newcomers. Witness, for example, the ways in which the community of medical practitioners closes ranks to resist the claims of alternative practices perceived as threats to its dominant position. Images of science and scientists are constructed and maintained because they are socially, politically and economically useful to those who do the constructing (the in-group). Where and how the demarcation lines are drawn is of great importance to both those included and those excluded; it can have enormously far reaching consequences, as Lakatos (1978a) reminds us.

> The problem of demarcation between science and pseudoscience has grave implications also for the institutionalization of criticism. Copernicus's theory was banned by the Catholic Church in 1616 because it was said to be pseudo-science ... The Central Committee of the Soviet Communist party in 1949 declared Mendelian genetics pseudoscientific and had its advocates ... killed in concentration camps ... the West also exercises the right to deny freedom of speech to what it regards as pseudoscience ... All these judgements are inevitably based on some sort of demarcation criterion. This is why the problem of demarcation between science and pseudoscience is not a pseudo-problem of armchair philosophers; it has grave ethical and political implications. (pp. 6–7)

While philosophers are justifiably cautious about stating clear criteria, science educators seem to be less so. Cobern and Loving (2001), for example, state that there is a sufficient level of consensus for a "standard account" to be discernible: "Science can be defined with sufficient clarity so as to maintain a coherent boundary for the practical purposes of school science curriculum development. That boundary excludes most forms of indigenous knowledge, if not all, just as it excludes art, history, religion, and many other domains of knowledge" (p. 64). Whether this expression of confidence is regarded as brave or foolhardy depends, I suppose, on how rigorously the proposed criteria are scrutinized, and from what perspectives. The criteria are arranged into three main groups (Cobern & Loving, 2001, pp. 60–61, all emphases removed):

> Science is a naturalistic, material explanatory system used to account for natural phenomena that ideally must be objectively and empirically testable.
>
> – Science is about natural phenomena.
> – The explanations that science offers are naturalistic and material.
> – Science explanations are empirically testable (at least in principle) against natural phenomena (the test for empirical consistency) or against other scientific explanations of natural phenomena (the test for theoretical consistency).
> – Science is an explanatory system – it is more than a descriptive *ad hoc* accounting of natural phenomena.
>
> The Standard Account of science is grounded in metaphysical commitments about the way the world "really is ..."
>
> – Science presupposes the possibility of knowledge about nature.

- Science presupposes that there is order in nature.
- Science presupposes causation in nature.

Nevertheless, what ultimately qualifies as science is determined by consensus within the scientific community.

A whole host of questions immediately springs to mind. What constitutes objectivity? Are scientific propositions always empirically testable? If science is about *natural* phenomena, where does that leave the many phenomena and events *created* by science? How is consensus reached? It is fairly easy to deploy sophisticated arguments from the philosophy of science, history of science and sociology of science to identify problems, flaws and lack of clarity in any such list, but we should bear in mind that these particular demarcation criteria are intended for use in school science, not for a course in philosophy. We should also ask about possible alternatives to this list. Should we seek another (necessarily) imperfect list, with a different set of flaws? Should we dispense with the idea of a set of demarcation criteria and conclude that we cannot distinguish science from non-science, or that the difference is obvious and not worthy of discussion?

Suchting (1995) suggests that a candidate for inclusion among the sciences should be judged by its "family resemblance" to those fields of study already recognized as science, an idea originally put forward by Wittgenstein and explored at length by Hacking (1996). While there is clearly no *one* scientific method, science does use particular "styles of reasoning" – not all of them at the same time and by all the sciences, but some of them by some of the sciences, some of the time. These styles of reasoning include experimental exploration and measurement, hypothetical modelling and use of mathematics; they involve the search for theories with broad scope, extensive explanatory power and high predictive capability. Family resemblances encompass the building of theories that are internally consistent, compatible with other domains of scientific knowledge (or, at least, not *in*compatible), capable of unifying seemingly unrelated phenomena, and productive of new lines of research. The diversity of scientific endeavours is underpinned and sustained by a family-wide commitment to rapid communication of findings, open and unfettered criticism, and willingness to change position in the light of good evidence and argument. Smith and Scharmann's (1999) position is that teachers should not attempt to teach a set of specific rules to demarcate science from non-science. Rather, they should aim for a set of descriptors that can be used to judge which fields of study or which activities are more scientific or *less scientific* than others. In other words, they acknowledge a degree of fluidity within the criteria, some fuzziness around the boundaries of science and a measure of overlap between areas of concern. Their position is that while there is no clear-cut universally applicable specification for science and non-science, scientists strive towards certain ideals. For example, they strive to be objective in the way they collect, interpret and report data – honestly and diligently, with no presuppositions or prejudice. In elaboration of their main point, Smith and Scharmann (1999) state: "objectivity is a poor candidate for an absolute criterion distinguishing science from nonscience. On the other hand, few educated individuals would argue that scientists who seek to increase their objectivity and decrease their subjective biases are not being more scientific than scientists who do not" (p. 499). Among the characteristics that the authors identify as making a field of study *more scientific* are: its empirical character (science deals with entities that can be measured and with relationships that can be quantified); testability/falsifiability (claims are accepted or rejected on the basis of evidence); replicability (experiments and observations have to be repeated successfully in order to rule out error, misconduct and divine revelation); tentative status (scientific knowledge changes over time as theories are modified or discarded, and new evidence deployed.[4] In addition, science puts high value on theories that exhibit (i) extensive explanatory power (i.e., can successfully address a wide and diverse range of phenomena and events), (ii) predictive capability, (iii) fecundity or fruitfulness (for example, raising new questions and providing new ways of looking at the world), (iv) open-mindedness, (v) parsimony, and (vi) logical coherence. In contradistinction, say Smith and Scharmann, fields of study are less scientific when they espouse a theological position, value authority over evidence and put reliance on faith rather than reason.[5] This position seems to sit very comfortably with an approach to NOS teaching based on the science teacher as anthropologist metaphor and the disciplinary characteristics outlined by King and Brownell.

Some 60 years ago, Robert Merton identified four "functional norms" or "institutional imperatives" that govern the practice of science and the behaviour of individual scientists, whether or not they are aware of it (Merton, 1973).[6] They are not codified and are not enforced by specific sanctions. Rather, they are transmitted in the form of precepts and examples, and are eventually incorporated as an ethos into what Ziman (2000) calls the "scientific conscience" of individual scientists. Not only do these norms constitute the most effective and efficient way of generating new scientific knowledge, they also provide a set of "moral imperatives" that serves to ensure good and proper conduct.

– *Universalism* – science is universal (i.e., its validity is independent of the context in which it is generated) because evaluation of knowledge claims in science uses objective, rational and impersonal criteria rather than criteria based on personal, national or political interests, and is independent of the reputation, race, religion and social class of the particular scientist or scientists involved.

– *Communality* – science is a cooperative endeavour and the knowledge it generates is publicly owned. Scientists are required to act "in the common good", to avoid secrecy and to publish their findings and conclusions so that all scientists may use and build upon the work of others.

– *Disinterestedness* – science is a search for truth simply for its own sake, free from political or economic motivation or strictures, and with no vested interest in the outcome. Because attempts to exploit the ignorance or credulity of non-scientists and/or fabricate results in pursuit of commercial or personal gain are strictly outside the code of approved scientific conduct, scientists have traditionally enjoyed a good reputation for ethical behaviour.

— *Organized scepticism* – all scientific knowledge, together with the methods by which it is produced, is subject to rigorous scrutiny by the community in conformity with clearly established procedures for judging matters such as methodological appropriateness, chain of argument from data to conclusions, and testability. These procedures ensure that all knowledge claims are treated similarly, regardless of their origin.

Two additional norms have been proposed by Barber (1962): *rationality* – science uses rational methods to generate and validate its claims to knowledge, and *emotional neutrality* – scientists are not so committed to an existing theory or procedure that they will not reject it or adopt a new one when empirical evidence favours it. Similar ideas underpin Garfinkel's (1967) "rules of interpretive procedure", which he contrasts with everyday reasoning: the rule of unlimited doubt asserts that scientists will not limit their scepticism by the kind of "practical considerations" that govern everyday life; the rule of *knowing nothing* allows scientists to suspend their own knowledge in order to see where it leads, while testing in everyday life proceeds on the basis of what can be taken for granted; the rule of *universalized others* enables scientists to trust the findings of others; the rule of *publicisability* requires that all findings are made public, regardless of personal motives and interests.

SCIENCE AS A COMMUNITY OF PRACTICE

Several of the research studies reported in Chapters 2 and 3 reveal that the social dimensions of science (both internal and external to the community of scientists) are the areas in which students' NOS understanding is weakest, least well- developed and most confused, and in which teachers are most reluctant to teach – ironically, the very areas that are central to the notion of critical scientific literacy and education for responsible citizenship. A principal strength of the science teacher as anthropologist metaphor and the King and Brownell list of disciplinary characteristics, and the modified version of Karl Popper's Three Worlds model used in Hodson (2008), is that the social dimensions of science cannot be ignored or marginalized. Because students are studying the characteristics of a *community* (of scientists), social issues are centre stage. They infuse discussion of science right from the outset. They provide a powerful counterpoint to traditional textbook accounts of theoretical developments that pay scant attention to the personal and social dimensions of scientific practice, neglect to consider the ways in which the decisions and actions of scientists are influenced by their worldviews, feelings, attitudes and prejudices, and fail to acknowledge how science is subject to a wide range of sociocultural and economic influences.

We need to make students aware that because of the theory-laden nature of observation and theory-impregnated nature of all the other processes of science, and the empirical under-determination of theories, what one "sees" as an observer, the proper conduct of experiments, the adequacy of a theoretical explanation, and so on, are all open to dispute, contestation and modification. Rather than attempting to reduce science to a cold, clinical, de-personalized method, and rather than presenting science as independent of the society in which it is located, we should be emphasizing the ways in which knowledge is *negotiated* within the scientific community by a complex interaction of imagination, experiment, theoretical argument, personal opinion and community-sanctioned norms of professional conduct. And we should be promoting the view that science is a theory-driven and creative endeavour, influenced throughout by social, economic, political and moral-ethical factors as they impact on the decision makers – that is, on the "gatekeepers" or guardians of the community's store of knowledge. As Robert Young (1987) says:

> Science is not something in the sky, not a set of eternal truths waiting for discovery. Science is practice. There is no other science than the science that gets done. The science that exists is the record of the questions that it has occurred to scientists to ask, the proposals that get funded, the paths that get pursued ... Nature "answers" only the questions that get asked and pursued long enough to lead to results that enter the public domain. Whether or not they get answered, how far they get pursued, are matters for a given society, its educational system, its patronage system and its funding bodies. (pp. 18 & 19)

Not only is science a social activity in the sense that society (writ large) determines "the science that gets done", but also in the sense that the rules of scientific procedure and the legitimacy of the "product" are determined by a community of practitioners. As Bereiter et al. (1997) remark, "If there is anything distinctive about science, it is not to be found in the workings of individual minds, but in the way scientists conduct themselves as a community" (p. 333). Again, there are two levels of community here. First, scientists rarely work entirely alone; most scientists in universities, research institutes and industry are members of a laboratory team working on closely related problems and extending over a number of years. Beyond the specific laboratory group an individual is also a member of the institution's team of scientists and a participant in the "invisible team" of scientists in all other institutions who conduct similar and related research. In a wider sense, scientists are dependent on one another for the overall research context within which they work, including availability of theoretical and procedural knowledge, investigative techniques, laboratory apparatus and instruments, communication channels, and so on. Second, the wider community of scientists determines what counts as acceptable scientific practice, exercises strict control of what is admitted to the corpus of accepted knowledge through its system of peer review, and maintains control over the education of future scientists and the initiation of newcomers into the community of practice. The rationality of science and the trustworthiness of the knowledge it generates is located in careful and critical experimentation, observation and argument, and in critical scrutiny of the procedures and products of the enterprise by other practitioners. The key point here is that science has a community-regulated and community-monitored rationality. Science is socially constructed through critical debate, and those involved in it have a commitment to maintain certain standards of debate. As Helen Longino (1990) points out, "it is the social character of scientific knowledge that both protects it from and renders it vulnerable to social and political interests and values" (p. 12). In Longino (1994), she identifies four conditions that a community of practitioners must meet if consensus is to count as *knowledge* rather than mere opinion.

— There must be publicly recognized forums for criticism – conferences, academic journals and the system of peer review, for example.
— There must be publicly recognized standards for evaluation of theory and practice – including empirical adequacy, appropriate research design, robust argument from evidence to conclusions, and so on.
— There must be uptake of criticism – the community needs to do more than merely tolerate dissent; it must act on it, such that ideas and theories change over time.
— There must be equality of intellectual authority – what is included or excluded must result from critical dialogue rather than the exercise of political or economic power. In a later work, Longino (2002) refers to a *"tempered equality of intellectual authority"* – that is, an equality that is tempered or qualified because of differences in intellectual abilities, educational background, research experience, and so on. However, she adds, the social position and economic power of someone in the scientific community ought not to determine the decisions that are made, and "every member of the community [should be] regarded as capable of contributing to its constructive and critical dialogue" (p. 132).

Longino (1994) notes that the specific criteria deployed in reaching consensus, or in mounting a serious challenge to a previously established consensus, may shift as the debate moves from the relatively private confines of the research laboratory to the public forum of scientific conferences and refereed publications. If it survives critical scrutiny by the community, using these public methods of evaluation and judgement, the knowledge item (model, theory, experimental procedure, instrumental technique, or whatever) becomes part of the written record of the scientific community and is made available to others. Because of this mechanism and the confidence that practitioners have in it, science is cumulative; current researchers utilize the knowledge generated by previous scientists and, in doing so, may develop it or discard it. Of course, the way in which the community of scientists exercises this public scrutiny at any one time is subject to a whole range of social, cultural, political and economic factors. For example, the likelihood of individual scientists being persuaded by a particular argument or influenced by particular evidence depends, in part, on their background knowledge, assumptions and values. The possibility that different perspectives will be brought to bear in the appraisal process is the reason for upholding the principle of academic equality and one of the guarantees of scientific objectivity. It is also a powerful argument for increased diversity within the scientific community (see later discussion). Taylor (1996) extends the notion of science as a social practice into the argument that the community of scientists collectively *identifies* the domain of science, *constructs* its procedures and standards, and *establishes* its demarcation criteria to reflect their personal, cognitive, technical and professional interests. Consciously or unconsciously, they construct definitions of science to "enhance their cognitive authority", accumulate and maintain professional resources (including research funding), establish a measure of control of science curricula and exclude what they *determine to be* pseudosciences or non-science. As noted above, communally generated standards constrain the production and legitimation of knowledge claims, the validity of which have to be argued before an audience of peers.

SCIENCE AS A THEORY-DRIVEN AND COMMUNITY-DIRECTED ACTIVITY

The important ideas of science cannot be acquired through everyday experience or through conversations with non-scientists. While students can look for and occasionally "discover" regularities in nature via observation and/or experimental investigations, it is absurdly naïve to expect them to be able to invent for themselves the abstract notions such as *gene*, *molecule* and *magnetic field* that scientists have developed over many years to describe, interpret, explain and predict.

> Learning science ... is seen to involve more than the individual making sense of his or her personal experiences but also being initiated into the "ways of seeing" which have been established and found to be fruitful by the scientific community. Such "ways of seeing" cannot be "discovered" by the learner – and if a learner happens upon the consensual viewpoint of the scientific community he or she would be unaware of the status of the idea. (Driver, 1989, p. 482)

Teachers have to shape students' views towards the inclusion of scientific notions in their *personal framework of understanding* (Hodson, 1998a). To do that, they must intervene. Thus, while everyday learning occurs spontaneously in the context of everyday experiences, the learning of science has to be organized via contrived situations. It has to be a purposeful *induction* (Lawrence Stenhouse's, 1975, term) that goes beyond personal experience and personal knowledge. It necessarily involves the provision of new experiences and the introduction of new ways of "seeing" and arguing. For example, the very first science lessons dealing with observation should be designed to convey the message that being a skilled observer means knowing what to look for, knowing where and how to look for it, having expectations of what will be seen, checking these expectations against reality, and knowing which observations are significant – as judged by the community of scientists. Therefore, in teaching young children how to make scientific observations, it is necessary to provide an appropriate and community-approved theoretical framework, emphasize distinctive and discriminating features, use examples, provide feedback on success, and so on. In examining a twig, for example, students will tend to concentrate on colour, texture and shape; they will not even notice leaf scars and growth rings unless the teacher points them out. Early on, it is important to bring children to a realization that objectivity in science does not consist in placing equal weight on all observations, something which children consider to be "fair" and which the traditional view of science sometimes seems to emphasize. What is needed instead is awareness that in learning to be a scientist students have to learn how to select relevant and appropriate observations and discard irrelevant, inappropriate and incorrect ones, a procedure for which they need a sound and well-understood theoretical framework. And, moreover, a theoretical framework developed and validated by the community of practitioners. The teacher's role is one of focusing attention on the *scientifically significant*, posing interesting and answerable questions, and guiding students towards relevant and significant observations by emphasizing theoretical issues and promoting good use of scientific language. There is an interesting paradox here: unless students know what to look for, they won't see anything, yet if they only look for what they expect they may miss the unexpected and the theoretically significant, or may misinterpret it. Students can be made aware of this point by asking them to consider how they might design an investigation to detect "Z-rays". Clearly, unless they speculate about the properties of Z-rays, they cannot design instruments to detect them, and the instruments will only detect Z-rays if they happen to have these particular hypothesized properties.

Hodson (1986) provides a fairly detailed discussion of how students, even very young students, can be made aware of the role and status of observation in science and the ways in which it entails "more than meets the eyeball", as Hanson (1958) puts it. Elements of the approach include the following: use of film and video material recorded with a moving camera, to make the point that the relative position and/or movement of the observer is a major factor influencing observations; work with optical illusions and other visual puzzles; use of instruments to extend observational possibilities (microscopes, telescopes, etc.); identification of objects by listening, smelling and tasting, to reinforce the point that observation requires assumptions to be made and questions to be asked as a guide to the collection of data; use of puzzle boxes, "black boxes" and "feely bags", followed by activities involving the use of maps and map construction. The purpose of exercises with maps is to show students that the knowledge contained in a map is one of several possible selections and representations of reality. The elements in a map are chosen, not imposed by Nature; no matter how useful a map may be, it is still a simplified and abstract representation of reality, just as scientific theory is a simplified and abstract representation of reality (see discussion in Chapter 6).

The "tricky tracks" activity devised by Lederman and Abd-El-Khalick (1998) is especially useful in introducing students to the distinction between observation and inference, and for emphasizing the importance of not claiming more from an observation than is justified. However, it is equally important for students to recognize, as soon as possible, that commonly used observational terms often include substantial theoretical inference. Unless a reasonable body of theory is taken-for-granted and theory-laden terms used for making observations, we would be unable to make progress. We would be constantly trying to return to simpler and supposedly theory-free terms. In the familiar exercise of observing a burning candle, all but very young children *know* that the liquid on the top of the burning candle is molten wax, because they have no doubts about the theoretical assumptions that impregnate such a view. To insist that they regard it as an inference is to be pedantic to a degree that can be counter-productive to good science. As Norris

(1985) says, "If you must doubt that the liquid at the top of candles is paraffin, how can you ever be expected to believe the far more profound theories and laws of modern science" (p. 830). The demarcation between observation and inference should be drawn at a point that is appropriate to the theoretical knowledge of the students, at the point where there is reason for students to doubt the theoretical assumptions that are involved. Insisting on a demarcation when there is no cause for doubt is to promote false notions about scientific observation. When theories are taken-for-granted they provide an observation language. So that with the general acceptance of a theory of solubility, for example, we see things *dissolve*, where previously we saw them *disappear*. Both scientists and non-scientists are able to observe solar eclipses, or craters on the Moon (rather than "dark patches"), because certain theories underpinning our views about the solar system and the cause of lunar features are not in dispute. Students who have studied some elementary physics can make sense of instructions to measure *voltage*, *current* and *resistance* because school science has led them to accept all the underlying theory assumed in designing and using ammeters and voltmeters. In the same way, sophisticated conceptual understanding permeates the observational language of practising scientists. In looking at a cloud chamber photograph, for example, an experienced scientist sees a record of sub-nuclear events whereas a non-scientist sees only confusing irregular lines. The point is that without a bold assumption of previously validated theories, scientific observation, and the theory-building that it informs and validates, would be impossible. Students can be made aware of this by repeating observational exercises from earlier in the course (or from a previous year) using more sophisticated observational language than previously. When heated in a Bunsen flame, certain materials can now be observed to melt, sublime or decrepitate, where previously they had just "changed". All three of these new terms include theoretical inference. There is much of value in conducting a discussion of the theoretical assumptions underpinning the design, construction and operation of common laboratory instruments, beginning with thermometers and microscopes, before proceeding to more sophisticated instruments such as ammeters, voltmeters and pH meters. While instruments such as microscopes and telescopes simply enhance ordinary human perception, instruments such as voltmeters and Geiger counters detect and measure phenomena that cannot be consciously or directly experienced by human beings. Their operation, and interpretation of the data they produce, involves deployment of a substantial amount of scientific theory. Martin (1972) sums up the situation particularly well.

> The idea that the best scientific observation is naïve like a child's, and that to become a good scientific observer one must unlearn all the ideas one has previously learned and return to the pristine purity of infancy, has little merit. Becoming a good scientific observer involves great sophistication and skill. The cure for bad scientific observation is not naïve observation, but more training and sophistication. (p. 112)

The point of this seeming diversion into the nature of scientific observation was to underline the point that science is community-generated knowledge, to which students have to be introduced by someone who is already expert (to a degree). In other words, science education is best regarded as enculturation into the knowledge, beliefs, language, values and code of conduct of the community of scientists, or the school versions of them. Each of Chapters 6, 7 and 8 is concerned with a particular aspect of enculturation into the communities of science and school science (namely, substantive structure, syntactical structure and language).

As these observation activities make clear, it is impossible to remove human influences from the practice of science. Decisions about priorities for scientific investigation, formulation of research questions, design and conduct of inquiries, collection and interpretation of data, and presentation of findings, are all made by people. Stories about science and scientists can show students something of the lives of the principal players in important episodes in science and identify the likely influences on their thinking and on their expectations and aspirations (see Chapter 10). Attention can be focused on the ways in which knowledge claims are subjected to rigorous scrutiny by other scientists and on the lengthy battles some scientists had to fight to get their ideas accepted. Stories are a very effective means of impressing on students that science is both a highly personal and a highly social activity.

As well as its own internal logic, science is shaped by the personal beliefs, education and political attitudes of its practitioners. The institutions of science, and the deployment of its practical results, reflect in part the history, power structures, and political climate of the supportive community. (Dixon, 1973, p. 71)

SUBJECTIVITY, BIAS, DISTORTION AND MISUSE OF SCIENCE

Echoes and reflections of the society that produced it can be detected in all aspects of science, even in the way that we conceptualize and theorize. The body of scientific knowledge exists as a tradition, with a history (item 4 in King and Brownell's list of characteristics, above) during which many different goals, attitudes and values will have left a mark. In Jane Gilbert's (2001) words, "it is clear that the neutral, objective, disembodied, raceless, classless, sexless subject of science is an illusion" (p. 295). Throughout history, for a variety of socioeconomic and political reasons, the majority of scientists have been men, so it is not surprising that there is substantial evidence of a strongly masculine influence on science. Students should be sensitized to it, and to consideration of what might be done to make science more gender inclusive.

As Harding (1991) points out, these male scientists have blithely assumed that they could tell "the one true story about the world that is out there, ready-made for their reporting, without listening to women's accounts or being aware that accounts of nature and social relations have been constructed within men's control of gender relations" (p. 141). The story of scientific theorizing about hominid evolution and early human social development can be used to illustrate Harding's assertion. Until the last 25 years or so the dominant accounts involved man-the-hunter and his crucial role in social evolution (Lovejoy, 1981). More recently, anthropologists and ethnobiologists such as Eleanor Leacock, Sally Slocum, Adrienne Zihlman and Nancy Tanner have focused on the central role of woman-the-gatherer and woman-the-toolmaker (Stanley, 1983; Alic, 1986; Hardy, 1986; Schliebinger, 1999; McGee & Warms, 2004). The key point to put to students is that both explanations accord with the evidence. Elizabeth Fee's (1979) account of craniology makes for interesting reading, too. She states that 19th century anthropologists "knew" that women are less intelligent than men but, as scientists, they needed the evidence provided by measurement of skulls. Measurement based on cranial volume indicates that elephants are more intelligent than humans, so volume had to be rejected as the principal indicator of intelligence. The ratio of brain size to body weight indicates that birds are more intelligent than humans, so other more complex measurements were taken, all in the drive to confirm the prior belief that men are more intelligent than women. Allchin (2004a) notes that the leading American craniologist, Samuel Morton, was highly selective in deciding which particular skulls to measure. This basic deviation from good scientific practice went undetected until Alice Lee and Marie Lewenz pointed it out, and used more sophisticated statistics than Morton's to reinterpret the data. Interestingly, they reached the opposite conclusion to Morton: women are more intelligent than men. It seems that scientific findings sometimes depend on who conducts the investigation! The field of primatology provides more compelling evidence that a change of perspective, from male to female, brings about a change in conceptualization. For example, Fedigan (1986) discusses what she calls the "baboonization" of primatology in the 1950s: savannah baboons, one of the most aggressive and male-dominated of all primates, was established as the preferred model for ancestral human populations despite knowledge of other, less aggressive primate populations. Subsequently, Haraway (1989) traced changes in the initial androcentric assumptions, language and theories of primatology, showing how a cohort of women scientists re-shaped accounts and explanations constructed by earlier scientists (exclusively male) to explain primate behaviour and social organization. In Helen Longino's (1990) words, "The existence of dominance structures in primate troops is 'obvious' until a different way of describing the interactions shows us that dominance is an interpretation of behaviour arising from the researchers' assumptions and expectations of social behaviour" (p. 221).

One final example should suffice to consolidate the argument. Traditional models of sexual reproduction (still the norm in school textbooks) posit a unidirectional causal relationship between sperm and egg: the notion of the "active" sperm penetrating and fertilizing the "passive" egg. In contrast to the large, passive egg, which is transported, swept or even drifts along the fallopian tube, the small, streamlined, active sperm delivers genes to the egg and thereby activates its developmental programme. By implying that the female biological role is less worthy and more dependent than the male role, such accounts may serve to reinforce gender stereotypes. The more complex, interactive and multi- factored models of reproduction developed by Ruth Herschberger (1948) and Emily Martin

(1991), which postulate equal causal contributions by the egg and the sperm, are better able to describe the role of females in reproduction and can, thereby, help to combat hierarchical gender stereotypes. However, Martin warns of shifting too far in the direction of an active egg that "captures and tethers the sperm with her sticky zona, rather like a spider lying in wait in her web" (p. 498). That way, she says, lies an image of woman as dangerous and aggressive, "the femme fatale who victimizes men" (p. 498). Arguing along similar lines to Emily Martin, Helen Longino (1994, 1995, 1997) advocates an approach in which traditional epistemic values, such as consistency with established knowledge, simplicity and scope, are replaced with values such as *novelty* (preferring theories that differ in significant ways from currently accepted theories with respect to the entities they postulate or the nature of the explanation), *ontological heterogeneity* (theories that invoke a range of causal entities are preferred to theories that postulate only one kind of causally efficacious entity or seek to use hierarchically-based reasoning to reduce explanations to just one causal entity) and *complexity of relationship* (theories are preferred when they regard relationships between entities as interactive rather than unidirectional, and multi-factored rather than single-factored).

In summary, it seems that culturally derived assumptions facilitate certain conclusions and exclude others, which is a good argument in itself for ensuring greater diversity within the scientific community. As Sandra Harding (1991) comments, exposing science to the scrutiny of those with alternative viewpoints, particularly those of oppressed and marginalized groups, would increase the objectivity of research by "bringing scientific observation and the perception of the need for explanation to bear on assumptions and practices that appear natural or unremarkable from the perspective of the lives of men in the dominant group. Thinking from the perspective of women's lives makes strange what had appeared familiar" (p. 150). In other words, it is easier to identify problematic assumptions underpinning science when one does not hold the same values as the scientist(s) in question. Intemann (2008) uses the reporting of research on oral contraceptives as an exemplar. In the drive to sensitize everyone to the need to reduce overall population growth, research data was presented in "the best possible light". For example, tests showing that oral contraceptives are effective in reducing the incidence of breast cancer were highlighted by the drug companies, while information concerning the increased risk of blood clots, strokes and cervical cancer were marginalized. The position adopted by Helen Longino and Sandra Harding is that if women had been the directors of the various research teams, they would have been much less likely to conclude that oral contraceptives present no health risks. Harding's remarks about the importance of including members of oppressed and marginalized groups are particularly significant in the context of my advocacy of a politicized science education directed towards issues of social justice. However, Pinnick (1994, 2008), Hekman (1997), Lennon (2004) and Landau (2008) urge caution: they point out the absurdity of assuming that membership of an oppressed or marginalized group automatically guarantees greater insight, and note that these groups are no more homogeneous in their views and values than any other group, including groups of white male scientists. Nonetheless, a scientific community with diverse contextual values will be *more likely* to recognize when the contextual values of an individual scientist have had negative influence on his/her research and *more likely* to take account of alternative perspectives on research priorities, investigative design, conclusions and publication of findings. A similar case is easily built for increased diversity among the community of *teachers* because of the additional or alternative perspectives they may hold with respect to science, education, learning, students, teachers and pedagogy, and for the insight they may have with respect to the affective and social dimensions of learning science in particular sociocultural contexts.

There is substantial evidence of gender bias in science: first, in the history of science and technology and the ignorance, neglect and suppression of women's contributions, and sometimes their false attribution to men; second, in the institutions of science, from which women were historically excluded and within which, even today, women are still under-represented (especially at senior levels), disadvantaged and made to feel unwelcome and uncomfortable – sometimes overtly, more often covertly; third, in the priorities for scientific research and development (which often fail to reflect women's interests, concerns and needs), and in the language of science, its concepts, theories, methods, criteria of judgement, forms of argument and underlying values. A "common sense" explanation for this gender bias runs along the following lines: since most scientists, for whatever social reasons, have been male there is good reason to expect that the science produced will be androcentric (male-centred and male-biased) simply because it reflects the outlook, interests, priorities and experiences of those who produced it. In addition to being profoundly influenced by their own individual preferences, experiences and feelings, these male scientists are subject to all the social, political, economic, religious, technological and moral-ethical forces that impact society and work towards the formation of particular attitudes, values and ways of thinking. Thus, they are "children of their time and place", just like everyone else, and if it is legitimate to refer to masculine ways of thinking then science could be expected to reflect them. For example, three of the "ideological pivots" identified by Smolicz and Nunan (1975) as impregnating science – anthropocentrism, quantitative methodology and analysis – could be regarded as masculine. The feminine equivalents would be biocentrism, qualitative methodology and holism. Some would regard science's fondness for dichotomies, hierarchies, linear reasoning and the search for clear and direct cause and effect relationships as masculine concerns; while women may be more concerned with webs of relationships, interconnections, interdependence and intuition. The underlying values of science relating to control and manipulation of the natural environment could also be considered masculine. Even the language of science and the style of reporting in the third person, past tense and passive voice has been labelled masculine. Furthermore, the argument goes, because science plays such a powerful ideological role in our society it has functioned to legitimate social inequality between the sexes – in other words, to discriminate against women. For many years, women were denied access to science, presumably in order to retain power and influence in the hands of men. In more recent years, with the removal of formal barriers

95

to access, the masculine face of science has functioned to dissuade, limit or re-strict access by making women feel uncomfortable or "out of place". In a society where one gender (or one race) is dominant there is likely to be a disproportionate distribution of resources, with the greater share going to the dominant group and the inequity being justified on the basis of presumed inherent differences between dominant and subordinate groups. It is likely, then, that science will have been used to benefit those in power and to exclude or disadvantage those already on the margins. These kinds of arguments have also been made, of course, with respect to technology, possibly to an even greater extent. Technology is seen as a masculine construct and masculinity is often described in terms of technical competence (Wajcman, 1995).

Laudan (1977) argues that when the basic premises of a theory are in conflict with the prevalent worldview, the theory is considered problematic and will stand a good chance of being rejected. When the premises and the dominant worldview are consonant, the problematic nature of the theory is resolved and acceptance is much more likely, provided that certain other criteria are also met. If we follow Laudan's logic, then because our society is sexist and gender biased, a sexist theory is unproblematic if it is androcentric (male-oriented), but problematic if it is gynocentred (female-oriented). A female-oriented science would be counter normative; it would conflict with the prevailing social climate; it would struggle to gain acceptance. However, Sandra Harding (1991) and Linda Shepherd (1993) point out that science is not always as rational, logical, analytical, objective, abstract, quantitative, hierarchical, competitive and morally neutral as the (mas-culine) stereotype would indicate, and that scientists often value the so- called "feminine" attributes of receptivity, subjectivity, multiplicity, nurturing, empa-thy, sensitivity, cooperation, intuition, qualitativeness and relatedness. Wallsgrove (1980) makes the point that it is irrational to pretend that everything you do, as a scientist, is rational: "If you don"t examine and come to terms with what you feel, your feelings will interfere anyway, but in a hidden and uncontrollable way" (p. 235). These issues will be explored further in Chapter 6.

Gender studies in science comprise a vast literature, which cannot be ad-dressed within the confines of this book.[7] Nevertheless, it is important to make the point that science teachers need to be aware of this literature, and to be aware of the equally vast literature concerning gender bias in science *education*, that is, in choice of curriculum content and context, teaching and learning meth-ods, textbook language, examples and illustrations, assessment and evaluation procedures, classroom interactions (among students, and between students and teachers), and what is deemed acceptable/unacceptable classroom behaviour. It is this complex of interacting factors that can make science in school unwelcoming, uncomfortable or even hostile to girls, neglectful of their experiences, interests and aspirations, and dismissive of their preferences for particular learning styles and assessment methods. In consequence, in many parts of the world, girls under-achieve in school science and women are under-represented in university science education (particularly at postgraduate level) and in careers in science and en-gineering. Part of the drive for universal critical scientific literacy necessarily includes establishment of a much more inclusive form of science education with respect to gender, ethnicity, physical disability, socioeconomic class and sexual orientation (Kahle & Meece, 1994; Gaskell & Gillinsky, 1995; Parker et al., 1995; Roychoudhury et al, 1995; Rosser, 1997; Solomon, 1997; Barton, 1998; Barton & Osborne, 1998; Mayberry, 1998; Rennie, 1998, 2003; Richmond et al., 1998; Rodriguez, 1998; Scantlebury, 1998; Mayberry & Rose, 1999; Brickhouse, 2001; Bianchini et al., 2002; Key, 2003; El-Hani & Mortimer, 2007; Klein et al., 2007; Scantlebury & Baker, 2007; Brotman & Moore, 2008). It is also important for *students* to be made aware of gender bias. A crucial element of critical scientific literacy is the detection of bias, of whatever kind, and students need experience of seeking evidence of it. Two fairly common sayings are: "You see what you expect to see" and "There is no smoke without fire". Perhaps both sayings are applicable to the issue of bias. First, the extent to which bias of any kind can be detected in any particular situation is, in some measure, a consequence of the knowledge, beliefs, values and attitudes of the person addressing it. We do not see things as they are; rather, we see things as *we* are. Second, individuals are unlikely to see anything amiss in what they have been socialized into regarding as normal. It is "invisible" to them. In that sense, bias is absent. Third, if bias is detected, even "between the lines", it must be there – whether or not the author intended it, or is even aware of it. This notion of *multiple realities* is a fascinating issue for students to confront. There is considerable value, also, in confronting students with a series of questions concerning an alternative gynocentred science (see Hubbard (1988) and essays in section 2 of Lederman & Bartsch, 2001). How might it be different from current science? Would we expect to see changes in areas of concern and re-search priorities, inquiry methods, data collection techniques, criteria for judging success, underlying values, or some other aspect of science? How much change could there be in any of these dimensions before the activity ceases to be science? And if it is no longer science, what is it? Does the term gynocentric science or *women's* science have any meaning? If science could be different, should it be different? Would a differently organized science lead to greater social justice or enhanced environmental awareness? If science should be different, how can we make it different? I am not suggesting that teachers try to provide answers to all these questions, but I am suggesting that teachers should ask them.[8]

It is here that discussion might turn to the ways in which bias can be so extreme that it results in the misuse of science for social and political ends. A prime example is the suppression of Darwinian ideas underpinning evolutionary biology in the Stalinist Soviet Union in favour of Trofim Lysenko's theorizing about Lamarkism, on the grounds that transmission of acquired characteristics was much more compatible with Marxist–Leninist ideology. Lysenko's argument was that all differences between individuals are due to environmental effects and, therefore, organisms can be radically modified by exposing them to environmental challenges. He claimed that the production of new crops and the adaptation of existing crops to new habitats need not involve the long processes of selective breeding, as claimed by scientists in the Capitalist world, but can be simply and quickly achieved by exposing seeds and young plants to suitable modifying conditions. Acting on this supposition, he introduced the "vernalization" method of seed adaptation to harsh climates and the grassland system of crop rotation, a policy that brought about a collapse of Soviet agriculture and led to widespread famine (Joravsky, 1970; Lewontin & Levins, 1976; Lecourt, 1977; Ayala, 1994). Of course, Lysenko avoided any serious and systematic experimental testing of his theories, preferring instead to employ crude experiments that could be interpreted in whatever way he chose. For many years, the abject and disastrous failure of Lysenko's agricultural practices was attributed to subversion by farmers and ene- mies of the State. The later deployment of these ideas by agronomist Luo Tianyu (a keen supporter of Lysenko) as part of Mao Zedong's "Great Leap Forward" initiative had even more disastrous results. Close planting, deep ploughing, ex- treme pest control measures directed against sparrows, insects and rats, banning of artificial fertilizers, insecticides, fungicides and selective herbicides, introduction of "innovative" farm machinery (that proved too heavy for all practical purposes), and inadequately built dams and irrigation channels (only natural materials such as earth, boulders and wood were allowed), had a disastrous impact on farming yields, soil quality and land stability. Again, there was widespread famine. Esti- mates of deaths in China from malnutrition during the period 1958 to 1961 vary between 20 million and 40 million (Becker, 1996).

Scientific racism is a term used to describe the ways in which science, scientific research and scientific argument are used to justify discrimination, dis- advantage and oppression of others. This is another area of scholarship that is outside the scope of this book, beyond urging teachers to include some consider- ation of scientific racism in the school curriculum. Suitable examples include the 19th century misuse of the Darwinian principle of natural selection to argue that white Europeans are superior to Africans in evolutionary terms (and so provide a spurious justification for colonization, and even slavery) (Fryer, 1984; Gould, 1981a,b, 1995) and the eugenics programme in the United States, United King- dom and Nazi Germany (Wegner, 1991). Seldon (2000) provides a richly detailed account of the ways in which knowledge of heredity was used in the United States in the 1920s and 1930s to promote the notion of "societal improvement" through selective breeding and sterilization, discourage inter-racial marriage and restrict immigration from Southern Europe.[9] Black (2003) also documents this history, together with the work of Francis Galton, Karl Pearson and Robert Rintoul in the UK and the full horrors of the pursuit of "genetic purity" in Nazi Germany. More recent examples of scientific racism include the low priority afforded to research on sickle cell anaemia (a disease that primarily affects those of African descent) and systematic misinformation about the condition to exclude African Americans from active flying duties in the US Air Force and prevent them achieving flight sta- tus with some commercial airlines (Michaelson, 1987; SSCR, 1987; Dyson, 2005; Howe, 2007), and the continuing misrepresentation of research by psychologists such as Eysenck and Jensen to claim the intellectual superiority of Caucasians (see Rushton, 1997, 2000; Rushton & Jensen, 2005). On this latter topic, Brush (1989) provides powerful food for thought in his description of the background to the Stanford–Binet IQ test, still widely used as a supposedly objective measure of intellectual capacity, and the basis of the Scholastic Aptitude Tests (SATs) for college entrance in the United States.

> When Lewis Terman tried out the first version of his intelligence test on white California schoolchildren, he found that the average score for girls was a little higher than for boys of the same age. This result was inconsistent with his preconception that males are at least as smart as females. So he balanced the test items on which boys tended to do better against those on which girls do better, in such a way that the average score would be 100 for each sex ... But when Terman and other psychologists administered the test to Blacks and found that they scored several points below Whites on the average, they did not make a similar revision to equalize the average scores; they were content with the "discovery" that Whites are smarter than Blacks. (p. 65)

Stanley and Brickhouse (1994) provide a salutary commentary on the continuing pernicious influence of a research tradition predicated on the assumption that intelligence is an easily defined human characteristic that determines educational attainment independently of sociocultural, political and economic factors – in itself, a strong case for including history of science in the school curriculum. As George Santayana (1955a) reminds us, "those who cannot remember the past are condemned to repeat it ... the condition of children and barbarians [is one] in which instinct has learned nothing from experience" (p. 82).

Before leaving this discussion, it is worth mentioning the criticisms that have been directed by Ruse (1979a), Janson-Smith (1980), Fausto-Sterling (1985), Birke (1986, 1992) and Hrdy (1986) towards sociobiology and its assertion that many human behavioural traits are a consequence of genetic inheritance and adaptation to environment – in particular, towards the work of E.O. Wilson (1975, 1981, 2004). Wilson argues that the human mind is shaped as much by genetic inheritance as it is by culture. In consequence, there are severe limits on the extent to which social intervention, education and environmental modification (including improvement of social conditions) can change human behaviour. Rose (1997) states that Wilson is guilty of circular reasoning: male-oriented views of society are projected onto the animal world, which then provides observational evidence that justifies reprehensible male behaviour in human society. At best, Rose argues, it is *bad science*; at worst, it is an endorsement of male promiscuity, polygamy, male domination and male violence, including rape, on the grounds that individuals who engage in this behaviour out-reproduce those who do not. And since this "masculine behaviour" is genetically-based, the argument continues, it is transmitted to these individuals' offspring. Evolutionary psychology is open to these same criticisms: that it takes today's social prejudices and projects them back into prehistory, thus elevating them to the status of timeless truths about the human condition. At the most extreme is Richard Dawkins" (1976, 1981) argument that our genes determine not only our biochemical and developmental pathways but also our cultural and social relationships. This mechanistic- genetic view of biological determinism leads to the conclusion that even our moral-ethical conduct is a consequence of the evolutionary fitness of individuals displaying a particular set of traits. This is certainly an interesting set of propositions to run past students.

ERRORS, FALSE TRAILS, MISCONDUCT AND FRAUD

Because science is carried out by people, with all their human frailties, it is subject to error. Errors are widespread and are to be expected, especially in the early days of a research project when new investigative methods and/or new instruments are being used. It takes time for technicians to "get a feel for the equipment", such that they can use it quickly and effectively, detect and avoid errors, and generate reliable data. The first difficulty centres on *recognizing* errors. Knowing which data to accept, which to reject and which to re-check is a matter of professional judgement or connoisseurship, and is acquired through experience. It also depends, of course, on the theoretical and procedural assumptions within which the scientist is working. Being human, scientists sometimes make wrong judgements, and errors can remain undetected for lengthy periods. There are occasions when errors of judgement lead to unconscious distortion of findings and even to the publication of false data and erroneous theories designed to explain them. For example, the supposed phenomenon of *bond stretch isomerism* became a stimulus for major experimental innovation and novel theorizing before it was eventually found to be a consequence of taking crystallographic data too literally (Parkin, 1992 – cited by Labinger, 1995). In the 1960s, two Russian scientists, Nikolai Fedyakin and Boris Derjaguin, announced that by heating water and letting it condense in quartz capillary tubes they had created "anomalous water", with a density almost 40% greater than water, a boiling point a little above 100°C and a melting point below 0°C. Over the next several years, many papers were published describing the behaviour of what came to be known as *polywater*. Its structure was elucidated and a theoretical explanation of its formation developed (Lippincott et al., 1969; Allen, 1970, 1971; Allen & Kollman, 1970; Linnet, 1970). Eventually the phenomenon was found to be a consequence of poor experimental technique: the water had become contaminated by particulate matter leached from the quartz capillaries and/or the grease used to seal joints in the apparatus (Allen & Kollman, 1971; Franks, 1981). In 1989, Martin Fleischmann and Stanley Pons of the University of Utah claimed to have achieved cold fusion by electrolysing a solution of 0.1M LiOD in 99.5% deuterium oxide/0.5% water using palladium electrodes, though they were somewhat vague about the exact procedure. Their explanation was that deuterium nuclei entered the crystal lattice of the palladium cathode, underwent nuclear fusion and released neutrons, which they detected by means of the characteristic secondary gamma radiation (Fleischmann & Pons, 1989). They also mentioned a substantial build-up of tritium. Instead of running a control experiment with water in place of heavy water (hydrogen is less fusible than deuterium by many orders of magnitude), Fleischmann and Pons concentrated on ways to increase the energy output of their experimental set-up. The great excitement generated by the prospect of colossal amounts of energy produced at virtually no cost died down when other research groups failed to reproduce the results, although vast sums of money had been invested in the efforts to do so (see Lindley (1990), Close (1991), Mallove (1991) and Gieryn (1992) for an account of the events).

Another intriguing false trail concerns claims of memory transfer through cannibalism. In 1955, James McConnell showed that planaria (flat worms) can be conditioned (though McConnell used the term "trained") to avoid light by simultaneously exposing them to a bright light and administering an electric shock. When cut in half, planaria will regenerate into two new individuals; it was found that *both* retained the light-shock association. McConnell subsequently fed "trained" planaria to others and claimed that these "untrained" animals were much quicker at acquiring the light-shock association than a control group fed "non-trained" planaria, which McConnell (1962) interpreted in terms of memory being stored chemically (in RNA). Few people were able to replicate his results (Bennett & Calvin, 1964). Following a rigorous attempt at replication, Hartry et al. (1964) concluded that McConnell's results are more plausibly explained by lapses in technique. McConnell's (1965) response was that the "worm trainers" used in these rival experiments were insufficiently skilled and had failed to appreciate the needs of the worms: "It seems certain that the variability in success rate from one laboratory to another is due, at least in part, to differences in personality and past experience among the various investigators" (p. 26). Elsewhere, he says that it is important to treat the worms "tenderly, almost with love" (cited by Collins & Pinch, 1993, p. 11). Even more surprising claims relating to memory transfer in mammals were made by Georges Ungar and his co-workers (1972). They claimed that light avoidance behaviour can be induced in untrained mice by injecting them with an extract of brain tissue from rats that have already been fear-trained to avoid the dark. Attempts to replicate the results by a Stanford University research team led by Avram Goldstein were unsuccessful (Goldstein et al., 1973). Once again, differences in experimental technique and careless work were cited by the original researchers as the source of the discrepancy. As Collins and Pinch (1993) remark, when only the original researchers can obtain positive results, which they term the "golden hands argument", there are good grounds for concluding that the underlying science is *unsound*. McConnell continued his research until 1971, when he was unsuccessful in gaining further research funding for his laboratory (see McConnell & Jacobson, 1974). Ungar continued his experiments with rats, mice and fish until his death in 1977. To date, no scientist has chosen to continue the research.

Errors can also lead to unexpected progress. A classic example is Pasteur's accidental discovery of what we now know as optical isomerism (in tartaric acid) as a consequence of unknowingly using exactly the right lab conditions for resolving the racemic mixture by means of crystallization. Chance seems to have played a substantial role in Woehler's synthesis of urea (he was trying to produce ammonium cyanate), Perkin's discovery of Mauve (during his attempt to synthesize quinine) and von Pechmann's discovery that polyethene is produced by heating diazomethane (a somewhat hazardous procedure in itself!). Physics has its fair share of allegedly chance discoveries, too: Herschel's discovery of infrared radiation, Becquerel's discovery of radioactivity, Röntgen's discovery of X-rays, and Seebeck's discovery of the thermoelectric effect. In biology, chance is attributed a major role in Minkowski's discovery of the pancreatic origin of diabetes, Richet's discovery of anaphylaxis, and Nagano and Kojima's discovery of interferon. From a student perspective, a particularly fascinating example of serendipitous discovery centres on research into anti-inflammatory drugs for treating angina and hypertension carried out at the Pfizer research laboratories in Sandwich (Kent, UK). One particular molecule was found in clinical trials to have entirely unexpected side effects. That drug is now marketed under the name *Viagra*.

The most widely known example of serendipity in science is Alexander Fleming's discovery of the bactericidal properties of penicillium moulds, in 1928 (reported in Fleming, 1929). When leaving for a vacation Fleming had failed to disinfect his cultures of bacteria properly; when he returned to the lab he found that they had been contaminated and that all the bacteria in the area surrounding the contaminating mould had been killed. The contamination was certainly accidental, but no-one could have been better positioned to make sense of this chance observation and realize its great significance. Twenty years previously Fleming had written a thesis on methods for treating bacterial infections; shortly afterwards he had observed the spectacular effect of Salvarsan on syphilis; during World War 1, he studied the usually fruitless attempts to combat bacterial infections with antiseptics. Most significant of all, in 1922, at a time when he had a cold, he introduced some mucus from his nose to a bacterial culture and observed that around the drop of mucus all bacteria were killed. In his report (Fleming, 1922) he concluded that the active component of nasal mucous, which he called *lysozyme*, was a key component in the body's natural protection against bacteria.[10] As Louis Pasteur is reputed to have said, "Chance favours only the prepared mind". It also only favours an *open* mind. Both are essential if the significance of information accidentally received is to be recognized, which goes some way to explaining why chance discoveries generally occur in a scientist's specialist field of research. As Wolpert (1992) observes: "Scientific research is based not on chance but on highly focused thoughts ... It is not by chance that it is always the great scientists who have the luck ... Advances in science, being a journey into the unknown, must inevitably confront scientists with the unexpected? It is the unexpected that provides the clues to guide further work ... We are surrounded all our lives by innumerable 'facts' and 'accidents'. The scientist's skill is to know which are important and how to interpret them" (pp. 79–84). In similar vein, Ziman (2000) comments that serendipity does not, of itself, produce discoveries; rather, it creates opportunities for making them:

Accidental events have no scientific meaning in themselves: they only acquire significance when they catch the attention and interests of someone capable of putting them into a scientific context. Even then, the perception of an anomaly is fruitless unless it can be made the subject of deliberate research. In other words, we are really talking about discoveries made by the *exploitation of serendipitous opportunities* by persons already primed to appreciate their significance. (p. 217, emphasis in original)

Dunbar (1995) also suggests that many of the findings labelled serendipitous are, in reality, the result of careful and systematic experimentation designed to reveal novel mechanisms. When surprising results are obtained, the investigator already has in mind a set of hypotheses and mechanisms to interpret them.

> When the control conditions produce unusual results, the scientist is already considering a host of potential mechanisms, and thus a surprising finding allows the scientist to focus on the aspects of his or her current conceptual structure that need to be changed or rejected. The surprising finding is genuinely surprising, but the use of controls and an already richly articulated conceptual structure enable the researcher to make sense of the findings and propose novel theories. (p. 390)

Thomas Kuhn (1977) identifies two prerequisites for serendipitous discoveries: first, "the individual skill, wit or genius to recognize that something has gone wrong in ways that may prove consequential" (p. 173); second, sufficient development of both instruments and concepts to make the emergence of anomalies likely and "to make the anomaly which results recognizable as a violation of expectation" (p. 173). Interestingly, one of the scientists interviewed by Wong and Hodson (2009) reports that he regularly searches the literature for surprising results and negative results as an aid to experimental design. In similar vein, Dunbar (1995) describes how the scientists with whom he has worked would frequently map an experimental problem with which they were struggling onto similar experiments that had proved successful, substituting any differences in approach into the previously unsuccessful experimental design.

Taking the view that one of the best ways to understand the nature of scientific practice is by studying episodes when it did not work exactly as it should, Allchin (2004a) argues that all science courses should, from time to time, focus on errors and where they led. "Teaching science without error", he says, "is like teaching medicine without disease or law without crime" (p. 944). Almost every scientist, at some point in her/his career, may have to admit to being wrong. Steven Weinberg (1994) argues that this salutary experience for individual scientists is what gives the scientific community confidence in the security of its knowledge: "It is the scientists' experience of being forced by experimental data or mathematical demonstration to conclude that we have been wrong about something, that gives us a sense of the objective character of our work" (p. 750). Errors can arise through (i) hasty, inattentive, negligent or careless data collection, (ii) inadvertent mistakes in preparing materials or taking measurements, (iii) reliance during the design and performance of experiments on theories that later turn out to be false or incomplete, (iv) failure to recognize other significant factors impacting behaviour, or (v) inadequate testing or even failure to test at all. It is dishonest of science teachers to pretend that scientists do not, from time to time, make these kinds of mistakes. It is also dishonest to deny, trivialize or remain silent about sharp practice, misconduct and fraud in science. In the highly competitive world of contemporary scientific research, with the constant need to secure research funding and generate publications, scientists are tempted to "cut corners", be less rigorous than perhaps they should, overstate their case, or make premature claims. Stealing data from other research groups, trying to mislead competitors by publishing erroneous or incomplete data, and using delaying tactics when reviewing articles for publication written by rivals, are more common than we might suppose or can feel comfortable about (Wong & Hodson, in press). How much of this less-than-savoury tale should we tell students? Should school science education address issues of scientific misconduct and fraud? By drawing attention to it, the authors of *Benchmarks for Scientific Literacy* (AAAS, 1993) seem to imply that the answer is "Yes".

> The strongly held traditions of science, including its commitment to peer review and publication, serve to keep the vast majority of scientists well within the bounds of ethical professional behaviour. Deliberate deceit is rare and likely to be exposed sooner or later by the scientific enterprise itself. When violations of these scientific ethical traditions are discovered, they are strongly condemned by the scientific community, and the violators then have difficulty regaining the respect of other scientists. (p. 20)

Borrowing from Allchin's (2004a) argument for teaching about errors in science, one could argue that studying misconduct in science is a powerful way of educating students about *proper* conduct as well as reinforcing their understanding of the sociocultural climate that seduces some scientists to violate the code. Misconduct involves such things as plagiarism (of ideas and data), failing to acknowledge co-workers as authors, simultaneous publication of the same data in different locations, failure to acknowledge involvement of (or funding by) commercial enterprises, use of threats and bribes to influence other scientists, falsification and fabrication of data, including concealment of "inconvenient data". It does not include errors in recording, selecting and analyzing data, or errors of judgement and interpretation. Clearly, it is very important to distinguish between what is intended and what is unintended. Thus, Wible (1992) draws a distinction between fraud and what he calls "replication failure", and Yoxen (1989) distinguishes between *hard* fraud ("invention of an entire set of data ... deliberate creation of substitutes for evidence, wholesale plagiarism of others' contributions") and *soft* fraud ("improving one's data, plagiarism of marginal ideas, poor statistical analysis, failure to cite important sources, and the prejudicial misrepresentation of others' work") p. 190).[11]

Fraud is potentially devastating to science because it undermines the trust that other scientists need to have in published work, on which they may choose to build their own research endeavours. Indeed, the entire scientific enterprise is based on trust: trust in the moral-ethical values that underpin the identification of research priorities and the daily conduct of scientists, trust in the validity of currently accepted scientific knowledge (until there is strong evidence to support its replacement), and trust that all practitioners accept and adhere to particular standards in the collection and presentation of data and the construction of scientific arguments (Norris, 1995; Harding & Hare, 2000; Gaon & Norris, 2001). In short, scientists need to know that they can trust the work of other scientists (including other members of their own research team) and that their own work will be accepted as trustworthy. Without such trust in each other the scientific enterprise cannot function effectively and productively. The independent justification of most of our beliefs is just not possible. Even those at the cutting edge of research must take on trust much of the knowledge they deploy, including the knowledge underpinning the design and utilization of the complex modern instrumental techniques on which so much contemporary research depends. As Lorraine Code (1987) points out, there is a complex network of interdependence that constitutes an *epistemic contract* among scientists.

> Scientists themselves must rely heavily for their facts upon the authority they acknowledge in their fellow scientists. They use the results of sciences other than their own and of other scientists in different areas of their own field, results they may feel neither called upon nor competent to test for themselves ... For this interdependence to be workable, there must be a tacit basis of trust and trustworthiness ... a sort of *epistemic contract*. (Code, 1987, p. 230, emphasis added)

Many research problems are located in multidisciplinary contexts, such that no one scientist can have the breadth of knowledge to solve it, nor the time and expertise necessary to collect and analyze the wealth of data that may be required. In consequence, scientific research is increasingly conducted in large teams, sometimes extending across international borders. Within such teams no one scientist can be privy to everything that occurs or be familiar with the details and nuances of all aspects of the work. Indeed, as Hardwig (1991) comments, for many multi-authored papers published in scientific journals, "none of the authors is in a position to vouch for the entire content of the paper" (p. 695). Trust among team members is crucial. Further, reviewers of papers submitted for publication usually have to trust the data that is presented. They do not have the time or opportunity (or even the inclination) to check it. Even when checks are carried out, it is difficult to detect "plausible and internally consistent fabrication" (Hardwig, 1991, p. 706). To deal with this situation, Hardwig proposes the *principle of testimony*: "If A has good reasons to believe that B has good reasons to believe p, then A has good reasons to believe p" (p. 697). In essence, he argues that A's good reasons depend on whether B can be regarded as truthful, competent and conscientious. Thus, the reviewer needs to evaluate the researcher's competence (with regard to the design and conduct of the inquiry and the quality of the analysis, interpretation and argument) and be vigilant for signs of vested interest. In short, it is not possible to draw a sharp distinction between evaluation of the research and evaluation of the trustworthiness of the researcher. Hardwig notes that the principle of testimony necessarily applies to the relationship between laypersons and "experts". Generally, neither members of the public nor journalists and documentary makers have access to original empirical data, or to details of research methods, and so must decide the extent to which they can trust particular researchers and research groups to be honest in their reporting.

Within the complex social networks operating within and between institutions, and in the wider scientific community, scientists are subject to all kinds of personal, social and economic pressures to win a research contract, get a promotion, publish a paper or complete a PhD thesis. These pressures can sometimes lead them to "cut corners" or engage in activities that fall short of the community's standards. In his classic work, *The Sociology of Science*, Merton (1973) claims that fraud is rare because the peer review system, and the importance all scientists place on experimental replication, quickly detects it.

> The virtual absence of fraud in the annals of science, which appears exceptional when compared with the record of other spheres of activity, has at times been attributed to the personal qualities of scientists ... a more plausible explanation may be found in certain distinctive characteristics of science itself. Involving as it does the verifiability of results, scientific research is under the exacting scrutiny of fellow experts. Otherwise put ... the activities of scientists are subject to rigorous policing, to a degree perhaps unparalleled in any other field of activity. (p. 276)

Writing some 20 later, O'Rafferty (1995) is less confident.

> In view of the number of cases of misconduct in science disclosed during the past 15 years or so, the statements of Merton ... now seem over-optimistic. Does the system of checks and balances take good care of fabricated evidence, plagiarism and falsification? Are dishonest scientists usually caught and discredited? To what extent does the nature of science keep scientists honest? The only possible answer to these questions is that we do not know. Events in recent years give reason to believe that the system of checks and balances within science is less adequate than many scientists had thought or would wish to believe. (p. 908)

O'Rafferty (1995) refers to three kinds of fraudulent data manipulation: cooking (the selection and exclusion of data to achieve a particular result), *trimming* (deleting outlying results, inflating values that are "too low" and reducing those deemed to be "too high") and *forging* (reporting observations that were not made). An interesting question with which to confront students concerns the circumstances in which it is acceptable/unacceptable to be selective about data use. A couple of historically significant experiments that feature in many school science courses can be used to address this question. It is now widely acknowledged that Robert Millikan was highly selective in the use of data in his famous oil drop experiments, rejecting data that did not fit the pattern he anticipated.[12] This careful selection of the data was a case of Millikan playing to his hunch about what the value of e (the charge on an individual electron) should be, allied to his intuition about when the experimental set-up was working well and not so well. To be fair to him, later improvements to his experimental technique did retrospectively justify his rejection of some data, and there is absolutely no evidence that he ever fabricated data. Holton (1978) talks about two contrary tendencies in Millikan's work.

One is the standard classical behavior of obtaining information in as deper-sonalized or objective a manner as possible. As every novice is taught, the graveyard of science is littered with those who did not practice a *suspension of belief* while the data were pouring in. But there is the other side of the coin, a strategy without which work of novelty could not get past those first hurdles whose exact nature can be identified in detail only after the fact. To understand this side of the researcher's behavior I introduce the notion of the *suspension of disbelief*; that is, the ability during the early period of theory construction and theory confirmation to hold in abeyance final judgments concerning the validity of apparent falsifications of a promising hypothesis. (p. 71)

Put more plainly, suspension of belief means avoiding firm conclusions when the evidence may be incomplete or difficult to interpret, or when consensus cannot be achieved; suspension of disbelief is continuing adherence to one's ideas in spite of contradictory evidence. Selectivity regarding data is a charge levelled at many scientists, often with justification.[13] Holton (1978) argues that it is a legitimate move and is part of scientific argumentation. It is, he says, "a mechanism for stabilizing belief in the efficacy of a hypothesis long enough to help it survive to the later stage of testing in public discussion" (p. 72). It also saves time that would otherwise be spent in the frustrating and, so the scientist believes, *fruitless* task of looking for the source of the discrepancies.

A strikingly different response to data variation was taken by Ernest Ruther-ford. In the now famous 1909 experiment conducted by Rutherford, Marsden and Geiger, a steam of alpha particles was fired at a thin film of gold foil. Almost all the particles passed through the foil and were detected by a photographic plate on the other side. Approximately 1 in 10,000 bounced back.[14] Marsden and Rutherford would have been justified in rejecting 0.01% of the readings as experimental error. They chose not to reject them, and their attempts to explain the "error" led to what we now know as the Rutherford model of the atom – a small dense nucleus of protons and neutrons, orbited by electrons.

Gould (1981a) uses the expression "unconscious finagling", rather than fraud, to describe situations in which scientists are so guided by their preconceptions, and so driven by their determination to discover, that inconvenient results, flaws in experimental design, miscalculations and sundry other errors are ignored or remain undetected, and are subsequently woven into the conclusions, as in the N-rays episode which Burdett (1989) has developed into a fascinating simulation for class use. In 1903, Rene Blondlot (Professor of Physics at the University of Nancy) was attempting to ascertain whether X-rays, recently discovered by Röntgen, were best regarded as waves or particles. If they are waves, Blondlot surmised, it might be possible to polarize them and to detect the polarized beam by means of an electric spark (which would be expected to glow more brightly at particular orientations of the polarized beam). The experiment appeared to con-firm the polarization, except that the beam was refracted by a quartz prism. Since X-rays were known not to be refracted, Blondlot concluded that he had discovered a new form of radiation, which he called N- rays.[15] He found that a Nernst lamp, which uses a metal oxide filament, was a particularly good source of N-rays. Further work convinced Blondlot that N-rays could pass through wood, paper and thin sheets of metal, but were blocked by water. His claims were accepted by others, and over the next year or two numerous articles on the generation and properties of N-rays were published, including accounts of their potential use in medicine (Klotz, 1980; Nye, 1980). The journal *Nature* commissioned physicist Robert Wood of Johns Hopkins University to visit Nancy to investigate Blondlot's experimental techniques and study his results, especially the rather bizarre claim that N-rays enhanced human vision. In a wonderful piece of theatre, Wood entered the dimly lit laboratory and surreptitiously removed the prism being used to diffract the N-rays. It made no difference to the results; the technician continued with his readings, unaware of Wood's subterfuge. Wood concluded that the researchers had been seriously misled by their investigations and were pursuing a false trail.

> After spending three hours or more in witnessing various experiments, I am not only unable to report a single observation which appeared to indicate the existence of the rays, but left with a very firm conviction that the few experimenters who have obtained positive results have been in some way deluded ... all the changes in the luminosity or distinctness of sparks and phosphorescent screen (which furnish the only evidence of n-rays) are purely imaginary ... not a single experiment has been devised which can in any way convince a critical observer that the rays exist at all. (Wood, 1904, pp. 531 & 532)

Surprisingly, Wood's report of the episode did not prevent research on N-rays continuing for some considerable time (especially in France). Eventually, however, when it was shown that Blondlot's published results displayed an accuracy that his experimental set-up could not have produced, and that the supposed physiological effects of N-rays were better explained in other ways, interest in further research came to an end and N-rays were recognized as a scientific mistake.

Sometimes, of course, false ideas persist for a long time because they accord with the observational evidence, make sense in terms of other accepted theories, and meet the needs and expectations of scientists. Such ideas are only discarded when a significant alternative becomes available, usually in response to a crisis of some kind, as described by Thomas Kuhn (1970) in *The Structure of Scientific Revolutions*. An example that might repay study in the school science curriculum is phlogiston theory – an idea that was introduced by Johann Becher in 1667, expanded by Georg Ernst Stahl, and persisted throughout most of the 18th Century. Allchin (1997) argues that the concept of phlogiston has considerable value in teaching about oxidation and reduction. Most science teachers would disagree, arguing that such an approach would take too long and could be counterproductive in terms of conceptual understanding. Handled carefully, however, the story of the overthrow of phlogiston theory by the modern theory of combustion is a superb illustration of how a theory that is perfectly adequate at a qualitative level is not necessarily so in quantitative terms (Talbot, 2000). The only way of defending phlogiston theory against the careful quantitative work of Antoine and Marie-Anne Lavoisier, on the weight gain of elements on combustion, was by postulating that phlogiston had negative mass (phlogiston, a component of inflammable materials, was supposedly released on burning). This absurdity pushed the theory into crisis. The story also shows how work by a scientist thinking within an established theoretical framework – in this case, work that led to the discovery by Joseph Priestley of the gases we now know as oxygen, nitrogen and carbon dioxide – can sometimes be reinterpreted by others (the Lavoisiers) and used in building a new theory that is capable of resolving a crisis and bringing about a scientific revolution. Antoine Lavoisier also brought about a revolution in the language and practice of chemistry. Prior to the 18th century the names of substances were based mainly on physical properties such as colour, taste and smell. In *Methode de Nomeclature Chimique*, published in 1787, Lavoisier and his co-workers proposed a system based on the names of elements (substances that cannot be decomposed into anything simpler), with the names of compounds deriving from the elements that comprise them (Crosland, 1962; Anderson, 1984). Adoption of this language shifted the primary concern of chemists away from surface characteristics to their elemental composition, thus transforming the discipline of chemistry from a practical art related to the work of dyers and glass makers into a science of molecular behaviour. Those students with a strongly developed sense of irony will note that the scientific revolution brought about by the Lavoisiers coincided with another revolution – the French Revolution and overthrow of the monarchy, during which Antoine Lavoisier himself perished on the guillotine.

Talbot (2000) argues that caloric theory (which regarded heat as a fluid) and its overthrow by the work of Rumford, Davy and Joule, makes a more fruitful case study of superseded ideas for school use than phlogiston theory. Caloric is certainly a less bizarre idea than phlogiston, and more in line with the kind of ideas that students themselves might hold.[16] Of course, students should also be asked to consider whether some of the ideas of contemporary science might seem just as outlandish to scientists a hundred years hence as phlogiston does to us. Considerable controversy surrounds the study of any discarded or superseded scientific ideas in the school curriculum. While some see their value in highlighting conceptual and methodological issues, and the social circumstances central to the events, others are concerned that the potential for confusing students and possibly compounding misconceptions more than outweighs the benefits. It is a fine line, and much depends on the students, the teacher and the resources available. Brown and Dronsfield (1991) and Schwartz (1995a) provide useful background for the phlogiston story; a case study of the events for school use, including some simple experimental activities, can be found in Solomon et al. (1994b, pp. 21–26). Useful background on caloric theory can be found in Roller (1950) and Fox (1971); Solomon et al. (1994b, pp. 27–30) describe some simple experiments that would enliven such a case study.

There might also be value in studying episodes in which perfectly sound ideas were initially rejected. Prime examples include the rejection of Alfred Wegener's (1915) theory of "continental drift" (van Waterschoot van der Gracht, 1928; Frankel, 1979; LeGrand, 1988; Suppe, 1998; Thompson et al., 2000) and rejection of the discovery by Oswald Avery et al. (1944) that DNA, rather than protein, is

109

the "transforming factor" responsible for transmission of hereditary characteristics (Ayala, 1994). In both cases the evidence was substantial, but Wegener was unable to provide a convincing explanation of how the continents could move, while Avery's finding (based on studies of pneumococcus bacteria) was rejected because erroneous knowledge about the structure of DNA suggested that it was too structurally simple to enable it to function as an "informative molecule". Mahoney (1979) describes a contrasting case in which erroneous data produced by a scientist of high reputation delayed a theoretical breakthrough concerning the structure of the universe. He describes how spiral nebulae were assumed to be relatively close (on the periphery of our own galaxy) on the basis of Adriaan van Maanen's measurement of their radial and rotational movements. Mahoney comments as follows:

> It was not until the work of Edwin Hubble, some 15 years later, that astronomers recognized that van Maanen's data were wrong. When his photographic plates were re-checked, a computer analysis suggested that he had made perceptual errors of minute proportions (about 0.002 mm) which – because they were systematically in favour of his expectations – yielded an aggregate distortion large enough to retard the acceptance of a theory. (p. 352)[17]

Regarding the earlier discussion of misconduct in science, it is worth considering some examples of overt scientific fraud for inclusion in the school science curriculum. Suitable candidates include the Piltdown Man forgery (see Weiner, 1955; Judson, 1980; Blinderman, 1986; Kohn, 1986; Williams, 1993; Johanson & Blake, 2001), the more recent *archaeoraptor liaoningensis* hoax (Sloan, 1999; Xing, 2000; Xing et al., 2000; Zhou et al., 2002) and the discredited work of the psychologist Cyril Burt (see Wade, 1976; Dorfman, 1978; Hearnshaw, 1979; Gieryn & Figert, 1986; Judson, 2004)[18] and cardiologist John Darsee (see Broad & Wade, 1982; Culliton, 1983a,b; Kohn, 1986; Judson, 2004). The Bell Laboratories affair involving Jan Hendrick Schon (described by Service, 2002, 2003) and the more recent scandal over Hwang Woo-suk's faked stem cell research (see Websites for *New York Times*, BBC, *Time* and *Nature*[19]) might also make good case studies. Lin (2007) describes an interesting case study in which undergraduates (both science majors and social science majors) were confronted with conflicting scientific data and opposing arguments concerning the authenticity of the Vinland Map (relevant sources include: McCrone, 1976; Cahill et al., 1987; Donahue et al., 2002; Whitfield & Dalton, 2002; Olin, 2003; Towe, 2004). Of course, any consideration of fraud begs a question of how many examples of scientific fraud have successfully escaped detection. Of significance, too, is the response of institutions to charges of misconduct or fraud. Judson (2004) comments that, except in the more spectacular cases of outright fraud, responses have been inappropriate: "Again and again the actions of senior scientists and administrators have been the very model of how not to respond. They have tried to smother the fire" (p. 6). John Ziman (2000) also comments on the marked contrast between the condemnation of fraudulent research claims as "instances of grave social deviance" and the "relatively lenient sanctions applied to those who perpetrate them" (p. 267).

It might be amusing and instructive for students to think about their own be-haviour: the frequency with which they have cheated, and the extent to which they have "massaged" data collected in experiments to fit the anticipated conclu-sion. Rigano and Ritchie (1995) describe several kinds of fraudulent behaviour or "fudging" of data by students: adjusting results to fit the data in the textbook or to match results obtained by other class members, excluding anomalous results, fabricating data or stealing results from others. Poor results in school practical lessons may be a consequence of shoddy equipment that fails to function properly, or they may arise because students misread or misunderstand the instructions, fail to recognize what is important (and so concentrate on something else), lack the skills necessary to collect good data, get bored and become careless, have an accident, or just fail to finish. Since there is rarely time to repeat the experiment, students "fudge" the results to ensure the right answer and, thereby, obtain a good mark for the write-up. The extent to which fudging occurs in any classroom is a consequence of how (and for what) marks are awarded, how teachers respond to less-than-perfect experimental results reported by students, and what they do to encourage students to identify the source of errors. At issue here is the extent to which teachers encourage students to adopt an *achievement* orientation (in which pursuit of marks for correct responses becomes paramount) or a *learn-ing* orientation (in which students pursue personal understanding). At issue, also, are the motives that teachers have for organizing practical work. For example, whether they see practical work as primarily for learning science content, learning about scientific investigations or engaging in such investigations (see Chapter 6 for further discussion of this issue).

NOTES

[1] I began my career as a classroom teacher in the United Kingdom, way back in 1968, long before the introduction of the National Curriculum for England and Wales. In those days there was not even a common curriculum for the Local Education Authority or the school, let alone the nation. So, in common with most other new teachers, my first task was to write my own chemistry curriculum for Forms 1–5 (ages 11–16). Such early experiences gave my generation of science teachers enormous confidence in their capacity to build curriculum (the positive aspect) and a long- standing distrust of centralization and a tendency to resist any mandated curriculum (which some would see as the negative aspect of the experiences).

[2] Essentially the same approach is envisaged in Cobern's (1995) description of science education as "an exercise in foreign affairs".

[3] See Hammond and Brandt (2004) for an extensive review of science education research that adopts an anthropological perspective.

[4] Smith and Scharmann (1999) go on to say that because science is accepted as fallible and repli-cation is considered essential, science is self-correcting. While this point has value, students also need to know that self-correction sometimes takes a while.

[5] The authors define *faith* as belief in the absence of supporting evidence. Of course, those with religion-based faith might argue that evidence of God's existence is all around us, and that a key article of faith is not demanding any other evidence.

[6] Merton first outlined this theoretical framework in a 1942 essay titled "Science and technology in a democratic order", published in the *Journal of Legal and Political Sociology*. The citations in this chapter refer to a subsequent work published in 1973 (see References).

[7] Useful starting points for teachers wishing to acquire some awareness of this literature are Harding and Hintikka (1983), Bleier (1984), Code (1991), Shepherd (1993), Keller and Longino (1996), Harding (1998), Kleinman (1998), Etkowitz et al. (2000), Lederman and Bartsch (2001), Miller et al. (2006) and Rolin (2008).

[8] Bleier (1986), Harding (1989, 1991), Keller and Longino (1996), Koertge (2004), Rolin (2004), Crasnow (2008) and the essays in section 6 of Lederman and Bartsch (2001) provide much useful background reading for teachers.

[9] Duster (1990) has warned of the eugenicist thrust pervading research in genetics, a feature that is even more apparent in the innovative work in biotechnology conducted in the years since he issued his warning.

[10] Interestingly, having shown the potential effectiveness of the *penicillium notatum* extract in combating scarlet fever, pneumonia, gonorrhea, meningitis and diphtheria, Fleming did not pursue the matter further. Howard Florey and Ernst Chain developed the first pencillin drug, which underwent successful clinical trial at the hands of Hobby, Henry and Meyer. Subsequently, Elizabeth McCoy produced a UV-induced mutant strain of *pencillium* that yielded almost 1000 times more penicillin than Fleming's strain. In 1949, Dorothy Hodgkin used X-ray crystallography to elucidate the molecular structure of penicillin, thus paving the way for the production of synthetic penicillin. Valuable background material for use in class can be found in McFarlane (1984), Jacobs (1985) and Brown (2004).

[11] An intriguing case cited by Kohn (1986, pp. 166–168) documents the (almost) simultaneous publication by the same scientists of contradictory results arising from the same investigation, with neither article (in *American Journal of Psychiatry* and *New England Journal of Medicine*, both of which are prestigious journals) making reference to the other.

[12] Holton (1978, 1986) reports that Milliken's notebooks reveal that he ran the oil drop experiment 140 times, but his published accounts list only 58 occasions.

[13] Mahoney (1979) suggests that Galileo, Newton and Mendel were all guilty of some degree of data manipulation in the light of theoretical expectation. Levere (2006) levels the same charges against Dalton and Liebig.

[14] Rutherford is reputed to have said: "It was quite the most incredible event that ever happened to me in my life. It was almost as incredible as if you had fired a 15 inch shell at a piece of tissue paper and it came back and hit you" (cited by Blatt, 1992, p. 93, and Rohlf, 1994, p. 168).

[15] N for *Nancy*, Blondlot's home town and seat of the university in which the work was conducted.

[16] See Duit (2006) for a comprehensive bibliography of published research on students' alternative conceptions in science, including their ideas about heat and heat transfer.

[17] Mahoney (1979) lists the following sources of information on this important episode in the history of astronomy: Hetherington, N.S. (1972), Adriaan van Maanen and internal motions of spiral nebulae: A historical review. *Quarterly Journal of the Royal Astronomical Society*, 13, 25–39; Hetherington, N.S. (1974), Edwin Hubble's examination of internal motions of spiral nebulae, *Quarterly Journal of the Royal Astronomical Society*, 15, 392–418; Fernie, J.D. (1970), The historical quest for the nature of spiral nebulae, *Publications of the Astronomical Society of the Pacific*, 82, 1189–1230.

[18] Studies by Joynson (1989), Fletcher (1991) and Gieryn (1995) have suggested that the judgement against Burt might have been overly hasty. Gieryn (1995) asks: "Was there intent to deceive (necessary, it seems, to sustain the charge of fraud) or was he simply a sloppy and negligent methodologist (reprehensible and even embarrassing perhaps, but hardly grounds for expulsion from science)" (p. 433). More recently, however, MacKintosh (1995) has reinforced Hearnshaw's original charges that Burt fabricated much of his data.

[19] Details of the Hwang Woo-suk story can be found at www.topics.nytimes.com, www.news.bbc.co.uk, www.time.com and www.nature.com.

CHAPTER 5
FURTHER THOUGHTS ON DEMARCATION

An interesting feature of life in the late 20th century and early 21st century has been the upsurge of interest in "New Age" and pseudoscientific beliefs such as the healing power of crystals, reflexology, aromatherapy, iridology, qigong,[1] feng shui, Tarot, and the like. Moreover, in this supposedly scientific age, traditional superstitions relating to black cats, broken mirrors, the number 13, walking under ladders and stepping on pavement cracks seem to be widespread, belief in telepathy, extraterrestrial visitation and clairvoyance is common, and interest in astrology is extensive (Walsh, 1992; Keranto, 2001; Newport & Strausberg, 2001). As Sagan (1995) reminds us, the seductiveness of these beliefs poses a considerable threat to science.

> Science arouses a soaring sense of wonder, but so does pseudoscience. Sparse and poor popularizations of science abandon ecological niches that pseudoscience promptly fills. If it were widely understood that claims to knowledge require adequate evidence before they can be accepted, there would be no room for pseudoscience. But a kind of Gresham's Law prevails in popular culture by which bad science drives out good. (p. 6)

In a survey of approximately 200 undergraduates in the University of South Australia (86% of whom were preservice elementary school teachers), Yates and Chandler (2000) found that only *four* students rejected all eight of the "New Age statements" categorized by the researchers and three of their colleagues as "totally unbelievable". Believers outnumbered skeptics in relation to UFOs, psychic séances and the predictions of Nostradamus; they were evenly distributed with respect to astrology and the healing power of crystals. If we take the view expressed by Culver and Iauna (1984) that belief in astrology is in inverse proportion to understanding of astronomy, we should be very concerned that so many people seem unable or unwilling to subject pseudoscientific ideas to the close critical scrutiny that would be expected to lead to their rejection. It is a substantial indictment of our current science education provision. It is also disturbing that so few children in the elementary schools of South Australia (and probably elsewhere, too) will, in the near future, be taught by a teacher capable of modelling informed skepticism about such claims. An enhanced level of critical scientific literacy among teachers and students is our defence against pseudoscientific claims and the seductions of outlandish New Age beliefs. Studying pseudosciences in the school curriculum is part of building that literacy.

To test just how widespread pseudoscientific beliefs and superstitions are among school-age students, Preece and Baxter (2000) used a questionnaire with a few follow-up interviews to study the views of some 2,000 students aged 11–18 attending secondary schools in South West England. Girls were found to be more gullible than boys, in all age groups, with 77% of 11 to13 year-olds believing in ghosts, 46% in astrology, 45% in palmistry and 35% in the healing powers of crystals. For boys, the percentages were 66%, 24%, 23% and 21%. Older students were more skeptical: 60%, 22%, 13% and 33% of girls continued to hold these four beliefs, and 39%, 15%, 11%, 19% of boys. Interestingly, and perhaps predictably, boys were found to be more likely than girls to believe that aliens had visited Earth at some time (54% boys; 42% girls). Coll and Taylor (2004) used the same questionnaire to structure interviews with 18 scientists based in New Zealand and the United Kingdom. Scientists were generally dismissive of all beliefs for which there is no potential theoretical explanation and, unlike non-scientists, they do not regard media reports or testimony by individuals as evidence. When there is a possibility of a theoretical explanation, however unlikely in terms of contemporary science, the scientists were remarkably open-minded. For example, some were prepared to believe in the possibility that crystals could improve health because it is known that crystals can exert electrical field effects. Exactly the same open-minded response was shown by some 16 year-olds responding to a question about whether talking to plants could help them to grow better. They speculated that exhaled breath could provide localized high concentrations of carbon dioxide that plants could use in photosynthesis (Leach et al., 1996). Probability arguments were used to entertain the idea that other life forms exist somewhere else in the universe and, therefore, it is conceivable that aliens had visited Earth in the past and may do so in the future. This is the kind of critical but open-minded stance that we should be pursuing with students. It can best be cultivated by subjecting pseudosciences to critical scrutiny as part of the school science curriculum. In making similar recommendations, Martin (1994) states that although a pseudoscience will have some "surface properties" in common with science, its underlying nature is unscientific. Specialist vocabulary, theories for which advocates claim there is supporting evidence, special training for practitioners, professional organizations and journals, and so on, serve to mask basic flaws and deflect attention from the fact that fundamental propositions are untested, untestable, or perhaps have already been falsified. Moreover, Martin says, there will be "attempts by the practitioners to prevent their theories being exposed to critical tests and evaluations, and the attempt to explain away any possible evidence ... Practice will also include attempts of the practitioners to isolate themselves from the mainstream of scientific inquiry and from critical interaction with the scientific community ... The attitude of the pseudoscientist will be dogmatic and slightly paranoid; he or she will be intolerant of all other theories" (Martin, 1994, p. 361).

Smith and Scharmann's (1999) characterization of fields of study as *more* scientific or *less* scientific, discussed in Chapter 4, has ready application to the classroom study of pseudosciences. Indeed, it has been used successfully by

Scharmann et al. (2005) to provoke preservice secondary school science teachers into more thoughtful consideration of NOS issues in evaluating the claims of evolution, intelligent design theory and umbrellaology.[2] In Smith and Scharmann (2008), the authors argue that the more scientific versus less scientific criterion is particularly well-suited to situations where students may hold views and values antithetical to those of science because it is less confrontational than a strict definition-based approach. Also of value is Thagard's (1980) description of a pseudoscience as a discipline that claims to be scientific even though its practitioners make little or no attempt to develop the underlying theory towards solution of the stock of unsolved problems in its domain, show no concern for attempts to evaluate the theory relative to other theories, and are highly self-serving in their views of what counts as confirmation and disconfirmation. What little research evidence is available on the success of teaching about pseudosciences is very encouraging. For example, Dagher et al. (2004) note that when they were asked to generate demarcation criteria for science and pseudoscience, many students identified experimental evidence, careful and systematic methodology, tentativeness and mathematical content. In a fascinating study designed by Craven et al. (2002), preservice student teachers were asked to make their own selections of written texts to exemplify scientific writing and pseudoscientific writing,[3] and to list what they perceive to be the defining characteristics. Student responses noted that pseudoscientific writing is characterized by unjustified assumptions, vague and unsubstantiated claims, inconsistencies, insufficient or poorly gathered evidence; scientific writing is based on systematic and careful research, which can be replicated by others. Furthermore, conclusions are based on evidence and have to be consistent with other well-established theories. Radner and Radner (1982) have proposed the following distinctive features of pseudosciences: inconsistency with established science, highly selective use of data, use of irrefutable hypotheses, obstinate refusal to revise ideas in the light of new evidence or argument, and appeal to myths, fictions and religious texts. This could provide a useful set of parameters for students investigating pseudosciences.

It might be instructive for students to conduct a survey of belief in pseudosciences and the paranormal, and seek to identify commonly held superstitions within the school and the local community, perhaps using an instrument adapted from Preece and Baxter's (2000) questionnaire. Different groups of students might then conduct a critical investigation of the four or five pseudosciences considered most plausible by respondents to the questionnaire,[4] making use of Radner and Radner's characterization (or similar criteria) and, of course, developing their own notions of what distinguishes science from pseudoscience. As an additional and separate case study, the class might address the question of whether acupuncture should be regarded as a science in terms of the criteria listed by Thagard (1980), Martin (1994), Hacking (1996) and Smith and Scharmann (1999). The lengthy description that follows draws very extensively on the published work of Douglas Allchin (1995, 1996), in which he points out that acupuncture has a history dating back two millennia.

There is no doubt that acupuncture works. As millions of successfully treated individuals will testify, it can quickly relieve pain, it is an effective treatment for insomnia, asthma, ulcers and vitamin E deficiency, it stimulates saliva production and dramatically increases the success rate of *in vitro* fertilization procedures. On a personal level, I can report that in a single visit to an acupuncturist my daughter was cured of a muscular problem that a physiotherapist had been unable to cure over the course of several visits. Acupuncture has also been used successfully in dentistry, surgery and drug dependency treatment. Like all traditional Chinese medicine, acupuncture treats the whole patient, not just the particular symptoms, on the grounds that the condition almost certainly results from a complex web of interacting relationships (physiological, psychological and social). Only rarely is there a simple cause and its resulting effect. The theoretical structure of acupuncture does not draw on Western science. Rather, it is based on the flow of qi (the life force) through the body via 12 major meridians (channels), each associated with a major organ (stomach, liver, gall bladder, etc.) and related to the Lunar cycle. Proper flow of qi maintains a balance between yin and yang and ensures good health. When flow of qi is impeded or becomes unbalanced, disease or organ malfunction may ensue, and pain may be felt. Acupuncture needles are used to promote or restrict the flow of qi, thus restoring the balance of yin and yang, with the correct position for insertion being determined by meridian maps that have been built up over several centuries of acupuncture practice (Baldry, 1993; Liao et al., 1994).

After a description of acupuncture, and possibly a talk by a practitioner, students could be challenged to decide whether they would classify it as science, pseudoscience or non-science. It has a specialist language, vast empirical base, theoretical structure, established methods for training practitioners, a long historical tradition and a progressive research agenda (as witness its extension to dentistry and surgery in the 1950s). For many years Western medicine regarded acupuncture as quackery, deceptive in its practice and fraudulent in its claims. Allegations were commonly made that patients would have recovered anyway, or that sleight of hand, distraction and diversion tactics are employed. It was said that success can be explained in psychosomatic terms. After all, acupuncturists screen their patients, on grounds that it is not considered an "appropriate treatment" for everyone. In other words, critics argue, only the gullible are selected for treatment, so success is guaranteed. Most damning of all, it is argued, there has been no systematic experimental verification of its success and its underlying theoretical structure does not make sense in terms of Western science. Teachers might ask students whether these are legitimate objections. To answer such a question, students need to know that experiment constitutes an ethical problem for traditional Chinese medicine. Withholding treatment from someone who needs it (in order to conduct a controlled experiment) is clearly not in the patient's interests and so violates their right to proper care. The demand for an explanation compatible with Western science is considered unreasonable. Acupuncture has a complex theoretical structure based on the flow of qi, and its proper application produces good results in patients. Therefore, an additional theory (expressed in Western terms) is unnecessary. In response to allegations of psychosomatic influences, defenders of acupuncture point out that it works equally well with horses, dogs and camels. What is also true is that if the needles are inserted in the wrong locations (for example, by someone who is untrained or incompetent), the treatment will be ineffective. In response to Western incredulity that the point for needle insertion is rarely at the seat of the pain, acupuncturists point out that Western doctors sometimes diagnose "referred pain", as in the condition known as angina pectoris. Cynical students might begin to suspect that acupuncture is resisted by the Western medical establishment because it is labour intensive, rather than capital intensive. There are no expensive drugs to sell, no sophisticated technology to be purchased and no fancy treatment centres to be built. Because it is "low tech" there is unlikely to be a large profit margin. Indirect justification for such an assertion could be found in the observation that pharmaceutical companies have responded to the upsurge of interest in traditional herbal medicine by conducting extensive research into the nature of the "active ingredients". If these substances can be isolated and identified, then they can probably be synthesized and marketed as a more convenient and "safer" alternative to chewing a few leaves gathered on the local riverbank.

Over the past 35 years or so, there have been numerous attempts to explain the success of acupuncture in terms of Western science. Perhaps the needles can block nerve impulses and so switch off the pain reflex. Perhaps this explains why the needles need to remain in place for several minutes before the patient notices their beneficial effect. Perhaps the needles trigger the release of endorphins or stimulate the release of cortisone. Experiments using naloxone, an endorphin suppressor, have shown that it prevents acupuncture needles from working properly (Pomeranz, 1987). If an explanation consistent with Western science can be found, acupuncture will undoubtedly gain greater respectability and will be more extensively recommended by Western doctors. Would this be sufficient for it to be classed as science? These demarcation questions can, and should, be put to students.

MULTICULTURAL AND ANTIRACIST SCIENCE EDUCATION

Demarcation criteria figure prominently in debate about multicultural and antiracist science education.[5] For some, the issue extends well beyond the argument that the school curriculum should acknowledge and celebrate the contributions of non-Western scientists to the origin, development and contemporary practice of science to encompass a demand for the radical reconsideration of what counts as science in the curriculum, who should make that decision, and who benefits and who is harmed or marginalized by it? In education, as in all spheres of human activity, decision-making necessarily reflects the perspectives of the decision-makers. Hence the selection of knowledge for the curriculum, and the science curriculum in particular, does not reflect a common heritage but one "drawn from the framework of those who have dominated society and educational discourse (i.e., mostly white, male, and middle class)" (Stanley & Brickhouse, 1994,

p. 388). That particular set of selection criteria results in the image of science, scientists and scientific practice so familiar to us through its depiction in the majority of school science textbooks and official curriculum documents, a view that has been and will continue to be heavily criticized throughout this book. Kawagley et al. (1998) argue that this "narrow view of science not only diminishes the legitimacy of knowledge derived through generations of naturalistic observation and insight, it simultaneously devalues those cultures which traditionally rely heavily on naturalistic observation and insight" (p. 134). Their view is representative of a body of opinion that argues for a substantial broadening of the science curriculum to accommodate knowledge about the natural world accumulated outside of conventional Western science, and variously described as traditional knowledge, Aboriginal knowledge, Indigenous knowledge and traditional environmental or ecological knowledge (often using the acronym TEK). Terms such as ethnoscience and folklore are also used from time to time. I have chosen to use the term traditional knowledge (TK) as indicative of knowledge accumulated by a group of people, not necessarily Indigenous, who by virtue of many generations of unbroken residence develop a detailed understanding of their particular place in their particular world. It includes, for example, the beliefs and practices of distinct social groups such as the fishing community in rural Trinidad described by June George (1999a) and the groups in South and East Asia designated "neo-indigenous" by Aikenhead and Ogawa (2007) (see below).

The substance of the debate is the politics of recognition: all human cultures have developed knowledge and belief systems about the natural world, and the principles of equality of respect and equality of worth that underpin multicultural and antiracist education demand equality of recognition (Taylor, 1994). At play, too, is a concern to redress the cultural imperialism of the past, through which traditional knowledge and practices were forcibly replaced by Western science and Western agricultural practices, often with untold damage to local ecosystems and destruction of the social fabric. Thus any challenge to the dominance of Western science in the curriculum carries extensive political baggage; it is a challenge both to the status of science and the political structures that sustain it. The argument runs along the following lines: science is primarily the accumulation of knowledge about the natural environment;[6] the priorities, procedures and products of science are shaped by social and cultural forces; there is considerable diversity of approach among the disciplines of science, with none having a claim to universality; every culture has, necessarily, amassed a store of knowledge about the natural environment in which it is located; these bodies of knowledge should be recognized as alternative sciences arising in different sociocultural and physical circumstances from those of Western science (see Stanley & Brickhouse, 2001). By making curriculum choices about what to include and exclude, we not only define and limit what counts and does not count as science, we also erect potential barriers that restrict access or make access difficult, thereby defining who has access to science and who is denied access. Clearly these are issues that teachers and curriculum designers need to address. But they should not be taken as an argument for including anything and everything in the curriculum under the banner of *science*. There are several very real and very serious dangers in adopting such a position. First, the danger of relativism: if the door is opened so wide that any claim to be counted as science is admitted, regardless of other considerations, then any view is as good as any other, without any need to attend to the circumstances of its production, the nature of its validation or the context of its deployment. By depriving students of the means to evaluate knowledge claims, relativism destroys the rationality underpinning the notion of critical scientific literacy. There would be no way to distinguish between good science and bad science or between science and non-science; there would be no means of confronting the claims and blandishments of snake oil salesmen, quacks and charlatans. Nor would there be any defence against the outlandish and unsubstantiated claims that appear in documents such as the *Science Baseline Essays* (Adams, 1990). In their very lively and spirited consideration of the Essays, and a number of similar publications, Loving and de Montellano (2003) provide some useful advice for teachers regarding selection of sound curriculum materials. While it is important to acknowledge that every culture has generated knowledge about the natural world, and that this knowledge has often been robust enough to function very adequately in the context in which it was developed, and that traditional knowledge may have some important things to teach all of us, it does no-one a service to label it as *science* if it does not fall within the criteria of definition, loose and fluid as they are.

Southerland (2000) argues that far from being a more egalitarian approach, the outcome of this blurring of the boundaries between science and other ways of knowing about the natural world is, in effect, a new form of scientism. Instead of acknowledging and celebrating the differences, it tries to force all forms of knowledge (however diverse) into the construct of "science". By redefining science to include all forms of knowledge about the natural world, we set no limits on the extent and authority of science and we reinforce the scientistic position that knowledge afforded the title of science is the *only* valid form of knowledge and understanding. Conversely, an approach that includes all forms of knowledge about the natural world under the banner of science is not so much about crossing borders as about pretending that the boundaries and barriers to access do not exist. There is also the implicit assumption that all members of a particular cultural group necessarily share, and *should* share, a particular set of views. Asserting and seeking to maintain collective identities can be restrictive for the individual and potentially damaging for the knowledge itself because it seeks to insulate it from any kind of criticism and to fossilize it in its current state (Irzik & Irzik, 2002). Striving to preserve cultural identity is a laudable goal, but we should be careful that it is not achieved at the cost of individual freedom, both now and in the future. Personalization of science entails the cultivation of respect for the autonomy of individuals (Appiah, 1994).

However, in the context of this book, my major concern is that the approach advocated by Kawagley et al. (1998) misrepresents both science and traditional knowledge. As discussed in Chapter 4, when we refer to a body of knowledge as "science" we invest it with some kind of privileged epistemic status. Hence there

119

is strong motivation among some scholars for according traditional knowledge this appellation and status. I believe these efforts to be seriously misguided. My position is as follows: to be classed as a science, a form of knowledge must meet the criteria for judging whether it is a science or not. There is much of value and benefit in traditional knowledge, and it clearly works well in the context in which it was developed, but it is not science simply because well-meaning people say it is science. No more than acupuncture is a science simply because it works well. Equating diverse systems of thought broadens definitions of science to the point of meaninglessness and denies the uniqueness of other forms of knowledge. In the words of Cobern and Loving (2001), "Diversity is lost. Meaning is lost. Communication is lost" (p. 61). It makes much more sense to value traditional knowledge for its own merits, for what makes it distinctive and powerful – a point to which I will return later in the chapter. In summary, traditional knowledge is legitimate and valuable in terms of its own epistemic and situational criteria; calling it "science" removes it from the context in which its legitimacy is located and requires that it be evaluated in terms of the criteria used to evaluate science. In the words of Cobern and Loving (2001), this is playing a game which traditional knowledge is bound to lose because it would have to compete on grounds where science is at its strongest and most persuasive. As El-Hani and Bandeira (2008) comment, by insisting on its designation as "science" we would be likely to devalue traditional knowledge in the very effort of advocating its curricular and sociopolitical value.

There is abundant research evidence identifying the barriers and difficulties faced by students of non-Western background in learning science (Jegede, 1995; Cobern & Aikenhead, 1998; Aikenhead, 2000, 2006). While those problems need not be recapitulated here, it is important to note that a clash of worldviews between local culture and the culture of science can sometimes figure prominently among them (Jegede & Okebukola, 1991; Baker & Taylor, 1995; Ogawa, 1998; Kawagley & Barnhardt, 1999; Koul, 2003a; Luykx & Lee, 2007; Malcolm, 2007) (Chapter 6 includes a discussion of worldview theory). Because a worldview includes fundamental beliefs about causality and about humanity's place in the world, it is fairly easy to see how it could be incompatible with the fundamental metaphysical underpinnings of science. For example, Richard Nisbett (2003) argues that East Asians are more likely to see humans and their surroundings as an integral part of a complex system, while Westerners tend to focus on individual actors. He argues that the widespread use of feng shui, the study of how structure relates to environment (including altitude, prevailing wind, compass orientation, proximity to water and mountains, etc.), shows that East Asians perceive the world as composed of complex relationships. In contrast, the Western tendency to solve problems through a specified series of steps indicates a commitment to rule-based, atomistic, universally applicable thinking. He concludes that Westerners are more analytic, paying attention primarily to the object and the categories to which it belongs, and using rules (including formal logic) to explain and predict behaviour. Westerners attend primarily to the focal object or person and Asians attend more broadly to the field and to the relations between the object and the field. Westerners tend to assume that events are caused by the object; Asians are inclined to assign greater importance to the context. When students with substantial exposure to, immersion in, and commitment to an alternative worldview are presented with a school science curriculum constructed in accordance with traditional views of science, they may experience great difficulty in effecting a border crossing into the sub-culture of science. Worldview theory is particularly helpful in distinguishing between thinking and knowing, and between understanding and believing. By thinking, one builds an epistemological case for particular understanding (a key element of learning *about* science), but believing is a metaphysical process by which one comes to accept as true that which one comprehends. Worldview theory helps us to unpack the metaphysical aspects of science (what it knows and assumes about the nature of being and existence) and to recognize when these features may be in conflict with a student's fundamental presuppositions about such matters. For example, Aikenhead (1997, p.220) describes how the interests of First Nations people in survival, coexistence and celebration of mystery are not in sympathy with the drive of Western science to achieve mastery over Nature through objective knowledge based on mechanistic explanations. Nor are the holistic perspectives of Aboriginal knowledge, with its "gentle, accommodating, intuitive and spiritual wisdom" in sympathy with reductionist Western science and its "aggressive, manipulative, mechanistic and analytical explanations". Border crossing is inhibited not so much by the cognitive demand of the learning task as by the discomfort caused by some of the distinctive features of science, features that are often exaggerated and distorted by school science curricula into a scientistic cocktail of naïve realism, blissful empiricism, credulous experimentation, excessive rationalism and blind idealism (Nadeau & Désautels, 1984).

Clearly, those with a pre-existing worldview that is in harmony with the scientific perspective will find it easier to learn science because it "makes sense" to them in terms of fundamental assumptions and underlying values. Science teaching will support and enhance their pre-existing worldview. Those whose worldview differs substantially may experience difficulties in learning science. As Cobern (1995) says, "One scientist trying to convince a colleague, or even a scientist from another field, is not the same as trying to convince those outside the scientific community" (p. 289). When there is a clash of worldviews the situation is, he says, like Charles Darwin presenting his book, *The Origin of Species*, to a public with very different, religion-based views about origins. Speaker and audience do not share the same fundamental ideas about the world and so may "talk past each other". Mutual incomprehension is, indeed, a poor basis for effective teacher-student relationships!

Malcolm (2007) addresses some of the collisions between the scientific worldview and the worldview of some traditional African cultures in the context of curriculum development in South Africa. Sometimes the sense of discomfort students feel can shift towards alienation, leading them to opt out of science altogether. Locust (1988), for example, notes that many Native American students will choose to fail a biology course that involves dissection rather than run the risk of "bringing bad luck" into their lives or the lives of family members. Numerous problems of this kind have also been identified by Kawasaki (1996) and Jegede (1998), but they are not restricted to the Japanese and African students these authors describe. They exist for Europeans and Americans, too, especially if they have strong religious, spiritual or aesthetic conceptions of the natural world (Cobern, 1996; Roth & Alexander, 1997). The problems run considerably deeper than cognitive or epistemological concerns. Science teaching may threaten, disrupt, overpower, marginalize and eventually displace long-standing beliefs and values that underpin some students' personal and cultural identity. In other words, students become assimilated into the dominant scientific cultural traditions at the expense of their other beliefs and values. Alternatively, students may resist this perceived attempt at displacing their worldview and decide that science is not for them. Interestingly, strongly held beliefs do not always prevent students from understanding and being able to use scientific knowledge that contradicts them. For example, Demastes et al. (1995) show that students can construct a perfectly adequate scientific understanding of evolution despite a clear rejection of its truthfulness that is rooted in strong religious beliefs (see also, Sinatra et al., 2003). Both knowledge structures (evolution and creationism) can be part of an individual's personal framework of understanding because they are used for different purposes. In some ways, this reflects the distinction between scientific *theories*, taken as "true" descriptions and explanations (in the sense to be discussed in Chapter 6) and scientific *models*, used for their predictive capability but not regarded as true. Similarly, as Ogunniyi et al. (1995) observe: "the Japanese never lost their cultural identity when introducing Western science and technology, because they introduced only the practical products of Western science and technology, never its epistemology or worldviews" (p. 822). Jegede (1995) has developed the notion of collateral learning to describe how individuals can hold and develop Western scientific thinking alongside traditional African knowledge and understanding. Regarding each individual as being in possession of a unique personal framework of understanding, and regarding each educational encounter as creating a unique learning context, allows for Jegede's description of four categories of collateral learning (parallel, simultaneous, dependent and secured) to be extended to all students in all classrooms (see Hodson, 1998a, for a full discussion of this notion).

It seems trite to point out that students fare much better when culturally relevant materials and methods are deployed. The question at issue in this chapter is whether this means that traditional knowledge should be taught as *science*. Brayboy and Castagno (2008) comment as follows:

If we consider the fact that Indigenous peoples in the Americas created toboggans to carry the heavy carcasses of deer and caribou, kayaks and canoes that were river and sea worthy, snow shoes and snow goggles; domesticated a wide range of plants including corn, potatoes, squash, beans and peanuts; built architectural masterpieces in which they live and ovens in which they cooked; used petroleum to create rubber and stars to successfully navigate the continent; and found ways to dry meat for storage and future use, it becomes evident that

Indigenous peoples have been scientists and inventors of scientific ideas for a long while. (p. 732)

The important question in the context of this book is whether this specific set of concerns, the development of innovative technologies, and the successful resolution of day-to-day problems, constitutes science according to the demarcation criteria thus far discussed. Similarly, Kawagley et al. (1998) maintain that Yupiaq traditional knowledge is based on a "massive set of scientific experiments continuing over generations" (p. 137). What they really mean is that this knowledge has been tested for its robustness in the field and has been judged successful. Whether this counts as experimentation is a different matter. They claim that the technology of the Yupiaq people (kayaks, river fish traps, and so on) is evidence of an underlying science: "technology does not spring from a void. To invent technological devices, scientific observations and experimentation must be conducted ... [it] could not have been developed without extensive scientific study of the flow of currents in rivers, the ebb and flow of tides in bays, and the feeding, resting, and migratory habits of fish, mammals and birds" (p. 136). And, therefore, this knowledge should be taught as science, or as an *alternative*, non-Western science. Even if we set aside the simplistic view that technology is merely (and only) the application of a pre-existing science, we still need to ask whether the underlying theory is abstract, internally consistent, formalized and systematized, broad in scope and application, rich in consequences, and productive of further research – criteria central to classification as science. In contrast to the position taken by Kawagley et al. (1998), Southerland (2000) draws a distinction between *instructional* multicultural science education (and the importance of accommodating the needs, interests, aspirations and values of diverse students) and *curriculum* multicultural science education (with its demands to redefine our conception of science), accepting the first and rejecting the second. The distinction has its merits, provided that it does not mean "business as usual" as far as critical scrutiny of curriculum content is concerned. My view is that students should scrutinize traditional knowledge in the same way that they scrutinize pseudosciences, by raising questions about what counts as science and in what ways traditional knowledge resembles and differs from science. Criteria would include such things as empirical content, and how it was established, conceptual underpinnings and capacity to construct explanatory models and theories, predictive capability, and the extent to which knowledge is quantified and precise relationships between concepts established. Sutherland and Dennick (2002) also suggest that the science curriculum should focus on epistemological differences between traditional knowledge and science. They report that the Cree students whom they interviewed about such matters were puzzled and disappointed that this was not already being done. In subsequent research, Sutherland (2005) reports that 11–14 year old Cree students who do draw a clear distinction between science and traditional knowledge are, in general, better motivated to learn science. They are also "more likely to attend school (despite enticement from peers to skip) and to invite challenges when learning" (p. 609).

What follows is not intended to be the definitive comparison of science and traditional knowledge. Such a task could only be carried out on a case-by-case basis by an individual with expertise in history, philosophy and sociology of science *and* lengthy immersion in the traditional knowledge under consideration. My purpose is to make some *general* points of comparison and contrast as the basis for further consideration of how best to deploy traditional knowledge in the school science curriculum. It addresses the very criteria of demarcation that I advocate using with school students and student teachers. Comparisons inevitably raise problematic issues. First, the tendency to over-generalize and stereotype. Just as it is mistaken to assume that there is one all-embracing description of science, so it is a gross over-simplification to assume that all forms of traditional knowledge have common underpinnings, purposes and values. Often there are concepts, ideas and principles embedded in a particular way of knowing that cannot be faithfully translated into another language or fully understood from a different cultural standpoint. For example, as Aikenhead and Ogawa (2007) point out, "the noun *knowledge* does not translate easily into most verb-based Indigenous languages" (p. 553). In consequence, these authors prefer to use the expression "Indigenous ways of living in nature" instead of "Indigenous knowledge". They also create the category "Neo-indigenous ways of knowing nature" to address the knowledge built up by long-standing, though not Indigenous or colonized communities of people – in their case, as a means of addressing traditional Japanese ways of knowing. Finally, the term "Eurocentric sciences" is used instead of "science" in an effort to avoid the kind of stereotyped, scientist and overly positivist views of science pervading many science textbooks. However, as van Eijck (2007) points out, this terms fails to acknowledge the complex origins, diverse roots and continuing cross-cultural borrowing that have led to contemporary science.

In general, traditional knowledge is accumulated over longer time periods than scientific knowledge, data are collected in natural settings rather than laboratories, propositions are tested by application in context rather than by systematic experimentation. Prediction is important in both, but the significance of *novel* predictions is peculiar to science. In traditional knowledge, validity and reliability are based primarily on the "test of time" – that is, how well the knowledge stands up to use in varying circumstances over many years; science does not look for continued confirmation. Both science and traditional knowledge are built on a mix of qualitative and quantitative data, though emphases may vary substantially from context to context. Whereas science aims for clarity and insists on unambiguous phrasing, traditional knowledge makes extensive use of metaphor, personification and symbolism to embellish or encode meaning and explanation.[7] Traditional knowledge often weaves observational data together with supernatural, religious and other subjective sources of information (including dreams) into an holistic explanation. For example, Kawagley et al. (1995) note that Yupiaq people view the world as being comprised of five elements: earth, air, fire, water and *spirit*. The authors contend that the incorporation of spirit resulted in "an awareness of the interdependence of humanity with the environment, a reverence for and a sense of responsibility for protecting the environment ... The shamans were the intermediaries between the spiritual and natural worlds. They informed the people what was appropriate or not in dealing with the earth" (p. 586). Meaning is often conveyed in narrative form, a practice that is uncommon in science. Traditional knowledge is located within an oral tradition (often including songs, prayers, dance and spiritual ceremonies), while science uses written text for communicating and recording knowledge. Traditionally, science puts considerable emphasis on its claim to be objective and value-free, though I have been at pains throughout this book to paint a somewhat different picture of science. Traditional knowledge is overtly value-laden; its purpose is not the provision of neutral information which individuals are then free to interpret and use as they see fit; its express intent is to impart useful information embedded in the values, moral rules, ethical guidelines and cultural traditions of the social group. Its role in maintaining social stability, order and cultural identity necessitates that it is widely acknowledged as immune to any criticism that might threaten social cohesion. Indeed, there can sometimes be very severe penalties for dismissing, ridiculing or even questioning established customs and traditional ways of knowing. In contrast, science puts great store on public scrutiny and on criticism, peer review, and so on. Whereas science is concerned with universal explanations, and seeks to apply theoretical systems more widely, traditional knowledge is rooted in, or connected to, place. This raises some interesting questions about the proper location for teaching traditional knowledge, or even teaching *about* it. Although theoretical explanations are used, there is rarely any interest in extending traditional knowledge beyond the context for which it was designed. Atran (1990) sees the fundamental difference in terms of traditional knowledge seeking to understand all aspects of the world in terms of the familiar and already experienced, while science seeks to explain the known in terms of the unknown through the creation of abstract theoretical concepts. The final demarcation between traditional knowledge and science that would be appropriate for students to consider is that traditional knowledge may not be freely available within the community. Factors such as gender, seniority, genealogy and aptitude may dictate who has access to it and who safeguards it and teaches it. In contrast, scientific knowledge is made public through formal publication, once it has been subjected to the rigorous procedures of peer review and critique. Aikenhead and Ogawa (2007) also point to a key difference in the historical-political context: for traditional knowledge, a contemporary context of sociopolitical and economic repression; for science, a continuing context of privilege and power. In similar vein, McKinley (2005) notes that the knowledge of many Indigenous peoples is under threat as a consequence of colonization, marginalization and cultural suppression: "Many indigenous peoples have become (involuntary) minorities in their own countries while others have been left with legacies of colonizing institutions and/or mindsets" (p. 228).

Using some of these criteria, George (1999b) views the relationship between school science knowledge and everyday practices based on traditional knowledge in terms of four categories: (i) situations in which science can explain traditional practices; (ii) situations for which science is currently unable to provide a satisfactory explanation; (iii) situations in which science and traditional knowledge advise similar practices, but explain the advice differently; (iv) situations where the two knowledge systems are in disagreement or conflict. The approach I am advocating goes much further than this kind of comparison of utility. It seeks to address key questions about the ways in which knowledge is constructed and deployed. The purpose of such an approach in school science is not to disregard, trivialize or marginalize knowledge constructed outside the realm of science but to recognize that such knowledge is based on different assumptions and values, and has different priorities, purposes and uses. Drawing attention to key issues concerning domains of application, the collection and interpretation of data, construction of explanatory systems, underlying assumptions and values, and so on, contributes substantially to learning *about* science. Perhaps the real value of the approach lies in identifying strengths and weaknesses, and recognizing ways in which different knowledge systems can complement each other. Finding value in traditional knowledge is not located in labelling it "science" but in recognizing its differences from science, valuing it on its own merits and for its distinctive virtues, and using these different perspectives as a means of critiquing the priorities, values and practices of science. For example, Snively and Corsiglia (2001) note that the spiritual basis of traditional knowledge, which is a major feature distinguishing it from science, incorporates "important ecology, conservation, and sustainable development strategies" (p. 23) – an argument also made by Christie (1991), Chinn (2007) and El-Hani and Bandeira (2008). The depth of ecological knowledge, its robustness over long periods of time and its context-specific detail within a holistic framework represent an invaluable framework for addressing current environmental crises. It is here that general works on traditional knowledge, such as *Wisdom of the Elders* (Knudtson & Suzuki, 1992), have an important role to play. It is here, too, at the intersection of Western science and traditional knowledge, that a recent curriculum initiative in South Africa is important. The goal is to integrate "indigenous knowledge systems" with school science throughout the entire school system (K-12).

> Indigenous or traditional technologies and practices ... reflect the wisdom of people who have lived a long time in one place and have a great deal of knowledge about their environment... Much valuable wisdom has been lost in South Africa ... and effort is needed now to rediscover it and examine its value for the present day. (Department of Education, RSA, 2002, p. 10)

Ogunniyi (2007a,b) reports that progress to date is encouraging, though many teachers are still uncertain about exactly what is entailed and apprehensive about their ability to engage in the radically different teaching and learning strategies that the new curriculum requires.

In non-Western contexts and in Western societies where some schools serve an Indigenous community (First Nations students in Canada, Māori in Aotearoa New Zealand, for example) there is a range of additional issues to consider, most of which are outside the scope of this book and outside my personal experience. Nonetheless, one or two points can usefully be made. Not only does the introduction of traditional knowledge into the curriculum play a major role in personalizing learning and assisting border crossings into science, it can be a crucial element in cultural stabilization or renewal. As noted earlier, much of this traditional knowledge is under threat. Indeed, much has already been lost. School science education can play a vital role in conserving that knowledge for future generations. The task is to find ways of incorporating traditional knowledge into the curriculum in order to celebrate and maintain important cultural traditions whilst attending to national development priorities that are rooted in contemporary science and technology. Gitari (2003, 2006) has undertaken this task with respect to traditional herbal medicine in rural Kenya (see also, Kithinji, 2000), Keane (2008) has done similar work with regard to traditional farming methods and views of nature in rural Kwa-Zulu Natal (South Africa), Chinn (2007) has shown how the Hawaiian science curriculum can incorporate traditional cultural practices focused on sustainability issues and development of a sense of place, and Palefau (2005) has conducted a study of how traditional knowledge in the Pacific Islands (specifically, the Kingdom of Tonga) can be preserved and incorporated into a science and technology curriculum concerned with economic development and responsible citizenship. Aikenhead (2002) describes how a group of teachers in Northern Saskatchewan worked together to design *Rekindling Traditions*, a grades 6 to 11 science course incorporating the local community's traditional knowledge, much of which was gathered by the students themselves through interviews with Elders and other knowledgeable people in the community. Aikenhead and Ogawa (2007) make reference to two other First Nations science education initiatives, *Lines and Circles* (Ahkwesahsne Board of Education, 1994) and *Forests for the Future* (Menzies, 2003), though they are highly critical of the way Western science is portrayed in the associated curriculum materials. Writing from a Japanese perspective, Ogawa (1998) argues that Western science should be taught and interpreted within the traditional cultural context of *Shizen* (a distinctive worldview that involves people in an emotional, spiritual and aesthetic engagement in the natural environment), for two reasons: first, to maintain indigenous Japanese cultural identity; second, to bolster movements for environmental protection. Watanabe (1974) also provides a helpful account of traditional Japanese views of nature. For a number of years, Mere Roberts (1998) taught an undergraduate course at the University of Auckland (New Zealand) through which Pasifika students with strong cultural knowledge, but little or no background in science, could gain some understanding of how science sheds light on the traditional environmental knowledge of the Pacific Islands region.[8]

It is noteworthy that *Science in the New Zealand Curriculum* (Ministry of Education, 1993) urges teachers to make science more relevant and more accessible to Māori students by acknowledging Tikanga Māori (Māori culture, beliefs and values), using and valuing Te Reo Māori (Māori language) and acknowledging and building on the experiences of Māori students. More significantly, the Ministry of Education (1996) has produced a parallel document *Pūtaiao: i Roto i Te Marauntanga o Aotearoa* for schools wishing to teach science through the Māori language, the goal of which is to teach science from a Māori worldview perspective, including Māori ways of understanding the living world and adopting "culturally relevant" pedagogical practices[9] (McKinley, 2000, 2005; Stewart, 2005, 2007; McKinley & Keegan, 2008; Wood & Lewthwaite, 2008). Although this curriculum is not without problems – for example, an unresolved tension between "science in Māori" and "Māori science" (Stewart, 2005) and a continuing problem over terminology (McKinley & Keegan, 2008)[10] – there is much to celebrate. The situation appears to be very different in Papua New Guinea, at least judging by Ann Ryan's (2008) account of the ways in which Indigenous knowledge and values have been systematically excluded from the science curriculum on the advice of overseas advisors and consultants. Sadly, the PNG situation is not uncommon. Ninnes (2003) has shown that even in societies with sizeable communities of Indigenous peoples, such as Australia and Canada, references to Indigenous identities and knowledge are rarely included in commonly-used school science textbooks, or they are presented in a stereotyped or historically outmoded way. Clearly there is an urgent need for efforts at both the preservice and inservice levels to equip teachers with the knowledge, skills, experiences and curriculum materials to address issues relating to traditional knowledge in a culturally sensitive and critical way. There are two dangers when teachers who are not members of the particular community in question attempt to "borrow" traditional knowledge for use in other contexts. One is the patronizing tokenism that can result from concentration of the "quaint and colourful" at the expense of rigorous academic considerations; the other is the spectre of cultural appropriation. As Brayboy and Castagno (2008) warn, "We must think about issues of ownership of knowledge and how to engage in a culturally responsive science education without engaging in theft, exploitation, or distortion of Indigenous knowledge" (p. 747).

CREATIONISM AND INTELLIGENT DESIGN THEORY

Although the idea of studying creationism or "intelligent design theory" in the school science curriculum would strike science teachers in many parts of the world as ludicrous, it makes eminently good sense to do so in the United States and possibly in Canada, too. Given the seemingly inexorable rise of Christian fundamentalist beliefs in the United States, there is great value in science teachers subjecting creationist beliefs to the same kind of critical scrutiny they would use to address the pseudosciences. This is not an argument for teaching creationism, it is an argument for teaching *about* creationism – a distinction that Pennock (2002) fails or refuses to appreciate in his argument against allowing creationism into schools. Teaching *about* creationism is the most effective way of denouncing it as a purported legitimate alternative to evolutionary theory, especially when set in the context of case studies of the events surrounding the publication of Darwin's *The Origin of Species* (Todd, 1984; Ruse, 1989; Duveen & Solomon, 1994; Solomon et al., 1994a,b) and events in the United States concerning demand for equal curriculum time. Resources for constructing suitable curriculum activities are considerable, though care should be exercised in scrutinizing such material for bias and prejudice. Three particularly valuable sources are Scott's (2005) comprehensive overview of the creation-evolution debate in the United States, which includes selections from an extensive literature and details of important legislation, Ruse's (2005) account of the origins of the debate in the Christian reformation and the crisis of faith it precipitated, and Pennock's (2000) critical account of the historical roots of the intelligent design movement and its underlying motives. What follows below draws very extensively on the writing of Paul Kitcher (1982), Michael Ruse (1988a, 2000), Ian Plimer (1994), Charles Taylor (1996) and Gregory Peterson (2002).

While there are variants of creationist doctrine, six general postulates are held in common: the origin of the universe in the work of a Creator; the alleged inadequacy of evolution as an explanation for the development of life forms; the occurrence of adaptive change only within fixed limits or originally created "kinds" (life-forms); separate ancestry for human beings and apes; explanation of geological history by catastrophism; the relatively recent origin of the Earth (between 6,000 and 10,000 years ago). Many creationists regard the account in *Genesis* as a factual representation of events, including, for some, a 6-day creation schedule. While most scientists regard creationism with the kind of bemused incredulity commonly directed towards "Flat Earthers", these beliefs are remarkably persistent, especially in the United States. Notable events in this long history include the 1925 State of Tennessee versus John Scopes trial (the so-called "monkey trial", later dramatized in the stage play and movie *Inherit the Wind*), the 1981 Arkansas "Balanced Treatment" trial, the 1987 US Supreme Court decision that it is unconstitutional for the State of Louisiana to require the teaching of creationism alongside evolution in public schools, the Kansas "Evolution Hearings" (2005–2007) in which demands were made for "equal treatment" of intelligent design theory alongside evolution, and the attempts by several school boards in Arkansas, Georgia, Louisiana and Oklahoma to require biology textbooks to carry stickers voicing a disclaimer about evolution (Borenstein, 2008).

An indication of the strength of support for creationist beliefs is evident in the findings of an NBC News opinion poll conducted at the time of the 1981 Arkansas "Balanced Treatment" trial: 76% favoured equal presentation of creationism and evolution in public schools, 18% favoured exclusive presentation of creationism (a figure that would have been much higher in 1925) and only 6% favoured evolution only (Nelkin, 1982). Similar surveys in 1986 and 1987 found that as many as 50% of science teachers supported inclusion of creationist doctrine in the classroom (Taylor, 1996). One former science teacher, an Australian by the name of Ken Ham, is now Director of the recently opened Creation Museum in Petersburg, Kentucky (www.AnswersInGenesis.org), an institution dedicated to promoting creationist beliefs. Also prominent among organizations promoting the creationist agenda are the Institute of Creation Research and the Discovery Institute. Numbers and Stenhouse (2000) describe the history of creationism in New Zealand and the events leading up to establishment of Creation Science (NZ) in 1992, which subsequently changed its name to Creation Science Foundation (NZ) to signal its close relationship to the Australian Creation Science Foundation. Taylor (1996) argues that creationism continues to be alive and well despite the response of

129

the scientific community, but perhaps also *because* of that response, which has often been ill-considered, arrogant, lacking in rhetorical expertise and politically naïve: "the demarcation of science accomplished in response to the emergence of creationism was, in many ways, a greater threat to science than creationism could ever hope to be" (p. 171). Taylor proceeds to identify a number of inappropriate or woefully inadequate responses by scientists. First, he says, "while the claim of creationism might well seem arcane, misguided, or just plain silly to many of us, it is unduly elitist to leave it at that" (p. 156). Neither denigrating those who hold creationist beliefs nor asserting the traditional authority of the scientific community (what Kitcher (1982) calls "credential mongering") is likely to be successful. It is simply not good enough to state that scientists "know best". Arguments that only scientists can or should make judgements about such matters are unacceptable, unless the criteria of judgement are made transparent and are justified by argument. Nor is it sufficient to state that evolution is a central plank of all science (not just biology).

> Evolution is one of the "great ideas" of science ... It is a tenet without which science as we know it could not continue to exist. (Gould, 1981b, p. 35)

> Nothing in biology makes sense except in the light of evolution? Without that light, it becomes a pile of sundry facts ... some of them interesting or curious but making no meaningful picture as a whole. (Dobzhansky, 1973, p. 128)

> Evolution is intertwined with other sciences, ranging from nuclear physics and astronomy to molecular biology to geology ... There is no basis for separating the procedures and practices of evolutionary biology from those that are fundamental to all sciences. (Kitcher, 1982, p. 4)

> The evolutionary hypothesis makes sense of the natural order ... In biosystematics and comparative zoology, the alternative to thinking in evolutionary terms is not to think at all. (Medawar, cited by Taylor, 1996, p. 158)

All this may be true, but does it make the case *for* evolutionary theory and *against* creationism? It smacks rather too much of "trust the experts ... we know what is best for you". As Taylor (1996) comments: "Initially, the vision of science constructed in the response to creationism was highly insular, holding that the conduct of scientific business, in any or all of its forms, including education, is an activity best organized and controlled by technical experts" (p. 172). In contrast, creationists made appeal to everyday observation, common sense and, of course, religion (in this case, Christian teachings). In a country (the United States) where some 70% or more identify themselves as Christian, many are likely to regard religion as better positioned than science to answer important questions, especially when their views of science may be incomplete or confused. Herein, of course, lies the case for enhanced levels of scientific literacy. Taylor (1996) summarizes the situation as follows:

By denigrating such an important and pervasive dimension of social life (implicitly or otherwise), the evolutionist response represented itself as unresponsive, even alien, to the needs (rational or otherwise) of the broader culture. Its denigration of alternative modes of rationality and its unreflective assumption of authority over traditionally social institutions only fan the flames of anti-intellectualism that spawn such reactionary movements as creationism. (p. 173)

Taylor may well be justified in his criticism of most scientists" response to the creationist lobbying for public support of their campaign for equal treatment alongside evolution in the school science curriculum, but my more immediate concern is whether students in school could do a better job. It is here that the NOS component of students' developing scientific literacy can test its mettle, especially since a number of creationists argue on scientific grounds, claiming that creationism is *better science* than evolutionary theory. For example, in his book *Evolution: The Fossils Say No!* Gish (1978) states: "For a theory to qualify as a scientific theory, it must be supported by events, processes, or properties which can be observed ... The general theory of evolution fails to meet these criteria" (p. 13). Scientists, Gish says, ask us to take their theory on trust, despite the lack of evidence. A particular target is the so-called gap in the fossil record: "It seems clear, then, that after 150 years of intense searching a large number of obvious transitional forms would have been discovered *if* the predictions of evolution theory are valid" (Gish, 1978, p. 49, emphasis in original). In contrast, says Gish, there is overwhelming evidence of design, and simple commonsense observation of animal species and of the fossil record lend much more credence to active creation than to development of diverse life forms by chance and natural selection: "What greater evidence for creation could we give than this sudden outburst of highly complex life forms ... The rocks cry out 'Creation!'" (Gish, 1985, p. 69).

Even quite young students should be able to identify inadequacies in the creationists' case. First, they might point out that the "simple commonsense observation" advocated by Gish does not exist. Teachers keen to cultivate NOS understanding will have ensured that students learn quite early in their science education that observation is inherently theory-laden and that making good scientific observations is a matter of using appropriate scientific knowledge. It is not the case that naïve and unsophisticated observation leads directly to secure knowledge of the world (see discussion in Chapter 4). Second, they might note a persistent misuse of the term *theory*. When creationists say, as they commonly do, that "evolution is *just* a theory", they imply that theories are little more than educated guesses or personal opinions, thus betraying a woeful lack of understanding of theory as complex explanatory structures comprising elaborate networks of inter-related concepts and conceptual relationships, underpinned by observational evidence and coherence with other accepted theories. This is precisely the point that needs to be made with respect to the significance of evolution in science: its centrality to and coherence with a range of well-established theories. Students will already be developing understanding that building, validating, criticizing and displacing theories is not a simple straightforward business, but one that demands particular procedures and upholds particular standards (see discussion in Chapters 6, 7 and 8). They will also have come to the realization that science is inherently tentative and uncertain, and that no amount of agreement (not even perfect agreement) between observation and theory, or between theory and prediction, would ever justify us in claiming that a theory is true. Of course, creationists would undoubtedly seek to exploit this open-mindedness as a weakness, arguing that scientists are trying to "paper over the cracks" of a theory that is unable to account for all the evidence.

A second response would be to look critically at the mass of data relating to estimates of the age of the Earth, the appearance of life forms and the emergence of *homo sapiens*. Creationists often respond by casting doubt on radiometric dating techniques, describing them as "mathematical tricks" or "arithmetic gymnastics". According to Ian Plimer's (1994) book, *Telling Lies for God*, creationists are also guilty of massive misrepresentation (not just misinterpretation) of scientific data. In recent years, proponents of intelligent design (ID) theory have accepted scientific evidence for the antiquity of the universe and the Earth (of the order of 13 billion years and 4.5 billion years, respectively) and acknowledged that life emerged on Earth about 4 billion years ago. Interestingly, ID theory does not claim that evolution never occurs or that natural selection does not play a role in the origin of some species, and some ID theorists may even accept the view that all organisms share a common ancestor (see Shanks, 2004). However, what ID theory does claim is that "intelligent causes are necessary to explain the complex, information-rich structures of biology" (Dembski, 1998a, p. 16). Dembski (1998a,b, 2003) also claims to have developed a mathematical means of detecting design, which he terms a "design filter", though it seems that design is not so much detected as imputed by the low probability of chance occurrence and Demski's assertion that there is "no acceptable alternative" to ID. Similarly, Behe (1996, 2003, 2004) argues that "designed structures" in the natural world can be recognized by their "irreducible complexity". Such irreducibly complex structures, unexplainable to Behe by appeals to chance or natural selection, include the mammalian eye and blood clotting systems. Put simply, anything that displays a high degree of complexity must be a product of intelligent design; so, too, any sudden changes in the structure of organisms revealed by the fossil record. Students should quickly recognize that ID theory fails to meet Karl Popper's criterion that theories should be *falsifiable*, at least in principle. ID theory can never be falsified because it attributes anything problematic to supernatural causes. Thus, it fails Popper's requirement for classification as science. Of course, creationists might argue that evolution is also unfalsifiable. Indeed, Karl Popper once described Darwinian evolution as a "metaphysical research programme" rather than a scientific theory, though he subsequently modified his position (Popper, 1988) by suggesting that its claims could, in principle, be falsified through critical scrutiny of "retrodictions" – the extent to which available evidence is consistent with the principles of evolutionary theory.[11]

Lakatosian perspectives may be more helpful than Popperian falsificationism, particularly the notion of science as a constellation of "research programmes": complex conceptual and methodological structures that guide the work of scientists by indicating potentially fruitful lines of development (the "positive heuristic") and specifying what cannot be done (the "negative heuristic"). According to Lakatos (1978b), the criterion for a satisfactory series of theories (a research programme) is that it should generate new facts and provide clear guidance for research. A research programme that is "progressive" has its hypotheses repeatedly corroborated, discovers unexpected novel facts, makes successful predictions, explains all the phenomena that competitor theories can explain, and explains phenomena and events that rival theories cannot explain. A research programme is said to be "degenerating" (and should be abandoned) when it loses its coherence and/or when it no longer leads to confirmed novel predictions. ID theory does not have any of the characteristics of a scientific research programme, as set out by Lakatos. There appears to be no research agenda and no stock of important questions awaiting an answer. Thus, students might conclude, as did Popper and Kuhn in regard to astrology, that creationism and intelligent design theory are nonscientific. Not because they fail to put their claims in a falsifiable form but because they have "no puzzles to solve and therefore no science to practice" (Kuhn, 1977, p. 276). The descriptions of pseudosciences advanced by Thagard (1980) and by Radner and Radner (1982), and noted earlier in this chapter, would be of especial benefit at this point. Peterson (2002) asks a whole raft of questions: "How frequently does design occur?" "Are human pathogens such as AIDS, cholera and malaria intelligently designed?" "Why does God seem to have a preference for insects?" Instead of addressing such questions, Behe (1998, p. 194) asks a different question: "What difference does intelligent design theory make to the way we practice science?" His answer is: "A scientist no longer has to go to enormous lengths to shoehorn complex, interactive systems into a naturalistic scenario". In other words, we do not have to struggle to answer difficult questions because ID theory allows us simply to by-pass them. Peterson (2002) observes that there is no attempt to build an explanatory structure for how and when design occurs and no attempt to explain *island biogeography*: the uniqueness of animal species and the distinctive patterns of diversity in isolated land masses such as the Galapagos islands and New Zealand – the very stimulus for Darwin's theorizing about natural selection.

> The challenge is the *patterns* of diversity that are found on islands. A history of the island in terms of evolution and natural selection is consistent with, and to a certain extent predictive of, the existence of large, flightless birds; high rates of endemic lizards and snails; and increased diversity on large islands. IDT, however, not only does not say anything about these patterns; it seems unable to. Any account that IDT could give would necessarily be ad hoc.[12] (Peterson, 2002, p. 18)

A scientific research programme stands or falls on its capacity to persuade other scientists of the validity of its claims. Persuasion resides in peer review of research findings and subsequent publication in academic or professional journals. However, the writing of proponents of ID theory, like those of old-style creationists, is directed largely at a non-scientific readership, many of whom may lack sufficient biological knowledge and understanding of NOS to engage in critical appraisal. Some unscrupulous creationists seek to exploit this knowledge deficit by creating confusion in the minds of readers, which they subsequently claim to resolve by the postulates of creationism. At the very least, such writing violates the Mertonian norm of *disinterestedness*: the requirement that scientists remain unprejudiced in their interpretation of data and argument. The bias or prejudice at issue here is, of course, the Christian fundamentalist orientation of many creationists. In arguing that such a dogmatic position is irreconcilable with science, Stephen Jay Gould (1981b) comments: "I cannot imagine what potential data could lead creationists to abandon their views. Unbeatable religious belief systems are dogma, *not science*" (p. 35). Donald Price (1984) concludes: "Creationism is simply one more branch of evangelistic apologetics sharing the same goal of preparing the ground for faith and conversion" (p. 167). I will leave the last word to Charles Taylor (1996):

> True scientific beliefs, on the other hand, are constructed as always open to revision in the light of new or more convincing empirical evidence ... [Creationists] seek to revise interpretations of particular *data* rather than their fundamental postulates, which are grounded in literalist biblical interpretation (p. 19, emphasis added)

In addition to emphasizing key features of NOS understanding, an exercise involving pseudosciences, traditional knowledge and creationism would help to clarify the *limits* of science and the domain of interest of science.

SCIENCE AND RELIGION

There are many human concerns, interests, needs and problems that fall outside the province of science. Part of critical scientific literacy includes knowing what science can and cannot do; what it aims to do and does *not* aim to do, and when to apply scientific ways of thinking and when *not* to do so. It includes the capacity to distinguish the ways in which scientific knowledge and scientific inquiry differ from religious, philosophical, aesthetic and psychological knowledge and inquiry.

> The scientific study of a work of art, say a picture, may give an exhaustive account of the chemical constitution of the pigments, the wavelengths of light they reflect, their reflection factors, masses and physical distributions. But such a scientific account has hardly begun to say much of interest to the viewer or to the artist. Aesthetic considerations, issues of meaning and matters of purpose are of far greater importance. A sociological study of the influences on artists' work will have similar limitations. It is not that pictures cannot be described in terms of chemicals, or mental activities in terms of brain functions – they can. What is wrong is to assert ... that these scientific accounts are the only valid ones there are. (Poole, 1996, p. 165)

One alternative and potentially conflicting system of thought is religion. The key point is that religion addresses questions that are outside the domain of interest of science. Science is the study of the natural world; religion asks whether there is anything *other* than the natural world. Thus, it is pointless to inquire of science about the supernatural; within the realm of science, such questions have no meaning. Religious knowledge and scientific knowledge are based on entirely different standards of validity, and are expressed in entirely different ways. Further, while science is required (at least by Mertonian norms) to be disinterested and dispassionate, religion concentrates on the spiritual, ethical and aesthetic aspects of the human condition. Carson (1998) uses this distinction/demarcation to argue against including creationism in the curriculum: "To include creationism in the curriculum is to communicate to students that it has met the standards of scientific validity, which it has not. The science classroom is not the place for airing theological viewpoints, any more than the pulpit is the place for delivering lessons on quantum mechanics or nuclear physics" (p. 1011). My own views on teaching *about* creationism should now be abundantly clear. Interestingly, Gauld (2005) expresses the view that scientific and religious "habits of mind" (as he calls them) are much closer than many people think.

> In both cases openness to argument and evidence, skepticism, rationality and objectivity are all held in high regard; in both, some ideas are more protected from attack while others are more open to challenge; and in both, at any time, there are various degrees of commitment to theories from skeptical rejection to passionate endorsement. Both habits of mind stem from the same scholarly attitude and any difference between them is probably due to differences in what are counted as appropriated evidence and good reasons. (p. 302)

Of course, there are some issues in which science and religion may both have an interest. In these circumstances it is pertinent to ask whether science and religion can co-exist. Barbour (2000) sees four possibilities:

- *Conflict*: science and religion are diametrically opposed and one "cannot be at home in both places".
- *Independence*: science and religion are different endeavours with different languages, concerns and epistemologies. There is no conflict because they serve different functions and purposes.
- *Dialogue*: critical dialogue between science and religion can benefit both, noting similarities and clarifying differences.
- *Integration*: variants on this theme include (i) seeking evidence of the existence of God in nature, (ii) reformulating religious beliefs in the light of science, and (iii) seeking a common conceptual framework.

Conflict between science and religion has been a recurring theme in the history of science, with Galileo's struggles with the Catholic Church being the most sensational (Brooke, 1991; Russell, 2002). Students may find much of interest, too, in the response of established religion to the work of Charles Darwin, much of it accessible through newspaper archives.[13] Peterson (2002) comments that in their campaign to present their ideas as science, the current crop of ID advocates have "obscured much of interest and importance in thinking about the relationship between God and the world. By radically polarizing and politicizing the science-and-religion dialogue, advocates of IDT stand to reverse two decades' worth of constructive dialogue and to reinvigorate the fractious ghosts of religion-science conflict" (p. 8). In startling contrast, as Margaret Wertheim (1995) points out, several prominent and influential scientists present their work as a "quasi-religious" quest for truth – most notably, of course, Albert Einstein: "Science can only be created by those who are thoroughly imbued with the aspiration towards truth and understanding. The source of this feeling . . . springs from the sphere of religion" (Einstein, 1940, p. 605). Einstein continues, "science without religion is lame, religion without science is blind" (p. 605). Elsewhere, in response to experimental data that initially seemed to threaten his theory of general relativity, Einstein commented: "The Lord is subtle, but he is not malicious" (in German, "Raffiniert ist der Herrgott aber boshaft ist er nicht"). Pais (1982) notes that in reply to a query about what he meant, Einstein replied: "Nature hides her secret because of her essential loftiness, but not by means of ruse" (p. vi). In summing up his objections to quantum mechanics, he said: "God casts the die, not the dice" (cited by Wertheim, 1995, p. 185) and "It seems hard to look in God's cards. But I cannot for a moment believe that he plays dice and makes use of 'telepathic' means (as the current quantum theory alleges He does)" (cited by Pais, 1982, p. 440).[14]

This quasi-religious theme has continued to permeate much of the writing concerning the quest for a unified field theory, popularly known as a "theory of everything". For example, in his introduction to Stephen Hawking's international bestseller, *A Brief History of Time*, Carl Sagan says: "The word God fills these pages. Hawking embarks on a quest to answer Einstein's famous question about whether God had any choice in creating the universe. Hawking is attempting, as he explicitly states, to understand the mind of God" (Sagan in Hawking, 1988, p. x). After detecting slight but persistent fluctuations or ripples in the cosmic microwave background radiation, and interpreting them as evidence of "the Big Bang", astronomer George Smoot is reputed to have said: "It was like seeing the face of God" (cited by Lederman, 1993, p. 400). In his 1993 book (written with Dick Teresi), Leon Lederman refers to the elusive Higgs boson, a particle crucial to a full understanding of the structure of matter, as "the God particle". In discussing the possibility of finally detecting the Higgs boson by means of a superconducting super collider, Lederman makes reference to what he calls "The Very New Testament": "And the Lord came down to see the accelerator, which the children of men builded. And the Lord said, behold the people are unconfounding my confounding. And the Lord sighed and said, Go to, let us go down, and there give them the God particle so that they may see how beautiful is the universe I have made (The Very New Testament, 11;1)" (Lederman, 1993, p. 24). Wertheim's (1995) comments on the book are worth quoting at length for her perspective on some authors' motives for using quasi-religious imagery in popular writing on physics.

Throughout his book Lederman refers liberally to God; he even calls one chapter on his own research "How We Violated Parity in a Weekend ... and Discovered God". The unmistakable implication is that particle physics is a direct path to a Deity.

Bolstering his religious subtext, Lederman quotes with approval fellow particle physicist Robert Wilson's association of accelerators with cathedrals. "Both cathedrals and accelerators," he writes, "are built at great expense as a matter of faith. Both provide spiritual uplift, transcendence, and, prayerfully, revelation." Lederman's invocation of accelerators as places of worship, his invention of a pseudo-religious document, and his naming of his own book are all, no doubt, attempts to make what is otherwise a rather dense and technical work more popularly appealing. Whether or not he truly believes the Hoggs boson is God's particle is impossible to tell. (p. 220)

Many scientists, philosophers and educators have written at length on the extent to which religion is or is not compatible with scientific thought. Relevant recent works on the topic, mostly in relation to Christianity and Judaism, include Davies (1984, 1992), Dawkins (1986, 1995, 2006), Barbour (1990, 1997, 2000), Brooke (1991), Gingerich (1994, 2006), Polkinghorne (1994, 1996, 1998, 2000, 2005), Tipler (1995), Haught (1995), Townes (1995), Mahner and Bunge (1996), Roth and Alexander (1997), Poole (1998, 2007), Watts (1998), Gould (1999), Ziman (2000), Peacocke (2001), Williams (2001), Bausor and Poole (2002), Russell (2002), Barr (2003), Koul (2003b), Colburn and Henriques (2006), Roughgarden (2006), Wilson (2006), Brigida and Falcão (2008) and Reiss (2008). A special issue of *Science and Education* (1996, volume 5, issue 2) addresses the question from the perspective of science education – see especially, Lacey (1996), Poole (1996) and Woolnough (1996). As Reiss (2007, 2008) reminds us, whether it is considered appropriate to address issues of religion in a science class (or in any other class, for that matter) will vary from country to country, from time to time, from school to school, and even from teacher to teacher. My own view, and my rationale for including this very brief discussion here, is that considerations of the nature of religious thought are enormously helpful in understanding the nature of science. The value of the activity depends, of course, on the willingness and capacity of teachers to present a reasoned, well-supported and critical perspective on all the issues arising, as Ruse (2005) urges with respect to the creation-evolution debate.

Given the threat that creationists pose to evolutionists of all kinds, it behooves evolutionists especially to start thinking about working together with Christian evolutionists, rather than apart. For a start, atheists like Dawkins and Coyne might consider taking a serious look at contemporary Christian theology (or the theology of other faiths, for that matter), rather than simply parroting the simplistic, schoolboy travesties of religion on which their critiques are founded. Conversely, Christians like Ward and Rolston might be encouraged to dig more deeply into modern, professional evolutionary biology and to start to get some understanding of its strengths and triumphs before they cast around for alternatives like self-organization. (p. 274)

Chapter 2 includes reference to a number of articles, written largely from an American perspective, that discuss pedagogical approaches suitable for teaching evolution as a controversial topic to students with strong religious beliefs.

Both Sardar (1988, 1989) and Loo (1996, 2001) have discussed the extent to which a distinctive Islamic science can be discerned and Yousif (2001) has examined the possibilities of incorporating it into the Brunei science curriculum as part of a programme of Islamization. Anees (1995) looks more generally at the ways in which science education in Islamic countries is impacted by a pervasive socio-religious context. In similar vein, Nandy (1988) and Nanda (1997) have investigated the potential of Hindu science as a key component in the Hinduization of education in India. In his appeal to Indian scientists to return to India's traditional science, Nandy (1988) says: "Instead of using an edited version of modern science for Indian purposes, India can use an edited version of its traditional sciences for contemporary purposes" (p. 11). In a spirited and colourful response, Loo (1999) accuses him and the contributors to his edited collection of essays, *Science, Hegemony and Violence*, of using flawed reasoning and misrepresentation to engage in "science bashing" – bashing of Western science, that is. While a review of this literature falls outside the scope of this book, the suggestion that such considerations might be included in the school science curriculum does not. Relevant here, too, are the many cross-cultural studies that look at teaching and learning science to students who have knowledge, beliefs, attitudes and values that conflict in some ways with the scientific worldview (see earlier discussion in this chapter, together with Loving & Foster, 2000; Shipman et al., 2002; Hansson & Redfors, 2008) and studies that examine the influence of teachers' religious views, and those of the surrounding community, on their teaching of science (Fysh & Lucas, 1998; Loving & Foster, 2000; Donnelly & Boone, 2007; Francis & Greer, 2001; Chuang, 2003; Stolberg, 2007; Mansour, 2008). As an interesting aside, relevant to discussion earlier in this chapter, it seems that people with strongly held religious beliefs are less likely than others to believe in the paranormal or to hold superstitious beliefs (Robinson, 1990). Another interesting aside is Smith and Scharmann's (2008) finding that a number of student teachers with strongly conservative Christian beliefs, who were initially opposed to the teaching of evolution in schools because, in their view, "it is not a legitimate scientific theory", concluded that evolutionary theory is *more scientific* than the ideas of intelligent design after engaging in a lengthy discussion-based activity focused on the rival scientific merits of paired enterprises such as astrology and astronomy, reflexology and acupuncture, philosophy and palmistry, intelligent design and reincarnation, and neurology and acupuncture.

MYTHS AND FALSEHOODS IN SCIENCE AND SCIENCE EDUCATION

Anthropologist Bronislaw Malinowski (1971) has written at length about the central role of socially constructed myths in legitimating and reinforcing particular ways of knowing, feeling, valuing and acting in all societies.

> Myth expresses, enhances and codifies belief; it safeguards and enforces morality; it vouchsafes for the efficiency of ritual and contains practical rules for the guidance of man. Myth is thus a vital ingredient of human civilization. (p. 19)

It would be surprising, then, if the community of scientists had not constructed some powerful cultural myths to enhance and codify its beliefs and practices. And it would be surprising, too, if the community of science education (teachers, teacher educators, researchers, curriculum developers and educational policy makers) did not have a related set of science curriculum myths to guide curriculum contruction and a set of pedagogical myths to guide its implementation. In his seminal work, *The Structure of Scientific Revolutions*, Thomas Kuhn (1970) provides some clues for how these myths come about. He argues that science does not proceed in an orderly, systematic and continuous way, but through a series of revolutions and periods of consolidation (what Kuhn calls *normal science*). In the beginning, in the early history of a science,[15] there is no agreement on how best to proceed. Each individual is busy defining and re-defining the field, establishing priorities, distinguishing significant from trivial problems, making statements about correct procedures, deciding what should count as evidence, what is an acceptable or unacceptable solution, and so on. The disorganized and diverse activities that characterize *pre-science*, as Kuhn calls it, become structured and directed when the community of practitioners reaches agreement on certain theoretical and methodological issues, that is, when the disciplinary matrix (the framework of theory and method, or *paradigm* as Kuhn terms it) becomes established and accepted. Once a particular paradigm has been established, scientists engage in *normal science*: they work within the paradigm, assuming that its basic premisses are valid; they follow its rules and procedures; they use its concepts and theories to explore, develop and extend the paradigm, widen its scope, solve its internal puzzles and problems (both theoretical and procedural), formulate quantitative relationships among concepts, and so on. Normal science "aims to elucidate the scientific tradition in which (the scientist) was raised rather than to change it ... The puzzles on which he concentrates are just those which he believes can be both stated and solved within the existing scientific tradition" (Kuhn, 1977, p. 234). In other words, the existing paradigm sets the agenda for thinking about the world. In Kuhn's (1970) words, "Work under the paradigm can be conducted in no other way, and to desert the paradigm is to cease practicing the science it defines" (p. 34). Because an accepted paradigm is highly resistant to falsification it will persist until a succession of unsolvable problems that strike at the very core of the paradigm precipitates a crisis that is resolved by the appearance of a new theory. Until that revolution, the basic premisses of the paradigm will be taken-for-granted. Here is a clue to how and why scientific myths become established. First, particular theoretical ideas become a key part in the education of newcomers. In the process of textbook production and lesson design, ideas become simplified, justifications become truncated, and what we know as contestable but evidentially well-supported ideas become presented as facts or truths. As "science-in-the-making" is transformed into established scientific knowledge ("ready made science"), the language of presentation changes, moving from forms such as "Jones has suggested that X is one of the factors inhibiting the proper function of the liver" to "the presence of X acts to inhibit liver function". Second, science itself is portrayed as a set of factual truths about the world, established through a powerful and all-purpose method of inquiry, and taken-for-granted by all scientists. After a scientific revolution there is much work to be done. All existing data not rendered irrelevant by the new paradigm has to be re-interpreted within the new theoretical framework; scientific instruments and research procedures have to be reviewed and appropriately modified. And, of course, new textbooks have to be written to ensure that the next generation of scientists is "properly" educated – that is, in the new way of thinking and acting. Sometimes this re-writing of science is so extensive and persuasive that all trace of preceding theoretical ideas are swept away, a phenomenon that Kuhn (1970) calls "the invisibility of revolutions".

> [The textbooks] have to be rewritten in whole or in part whenever the language, problem-structure, or standards of normal science change. In short, they have to be rewritten in the aftermath of each scientific revolution, and, once rewritten, they inevitably disguise not only the role but the very existence of the revolutions that produced them. (p. 137)

This is the mechanism of myth making: data become explanations, explanations become facts, tentative judgements become definitive conclusions, and science is presented as truth. Unfortunately, the careful and clear presentation of scientific concepts and ideas demanded by journal editors and textbook publishers can have the side-effect of making them seem both simple and certain. Although the notion of science as absolute would be rejected on philosophical grounds by most scientists and science teachers, this particular myth still plays a central role in scientific practice and science teaching. We act as though science is truth. As Broad and Wade (1982) comment, "The myth of science as a factual/logical process, constantly reaffirmed in every article, textbook, and lecture, has an overwhelming influence on scientists' perceptions of what they do" (p. 126). About a quarter century ago, as part of a major survey of Canadian science education conducted by the Science Council of Canada, Nadeau and Désautels (1984) identified what they called five "mythical values stances" suffusing science education:

- *Naïve realism* – science gives access to absolute truth about the universe.
- *Blissful empiricism* – science is the meticulous, orderly and exhaustive gathering of reliable data.
- *Credulous experimentation* – experiments can conclusively verify hypotheses.

- *Excessive rationalism* – science proceeds solely by logic and rational appraisal.

- *Blind idealism* – scientists are completely disinterested, objective beings.

The cumulative message is that science has an all-purpose, straightforward and reliable method of ascertaining the truth about the universe, with the certainty of scientific knowledge being located in objective observation, extensive data collection and experimental verification. Moreover, scientists are rational, logical, open-minded and intellectually honest people who are required by their personal commitment to the scientific enterprise to adopt a disinterested, value-free and analytical stance. In Cawthron and Rowell's (1978) words, the scientist is regarded by many of those responsible for the school science curriculum as "a depersonalized and idealized seeker after truth, painstakingly pushing back the curtains which obscure objective reality, and abstracting order from the flux, an order which is directly revealable to him through a distinctive scientific method" (p. 32). More recently, Hodson (1998b) has identified ten common myths and falsehoods promoted, sometimes explicitly and sometimes implicitly, by the science curriculum. They are reproduced below (Table 5.1), alongside a broadly similar list of falsehoods generated by McComas (1998) from his critical reading of science textbooks. Further detailed criticisms of the distorted, incomplete and misleading views about science and scientific practice commonly found in school science curricula, textbooks and non-fiction "trade books" can be found in Pappas (2006), Yip (2006), Abd-El-Khalick et al. (2008), van Eijck and Roth (2008), Kosso (2009) and Schroeder et al. (2009).

As Milne and Taylor (1998) point out, the all-too-ready acceptance of naïve realism by teachers leads to the adoption of authoritarian, didactic methods, with undue emphasis on "correct definitions" dispensed by the teacher or located in the textbook. Its adverse impact on the way teachers design practical work activities can be summarized in terms of four "pedagogical myths": (a) all students observing a particular phenomenon or event will make identical observations, if they are sufficiently careful; (b) because observations "speak for themselves" there is no need for discussion of the results (and probably not for the purpose and procedure, either); (c) theoretical understanding will arise naturally and inevitably from careful scrutiny of observational data; and (d) the lab report is an objective record of the activity and its findings. The overriding conclusion is that collecting data and conducting experiments are both quick and unproblematic. Furthermore, the success of school science experiments in generating the "correct data" and/or reaching the "right conclusion" reinforces myths about the certainty of scientific knowledge (an issue to be explored further in Chapters 6 and 7). As I have argued a number of times in this book, and in Hodson (2008), progress towards critical scientific literacy for all students is dependent, in the first instance, on teachers themselves (and the curriculum materials they deploy in class) adopting a more realistic and authentic view of science, scientists and scientific practice. Although not all students will internalize all the misunderstandings of science embedded in less enlightened science curricula, there is ample evidence, as discussed in Chap-

TABLE 5.1

Myths and falsehoods about science.

Hodson (1998)	McComas (1998)
Observation provides direct and reliable access to secure knowledge.	Hypotheses become theories that in turn become laws.
Science always starts with observation.	Scientific laws and other such ideas are absolute.
Science always proceeds via induction.	A hypothesis is an educated guess.
Science comprises discrete, generic processes.	A general and universal scientific method exists.
Experiments are decisive.	Evidence accumulated carefully will result in sure knowledge.
Scientific inquiry is a simple algorithmic procedure.	Science and its methods provide absolute proof.
Science is a value-free activity.	Science is procedural more than creative.
Science is an exclusively Western, post-Renaissance activity.	Science and its methods can answer all questions.
The so-called "scientific attitudes" are essential to the effective practice of science.	Scientists are particularly objective.
All scientists possess these attitudes.	Experiments are the principal route to scientific knowledge.
	Scientific conclusions are reviewed for accuracy.
	Acceptance of new scientific knowledge is straightforward.
	Science models represent reality.
	Science and technology are identical.
	Science is a solitary pursuit.

ter 2, that many students do leave school with a confused, confusing, deficient or distorted view of the nature of science and the activities of practising scientists.

Of course, we should not replace myths about the certainty and absolute truth of scientific knowledge with the equally damaging view that all scientific knowledge is tentative and uncertain. Much of the scientific knowledge that students learn in school is sufficiently well-established for scientists to take it for granted as the starting point for further exploration and theorizing, a matter that is explored a little more fully in Chapter 6. We should also ask why false or confused NOS knowledge constitutes a major problem for science education. In short, why does it matter what image of science is presented and assimilated? It matters insofar as it influences career choice, and so may have long-term consequences for individuals. It matters if the curriculum image of science is such that it dissuades creative, non- conformist and politically conscious individuals from choosing to pursue science at an advanced level. It matters if the image of science is such that it dissuades women, members of visible minority groups and students from lower socioeconomic status homes from entering science-related careers or seeking access to higher education in science and engineering because they do not see themselves included and represented in the science curriculum. It matters if our politicians, public servants and industrialists are so ignorant of scientific and technological issues that their decision-making is ill-informed and uncritical. It matters if the general population is unable to respond knowledgeably and critically to the claims and proposals of those in society who might use scientific arguments (and sometimes pseudoscientific or scientifically spurious arguments) to persuade, manipulate and control. It matters if a significant part of humankind's cultural achievement is poorly understood. Failing to provide every student with an adequate understanding of the nature of science runs counter to the demand for an educated citizenry capable of responsible and active participation in a democratic society. As I argued in Chapter 1, a proper understanding of science and the scientific enterprise is just as essential as scientific knowledge (i.e., conceptual understanding) in ensuring and maintaining a socially-just democratic society.

> [Scientific literacy] should help students to develop the understandings and habits of mind they need to become compassionate human beings able to think for themselves and to face life head on. It should equip them also to participate thoughtfully with fellow citizens in building and protecting a society that is open, decent, and vital. (AAAS, 1989, p. xiii)

There are encouraging signs that some teachers, in some schools, are able to ensure that students develop a more authentic view of scientific practice, sometimes as a consequence of an explicit programme of study in the history, philosophy and sociology of science, sometimes as a consequence of practically oriented experiences in laboratories, in the field or in zoos and museums. The remaining chapters of this book seek to explore a variety of approaches to consolidating students' NOS knowledge and assisting them in deploying it effectively in real world situations.

SCIENCE, TECHNOLOGY AND PRACTICAL KNOWLEDGE FOR ACTION

An intriguing study by Layton et al. (1993) addresses the notion of "local sciences" – that is, the kind of knowledge accumulated and deployed by non-scientists to address specific needs, interests and issues, and to solve problems in everyday life. The study focused on four groups: (i) parents of children with Down's Syndrome; (ii) elderly people coping with domestic energy problems (and seeking to reduce their power bills); (iii) local government officials responsible for waste disposal; and (iv) people working at, or living close to, the Sellafield nuclear processing facility in West Cumbria (UK), who might be considered at risk from potential radiation leaks. What is striking about the findings is the way in which concerned citizens build "practical knowledge for action". Applicable only in the particular situation, this cluster of knowledge and skills often constitutes understanding that is very different from the scientific knowledge normally presented in school, especially in terms of its sophistication. This was especially evident in the case of the Downs Syndrome parents and the West Cumbria residents. Epstein (1995, 1997) reports a similar situation in which AIDS activists, with virtually no formal education in science, acquired sufficient expertise to become effective

143

and respected participants in the design, conduct, interpretation and reporting of clinical trials of a range of AIDS-therapy drugs. At the school level, Alsop (2009) reports on the ways in which a group of students at the lower secondary school level acquired complex and socially valuable knowledge about severe respiratory problems and sleep disorders, triggered by the illness of a close relative of one of the students.

Practical knowledge for action may include fragments of scientific knowledge adapted to address specific purposes and problems more directly, together with *alternative* scientific knowledge and highly idiosyncratic judgements deriving from personal experience. For example, in the study by Layton et al. (1993), the knowledge commonly deployed in addressing domestic energy problems had more in common with caloric theory than kinetic theory, and many of the elderly people who were interviewed perceived *cold* as an entity with distinct properties of its own, rather than recognizing it as the absence of heat. In other words, they developed an explanatory system rooted very firmly in personal experience of living in a draughty home. As Jenkins (2000) observes, the scientific knowledge learned in school is often irrelevant or no more than marginal to decisions about practical action: "Addressing satisfactorily the coldness of a room by closing a door, double glazing one or more windows, or insulating the walls and ceiling does not require an understanding of cold as the absence of heat conceptualized in terms of molecular motion" (p. 210). Scientific knowledge delivered by experts in the form of leaflets and advice from medics was often seen as unhelpful. In the case of the Downs Syndrome parents, this knowledge was too often "a message of despair when they were desperate for one of hope. Knowledge was offered in the wrong form, reflecting priorities different from those of practical action: in the wrong way, discounting understandings which parents had wrought from experience; and, often, at the wrong time, serving the convenience of donors, ignoring emotional traumas which parents might be undergoing, and undiscerning of the moment of need" (Layton et al., 1993, pp. 57–58). Thomas (1997) cites a study reported by Irwin et al. (1996) in which residents in housing complexes located close to potentially hazardous industrial sites responded negatively to expert scientific knowledge. Because it was couched in inaccessible language and seemed unable to answer their most pressing questions, it simply promoted dissatisfaction, elevated anxiety levels and exacerbated feelings of powerlessness.

In evaluating the usefulness of scientific knowledge, non-standard criteria are often employed. For example, open fires were often preferred to more efficient forms of heating because they are "cosy" and draught excluders were rejected on grounds of social acceptability ("they are naff!"). It is crucial, therefore, that those who seek to raise levels of scientific literacy in the wider community, and to assist people in making use of scientific knowledge in their daily lives, are cognizant of these other, largely social agendas. This applies at the community level as well as the individual level. Scientific knowledge can be rejected or re-garded with deep suspicion by social groups because it is not tailored to their specific needs, interests and social circumstances, or because it fails to take ac-count of other agendas. It may also be rejected because the "experts" who deliver scientific knowledge are sometimes regarded as not entirely trustworthy because of a perceived bias. For example, in Wynne's (1996) study of Cumbrian sheep farmers in the aftermath of the Chernobyl disaster, most of the farmers maintained their belief that radioactive contamination was attributable to the local Sellafield nuclear processing plant, and not to Chernobyl, despite assurances from scientists that they had unambiguous "fingerprints" that proved it was from Chernobyl. Also, the farmers understood the scientists' statement that radioactive caesium is flushed from lambs more quickly when they graze on valley grass rather than on the hills but, as Wynne (1991) comments, they also knew that "valley grass is a precious and fragile resource whose loss by intensive grazing can have damaging consequences for future breeding cycles" (p. 114), and so they chose to ignore the scientists' advice. Wynne (1995) refers to episodes like this as the *social construction of ignorance*: the deliberate avoidance of scientific knowledge because it is perceived as contrary to one's interests or too much in the other party's interests (as in, "they are just trying to sell us something"). Workers at the Sellafield nuclear processing plant told researchers that they avoided scientific knowledge that could potentially have helped them to assess health risks more effectively because: (i) trying to resolve the various controversies would be too time consuming; (ii) being too conscious of risks would raise anxiety levels; (iii) they did not want to signal mistrust of the staff whose job it is to assess risks and institute safety procedures. Wynne notes that the workers were not so naïve that they simply took existing arrangements on trust, but instead of devoting time and attention to acquiring specific scientific knowledge they concentrated on a critical assessment of the extent to which they could trust, or should distrust, the plant management and the particular individuals charged with responsibility for safety. In a study conducted by Lambert and Rose (1996), most of the patients diagnosed with *familial hypercholesterolaemia* (a genetically transmitted inability to metabolize lipids that greatly increases an individual's susceptibility to cardiac arrest) constructed personal knowledge for action that sought to balance scientific knowledge concerning above-average risk of premature death with consideration of the implications of a more restricted existence, opting for a compromise between risk-reducing action (particularly, strict dietary control) and maintaining an enjoyable social life. Many of the Cumbrian sheep farmers interviewed by Wynne declined invitations to undergo whole-body radioactivity scans on the grounds that they would "do nothing but worry" if high levels were revealed. However, they commented that their requests for water analysis were ignored, although water supplies could have been changed if analysis had shown a need to do so. As Wynne (1991) comments, "from this group's perspective, useless knowledge was offered, while useful knowledge was denied" (p. 117).

What these studies show is that people faced with making important decisions in everyday life may not always use "pure" scientific knowledge. They may use restricted or adapted scientific meanings; they may incorporate knowledge from areas outside science; they may rely heavily on hunch, intuition, personal experience and testimony from other non-scientists. This complex of knowledge is assembled into a highly personal and context-specific means of thinking about issues, solving problems and reaching decisions. I believe that there is enormous value in studying some carefully selected examples of *practical knowledge for action* as a way of sharpening NOS understanding, addressing demarcation criteria and appreciating both the strengths and limitations of scientific knowledge in addressing real world problems. I also wish to reiterate the point that enhanced NOS understanding is a crucial element in selecting appropriate scientific knowledge, appraising it in the context of use, seeking additional knowledge resources, interrogating expert opinion and reaching informed decisions. In other words, sophisticated NOS knowledge is essential to proper consideration of science and technology in the real world, and addressing real world issues enhances NOS understanding. These matters will be explored further with respect to the distinction between theories and models in Chapter 6. A subsequent book will address their significance in the confrontation of socioscientific issues, politicization of students and preparation for sociopolitical action. What the studies reviewed in this section make clear is that practical knowledge for action is closer to technological knowledge than to scientific knowledge, both in terms of concepts and criteria for judging acceptability. Consideration of such differences and similarities among science, technology and practical knowledge for action is an important element of learning *about* science. Jenkins (2000) sums up the situation as follows:

> Science, as encountered by most people in their everyday lives, is rarely objective, coherent, well-bounded and unproblematic ... [a situation] firmly at odds with the view of science commonly presented at school, where the discipline is well-bounded, answers are secure, and uncertainty, doubt or debate are not admitted, save under severely limited conditions. (p. 209)

Sometimes scientific knowledge needs to be reclassified or re-coded to make it more useful in practical situations. To be useful in practical contexts, abstract, idealized science has to be adapted and modified to take account of the complexity and non-uniformity of the real world. As Layton (1991) points out, chemists classify substances in terms of molecular structure and functional groups, while pharmacologists do so in terms of human responses to chemicals – that is, as stimulants, depressants, decongestants and analgesics. While physicists are insistent on the principle of conservation of energy, heating engineers are concerned about heat *loss* and finding ways to conserve energy. The scientific knowledge needed for practical purposes is different in many respects from what we might call "academic science". While school science courses deal with *frequency*, *wavelength* and *amplitude*, sound engineers and people living close to stadiums used for rock concerts or speedway racing are more concerned with noise level, noise pollution and noise damage (to buildings and people). The biological classification of diseases in terms of causal agents such as viruses, bacteria and protozoa is much less useful to ordinary citizens than a classification in terms of environmental transmission characteristics.

> Information that is generated within one system exists in a particular coded form, recognizable by and useful to participants in that system. If it is to be transferred from one system to another – say from science to technology, or

from technology to the economy, or in the reverse direction – it has to be translated into a different code, converted into a form that makes sense in a world of different values. (Aitken, 1976, p. 18)

Interestingly, Mulkay (2005) argues that these key differences between science and technology cast doubt on the commonly deployed arguments that success in technology, engineering and medicine provide convincing evidence of the objective reality of scientific knowledge. If, as argued here, scientific knowledge for technological purposes is constituted differently, in accordance with interests, purposes and criteria of adequacy, we have to conclude that it is not so much that practical utility furnishes an objective criterion of the universal validity of scientific propositions, but that judgements of cognitive adequacy vary with social context.

Although school textbooks often describe technology as the *direct* application of scientific knowledge, with scientific knowledge being seen as an essential precursor to technological innovation, Derek de Solla Price (1965, 1969) observes that much of science has no technological application and much of technology has no input from science. Rather, he argues, science seems to grow on the basis of past science, technology seems to grow on the basis of past technology. In reality, the precise relationship between the theoretical knowledge of science (knowing *that*) and the practical knowledge of technology (knowing *how*) is neither simple nor straightforward, and varies from case to case (see discussion in Chapter 1). The history of science and technology reveals four possibilities: science precedes technological application (the common perception of technology as applied science); technology precedes the science that eventually explains it (as in most traditional and long-standing technologies such as cheese, wine and glass making, selective breeding of plants and animals, and most traditional medicine and healing practices); scientific and technological development seem to run on entirely independent lines; scientific and technological development are mutually dependent and interactive (each informs and stimulates developments in the other). The history of radio is a fascinating story of how early scientific work by Faraday, Maxwell and Hertz made radio transmission a feasible goal, but subsequent work by Marconi, Lodge, Muirhead, Fleming and DeForest was not science-driven.

> While science played an essential role in making radiotelegraphy possible, it contributed little to the development of the technology thereafter ... Even Fleming's diode valve, the strategic invention that ushered in the second phase in the history of telecommunications, required no new scientific knowledge for its discovery. The so-called "Edison effect" in electric light bulbs had been observed many years before ... The same generalization can be made with reference to DeForest's triode vacuum tube, a device of major technological importance whose principles of operation the inventor himself did not understand and which certainly called for no new inputs of information from science. (Aitken, 1976, p. 326)

Sometimes the boundaries between science and technology are clear; sometimes they are unclear, fuzzy and ambiguous; sometimes they are permeable, with knowledge and people crossing and re-crossing the borders. Layton (1988) makes the point that even when technology is "just applied science", there is still lots more work to be done. He gives the example of W.H. Perkin's development of Mauve, and says that its translation into a commercial product required "knowledge, skills and personal qualities very different from those needed for the test tube oxidation of aniline" (p. 371). There were several early problems: Mauve would not take evenly on large batches of cloth, there was no suitable mordant for cotton, raw materials were not readily available, handling concentrated nitric and sulphuric acids on a "factory scale" presented all manner of engineering problems, there were problems of marketing associated with consumer reluctance, and so on. Similarly, extensive work by bacteriologists, crystallographers, clinicians and engineers was needed in order to "convert" Fleming's observation of the bactericidal properties of pencillium moulds into a commercially viable drug. Notions such as optimization, feedback modelling, systems analysis, critical path planning and risk assessment have to be included whenever science is applied to real world situations. Real-world problems rarely exhibit the simple cause and effect relationship portrayed in traditional science curricula; rather, attempts at solution often reveal layers of increasing complexity and uncertainty that cannot be contained within a particular disciplinary framework. Issues and problems become inextricably linked with considerations in economics, politics, aesthetics and moral philosophy. Criteria of acceptability in science include empirical adequacy, internal consistency and consistency with other accepted theories. It helps also if an explanation is elegant and parsimonious, though those criteria may not be considered essential. In technology, a solution has to work, of course, but it also has to be efficient, cost-effective, reliable and durable, easy and safe to use, possibly aesthetically pleasing, and so on. There may also be critical considerations relating to government regulations, legal implications, moral-ethical issues, cultural appropriateness, social and environmental impact, and client deadlines. Because technological solutions are often incomplete, tentative and provisional they may carry a measure of risk and hazard. They may have adverse side effects, some of which may not become apparent for some time. Thus, technologies are rarely "good" in an absolute sense. Rather, they are good from some perspectives, less good (or even undesirable) from others. In that case, whose perspective is to count, whose interests are to be served, whose values are to be upheld? One person's acceptable risk or cost is another person's intolerable hazard, social disruption or cultural insensitivity. The deployment of a new technology may have entirely unanticipated consequences in terms of benefits, risks, hazards, and social and economic costs, and may have influences that extend well beyond the problem it was designed to solve, the needs it was intended to meet and the situation into which it was introduced. Moreover, in the modern world, the development, manufacture and operation of new technologies bring with them a complex network of inter-related technologies. For example, the invention of the internal combustion engine and the manufacture of motor cars precipitated the development of technologies for the detection, extraction and refining of petroleum, the establishment of filling stations and car repair shops, the production of inflatable tyres, the design of bigger and better roads, the establishment of a system of driving laws and regulations governing driving and the licensing of vehicles, and the growth of a wide range of leisure industries (including tourism) that would have been out of the reach of most people until the availability of rapid and relatively cheap travel. In short, technology is inescapably a social activity: its priorities are determined by, and decision making on rival solutions is impacted by, the prevailing distribution patterns of wealth and power. The curriculum should acknowledge these realities.

Before leaving this topic it should be noted that biotechnology, defined here as any technique that uses biological knowledge, procedures and/or systems to solve problems or create/modify products for the benefit of people and/or the environment, is potentially the most fruitful focus for addressing the relationship between science and technology. Biotechnology has been part of human activity for millennia. For example, using bacteria to make yoghurt and cheese, fermenting grapes to make wine and selectively breeding plants and animals to create organisms that better suit our needs and purposes. In recent years, genetic manipulation has opened up enormous possibilities that have far reaching economic, environmental, sociopolitical and moral-ethical dimensions. Consideration of these matters, and any sustained study of demarcation issues, in whatever specific context, will throw up four major topics for discussion: mode of rationality and argumentation; language issues; modelling and theory building; worldview, values and ethical practice. Each of these warrants special consideration in the curriculum as part of the explicit and reflective teaching of nature of science, and is afforded such consideration in Chapters 6 to 10. A major problem for me has been to decide on the most suitable order for these chapters. Each has implications and major points of overlap with the others, and readers may choose to read Chapters 6, 7 and 8 in a different order.

NOTES

[1] Qigong is a technique, associated with some martial arts, for coordinating different breathing patterns with various physical postures and motions of the body in order to achieve and maintain good health and enhance mobility and stamina.

[2] In an exercise designed by John Somerville (1941) to focus attention on demarcation criteria, students are invited to consider whether umbrellaology (dealing with colour and weight of umbrellas, number of umbrellas per household, age and sex of owners, and making extensive use of statistical data), qualifies as a science.

[3] No particular literature sources were recommended. Students chose to consult sources ranging from high status publications such as *Scientific American*, *Discovery*, *The New York Times* and *Newsday* to supermarket tabloids like *The National Enquirer* and *The Globe*.

[4] I would restrict the investigations to pseudosciences on the grounds that they provide much more scope than the paranormal or common superstitions for addressing demarcation issues.

[5] Some readers may prefer the label *intercultural* or *cross-cultural education*, *culturally responsive schooling* or *education for cultural diversity* to the somewhat dated terms multicultural and antiracist education. Pomeroy (1994) has identified nine interpretations of the notion of science education for cultural diversity and has traced the underlying beliefs, goals, pedagogical strategies and research agendas: science and technology career support systems for under-served groups; localized contexts for science curriculum; culturally sensitive teaching and learning strategies; revised and authentic

149

history of science; demystification of science and scientist stereotypes; science communication skills for language minority students; inclusion of traditional knowledge; bridging students' worldviews and scientific worldview; comparing and contrasting epistemologies of Western and non-Western knowledge of the natural world.

[6] Of course, science also studies the built environment, together with phenomena and events *created* by science.

[7] Merton (1973) states that up until the 19th century it was common for scientists to report their discoveries in the form of anagrams, thus simultaneously claiming priority and concealing important ideas from rivals until further investigation could be completed.

[8] The course, titled Indigenous science and Western science: Perspectives from the Pacific, was located in the Centre for Pacific Studies. Although enrolment was open to all students taking a General Arts BA degree, most of the students were Pacific Islanders. Sadly, when Dr Roberts left the university, the course was discontinued.

[9] Ladson-Billings (1995) regards "culturally relevant pedagogical practices" as those that enable all students to experience academic success, develop cultural competence and acquire the critical consciousness needed to question and challenge the socio-political status quo. This latter criterion fits well with my personal commitment to the politicization of students through the science curriculum. Lee (2002) and Luykx and Lee (2007) use the term "instructional congruence" to refer to the continuities and discontinuities between the demands of school science and students' existing cultural and linguistic experiences.

[10] McKinley and Keegan (2008) note some of the complications that can arise in choosing appropriate vocabulary when scientific meaning is not entirely coincident with traditional usage. For example, the Māori word for fish [ika] traditionally referred to all creatures living in the sea, not just vertebrates with fins and gills. The Māori word *ngarara* is used in the *Pūtaiao* document for insect, although the word traditionally includes all crawling animals, such as lizards and skinks. The point is that classifications are constructed to serve particular purposes, and vocabulary reflects that purpose. McKinley and Keegan note that even when Maori terms are still available (and many are not), they "lack meaning without the associated experiences and activities" (p. 144). Writing from a North American perspective, Davison and Miller (1998) urge that First Nations and Native American students study plants indigenous to their area, using traditional names and classifications based on their medicinal properties and their role in traditional culture.

[11] The processes of evolution occur over such a vast timescale and are so highly sensitive to both initial conditions and new conditions arising during the course of evolution that it is impossible to predict with any degree of accuracy what mutations will arise. Retrodiction involves "making predictions about the past" – that is, noting the extent to which data that has already been gathered (about existing species and species detected via the fossil record) is consistent with what theory would have predicted (see Popper, 1980, 1988).

[12] Generating *ad hoc* hypotheses to explain "special cases" is another move that Popper declares to be unscientific.

[13] The controversy surrounding Darwin's work extended well beyond religious beliefs; it impacted profoundly on economics, politics, geology, psychology and history; it raised questions about the ways in which human society should be organized and why particular kinds of social action ought to be implemented (see Hull, 1973; Ruse, 1979b; Bowler, 1985, 1996).

[14] Werner Heisenberg (1971, pp. 82–92) gives details of a fascinating conversation on science and religion involving himself, Wolfgang Pauli and Paul Dirac (in which they paid special attention to the views of Albert Einstein and Max Planck), and reports on a subsequent conversation on the same subject with Niels Bohr.

[15] Note that Kuhn discusses the development of sciences, not science as a whole. For Kuhn, there are significant differences among the sciences – a point of some importance in the context of this book.

CHAPTER 6
THE SUBSTANTIVE STRUCTURE OF SCIENCE

A strong case can be made for beginning discussion of curriculum issues relating to the King and Brownell disciplinary characteristics with consideration of the substantive structure of science (item 6 in the list). Firstly, because covering content almost always has the highest priority with teachers and students, and can often divert attention from the pursuit of other curriculum goals. Issues relating to content command the most curriculum time, dominate the approach to assessment and evaluation design, and usually provide the rationale for structuring, organizing and sequencing school science textbooks and other curriculum resources. Secondly, because the drive to construct theoretical explanations is widely regarded as the major characteristic distinguishing science from traditional knowledge, everyday commonsense knowledge and practical knowledge for action (see discussion in Chapter 5). Indeed, it could be argued that theorizing is the most distinctive feature of scientific thinking.

While debate continues about precisely what knowledge, skills, attitudes and dispositions a person designated as scientifically literate should possess and exhibit, there is universal agreement that science content comprises a significant component. It can safely be said that nobody would argue for a definition of scientific literacy that did not include understanding of *some* of the fundamental ideas, principles and theories of science. Discussion about the particular content that constitutes "some of the fundamental ideas, principles and theories of science" is outside the scope of this book. Here the focus is on understanding the nature of scientific knowledge – in particular, the reasons that motivate its construction, the ways in which it is constructed and validated, and the role and ontological status of the knowledge so produced. Part of what I have in mind is a second order activity that directs attention to the thinking processes that scientists use in developing the content knowledge students are required to learn if they are to be regarded as scientifically literate. In addition, students should know something about the sociocultural circumstances and, of course, the cognitive circumstances that led to particular theoretical explanations, and the changed circumstances that led to their subsequent rejection or replacement (if, indeed, they have been superseded). Issues concerning the extent to which scientific knowledge is socially constructed will inevitably arise.

Because scientific knowledge is produced in a social context it is necessarily impacted by the personal and professional goals of scientists, the interests and priorities of funding agencies, and the cluster of economic, political and moral-ethical influences that impregnate the sociocultural context in which scientific practice is located. In other words, scientific knowledge is socially constructed through the practices of the scientific community in response to the demands, needs and interests of the wider community that surrounds and supports it. Furthermore, the procedures of science (experiment, observation, etc.) are based on conventions that are, themselves, human constructs. Some sociologists of science take this to mean that the knowledge produced by scientists is *no more than* a social construct and could have been constructed differently. In other words, scientific knowledge is "what scientists say it is", for the reasons they choose. It has no special status or claim to validity. Science is "no better" than any other knowledge about natural phenomena and events; it is just "different".

It is important that the school science curriculum achieves a sensible balance between the view that science is absolute truth, ascertained by value-free disinterested individuals using entirely objective and reliable methods of inquiry (see discussion of common myths about science in Chapter 5), and the relativist view that "scientific truth" is any view that happens to suit the prevailing cultural climate or reflect the interests of those in positions of power. There is no doubt that scientific practice is profoundly influenced by the social context in which it is located. The key point at issue is the *extent* of this social influence or, as Giere (1988) puts it, "The real issue is the extent to which, and by what means, nature constrains scientific theorizing" (p. 55). For those at one extreme, social factors are acknowledged to influence research priorities, but little else; for those located at the other extreme, scientific knowledge is regarded as social in *content* as well as in origin. In other words, the second position is that scientific knowledge is *no more than* a social construct and, therefore, has no more legitimate a claim to describe reality than any other form of knowledge. One response to this claim is to ask for a clear example of how social conditions *determine* scientific knowledge, how the social context leads unerringly and inevitably to particular scientific knowledge. Slezak (1994) asks what elements in the social circumstances surrounding Isaac Newton's work led directly to the inverse square law? In different social circumstances, would he have formulated the inverse *cube* law of universal gravitation? Similarly, we might ask whether Newtonian physics was replaced by Einsteinian physics simply because the social climate changed or because the new theory explained more and solved some outstanding problems. We might also ask why we continue to believe in theories that were produced in social circumstances very different from those of the present day.

It would be grossly misleading to believe that there is only one conceivable representation of the world, or that we can know what the world is like independently of our conceptual structures. But this is not to say that the world is merely a construct of the human mind, that our knowledge is purely arbitrary, or that individuals are free to fabricate any view of the world that happens to suit them. The admission that observation is theory-dependent and that theories are created by individuals does not mean that science loses all objectivity. The admission that theoretical explanations could be different does not reduce science to mere fashion, prejudice or social convention. Science cannot be anything that important scientists choose to say it is. The world does limit us in some ways. Scientific knowledge is a product of the interaction between the external real world and our intellectual needs and capabilities. Of course, what we (as scientists) contribute to that interaction changes over time in response to social, cultural and political change, technological innovation and new theorizing. While social forces do not determine how the natural world is constituted or how it behaves, they do "open the eyes" of scientists in particular ways, direct their attention to particular phenomena and events, and impact on the ways in which they make sense of them using procedures devised and sanctioned by the community of scientists. It is also the case that, from time to time, science and technology create phenomena and objects that did not previously exist – for example, Dolly the cloned sheep, laser beams, and all the new molecules and materials produced by chemists and materials scientists. Once created, however, these things have an objective existence that is, in some respects, independent of the scientists or engineers who created them.[1]

Giere (1997) provides a diagrammatic overview of the relationship between a theoretical structure (for which he uses the term "model") and the real world (Figure 6.1) that might have value in the context of this discussion. Scientists construct theories and models they believe correspond in some important respects to features of the real world. By using the model/theory, they can make specific predictions of what will be observed under certain circumstances and look for evidence that the prediction is upheld, or not. The extent of agreement increases or reduces confidence in the model/theory. Clearly, the procedures of model building and testing do not comprise a simple step-by-step algorithm. Rather, there is a cluster of interacting, overlapping and recursive steps involving (i) collection of data via observation and/or experiment, (ii) reasoning, conjecture and argument, (iii) calculation and prediction, and (iv) critical scrutiny of all these matters by the community of practitioners, leading to a decision about acceptance, rejection or further modification of the model/theory (issues that will be raised in discussion of the syntactical structure of science, in Chapter 7, and the nature of scientific argumentation, in Chapter 8.

Larry Laudan (1977) describes scientific practice as primarily an exercise in problem solving. He divides problems into two broad categories: *empirical* problems (any feature of the natural world that is in need of explanation) and *conceptual* problems. Conceptual problems can be *internal* (logical inconsistencies, ambiguities, circularities, etc.) or *external* (conflicts with other theories), while empirical problems can be categorized as *solved problems* (those satisfactorily explained by the theory), *unsolved problems* (those that fall outside the scope of the theory) and *anomalous problems* (those not solved by the theory, but solved by a rival theory). Solved problems are important because they constitute a substantial part of the warrant for belief in the theory's central premises (what Lakatos (1978b) calls the "hard core") and are especially important if rival theories are unable to solve them. However, as history reminds us, solved problems are no guarantee of truth: false theories can work just as well as true ones, and there

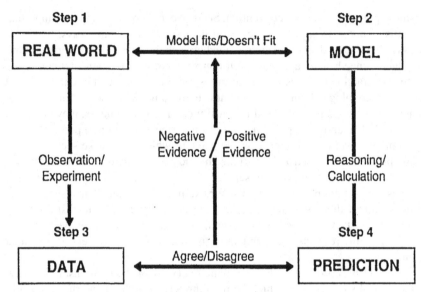

Figure 6.1. Giere's Model of the relationship between data, reasoning and argument (as modified by Driver et al., 2000).

are many examples of discarded theories that were previously accepted as true because they "worked well". The goal of the researcher is to maximize the number and scope of solved problems and minimize the number and scope of anomalies and conceptual problems (compare the Lakatosian notion of a "positive heuristic"). The theories that best solve problems will contribute most to meeting the aims of science: making accurate and reliable predictions, establishing manipulative control, increasing the scope and precision of explanatory systems and, whenever possible, integrating and simplifying the various components of those explanatory systems. Laudan (1984a,b) makes the point that there is a constant and complex process of mutual adjustment and mutual justification among aims, methods and theories. Aims justify methods and must harmonize with theories; methods justify theories and are chosen to meet aims; theories constrain methods and must harmonize with aims (Figure 6.2). The components are mutually interdependent: methodological rules determine the range of possible theories, new theories may radically alter existing aims, and so on. This view accords well with arguments to be used in Chapter 7 concerning the contextual and conceptual contingency of scientific inquiry. Scientific theories and models are constructed in accordance with procedures legitimated by the community of scientists in order to serve two functions: *explanation* and *exploration* – that is, to make sense of what we already know and to guide future inquiry and investigation. After years of scientific endeavour, both the conceptual and procedural functions of theoretical structures are now highly context specific. The days when scientists in different sub-disciplines engaged in broadly similar activities are now long gone – if, indeed, that was ever the case (see discussion in Chapter 7).

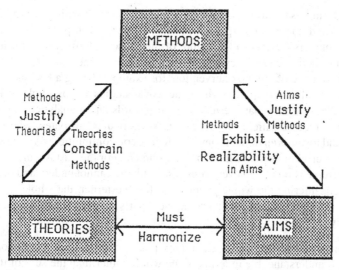

Figure 6.2. Interactions among aims, methods and theories, after Laudan (see Duschl, 1990).

Theory acceptance is not just a matter of solving empirical and conceptual problems but of reconciling the proposed explanatory structure with our broad framework of assumptions about the world. We have expectations about what the world is like, how we can reason about it and make sense of it. Science has to fit with those expectations! Stephen Pepper (1942) refers to these fundamental beliefs as *world hypotheses*,[2] though worldview has become a more widely used term in recent years. A simple and precise definition of worldview is not possible. The best I can manage is to describe it as a set of presuppositions (assumptions that may be true, partially true or entirely false) that we consciously and/or unconsciously hold about the world, sometimes consistently and coherently, sometimes inconsistently and erratically. Kearney (1984) refers to worldview as "culturally organized macrothought: those dynamically inter-related basic assumptions of a people that determine much of their behavior and decision making, as well as organizing much of their body of symbolic creations ... and ethnophilosophy in general" (p. 1). He identifies seven major components of a worldview: the Self, the Non-Self or Other, classification, relationship, causality, space and time. Kearney refers to these items as "universals" because they are, he argues, universal in their significance across cultures, though varying in nature and detail. While each component is distinctive in some way, it is the interactions among the universals that most effectively sum up the way a particular cultural group sees the world. For example, how a society functions is crucially dependent on culturally-determined notions of the nature of Self, setting boundaries of "who I am" and "what I am". Non-Self or Other is everything in the universe except the Self. Every Self interacts with the Other, both living and non-living; the *relationship* universal describes the ways in which people view these interactions – in terms of harmony, competition, control, negotiation, etc. The ways in which we classify (the third universal) says a great deal about the presuppositions we hold about the world and about our purposes in interacting with it in particular ways. The *causality* universal refers to the nature of the cause and effect relationship used to explain phenomena and events. It may be seen as mechanistic, teleological or a consequence of Divine intervention, for example. Using a fascinating blend of sociocultural history, philosophy, educational studies and psychology, Richard Nisbett (2003) argues that Westerners tend towards an an analytic view that focuses on perceived salient objects and their particular attributes, whereas East Asians tend towards an holistic view that focuses on continuities in substances and complex interrelationships within the environment. For the East Asian, everything in the universe is related to everything else, and one cannot understand the pieces without considering the whole picture. For the Westerner, the whole is best understood through analysis into its component parts, each of which can be studied separately.

> To the Asian, the world is a complex place, composed of continuous substances, understandable in terms of the whole rather than the sum of its parts, and subject more to collective than to personal control. To the Westerner, the world is a relatively simple place, composed of discrete objects that can be understood without undue attention to context, and highly subject to personal control. (p. 100)

He further argues that Westerners have a much greater tendency to categorize objects than do East Asians, find it easier to learn new categories by applying rules about properties to particular cases, and make more inductive use of categories – that is, they have a greater tendency to generalize from particular instances of a category to other instances or to the category as a whole. For Easterners, explicit modelling or rule making is less characteristic of causal explanations; they are less likely to use rules to understand the world, less likely to make use of categories, and find it hard to learn categories by applying explicit rules. The *space* and *time* universals complete an individual's or a society's collective worldview. Differences can be substantial. For example, space (and distance) are perceived very differently by city dwellers and those who live in the countryside, and by those who live on small islands and those who live in the centre of a large land mass. To compound matters, science presents all manner of challenges: exceedingly small entities (sub-atomic particles, for example), very large distances (e.g., remote galaxies located many light years away), and the notion of curved space. Time is equally complex and varied, in that some people (and the societies in which they live) can sometimes be described as predominantly future-oriented, present-oriented or past-oriented. For some, the past is last week and for others it is a thousand years ago; for some, the future may be the next decade or even further away, for others it is tomorrow. It is evident from everyday conversation that some people see time passing slowly while others see it passing all too quickly. While Western societies tend to regard time as linear, unidirectional and irreversible, some African cultures have an oscillating image of time, that is, time runs in circles or zig-zags. Horton (1967) notes that there are different time scales for different occasions and situations. For example, the cycle of life and death, the seasonal cycle, connections between current events and the lives of ancestors or the establishment of major community institutions. Peat (1995) writes at length about both the metaphysical and sociocultural aspects of time as conceived by some Native American peoples. On some occasions, it may be important to combine past, present and future into a "fourth state" as a way of addressing important issues (Armah, 2005, cited by Keane, 2008, p. 601). Science invites us to recognize some substantial differences in the "scale" of time, ranging from processes taking no more than a few nanoseconds to those taking billions of years, while post-Einsteinian physics envisages a complex interaction of space, time, mass and energy, and contemplates the notion that time moves more slowly at velocities approaching the speed of light. Einstein is reputed to have said: "people like us who believe in physics, know that the distinction between past, present, and future is only a stubborn, persistent illusion" (cited by McFarlane, 2002, p. 126).

Capra (1983) describes the Cartesian/Newtonian worldview underpinning modern science and most of school science as comprising five main ideas: (i) there is a fundamental division between mind and matter; (ii) nature is a mechanism that works according to exact and universal laws; (iii) reality comprises a multiplicity of building blocks, which collectively constitute the whole; (iv) the scientific method is assumed to be the only valid way to ascertain the true nature of phenomena; and (v) the goal of science is to understand nature so that it can serve human ends. A number of questions spring to mind: First, is this any longer a true picture of science? As noted above, contemporary physics may paint a somewhat different picture. Second, is this view promoted through the science curriculum? Discussion in chapter 5 suggests that it is still the predominant view of science projected by the curriculum (Nadeau & Désautels, 1984; Hodson, 1998b; McComas, 1998). Third, is it a good or bad thing to promote? Would science be better served by a reorientation of its underlying values? Would society be better served by a re-constituted science?

REALISM OR INSTRUMENTALISM?

As far as school science is concerned, teachers need to address the role and status of scientific knowledge and reach a decision on where they wish their students to stand on the issue of realism versus instrumentalism. *Naïve realism* is the view that scientific knowledge constitutes true knowledge of the world; *instrumentalism* (one of several variants of "anti-realism"[3]) is the notion that theories are simply convenient devices ("fictions") that enable us to predict, manipulate and control. Created entities like atoms, electrons, genes, gravitational fields and black holes are built into theoretical structures designed to account for "the world beneath surface appearances" and provide scientists with a means of interacting with the real world in predictable ways.[4] Provided that we solve our problems quickly and make predictions reliably and accurately it really does not matter whether the theory we employ is true or not, or whether the entities postulated by the theory really exist. Truth is irrelevant. *Utility* is the significant criterion.

Both naïve realism and instrumentalism are beset with enormous difficulties. Naïve realism sets standards for the acceptability of knowledge claims that simply cannot be met. We cannot gain direct access to truth about the world because, in a very real sense, we are "prisoners" of our woefully inadequate senses (our physical frailty) and our capacity to theorize (our intellectual frailty). We have no way of stepping outside our minds and bodies to see the world in a theory-free way, assisted by perfect senses. Our success in building a description and explanation of the world cannot be checked against the reality we are theorizing about. Even the instruments we build to give us greater access to information about the world are "contaminated" by the theoretical assumptions we make during their design, construction and use. Moreover, whatever we choose to investigate could be changed, in some possibly unknowable way, by the interaction – that is, we risk *contamination* and *distortion* through the very act of investigation. In consequence, agreement with the "facts" (observable evidence) does not mean that a theory is true. Firstly, the facts may be wrong. After all, they are interpreted responses to sense data collected with severely limited "tools" (our senses, or our instrumental extensions of them). Secondly, there may be other theories that also agree with the facts. The observational evidence may be susceptible to interpretation in more than one way, to suit more than one theory.

If two theories can provide contradictory accounts for the same observable phenomena, yet make similar predictions about observable events, they are equally well corroborated by the available evidence. However, they cannot both be true, in the sense that the theoretical entities they utilize in explanation exist independently of our theorizing about them. We cannot assume truth from explanatory power. Anti-realists might also argue that the methods by which scientists investigate knowledge claims are so extensively theory-dependent and theory-driven that they are not so much a means of discovery as a means of constructing and re-constructing knowledge, and so cannot tell us anything about the world outside those constructions. In other words, if the ideas and procedures that scientists use to investigate the world are social constructs then what they tell us about the world is also a social construct, unless they "hit upon the truth" by accident. Therefore, it makes sense to admit that theories are simply fabrications of the human mind, invented to give us a measure of control and predictive capability. Indeed, Laudan (1984b) argues that because the realist goal is unattainable, and because there is no means to measure how close a theory is to the truth, scientists who pursue realist goals are acting irrationally. Cobern and Loving (2008) make the commonsense point that while we "cannot know with certainty that perceptual and experiential experiences are significantly grounded in ontological reality … it is equally true that we cannot know for certain that perceptual and experiential experiences are *not* significantly grounded in ontological reality" (p. 441, emphasis added). A major counter argument to instrumentalism is that it provides no incentive for the scientific endeavour. Once in possession of a satisfactory explanation that adequately accounts for the data, and enables accurate predictions to be made, there is no reason for an instrumentalist to seek alternatives and to develop theories further. Why bother to do more? Only a realist, someone highly motivated to find out what the universe is really like, would continue the effort. Like Holton (1993), Ayala (1994) and Newton (1997), I have good reason to believe that most scientists are realists and that the prime motivation for "doing science" is to construct plausible, intellectually satisfying and productive explanatory frameworks, as many of the scientists interviewed by Wong and Hodson (2009) confirmed.

A major problem for instrumentalism is that it cannot easily explain why theories are successful and, in particular, why they have predictive capability, nor how an idea introduced expressly to solve a particular problem can sometimes lead to the solution of quite different problems. If scientific theories are not (approximately) true, how is it possible for them to solve diverse problems and to yield such accurate (and sometimes very surprising) observational predictions? The continuing success of a particular theory in solving both theoretical and practical problems, its ability to provide common explanations for diverse phenomena, and its capacity to grow, develop and extend its domain, are good grounds for adopting a realist position. In other words, a theory's success is evidence that its Lakatosian "hard core" is approximately true. How can a theory be no more than a fictional calculating device if it can predict real and surprising phenomena such as the bending of light in the vicinity of an intense gravitational field, and make such predictions with surprising regularity? How can instrumentalist theories, which are supposed to be mere calculating devices, lead to the discovery of new phenomena using concepts that are supposedly no more than theoretical fictions? In Putnam's (1975) words, "the positive argument for realism is that it is the only philosophy that doesn't make the success of science a miracle" (p. 73). It is also difficult for instrumentalism to account for those occasions when science creates phenomena and events that do not exist in the natural world, especially when they go on to form the basis of significant technological artifacts. Arguing in this vein, Hacking (1983, 1991) declares that engineering success constitutes the strongest case for realism. Enhanced instrumental capability is an indication that our theories are getting closer to the truth (to reality). Perhaps electrons were once merely theoretical fictions, but now that we can manipulate them in a controlled way we are justified in believing that they are real. Tala (2009) makes the same point with respect to successes in nanotechnology: "If the nanostructures under changing macroscopically controlled conditions behave as expected, that is, if the nanoconveyors and minimachines manage to cause expected macroscopic effects, then the entities behind these causes must be real" (p. 285). Put in another way, as our theories more sharply and more accurately focus on reality, their instrumental power can be expected to increase. We might also anticipate increased conceptual coherence.

Anti-realists argue that empirical success and predictive capability are no guarantee that the particular entities comprising a theory actually exist. As Laudan (1981) points out, there is a long list of now abandoned theories that once had predictive power, including the crystalline spheres of ancient and medieval astronomy, the phlogiston theory of burning, the caloric theory of heat and various theories of spontaneous generation. We no longer believe that the entities in these theoretical structures exist. Therefore, he argues, we have good reason to believe

159

that the entities we currently accept as real will also eventually be discarded as untrue. Realists can readily counter this argument on the grounds that theory replacement is a case of theoretical understanding being brought closer into conformity with the real world, as an occasion when science learns from its mistakes and gets closer to the truth. The fact that theories have anomalies, and that from time to time we abandon entire theories (and even the entities that comprise them), tells us that there is "something out there" that is different from our current conception of it. Our desire to ascertain the precise nature of that "something" is the impetus for further study, further investigation and further theory building. Duschl (2008) makes essentially the same point when he asserts that "most of the theory change that occurs in science is not final theory acceptance but improvement and refinement of theory" (p. 274).

One way of solving the dilemma concerning choice between realism and instrumentalism is to regard the status of scientific knowledge as related to our purpose in seeking it, and to the role it plays once it has been acquired. Sometimes scientists aim at a true description of the world and a true explanation of observable facts; sometimes a convenient predictive instrument is all that is required. When they wish to explain "how things are", when they are trying to make sense of it, scientists develop *theories* – our current "best shot" at the truth. Each theoretical development takes us a little nearer to the truth about the universe, though we recognize that we are unlikely ever to achieve that goal and would not know if we had. When we simply want "to get the job done" – that is, to make a prediction, perform a calculation or achieve a measure of control – we invent *theoretical models*. We also invent models when phenomena are too complex to be susceptible to theorizing, though we continue to hold that as a future goal. Phenomena may be, at present, inaccessible to our senses, they may be conceptually difficult or entirely novel. In such circumstances, our models are "useful fictions" that enable us to make interim sense of phenomena. As Barbour (1974) says, "they are partial and provisional ways of imagining what is not observable; they are symbolic representations of aspects of the world which are not directly accessible to us" (p. 6). Viewed this way, models and theories have significantly different purposes, with models functioning as a bridge between scientific theories and the real world they purport to describe. They provide a simplified description of real world phenomena and events, thereby assisting ready manipulation of objects, quantification of conceptual relationships, calculation and prediction. Thus, scientists play both a realist game and an instrumentalist game: they develop theories when they aim to explain and describe the real world; they use convenient models when they are content with a quick and accurate calculation or prediction, or when they are *building* complex theories. This position, which some have termed *critical realism* (Putnam, 1987; Bhaskar, 1989), enables scientists to be realist about some theories (those they consider to be genuine attempts to uncover the truth) and instrumentalist about others (those they find useful but do not accept as true descriptions). I believe there is enormous value in referring to the former as *theories* and the latter as *models*. In contrast, instrumentalists are always instrumentalist and blur the distinction between theory and model. This distinction between models and theories is not commonly drawn. Indeed, Sensevy et al. (2008) report that the teachers with whom they have worked regard teaching about modelling as a sufficiently difficult task in itself, without introducing this distinction.

It is interesting that a model originally designed simply to achieve a short term instrumentalist goal may sometimes be developed, over time, into a realist theory. Models often function as the means of building progressively more complex (and realist) theories. They function as "scaffolding devices", by means of which scientists make interim sense of observational data as they develop and extend their thinking. They play a key role in the design and conduct of experiments and in thought experiments. In short, they are "tools for thinking" or "working devices" for building and extending theories (Nersessian, 1992). In Barbour's (1974) words, "a model is a symbolic representation of selected aspects of the behaviour of a complex system for particular purposes. It is an imaginative tool for ordering experience, rather than a description of the world" (p. 6). In other words, models are approximations of reality; they may omit a number of details, including an account of variables known to be relevant overall but considered minor in the specific context under consideration. But what is lost in terms of veracity is compensated by increased clarity and accessibility.

> In modelling we sacrifice perfect truthfulness because the object is distorted by simplification. This is, however, more a strength than a weakness because the model gains greatly in transparency and conceptual clarity. It deliberately neglects the less relevant details and emphasizes the most important facets ... A reliable model exhibits the essence of the property or a phenomenon under study, since attention is focused only on selected and dominant features (effects). (Maksić, 1990, p. xiv)

> Representational forms and particular representations are simultaneously illuminating and limiting. They cannot perfectly represent their objects because they do not display all the features of the thing represented. Therefore, they must be judged, at least in part, in terms of their usefulness. (Mitchell, 2003)

Cartwright (1999) likens the way in which a model effects the transition from the abstract concepts of scientific theories to the concrete realities of daily experience to the role of fables in relating real life to abstract moral principles: "Fables transform the abstract into the concrete, and in so doing, I claim, they function like models in physics" (p. 36). Woody (1995) notes that while some models are only to be used in accordance with strict rules, others are "projectable" – that is, they do not have well-defined or fixed boundaries and can be used to extend thinking in new directions.

Like scaffolding, a model may be discarded once it has served its function. Indeed, it may be that a model used early in a theory's developmental history proves incompatible with the premises of the final refined theory, despite its pivotal role in theory construction.[5] Thomas Kuhn (1970) points out a somewhat different purpose for the various models that he regards as embedded within (or derivable from) a particular theoretical structure. During periods of normal science, he says, scientists may use a range of such models in problem solving, prediction and further theory articulation. Indeed, within all significant theoretical frameworks there is a "family of models" (Giere, 1988, 1999) that scientists may deploy for solving short-term, practical problems, making calculations, generating specific hypotheses for testing and formulating predictions. Because they encapsulate and communicate knowledge about the design, materials, apparatus and practical conditions necessary to conduct scientific investigations, Baird (2003, 2004) extends Giere's notion of a family of models to include laboratory equipment, instruments for measurement and experimental set-ups. They can be regarded as "models" because they enable us to control and manipulate the world, and even to create some phenomena and some entities, that is, they are devices created specifically to assist problem-solving and theory-building.

There is nothing that compels us to treat all the entities and claims within a theoretical structure in the same light. Some have realist purpose, some have instrumentalist purpose (Slater, 2008). Instrumental models are used when the full theory is too complex or too abstract for direct use in problem solving. The full theoretical force of the theory does not have to be engaged if instrumental models within it can suffice. Nagel (1987) expresses broadly similar views when he describes models as being *intermediaries* between the complex abstractions of theoretical structures and the concrete actions of experiments, thereby assisting scientists to make predictions, design inquiries, summarize data, justify outcomes and facilitate communication. Eventually, when scientists encounter insurmountable problems with a particular theory, find severe limitations to it, or construct a theory that is more satisfying, the old theory is rejected and reverts to the status of a model. Although we no longer believe that it *explains* matters, it may still be useful for making predictions. From the critical realist position it is not illogical to retain a falsified or superseded theory in an instrumental capacity, provided that its status is recognized and acknowledged. It may be that within a restricted domain of application, a falsified theory (in its new status as model) is more useful than the currently accepted theory simply because it is easier to use. Newtonian physics is certainly much easier to use in addressing real world problems than is relativity theory, and is perfectly adequate for most purposes. Moreover, within the critical realist position it is not illogical to utilize alternative instrumental models, even incompatible or contradictory ones, to deal with different aspects of the same phenomenon. For example, we can use a wave model of light to solve problems relating to interference patterns and a particle model of light to solve a problems relating to photoelectric effects. In chemistry, we may switch from a Lowry–Bronsted model of acids and bases to a Lewis model as the specific nature of the problem under consideration changes. The use of diverse and sometimes mutually contradictory models is fairly common practice in engineering and in practical knowledge for action (see Chapter 5). So, too, is the use of models that function only within a narrow range of conditions and models that have general guidelines for use and wide parameters of application, rather than precise and exact conditions. As Bunge (1966) comments: "Think of the variable loads a bridge can be subjected to or of the varying individuals that may consume a drug. The engineer and the physician are interested in safe and wide intervals centered in typical values rather than in exact values. A greater accuracy would be pointless since it is not a question of testing. Moreover, such a greater accuracy could be confusing because it would complicate things to such an extent that the target – on which action is to be focused – would be lost in a mass of detail. Extreme accuracy, a goal of scientific research, not only is pointless or even encumbering in practice in most cases but can be an obstacle to research itself in its early stages" (p. 334).

As Chinn and Samarapungavan (2008) remind us, models designed to explain a particular phenomenon can sometimes be pitched at different levels of analysis: "For example, in biological models of disease, many phenomena can be simultaneously modeled at the macro-anatomical level (e.g., asthma results from the constriction of bronchial airways in response to allergens like pollen), at the cellular level (e.g., asthma is caused by the activation of lymphocytes), and at the molecular level level (e.g., the activation of lymphocytes depends on molecular signals from dendritic cells)" (p. 202). Perkins and Grotzer (2005) point out that models may also vary with respect to: (i) sophistication in the level of mechanism ascribed to the phenomena and events being modelled (from observable surface features to abstract idealizations); (ii) *interaction patterns* invoked (from simple push, pull and resistance to interactive causality, feedback loops and constraint-based systems); (iii) use of *probability* (from deterministic models to those based on chance occurrence, fundamental uncertainty or chaos theory); and (iv) *agency* (whether there is a central "agent" as causal actor or more complex possibilities such as additive causes, self-organizing systems and emerging properties).

Elsewhere I have argued for a critical realist position for the school science curriculum on the grounds of *authenticity, flexibility* and *pedagogical utility* (Hodson, 2008). I argue for the *authenticity* of critical realism on the grounds that most scientists are realists of a kind, for whom the principal goal of science is to obtain knowledge of maximum clarity, validity, reliability and predictive capability. Science assumes that there is regularity in nature and that whatever underpins this regularity can be ascertained. Of course, until theoretical controversies are settled and consensus established, scientists may continue to regard explanatory structures as tentative instrumental models. My claim for *flexibility* rests on the point that critical realists can be realist about those conceptual structures they regard as genuine attempts to uncover the truth and instrumentalist about those that have utilitarian value (for prediction, control, etc.) but make no claim to truth. This position also accommodates both the tentative nature of scientific knowledge and the sociocultural embeddedness of scientific practice. On the first point, it acknowledges that scientific observation can be mistaken, experimental design flawed, interpretation suspect, and so on, but that these shortcomings can be recognized and ameliorated through critical scrutiny within the scientific community. On the second point, it acknowledges that scientists are necessarily impacted, to some extent, by personal predispositions, values, attitudes and prejudices, sociocultural influences, economic pressures, power imbalances and political imperatives. It is for these reasons that Giere (1999) coined the term "*perspectival* realism". Scientific theories cannot capture the "totality of reality"; rather, they provide us with particular views of it. Matthews (1998) argues that the critical realist position (or what he calls "modest realism") enables students to recognize that "science is a human creation, that it is bound by historical circumstances, that it changes over time, that its theories are underdetermined by empirical evidence, that its knowledge claims are not absolute, that its methods and methodology change over time ... that its research agendas are affected by social interests and ideology" (p. 166), while still claiming that in many respects it describes the real world. My criterion of *pedagogical utility* is based on the assumption that critical realism is more likely than instrumentalism or naïve realism to motivate students. Students who are critical realists are more likely to think that they have the ability to build understanding and explanations for themselves through careful and critical reflection on their experiences. A critical realist position enables students to construct their own models to solve short-term problems, where initially the only criterion of acceptability is "what works". By comparing and contrasting their model with models constructed by others, students see the importance of robustness, breadth and scope, and predictive capability. Later, they might come to appreciate how further rounds of inquiry and reflection can result in these models being developed into realist theories.

HOW TENTATIVE IS TENTATIVE?

Concerns may also arise for students faced with the proposition that scientific knowledge is tentative while simultaneously being presented with authoritative statements concerning current theoretical understanding. It is important for students to recognize that scientific knowledge is tentative because it is based, ultimately, on empirical evidence that may be incomplete, and because it is collected and interpreted in terms of current theory, which may eventually be changed as a consequence of the very evidence that is collected. In all these endeavours, the creative imagination of individual scientists is impacted by all manner of personal experiences and values. Moreover, the "collective wisdom" of the scientific community that supports the practice, scrutinizes the procedures and evaluates the products, is also subject to complex sociopolitical, economic and moral-ethical forces. In consequence, there can be no certainty about the knowledge produced. There is also much of value in encouraging students to question, and to express whatever doubts they may have about the ideas we present in class, but it can be counter-productive for students to believe that everything in science is tentative or the scientific community's well-established and taken-for-granted theories can be falsified by a few moments of data collection and argument in a school science class. School science experiments cannot refute established scientific knowledge, though they can provide some evidence for accepting it. It is here that Harding and Hare's (2000) notion of *open-minded realism* is particularly valuable. All theories are tentative when first proposed. As supporting evidence builds up and theories are deployed successfully in a range of situations, confidence in them increases. Without such confidence, further research and theory development would be impossible. In other words, scientists have good reasons to accept some theories as true, or close to the truth, but they do so on the understanding that they can subsequently change their minds if new evidence is obtained or new ways of looking at existing evidence arise. Harding and Hare (2000) sum up their notion of open-minded realism as follows:

> When compelling evidence has accumulated, scientists will come to regard the theory as settled and established. The claim in question, however, continues to rest on available evidence, and one's open- mindedness consists not in suspending judgment where there would seem to be compelling evidence, but in being willing to reconsider present conclusions in the light of whatever evidence or rival interpretation appears. (p. 230)

Well-established scientific knowledge is used to construct larger, wider-ranging and more ambitious theoretical structures, such that the whole may be considered tentative and "under construction" while many of the components are taken-for-granted as secure knowledge. To admit that absolute truth is an impossible goal is not to admit that we are uncertain about everything. We *know* many things about the universe even though we recognize that many of our theoretical systems are still subject to revision, or even rejection. There are several closely related issues to consider. First, very specific claims about phenomena and events may be regarded as true even though the theories that account for the events are regarded as tentative. Because the whole necessarily extends beyond the parts of which it is comprised, the whole may be seen as tentative while the parts (or some of them) are regarded as certain. Most theories are tentative when first developed, but are accepted as "true" (in a scientific sense) when they have been elaborated, refined and successfully used, when they are consistent with other theories, and are strongly supported by evidence. Teachers make a grave mistake when they encourage students to regard all science as tentative. Indeed, if scientists did not accept some knowledge as well-established we would be unable to make progress.

Millar (1997) asks whether it is in the best interests of students and teachers to portray demanding and difficult ideas as tentative conjectures. He asks: "Is this likely to increase students' motivation to undertake the challenging mental work of coming to terms with these ideas, so that they can use them as tools for their own thinking about new problems and situations?" (p. 91). He goes on to state that there is a moral issue here: for example, "the germ theory of disease is intended to provide students with understanding which can be practically useful in taking decisions about their own health, is it responsible to portray generally accepted core knowledge as tentative conjecture?" (p. 91). Winchester (1993) points out the absurdity of teaching that there is no truth in science but there *is* truth in the philosophy of science, history of science and/or sociology of science when practitioners in these fields suggest that all scientific knowledge is uncertain. These areas of scholarship rest on much shakier ground than most of our chemistry, physics and mathematics! The key point is that we do want students to be aware that many

165

theoretical structures are well-established and no longer subject to doubt, but we want them to be aware that theories achieve this status only when they have been intensely and systematically scrutinized by the community of practitioners. In scientific practice, the construction of knowledge claims and the critiques of those claims are inter-dependent aspects of theory-building. Students need to know the warrant for believing a particular theory – that is, the evidence and arguments that were used in reaching decisions about the theory's appropriateness and accept-ability. Millar (1997) makes the point that scientific consensus arises from "the inability of those working with the phenomena in question to entertain any other way to make sense of it" (p. 99). He continues as follows: "That, of course, is what we normally mean when we say we are "certain" of something. The meaning of words like "certain" and "sure" (and "true" and "real") is sustained by their everyday usage" (p. 99).

CONCEPTS, MODELS AND THEORIES

As argued earlier, critical scientific literacy entails a robust and wide-ranging understanding of scientific ideas, principles, models and theories. Achieving this kind of understanding is not a simple task for which a single self- contained teach-ing and learning approach can be devised. Students need to know the nature, scope and limitations of these ideas, principles, models and theories, and they need to be provided with opportunities to deploy them in a range of situations – the *learning science* component of the curriculum. They need to know something of the ways in which a complex of cognitive problems and factors related to the sociocultural context influenced the development of key ideas over time, and they need to un-derstand the role of models in the design, conduct, interpretation and reporting of scientific investigations – the learning *about* science component. They need to experience model building for themselves, and to give and receive criticism in the quest for "better" models – the *doing* science component. Finally, they need to experience and gain a measure of expertise in making decisions about the suitability of particular models and theories for addressing socioscientific issues, formulating a position on controversial matters, and taking effective and appro-priate action. This is a very tall order. It is both an enormously ambitious and an extremely complex curriculum endeavour, but one that is attracting considerable research attention. The nature of mental models has long been an area of research in cognitive psychology, dating back to the seminal work of Johnson-Laird (1983) and Gentner and Stevens (1983), but in recent years the topic of models and modelling has generated considerable interest among science educators (Gilbert & Boulter, 1998, 2000; Gilbert et al., 1998a,b; Franco et al., 1999; van Driel & Verloop, 1999; Greca & Moreira, 2000, 2002; Coll & Treagust, 2002, 2003a,b; Justi & Gilbert, 2002a,b,c, 2003; Treagust et al., 2002; Davies & Gilbert, 2003; Saari & Viiri, 2003; Taber, 2003; Taylor et al., 2003; Erduran & Duschl, 2004; Gilbert, 2004; Gobert & Pallant, 2004; Halloun, 2004, 2007; Hansen et al., 2004; Kawasaki et al., 2004; Coll & Taylor, 2005; Coll et al., 2005; Justi & van Driel,

2005; Lehrer & Schauble, 2005; Perkins & Grotzer, 2005; Khan, 2007; Koponen, 2007; Lopes & Costa, 2007; Shen & Confrey, 2007; special issue of Science & Education, 2007, 16, issues 7-8; Clement & Rea-Ramirez, 2008; Hart, 2008; Nersessian, 2008; Russ et al., 2008; Maia & Justi, 2009). This interest can be categorized into three principal areas of concern.

- The particular models and theories produced by scientists as explanatory systems, including the history of their development.
- The ways in which scientists utilize models as cognitive tools in their day-to-day problem solving, theory articulation and theory revision.
- The role of models and modelling in science pedagogy.

In addressing these three areas there may be some advantage in drawing distinctions among four different kinds of models: *mental* models, *expressed* models, *consensus* models and *teaching* models (Gilbert et al., 1998a). However, the distinction I drew earlier between (realist) theories and (instrumentalist) models will not be used in this discussion. First, because the existing science education research literature does not adopt this convention. Second, because in the context of a discussion about learning, the role and status of a particular conceptual structure is not immediately apparent unless the "learning history" of the students is known in detail. There are two further uses of the term "model" that are relevant to the matters addressed in this book: first, attempts to develop a model of (or for) scientific practice – that is, a generalized and theorized view of how scientists conduct the day-to-day business of scientific investigation, how they investigate phenomena, build explanatory systems, and so on; second, the notion of a "model scientist" – that is, someone who works assiduously and diligently in pursuit of worthy scientific goals and upholds the standards, norms, values and codes of behaviour of the scientific community. This latter concern is a focus for attention in Chapter 10.

Mental models are the cognitive constructions individuals create for themselves to describe and explain phenomena and events that cannot be experienced directly. They may be highly idiosyncratic, incomplete, unstable, internally inconsistent, highly influenced by sociocultural circumstances, and sometimes of little use to anyone else, but they are central to personal understanding, as discussed by Hodson (1998a). Expressed models are the mental models individuals choose to place in the public domain in concrete/material form or through speech, text/writing or other symbolic forms of communication. Of course, the very act of talking, writing or drawing is likely to change the mental model in some way (see Chapter 9). Consensus models are expressed models that have been subjected to critical scrutiny by the scientific community or a sub-community of scientists, or even by a group of students, and have thereby gained a measure of social acceptance. The term *scientific* model might be used for models that have achieved consensus within the wider community of scientists and have been afforded the status of "accepted knowledge". As noted earlier, it is possible for incompatible consensus models to co-exist (for example, the particle model and wave model views of light), with each model being a focus for research. And, of course, a consensus model may subsequently be superseded by a more robust and success-ful model. It might then be designated an *historical* model. Teaching models are expressed models used by teachers to assist students in gaining an understanding of a consensus model (or the school curriculum version of it) and/or deploying that model in explanation and problem solving. Hart (2008) lists the following criteria for an effective teaching model: it must be intelligible and meaningful to students in terms of their existing knowledge and understanding; it must allow commonly occurring conceptual difficulties and misconceptions to be articulated and addressed; it must engage students' imagination and intellect; it must enable students to make progress towards the target scientific consensus model; ideally, it should promote discussion and provide insight into the nature of models and modelling in science (i.e., it should contribute to NOS understanding). Teaching models are necessarily simpler than the target scientific consensus models, but this simplicity can itself be a source of misunderstanding and confusion. Justi (2000) uses the term *hybrid* model to refer to teaching models that purposefully or inadvertently merge the characteristics of more than one historical model or scientific consensus model. On occasions, such models can be very confusing and misleading. As Gilbert (2004) points out, expressed models can be placed in the public domain – that is, the classroom, the community of scientists or the wider social domain – in (i) concrete form (e.g., a scale model of an organism or technological artifact, ball and spoke model of a molecule or polystyrene model of a crystal structure), (ii) verbal or written form, (iii) iconic or symbolic form (for example, chemical formulae and equations), (iv) mathematical form (such as $PV=nRT$, $F=ma$), (v) visual form (via graphs, diagrams and animations), (vi) ges-tural form (through dance and mime), and/or (vii) any combination of these forms (as in multimedia presentations and simulations).

It is important for teachers and students to keep these distinctions clear. There is much of value in creating opportunities for students to explore and develop their own mental models, and to present them in appropriate form for critical scrutiny by others. Through such activities a group of students may reach consensus, though subsequent investigation and data collection may result in further phases of criticism and revision, resulting in a new consensus model. Through such ex-periences students gain insight into the ways in which scientists utilize models as cognitive tools in theory building (see later discussion). However, the models that have gained consensus within the scientific community (the accepted knowledge store of science) are not likely to be seriously challenged by the critical scrutiny of school students. They are too well-established to be vulnerable to data collected in school science investigations. However, it is crucial that students know the rationale, arguments and evidence that led to their acceptance by the community, and are aware of the strengths, weaknesses, scope and limits of applicability of the various models they study. It is important to note that knowing what a model is intended to achieve entails knowing what it does *not* address. What I am arguing here is that teachers need to be clear which "game" is being played at any one time. Are students seeking to acquire and learn to use the established knowledge of the scientific community or are they engaging in model building and its associated critique? Being clear about this distinction may help teachers to avoid many difficulties and confusions.

Stewart and Rudolph (2001) urge teachers not to restrict lab-based or field-based work to the usual kind of *model-data fit* problems, or what Laudan (1977) calls "empirical problems": those problems in which students use an existing scientific model to solve problems for which the model is known (usually by the teachers but possibly by the students, too) to be perfectly adequate or try to revise the model in order to accommodate anomalous data. They should also engage students in confronting what Laudan calls "conceptual problems": (a) *internal* problems, concerned with identifying and correcting logical inconsistencies, self-contradictions, conceptual ambiguities or circularity; and (b) *external* problems, concerned with recognizing and attending to inconsistencies with other accepted models and theories. Addressing conceptual problems is an integral part of mod-elling and learning about modelling (see later discussion), tackling model-data fit problems (empirical problems) is oriented more towards learning and using particular scientific models and theories. Modifying and developing models and theories in the light of anomalous data may be part of both activities.

With regard to learning the models and theories produced by scientists, it is clear that students will not acquire the necessary breadth and depth of under-standing unaided. What they need is access to a teacher (or other knowledgeable adult) who is willing and able to provide advice, guidance, encouragement and support through the careful arrangement of appropriate curriculum materials and learning tasks. My own views on teaching and learning science content have been expressed at length in Hodson (1998a) and space precludes any extended discussion here, save to note the importance of *personalization* of learning – that is, attending to the particular needs, interests, experiences, aspirations and values of every learner, and to the affective and social dimensions of learning environ-ments. The key lies in the creation of a supportive and emotionally safe learning environment for all students. The notion of *scaffolding* is particularly helpful (Wood et al., 1976; Collins et al., 1989; Stone, 1993, 1998; Hogan & Pressley, 1997). Scaffolding involves the teacher (or a knowledgeable "other") adjusting the complexity of the learning task so that the learner is able to solve a problem, perform a task or achieve a goal that would be beyond their unassisted efforts. Scaffolding should not alter the overall structure of the learning task. Rather, it should adjust the precise nature of the learner's participation as the teacher as-sumes responsibility for those aspects of the task that require knowledge or skills that the learner does not yet possess. In a scaffolded task such teacher assistance is only considered productive if the learner has fully comprehended the purpose and structure of the task, understands why the particular strategies were employed, and appreciates how the conclusions have been reached. Only in these circumstances will assistance be educative, criticism productive and feedback effective. These mattters are discussed at greater length in Hodson and Hodson (1998a,b).

The most appropriate form of scaffolding depends, of course, on the nature of the learning task, the particular learners involved, and the specific educational context. It also changes with time. As learners gain experience with the task or other similar tasks the amount of support can, and should, be decreased (the notion of "fading"). Fading the scaffolds encourage students to apply what they have learned from previous scaffolds to the current learning task. Early scaffolding steps are largely a matter of finding connections between what learners already know and have experienced, establishing a context that is meaningful and relevant to the students, gaining their interest and commitment, identifying similarities and differences between the new situation and the old, and ensuring that terminology is properly understood. Second phase scaffolding is largely a matter of ensuring an appropriate level of cognitive demand – that is, locating learning tasks in the *zone of proximal development*[6] and reducing the number of steps in the learning task to a manageable number. Once complex tasks have been broken down into more manageable chunks, the teacher might carry out some of them, the student carry out some, while the remainder are shared. During this phase the teacher highlights and directs attention to important features to be learned, emphasizes key vocabulary and introduces new terminology, functions as a kind of "external memory" by providing appropriate information as it is required, assists the learner in analyzing tasks, provides guidance and advice on learning strategies, models and demonstrates critical steps in procedures and furnishes evaluative feedback on student performance. Other aspects of scaffolding include reducing linguistic complexity, eliminating sources of distraction and reducing the level of "noise" associated with hands-on activities (see below). The first step in stage 3 scaffolding – ceding control and responsibility for learning to the learner – begins when students are encouraged to elicit the help they feel is needed for task clarification or completion. Instead of learners simply receiving teacher guidance as and when the teacher determines, there is a negotiation of the nature and extent of teacher participation. The learner assumes increasing control, with the teacher acting as support as and when required, and what was previously guided by the teacher is now guided and directed by the student. Through increasingly skilful questioning (something else that has to be taught to students, of course), students assume responsibility for structuring the task and allocating specific sub-tasks. In other words, students assist the teacher to assist, and teachers and students co-construct both the learning activity and the knowledge that is developed through engagement in it. Knowing what kind of teacher assistance is required, and when it is required, is an important part of learning and is the basis of self-knowledge and metacognitive awareness. Teachers should be aware that providing assistance at too high a level can sometimes be counter-productive, and that assistance given without request may be perceived by some students as interference. Chinn and Samarapungavan (2008) provide some useful advice for teachers in deciding where and when to intervene in their very detailed analysis of the kind of obstacles likely to prevent students from developing good understanding of models, including (i) poor understanding of epistemological issues (NOS knowledge), (ii) clashes with existing conceptual understanding, (iii) incompatible views of learning (principally, an achievement orientation rather than orientation towards understanding) and (i) adoption of inappropriate learning strategies.

"Hand-over" is complete when students no longer need teacher assistance and are capable of self-direction. How quickly this happens varies enormously from student to student and context to context. Reigosa and Jiménez-Aleixandre (2007) point out that transfer of responsibility from teacher to student is hindered by excessive task difficulty (but note that inadequate task difficulty will not facilitate development) and by an overall school culture that encourages conformity, fosters a task or achievement orientation at the expense of a learning orientation, and promotes a teacher-centred curriculum approach. Problems can also arise when the social climate is such that more able students have no interest in sharing their ideas or when they take personal control of group tasks, thereby marginalizing other group members. Before leaving this discussion, it is worth noting that it may take a considerable time for students to develop a full understanding of some of the so-called "big ideas" of science, that is, the content that collectively constitutes a major component of critical scientific literacy. Evidence and arguments for a particular view may be accumulated over several lessons (or several topics) as students use their developing understanding to explain new and increasingly complex phenomena, make predictions in novel circumstances, and so on. Krajcik (2008) notes that experience of the increasing explanatory and predictive power of a particular model constitutes the *fruitfulness* component of the well-known model of conceptual change devised by Posner et al. (1982) (see discussion in Chapter 3).

PRACTICAL WORK

Before leaving this discussion of teaching and learning strategies it may be helpful to comment on the use of practical work.[7] I suggested in Chapter 1 that we could regard science education as comprising four major components: (i) *learning* science and technology – acquiring and developing conceptual and theoretical knowledge; (ii) learning *about* science and technology – developing an understanding of the nature and methods of science, appreciation of its history and development, awareness of the complex interactions among science, technology, society and environment, and sensitivity to the personal, social and ethical implications of particular technologies; (iii) *doing* science and technology – engaging in and developing expertise in scientific inquiry and problem-solving, developing confidence in tackling a wide range of "real world" tasks and problems; and (iv) *engaging in sociopolitical action* – acquiring (through guided participation) the capacity and commitment to take appropriate, responsible and effective action on science-related and technology-related matters of social, economic, environmental and moral-ethical concern. It makes good sense for teachers to draw a clear distinction between the first and the third categories when designing practical work, regardless of whether it involves teacher demonstration or hands-on work by students.

When the purpose of the activity is to bring about particular conceptual understanding it is important that results are obtained that lead directly and unerringly to that understanding. While wayward results in one investigation do not necessarily undermine the teacher's case for the target understanding, they do constitute a threat to students' belief in the rationale of science because the warrant for belief has been seen to shift from personally gathered data to teacher authority. Teachers using practical work to bring about clearly identified conceptual understanding can increase the likelihood of students "getting the right results" by simplifying the activity: reducing the number of steps, restricting the number of variables, using simpler apparatus and laboratory procedures, pre-assembling apparatus, pre-weighing materials, and using computers to capture and present data, or to monitor and control experiments. Collectively, these actions can be seen as re-ducing the level of "noise". If the "right results" still cannot be obtained, my response would be to *conjure* it – that is, use sleight of hand and other kinds of secret manipulation of equipment and materials to ensure the data match ex-pectations, as described by Nott and Smith (1995). Alternatively, teachers should use a computer simulation[8] or multimedia material, access data from a reputable source (for example, the library of microscopic images made available through websites at UMIST and the University of Adelaide), or replace the activity with some alternative active learning method. I take the view that practical work de-signed specifically for learning science (as defined above) is theatre. Illusion and make-believe are prominent and effective features of theatre; I see no reason why they should not be prominent and effective ways of promoting conceptual understanding.[9]

When a practical activity is primarily concerned with giving students ex-perience of *doing* science, obtaining the "right answer" is not an issue. In fact, generating "wrong answers" can sometimes help to focus attention on the complexities of good experimental design. Experimentation is not a simple, straightforward business, and scientists often need to go through several revisions of the experimental design before they find a robust and reliable approach. It is often the case, too, that experiments are adjusted and modified as the investigation proceeds, as it becomes apparent that events are not unfolding as planned. All of this experience constitutes learning *about* science as a consequence of doing science (see Chapter 7 for further discussion). In *doing* science, students have responsibility for posing questions, devising methods of inquiry, analyzing and interpreting data, reaching a conclusion, constructing a convincing argument for that conclusion, and communicating their methods, findings and conclusions to others. Because the questions make sense to the students, and are grounded in what they know and care about, these activities generate much higher levels of student motivation, interest and involvement. Student teachers are often told that motivation is the confluence of "the 4 Cs": control, choice, challenge and collab-oration. *Doing* science activities, especially challenging activities conducted in groups and designed by students, would appear to be among the most highly mo-tivating experiences we can provide. Through such experiences students become aware of the interplay of theory, design and evidence in ways that rarely occur with standard laboratory exercises. The effectiveness of practical work would be immeasurably improved if *students* were made aware of whether the practical activity in which they are engaged is a learning science or a *doing* science activity. Different purposes engender different attitudes to the activity and different responses to the data. Students need to know, in advance, whether the purpose of the activity is to experience phenomena and events, demonstrate an idea, principle or theory, acquire a specific bench skill, make a measurement, test a hypothesis, design and conduct an investigation, manipulate some variables, collect lots of data to see if a pattern can be detected, or "just to see what happens". These matters are explored at length in Hodson (1990b, 1993b, 2005). Evidence of a serious mismatch between teachers and students in views about the purpose of practical work can be found in Wilkinson and Ward (1997), while the importance of clarity about the learning goals of particular lessons involving practical work is emphasized by Nakhleh et al. (2002) and Séré (2002).

In the context of the earlier discussion of models and theories, it is important to emphasize that *learning* science involves students acquiring a robust understanding of, and ability to use appropriately, the models and theories that comprise important scientific knowledge. Learning *about* science involves students reaching an appreciation of models in theory building and understanding the history and development of particular items of scientific knowledge. *Doing* science involves students creating, criticizing and testing their own models and seeking to reach consensus in class. In Tsai's (1999) study at the grade 8 level, students with constructivist views of science, as measured by a Chinese version of Pomeroy's (1993) questionnaire, were better able to negotiate meaning with their peers during laboratory activities than students with an empiricist orientation.[10] They engage more frequently and more effectively in interpreting and explaining experimentally generated data, investigate theoretical aspects more fully, and express a strong preference for open-ended inquiry (rather than closed-ended, teacher directed activities) and may lose motivation when their preferences are not met. In contrast, empiricist oriented students see practical work mainly as a means of confirming facts presented by the teacher or located in the textbook. In a direct response to the learning science, learning about science and doing science categorization, Tsai (1999) concludes: "Constructivist students believe that laboratory experiences help them understand the scientific concepts involved in a richer or more concrete manner (the first purpose), understand the processes of science and where scientific knowledge came from (the second purpose), and how scientists do/did science (the third purpose) ... Empiricist learners seemed to not perceive such purposes" (p. 671).[11] It seems, as expected and discussed in Chapter 2, that NOS views inform students' responses to practical work, while their involvement in practical work (especially open-ended inquiries) impacts their NOS views – a matter to be explored further in Chapter 7.

A ROLE FOR HISTORY OF SCIENCE

It has been widely argued that historical case studies can assist concept acquisition and concept development (see Matthews, 1994; Bevilacqua & Giannetto,

1995; Leite, 2002; Stinner et al., 2003). The argument goes along the following lines: (i) a thorough understanding of conceptual niceties can only be acquired by putting oneself in the same position as the "pioneers" of the discipline and appreciating the specific problems with which they grappled; (ii) students learn from recognizing, analyzing and correcting their own mistakes in the light of descriptions of the mistakes of others; (iii) individual learning histories in science sometimes mirror the historical development of scientific concepts. If the third argument is valid, it follows that knowledge of the historical development of a discipline might help teachers to anticipate, understand and deal with some of the conceptual difficulties students are likely to encounter and the kind of misconceptions they are likely to hold (Wandersee, 1985, 1990; Steinberg et al., 1990; Barker, 1995; Halloun, 2007). Studying episodes in the history of science will indicate the kind of questions teachers should ask, the observational evidence they should seek, and the experiments they should conduct in order to enhance student learning. Studying episodes of strong resistance to new ideas, both within the scientific community and in the wider public arena, might help teachers to anticipate likely sources of resistance among students. In that sense, history of science is just as much for teachers as it is for students. It is both intriguing and comforting for students to discover that eminent scientists of the past had views similar to their own. They realize that it was a perfectly good and sensible view to hold, even though it later turned out to be incorrect, and so they are better placed, both cognitively and emotionally, to engage in critical confrontation of other ideas they currently hold.

In developing their notion of *Large Context Problems* (LCPs), Stinner and Williams (1998) add a motivational element to claims for the educational benefits of including historical studies in the curriculum: the excitement of sharing "the frustrations and rewards of the intellectual struggles of those who have made important scientific discoveries" (p.1029). LCPs are "science stories" built around powerful unifying ideas; they are chosen for their curriculum saliency. As Stinner (1992, 1995, 1996, 2006) reminds us, setting science learning in context involves consideration of three contexts: the historical context (including the prevailing sociocultural and economic situation), the conceptual context (the kind of conceptual questions and practical problems addressed), and the procedural context (the investigative methods available or developed in response to a particular inquiry). Each of these contexts is complex and multilayered, so that in addressing an LCP students may have to pose, address and solve many other subordinate problems (see Winchester (2006) for a discussion of LCPs and the history of similar approaches, including the PBL approach pioneered by the McMaster University medical school). Not surprisingly, as Wang and Marsh (2002) report, teachers using history of science in class devote most attention to the historical context and least attention to the procedural context. Elementary school teachers tend to have particular difficulty with the procedural context. Of course, as Klassen (2006a) reminds us, there are two other contexts for teachers to consider. First, the social context in which the students confront an LCP, and the extent to which that might be designed to reflect the ways in which scientists themselves confront problems, both within research groups and within the wider community. Second, the affective context and the need to engage students' interests, feelings and emotions. It is here that narrative has a powerful role to play (see Chapter 9).

Stinner and Williams (1993) argue for the value of LCPs in teacher education programmes on two grounds: first, well-designed LCPs make very effective curriculum resources; second, their construction can play an effective role in sensitizing teachers to the value of historical studies and awakening awareness of students' alternative frameworks of understanding. An interesting variation on this theme is Lochhead and Dufresne's (1989) idea of using "dialogues" between important scientists – for example, between Aristotle and Galileo – as a way of elucidating conceptual difficulties and illustrating the chain of argument connecting theory and observation. More recently, Hoadley and Linn (2000) have described an approach simulating "historical debate" (between Kepler and Newton) using online asynchronous discussion.

It is important not to neglect the key learning *about* science goals associated with concept acquisition and development, no matter what pedagogical strategies are employed. Principal among them is an understanding of the ways in which expressed models achieve the status of consensus models through the complex processes of debating and testing within the community of practitioners. A second important aspect of learning *about* science concerns the history and development of particular ideas in science. How did concepts arise and develop? Why were they displaced by others? What instrumentalist and realist episodes were involved? What evidence was generated? What arguments were employed? For example, Justi and Gilbert (1999) have identified eight distinctive models that played a part in the history of chemical kinetics, while Borges et al. (1998) have delineated five historically significant models in the study of magnetism. A third important aspect of learning *about* science focuses on the ways in which individual scientists utilize models as tools for thinking (Hesse, 1966; Gilbert & Mulkay, 1984; Giere, 1988; Stavy, 1991; Nersessian, 1992, 1995; Ogborn & Martins, 1996; Penner et al., 1997; Ault, 1998), a matter that is addressed in the following section. By studying episodes in the history of science, especially through original papers, and by reading biographical and autobiographical material, students can gain insight into the ways in which particular scientists thought about conceptual and procedural issues, designed investigations and went about solving problems. They are given graphic illustration of the interplay of theory and experiment: scientists design new experiments to counter an opponent's theoretical position and the opponent counters with a modified theory, a new experiment, or a technological innovation capable of making more accurate readings or gathering an entirely new kind of observational evidence. There may be value, once in a while, in students replicating experiments that were highly significant in those respects (Kreuzman, 1995; Heering, 2000, 2003; Hottecke, 2000; Binnie, 2001; Cavicchi, 2008). For example, Kipnis (1996) uses an approach based on guided discovery pedagogy that he calls the "historical-investigative" approach. Fowler (2003) advocates the use of original texts, though he warns teachers to be vigilant in ensuring that the mathematical complexity of the original text is not too daunting for students.

Other problems relate to language use, such as unfamiliarity with archaic vocabulary and forms of expression no longer in use, and discrepancies between the original and contemporary meanings of technical terms (see Stinner (1994) for an interesting discussion in relation to the concept of *force*).

In studying how particular conceptual models and theories were developed and established, students quickly realize that the supposedly unambiguous facts recorded in textbooks have not always been so unambiguous, nor have our taken-for-granted explanations of phenomena and events always been so well-established. Facts are not individual pieces of data but interpretations of observations linked in meaningful ways into explanatory systems, and underlying and underpinning these complex conceptual frameworks is a host of assumptions and a wealth of interpreted data, many of which are not apparent on first encountering them. Some models and theories are so familiar that their fundamental assumptions remain hidden, their inadequacies are glossed over, their layers of embedded meaning remain unexplored and their misleading aspects ignored. Historical studies can be used to show that scientific claims always hinge on evaluation of evidence, and that evaluation is a matter of argumentation and criticism both in individual laboratories and in the wider community. Kolstø (2008) argues that changing the common conception of students that science deals with facts to an understanding that it rests on evidence and argument is of paramount importance for democratic debate because the idea that science is factual implies that lay people have "no right to doubt" (p. 986). Historical studies enable theories to be "unpacked" to reveal their components and the influences that led to their construction, thereby leading students to a deeper understanding of how theoretical systems are constructed and what criteria must be satisfied in order for them to be validated (Hagen & Kugler, 1992; Castro & Carvalho, 1995; Weck, 1995; Justi & Gilbert, 1999; Niaz, 2002). Many important questions are raised: What constitutes a well-built theory? How are convincing arguments made? What factors are likely to influence acceptance and rejection by the community?

LEARNING MODELLING AND LEARNING ABOUT MODELLING

It seems that experts (whether in science or in some other field) not only know more than novices, they also have more accessible and usable knowledge because it is differently (better) organized and differently (more productively) deployed (Larkin, 1983; Carey, 1986; Chi et al., 1981, 1988; Feltovich et al., 1997). While novice problem solvers access concepts, procedures and equations one by one, experts access related clusters of relevant knowledge; and while novices tend to address superficial features of a problem, experts are able to use more powerful overarching principles to address the fundamentals. While novices use means-ends analysis and attempt to employ previously learned formulae and algorithms to situations they recognize as similar, experts are more holistic and attempt to solve problems from first principles, drawing on their rich base of domain knowledge and stock of experiential knowledge for exemplars. In other words, experts function at a more general level than novices. Experts are fast, accurate and flexible in solving problems in their chosen domain because many skills have become automatic, freeing up cognitive capacity for processing other aspects of the task. They also exercise strong metacognitive control over their reasoning (Glaser & Chi, 1988; Ericsson & Smith, 1991; Ertmer & Newby, 1996), spend a great deal of time analyzing and representing a problem before they start trying to solve it, and continue to monitor, evaluate and reflect as they proceed. From her analysis of important episodes in the history of science, Nersessian (1995) postulates that *modelling* is a crucial part of expert scientific thinking: scientists construct qualitative models, sometimes a series of intermediate models that approximately represent the target problem, as a means of deriving quantitative relationships and equations.

> Having expertise in physics requires, in addition to domain knowledge, facility with the domain-independent practice of "constructive modeling" ... a dynamic reasoning process involving analogical and visual modeling and mental simulation to create models of the target problem where no direct analogy exists. (pp. 204 & 207)

From a questionnaire-based study of 24 leading scientists, in a range of disciplines, van der Valk et al. (2007) conclude that practising scientists see models largely in the ways described earlier in the chapter. For example, a model is always related to a particular object, phenomenon, event, process, system or idea, and is created for a specific purpose. That purpose may be to describe phenomena, make predictions, assist experimental design and data collection, aid decision-making or stimulate theory development. Different models for the same phenomenon may co- exist, depending on the specific purpose in mind. There can be conflict between the need for a complex model that closely resembles the target and a simpler model that is easier to handle. In contrast to novices and non-scientists, experienced scientists seek to build coherent and inclusive models, with no significant gaps or internal uncertainties and inconsistencies. Non-scientists are often content to use incomplete explanations. Scientists subject the model to rigorous scrutiny and test by actively seeking counter evidence and contrasting instances, detecting flawed evidence and experimental error, and considering alternative interpretations. Once they have found a way to make sense of some aspects of the world, non-scientists may be reluctant to consider alternatives. Conversely, they may want the model to include everything they know, even if some elements are mutually contradictory. Of course, once they have become committed to a particular explanation, scientists may strongly resist attempts by others to falsify it (see Hodson, 2008, for a discussion of the factors influencing theory change that is appropriate for school science).

Regardless of one's position with respect to the argument that students should be encouraged to learn science via activities that mimic the thinking processes of creative scientists, there is considerable value in having students learn something about the ways in which real scientists utilize models in their thinking through case studies in the history of science, though I am certainly not suggesting that teachers use this approach for every topic in the curriculum. In their interview-based study of 12/13-year olds and 16/17-year olds, Grosslight et al. (1991) revealed three levels of student understanding of models: at level 1, there is a 1:1 correspondence between models and reality, with models regarded as toys or as incomplete copies of reality; at level 2, models are regarded as serving a specific communications purpose, in pursuit of which some aspects of reality are deliberately omitted, minimized or enhanced; at level 3, they are recognized as tools for thinking, for testing and developing ideas. None of the students in the study reached level 3, though Keys (1995) and Raghavan and Glaser (1995) show that experiences involving experimental design and computer modelling can assist most students in moving much closer and more rapidly to level 3 understanding.[12] Interestingly, 13 of the 14 student teachers in the study conducted by Crawford and Cullin (2004) were classified no higher than level 2 at the end of a purpose-built instructional module. Although there were some promising signs that these teachers would eventually gain a more complete understanding of the role of models in science (and in science education), by the end of the study only one had articulated the "informed view" that models play a key role in theory building and that scientists change models in the light of new data. In a survey of 39 Brazilian teachers by Justi and Gilbert (2003), elementary school teachers and secondary school teachers of biology fared no better than high school students, although secondary school physics and chemistry teachers had views much more in accordance with the accepted scientific viewpoint, attributed by the authors to the more extensive use of models in physics and chemistry.

Research suggests that many teachers lack background knowledge (particularly about the role and status of models in science) and do not have sufficient pedagogical content knowledge to teach about models and modelling success-fully (van Driel & Verloop, 1999; France, 2000; Justi & Gilbert, 2002b, 2003; Justi & van Driel, 2005; Oliva et al., 2007). For example, Justi and van Driel (2005) noted that the five preservice secondary teachers with whom they worked did not have comprehensive knowledge about models and modelling, and when and how to deploy them, though a purpose-built course on models and modelling did enhance their ability to engage students in modelling activities, and bolstered their confidence in doing so. A study of Dutch science teachers by van Driel and Verloop (2002) concluded that teachers can be divided into two groups: one group focuses on the conceptual content of specific models, mainly via a teacher-directed approach; the other group pays more attention to the nature, design and development of models, using more student-directed learning activities. A subse-quent study by Henze et al. (2007a,b) with teachers drawn from the same overall population, but following the introduction of a new curriculum that emphasizes NOS understanding and encourages teachers to adopt more student-driven and media-oriented learning methods, identified *three* sub-groups of teachers: type 1 teachers combined modelling as an activity undertaken by students with the learning of specific model content; in type 2 teaching, the learning of model content was combined with critical reflection on the role and nature of models in science; in type 3 teaching, the learning of model content involved both students' production and revision of models and critical examination of the nature of scientific models in general. The authors conclude that the new curriculum (*Public Understanding of Science*) has effected a substantial shift in classroom practice, although the authors note that there is still a need for professional development programmes (including opportunities for group-based critical reflection), purpose-made curriculum materials, and appropriate forms of assessment and evaluation.

It is clear that the complex knowledge and understanding about models, modelling, theories and theorizing encapsulated in the notion of critical scientific literacy is unlikely to arise from a single kind of learning experience. For example, in a study with A-level students (17–18 year olds), Leach et al. (2003) report that although two purpose-built lessons focused on the relationship between models and data (one in biology, one in physics) did enable some students to reach a very tentative view about the role and status of models, a significant number did not appear to know what the lessons were about and made little or no progress. Many were still unable to differentiate between "models based on theoretical ideas and data collected through experimental measurements" (p. 839). Kuhn et al. (1988) report that when grade 3 and grade 6 students were asked to provide evidence for a theory, or for their proposed model, they simply restated the model/theory. Moreover, when evaluating theories in the light of evidence, they frequently ignored or marginalized disconfirming evidence. Chapter 2 discusses some more encouraging research findings relating to students' understanding of models and theories. What these studies make clear is that teachers need to direct students' attention to the key NOS features (such as the relationships among observation, inference, prediction and explanation) and provide ample opportunities for reflection and criticism. Case studies of historically significant episodes in science and of contemporary scientific problem solving can play a crucial role, but they are insufficient in themselves. Students need practice in using models for specific purposes and opportunities to reflect on the relationship between purpose and design; they need experience of revising and adapting models to suit new situations and/or overcome any shortcomings in the model. They should be provided with opportunities to build, test and argue for their own models, and to gather evidence in support of them through student-led inquiry, something that is still somewhat uncommon in school science education. Part of this experience involves making a case for a particular model, giving and receiving criticism, reaching consensus and communicating conclusions to others – parents and family members, teachers, "friendly scientists", other students, and so on. If students are to feel satisfied with a particular model, or to express a preference for one model over another, they need some robust criteria. How well is the model supported by the evidence? How much evidence can be regarded as sufficient? Is that same evidence capable of supporting a different model? Is the model internally consistent and coherent? Does it make sense in terms of accepted theories in closely related areas? Does it have predictive power? There are two principal learning goals here: first, enhanced conceptual understanding through critical discussion of the evidence supporting (or not) a particular view; second, NOS understanding concerning the need for evidence and its rigorous appraisal.

One question that immediately springs to mind is whether the creative aspects of model construction can be taught. Not according to Morrison and Morgan (1999):

> When we look for accounts of how to construct models in scientific texts we find little on offer. There appear to be no general rules for model construction ... Some might argue that it is because modelling is a tacit skill, and has to be learnt not taught ... it is, some argue, not only a craft but also an art, and thus not susceptible to rules. (p. 12)

Nevertheless, as Justi and Gilbert (2002a) argue, there are some general principles that may be of value in assisting teachers to provide structured experiences for students. First, there needs to be clarity about the purpose of the model. Is it intended to provide a qualitative description of a phenomenon or event, establish a precise relationship between postulated entities, ascribe cause and effect, make a prediction or achieve some other purpose? Second, students need to be aware of their options in choosing the components of the model. Some models utilize only the readily observable (such as the heliocentric model of the solar system) or use analogy with the already familiar. With sufficient support, students can engage in building models like these at first hand (see below). On other occasions, modelling makes sense of observable phenomena by introducing unobservable entities such as *atoms* and *genes*, or postulating abstract concepts such as *force* or *energy*. There is little prospect of school students making these kinds of creative leaps, though well-chosen case studies can illustrate how scientists made these kinds of creative breakthroughs. Deciding on mode of expression comes next: should the model be expressed in concrete, visual, verbal or mathematical form? Valuable learning accrues from exercises in "translating" a model from one mode of expression to another, an activity that is enormously enhanced by contemporary computer software – "mindtools" as Jonassen and Carr (2000) call them.[13] The ability to "translate" is particularly important in chemistry, where students gain knowledge of chemical behaviour through hands-on activities in the lab or in fieldwork, interpret their findings in conceptual terms (using notions such as atoms, molecules, electron configurations and energy changes, for example), and further articulate their explanations with chemical equations and 3-D models of molecular structures and/or crystal lattices (Johnstone, 1991; Kosma & Russell, 1997; Tsaparlis, 1997; Hinton & Nakhleh, 1999; Kosma et al., 2000; Robinson, 2003; Erduran et al., 2007). Representational devices such as data tables, graphs, formulae and equations do not simply communicate thought and ideas, they do so in ways that enhance and develop those thoughts and ideas. They shape thought; they open up new possibilities. Consequently, acquiring the language of inscriptions and notations is a key element in critical scientific literacy, and utilizing this vocabulary is an essential step in modelling. Of course, using the language of mathematics is also a particularly important element of model building. However, many teachers delay the mathematization of scientific ideas, believing that it might encourage an emphasis on computation rather than understanding. Although I agree that it is often desirable to consolidate qualitative understanding before moving to a quantitative representation, there are occasions when quantitative relationships are easier to comprehend and more susceptible to study. Much depends, too, on the kind of mathematics teaching the students have experienced. Lehrer and Schauble (2000, 2003, 2006) have demonstrated how students' use of increasingly mathematical representation of ideas relating to plant growth opens up new opportunities for creative thought.

> Once students understood the mathematics of ratio and changing ratios, they began to conceive of growth not as simple linear increase, but as a patterned rate of change. These transitions in conception and inscription appeared to support each other, and they opened up new lines of inquiry. Children wondered whether plant growth was like animal growth, and whether the growth of yeast and bacteria in a petri dish would show a pattern like the growth of a single plant. (Lehrer & Schauble, 2006, p. 183)

Once the basic model is built, it has to be subjected to rigorous scrutiny – an activity that is best conducted in groups. When students give and receive criticism, and build on the ideas of others, they are mimicking the early stage of model building within the community of scientists, that is, the phase when individual scientists discuss their embryonic ideas with their immediate colleagues in a non-threatening and supportive environment (see Woodruff & Meyer, 1997). Such activities may not lead students directly to established scientific knowledge, and will certainly not lead them to generate *new* scientific knowledge, but they can lead to ideas that are new to the participants and that the students recognize as superior to their previous understanding in some important respect(s). This is the kind of classroom discussion that Carl Bereiter (1994) describes as "progressive discourse". At issue here are matters of clarity, accuracy, applicability, internal consistency, explanatory power (whether the model is in agreement with available data and can make successful predictions), and whether the model is consistent with what the students already know. Next is the stage of empirical testing. Students use the model to design an investigation to collect data capable of supporting or refuting the model, or aspects of it. If the model fails this testing stage, it should be revised and re-tested – provided, of course, that the design of the investigation was appropriate, the methods of data collection were valid and reliable, and the students had the necessary bench skills to collect good quality data. This is a timely reminder to students that it is not only the model or theory that is under test during an empirical investigation, but also the means by which data is collected and the chain of argument deployed.

If the model has survived thus far, students can proceed to articulate it in a suitable form for communication to others. They need to formulate a convincing argument for the validity and usefulness of the model (and, possibly, the rejection of others), including a clear statement of the evidence that constitutes the warrant for belief; they need to outline the scope and limitations of the model, and present their arguments in a way that conforms to the conventions of the community of

181

scientists. This entails familiarity with the form of scientific arguments, the forums in which they are presented (conferences, journals, etc.) and the language in which they are expressed. Chapter 8 provides a lengthy discussion of how Stephen Toulmin's model of scientific argumentation can be used to assist students in organizing the evidence used, the warrants and backings invoked, the qualifiers employed and the rebuttals raised by opponents or anticipated by the proposer in the construction of explanations, theories and models. Reiner (2000) and Reiner and Gilbert (2000) remind us that the processes of refining models, using models in problem solving and arguing in favour of a particular model can sometimes be enhanced by engaging in thought experiments.[14] As Reiner and Burko (2003) comment, thought experiments "force a learner to access tacit intuitions, explicit and implicit knowledge, and logical derivation strategies, and integrate these into one working thought process" (p. 380). Unlike scientists, who often plan the use of coherent and elegant thought experiments as a formal argumentation device, students intuitively and spontaneously engage in thought experiments as a way of solving their immediate problems as they struggle to build and refine models and theories (Reiner, 1998, 2006). If models do not survive this kind of testing there is little point in proceeding to an empirical testing phase.

Dynamic modelling tools, including spreadsheets, expert systems, systems modelling tools and microworlds can also play a central role (Jonassen & Carr, 2000). Newly developed (and developing) computer-managed, multimedia packages provide enormous scope for enhancing both student and teacher understanding of models and their capacity to build and use models. These packages enable phenomena and events to be witnessed (albeit at second hand) via video; they can present models of these phenomena and events in conventional 2-D form (through diagrams), "virtual 3-D form" (via dynamic computer graphics) and in graphical and mathematical form; they provide opportunities for teachers and students to manipulate and juxtapose these images, expand them, alter them and even create new ones (Buckley, 2000). In the early stages of model building, computer programs such as SemNet, Learning Tool, Explanation Constructor, Thinkertools, Artemis, Inspiration, Mind Mapper, MindMan, Worldwatcher, AxonIdeaprocessor, VisiMap and ActivityMap enable learners to identify important concepts, graphically inter-relate them in multidimensional networks, and label the relationships between them (White & Frederickson, 1998, 2000; Barab et al., 2000; Jackson et al., 2000; Krajcik et al., 2000; Edelson, 2001; Reiser et al., 2001; Fretz et al., 2002; Kim et al., 2007). For example, Sandoval and Reiser (2004) describe the use of an electronic journal, *ExplanationConstructor*, to assist students in compiling a detailed and critical record of their investigations and support the construction and evaluation of explanations. As a scaffolding tool, *Explanation-Constructor* supports students in their attempts to build coherent, causal accounts of the data they collect via a series of prompts concerning the need to make systematic use of data in seeking to establish cause and effect, recognize possible limitations, and rule out alternative explanations. Similar prompts assist students in criticizing explanations developed by peers and in developing counter claims. Reflection on model building activities and explicit comparison with the work of practising scientists (via case studies) ensures a strong measure of authenticity in the curriculum, as discussed in Chapter 7. Conversation tools, such as online discussion forums and email, enable students to build effective learning communities as they clarify and establish shared understandings and representations.

ENSURING BETTER LEARNING

While the acquisition of modelling expertise might best be described as "a lengthy and somewhat haphazard process, characterized by a high level of uncertainty, partial success, and even failure" (Justi & Gilbert, 2002a, p.374), there is some evidence that students do improve their model building abilities with experience and carefully sequencing teaching (Webb, 1994; Raghavan et al., 1998a,b; Lehrer & Schauble, 2000; Smith et al., 2000; Saari & Viiri, 2003; Taylor et al., 2003; Acher et al., 2007; Kawasaki et al., 2004; Lin & Chiu, 2007; Prins et al., 2008; plus Duschl et al. (2007) for a detailed review and critique of relevant research). This latter point regarding carefully sequenced teaching is crucial: understanding the nature of models and modelling will not occur unaided; it requires purposeful and sustained interventions. A number of factors seem to be significant. First, as noted above, clarity of purpose – that is, whether the model is being constructed to describe observable behaviour, establish a cause and effect relationship, or for some other purpose. Second, familiarity with and interest in the relevant phenomena and events. Third, adoption of a learning perspective, rather than a performance or achievement perspective. Fourth, sensible sequencing, so that students gain experience with simple model forms before proceeding to more complex and abstract models. Fifth, opportunities for guided reflection on key aspects of models and modeling. Sixth, extensive use of student models for solving problems and making predictions. Seventh, focus on limitations as well as productive features of models. Eighth, a non-threatening and supportive classroom climate. Ninth, a sense of ownership. Tenth, sufficient time. It almost goes without saying that the use of *authentic* modelling activities in which the development and critique of models is linked to real world problems and socioscientific issues is both an invaluable way of reinforcing the NOS learning goals and a powerful aid to building motivation through a sense of ownership. However, as Prins et al. (2008) point out, real world problems sometimes demand a level of conceptual understanding beyond that available to school-age students.

Schauble (2008) comments: "There appears to be no one point when model construction begins, and no point when we would agree that it is completed. Instead, it is a very complex form of thinking, one that includes aspects that develop only over the long term and only in situations where they are valued and supported" (p. 55). Because the thrust of model building activities is the development and elaboration of students' ideas, and clarification of the reasons for building them, it is important that there are regular opportunities to expose those ideas to critical scrutiny, both within and among student working groups. By making their ideas explicit, and by giving and receiving criticism, students concretize, clarify and systematize their ideas, recognize shortcomings and areas of confusion and uncertainty, and gain support, encouragement and advice for further refinement. In Schauble's (2008) words: "Students who are asked to explain and justify their explanation of phenomena, and to be accountable for restating, elaborating, and constructively critiquing the contributions of their classmates, are likely to understand over time that not everyone interprets the world in the same way" (p. 55). When students are required to give reasons and evidence for the models they develop, and to present the rationale in a clear and straightforward way, attention can be focused on the nature of scientific arguments (see Chapter 8). These activities also help students to develop their ability to communicate effectively and purposefully, and to develop a sense of audience – matters that will be addressed in Chapter 9. Maia and Justi (2009) point out the crucial role played by the teacher: reminding students to take account of other accepted knowledge and to acknowledge and respond to the ideas of other students; providing new information and new contexts against which students can test their developing ideas; urging students to clarify meanings, consider alternative modes of data representation, and so on. Given these conditions, and this kind of classroom climate, it seems that even elementary school students can build models that extend beyond the observable, and can recognize these models (and the models of scientists) as conjectural and provisional. Even so, it seems that students (and adults, too) always prefer simple models to more complex ones. Even those high achieving students who successfully use sophisticated models in examinations will opt for simpler but less adequate models for what we might call "everyday explanations" of chemical phenomena (Coll & Treagust, 2003a). Similarly, Perkins and Grotzer (2005) note that although students may use a complex relational causal model in most of their reasoning about air pressure, they may still revert to using a simple linear model when accounting for hurricanes. And students of all ages, and across a variety of cultures, express a preference for simple "realistic-looking" models (e.g., space-filling and ball-and-spoke 3-D models of molecular structures and simple octet rules for chemical bonding) (Coll, 2006).

Treagust et al. (2002) have developed a useful 27-item questionnaire, *Students' Understanding of Models in Science* (SUMS), to assess understanding in terms of five themes: models as multiple representations, models as exact replicas, models as explanatory tools, how models are used, the changing nature of models. Their studies with some 200 students in Years 8 to 10 at two high schools in Perth (Western Australia) reveal some curious inconsistencies in student understanding. For example, although 60% of students recognized that a range of scientific models can be developed to deal with different aspects of a phenomenon, some 43% believed scientific models are exact replicas of reality. As with many other aspects of NOS, students' understanding of the instrumental role of models generates both encouraging and discouraging data. Treagust et al. (2002) report that while many students at grades 8 and 10 appreciate the value of models in assisting their own learning (by providing different views and perspectives of an object, phenomenon or event), substantially fewer recognize that scientists use these same features of models as tools for thinking and for building theories.

While it is encouraging that half the students see the usefulness of models in making predictions, it is disturbing that up to 43% hold the naïve realist view that scientific models are exact copies of reality, a view held by most of the student teachers in Smit and Finegold's (1995) study.[15] However, in the study by Coll and Treagust (2003a), students were often well aware that their mental models of metallic bonding do not describe the real world, but they continue to use them because they are adequate for the task of predicting metallic behaviour. They recognize that an elaborate realist theory exists, but in these circumstances it is not necessary to invoke it. Although it may need further refinement before it is robust enough to link understanding directly with curriculum experiences, the SUMS instrument has considerable potential in assisting teachers to design and implement more effective ways of enhancing students' knowledge about models and their ability to engage in modelling activities, especially when SUMS data are used in conjunction with findings from insightful studies of the development of teachers' content knowledge, curricular knowledge and pedagogical content knowledge about models and modelling, such as those conducted by Justi and van Driel (2005). Interestingly, a study by Chittleborough et al. (2005), using a six-item questionnaire instrument, *My Views of Models and Modelling in Science* (VOMMS), developed by Treagust et al. (2004) from Aikenhead and Ryan's (1992) VOSTS item bank, showed much improved understanding of models among students as they proceed through grades 8 to 11: 25%, 34% and 23% of students in grades 8, 9 and 10 described models as "accurate duplicates of reality" but only 8% of grade 11 students did so. This may be very encouraging evidence that the recent curriculum emphasis on understanding the nature of science is beginning to pay dividends. Because VOMMS is designed to enable students to explain why they opt for a particular response to questionnaire items, the authors were also able to gain valuable information about the way students' understanding of the changing nature of models in science, and their understanding of why there are sometimes different models for the same phenomena and events, develop as they are exposed to NOS-oriented instruction. It seems that such experiences also enable students to develop an increasingly sophisticated understanding of the role of models in teaching and learning science. It is worth noting that the experienced science teachers interviewed by van Driel and Verloop (1999) generally held the view that models are "simplified representations of reality" (p. 1150), used for specific purposes such as describing, predicting and developing explanations. In a subsequent study of nine experienced teachers, Henze et al. (2007a,b) found a combination of two potentially conflicting views: belief that a model is a sim-plified reproduction of reality co-existing with recognition that models are the products of human thought and creativity. Whether these views are conflicting or not, and whether teacher educators need to be concerned about it, hinges on exactly what teachers understand by the term "representations of reality".

WORKING IN GROUPS

In traditional classrooms, all classroom talk is orchestrated by the teacher. Students speak only when they have been granted permission to do so (usually in response to a raised hand) or have been nominated by the teacher. There is little or no talk between and among students, or initiated by students. The teacher decides the focus for discussion and is the final arbiter of what views should be accepted. Such a situation is not conducive to the exploration of ideas and the engagement in criticism and argument that constitutes model building and model appraisal. If students are to understand how scientific knowledge is negotiated within the community of practitioners they need some direct experience of critique and negotiation. They need the opportunity to construct, discuss and debate the merits of their ideas with others. If they are to achieve the intellectual independence we seek, students have to be afforded a substantial measure of responsibility for their own learning. Student-led discussion in small groups is ideal for supporting students as they generate theoretical explanations, build on each other's ideas and subject ideas to rigorous criticism. Subsequent large group discussion or formal presentation encourages clarity of expression and careful consideration of possible counter arguments. Van Zee and Minstrell (1997) contrast traditional teacher-dominated classroom talk with what they call "reflective discourse", during which three conditions are met: "(i) students express their own thoughts, comments and questions, (ii) the teacher and individual students engage in an extended series of questioning exchanges that help students better articulate their beliefs and conceptions, (iii) student/student exchanges involve one student trying to understand the thinking of another" (p. 209). Following a collaborative research and professional development project involving teacher educators and elementary school teachers, van Zee et al. (2001) concluded that students ask more questions, and more perceptive questions, when they are invited to do so and feel comfortable and well-supported in doing so, when they are familiar with the content and context of the discussion, and when teachers allow them to work together without direct supervision.

Mercer (1995, 2000) identifies three distinctive categories of student talk in collaborative groups: *disputational, cumulative* and *exploratory*. Disputational talk is simply an exchange of opposing views in which disagreements are emphasized. There are no apparent attempts to pool ideas, reach decisions or engage in constructive criticism of ideas raised by others. In cumulative talk, students build positively but uncritically on what others say. It typically features repetitions, confirmations and elaborations. In exploratory talk, students engage critically with each other and support each other in collaborative reconstruction of ideas. When challenges are made they are supported by arguments; alternative viewpoints are suggested. There is still a crucial role for the teacher: as scientific expert, guide, critical friend and facilitator. First, laying down guidelines and rules concerning how to participate successfully in group discussions: how to argue and negotiate, how to listen, how to give and receive criticism, and so on.[16] Second, pushing students to ascend the hierarchy of talk categories (disputational, cumulative and exploratory). Third, modelling what constitutes a good argument and providing clear advice on how to present a carefully reasoned and well-articulated argument, with appropriate evidential, methodological and theoretical support – matters that will be revisited in Chapter 8. At times, teachers may need to solicit ideas, guide students in appraising, comparing and contrasting them, and assist them in resolving differences and incompatibilities. From time to time, they may need to supply additional data and ideas. It is important that teachers point out any deficiencies in student reasoning. Ford (2008a) expresses this position particularly well: "The teacher has a voice, and a privileged one, but it is through the role of critiquer that the students themselves are expected to play and are learning to play. Thus, the teacher's critiques should function not only to identify errors, but also to model the kind of thing that students are expected to do with their peers ... and with their own knowledge claims" (p. 420). Chapter 9 deals with the issue of classroom talk at much greater length.

It is reasonable to suppose that metacognitive skills, social skills and organizational skills could be factors influencing the success of model building activities, and that students with well-developed skills in these areas will participate more effectively in the processes of reasoning, arguing, persuading and reaching consensus. Research evidence is moderately encouraging on this matter. For example, Hogan (1999) developed an intervention strategy for grade 8 students called *Thinking Aloud Together*, which aims to help *individuals* understand the nature of collaborative reasoning (and its significance to scientific practice), the difficulties it sometimes presents, and the tentative nature of its conclusions, and to help *groups of students* plan, monitor, regulate, reflect on, and evaluate their activities. The experimental group did achieve a better understanding of the nature of collaborative reasoning and were able to articulate the processes and procedures in which they had engaged, though they were no more successful than the control group in applying their conceptual knowledge to a novel situation or using scientific reasoning to address an ill-defined problem. In an earlier study, Hogan (1998) identified a number of roles consistently adopted by particular students during model building activities. Four roles promoted the group's reasoning processes and led to successful outcomes: promoter of reflection; contributor of content knowledge; creative model builder; mediator of group interactions and ideas (Hogan's terminology). Four roles had little or no impact, or had an inhibiting effect: promoter of acrimony; promoter of distraction; promoter of simple task completion or unreflective acceptance of ideas; and reticent participant. Clearly the social and interpersonal skills that students bring to the group can be just as influential as their scientific knowledge and cognitive abilities. Individuals will vary enormously in terms of what they can and are willing to contribute: conceptual knowledge, NOS knowledge, other beliefs, values, attitudes, commitments and dispositions, cognitive skills, metacognitive knowledge and skills, motivational attributes, organizational and regulatory abilities. Groups will be effective to the extent that group members are able and willing to contribute and to utilize the contributions of others. In any discussion group there will be students who look to dominate and exercise control, those who rarely if ever contribute, those who slavishly accept the views of socially prominent or intellectually capable individuals, and those who seem unable to listen to others or consider their views. Bianchini (1997, 1999) presents some disturbing evidence of students who are perceived as low status by their peers being denied access to materials and excluded from group discussions. Alarmingly, these are predominantly students from groups that have historically under-achieved in science and are generally under-served by science education – that is, in the American context in which Bianchini works, female students, Latinos and African Americans. Herrenkohl et al. (1999) report improved conceptual understanding of the problem under consideration and enhanced ability to build explanations and solve specific problems (in the context of a topic on floating and sinking) when teachers use a combination of (i) explicit teaching of three common reasoning practices (predicting and theorizing, summarizing results and relating predictions to theories), (ii) explicit guidance on roles that students can take in class discussions to help them monitor and encourage their own and others' thinking (such as equitable task distribution, questioning each other and providing feedback), and (iii) opportunities for students to publicly document their changing views and ideas. However, there was no enhancement of students' overall understanding of the nature of scientific theories.

Before leaving this discussion it is important to be reminded that changing one's mind (in this case, with respect to a model or theory) is not always a simple, straightforward business. There can be very significant emotional, attitudinal and social barriers to be overcome, and there can be quite severe emotional and social consequences attending a change of position. As Lemke (2001) comments, "It is not simply about what is right or what is true in the narrow rationalist sense; it is always also about who we are, about who we like, about who treats us with respect, about how we feel about ourselves and others. In a community, individuals are not simply free to change their minds. The practical reality is that we are dependent on one another for our survival, and all cultures reflect this fact by making the viability of beliefs contingent on their consequences for the community" (p. 300). Unfortunately, says Lemke (2001), many science educators fail to recognize these realities: "An apparent assumption of the conceptual change perspectives in science education is that people can simply change their views on one topic in one scientific domain, without the need to change anything else about their lives and their identities" (p. 301). Both teachers and students need to recognize that each of us, whether student, teacher, scientist or whatever, is a member of a number of sub-cultures (see discussion in Chapter 4), and that we necessarily align with particular views and adopt particular positions with respect to social and cultural conflicts that extend far beyond the classroom, but may have classroom manifestations. Discussion of the affective and social dimensions of learning is outside the scope of this book, though it is important to acknowledge their significance (see Pintrich et al., 1993; Sinatra & Pintrich, 2003; Alsop, 2005; Sinatra, 2005; Johnston et al., 2006). When scientific knowledge and scientific reasoning are used to address socioscientific issues, the significance of affective and social factors becomes even more pronounced. So, too, do issues of power and privilege. As Delpit (1988) reminds us, a "culture of power", with codes and rules, pervades all social institutions, including schools. She notes that these rules reflect the norms, conventions and values of those who hold positions of power. Delpit also notes that those with power are often unaware of this culture of power, or they are unwilling to acknowledge it, while those with less power are often uncomfortably aware of its existence. Making the rules explicit and discussing strategies for using them, criticizing them and (possibly) changing them are key elements in fostering upward social mobility.

THE ROLE OF MODELS IN SCIENCE TEACHING

Students can also learn much about modelling from teachers' deployment of teaching models, especially when accompanied by opportunities for reflection and critical questioning. Teachers use a variety of 2-D and 3-D concrete models, diagrams, metaphors and analogies to render the abstract and complex ideas of science more tangible and accessible, organize and codify what would otherwise be a collection of disconnected facts, help students to see the relevance of their experiences to the concerns of the theory (or vice versa), and/or adjust the level of cognitive demand of a topic to suit a particular group of learners (see Harrison & Treagust (1998, 2000) for a useful typology of teaching models). The success of concrete models resides in the students' spatial awareness, which Barnea (2000) characterizes in terms of a three-item hierarchy: *spatial visualization* (the ability to understand 3-dimensional objects from 2-dimensional representations), *spatial orientation* (the ability to imagine what a representation will look like from a different perspective) and *spatial relations* (the ability to visualize the effect of operations such as rotation, reflection and inversion – that is, to mentally manipulate objects). Clearly, such skills are essential to a proper understanding of, capacity to use, and ability to create concrete models. They are also central to an understanding of other kinds of teaching models. How students develop these skills, and can be provided with compensatory learning experiences when progress is slow, is a matter in dire need of further research. One major area of research interest focuses on computer-based visualization tools. For example, the MacSpartan software enables students to view and rotate molecules, measure bond angles and lengths, modify molecules and construct new ones, thereby helping students to understand chemical concepts that are difficult to convey using static 2-D and 3-D models (Crouch et al., 1996).

Substantial research attention has been directed to other aspects of teaching models. For example, much has been written about (i) the characteristics of good analogies and metaphors[17] (Zeitoun, 1984; Dagher, 1994, 1995 a,b; Duit, 1991; Thagard, 1992; Nashon, 2004; Aubusson et al., 2006; Orgill & Bodner, 2006; Hart, 2008), (ii) the best tactics for their classroom deployment (Glynn, 1991; Treagust et al., 1992, 1998; Clement, 1993; Harrison & Treagust, 1993, 1996, 1998, 2000; Harrison & Treagust, 1993, 1996, 1998, 2000; Treagust, 1993; Venville et al., 1994; Newton & Newton, 1995; Ogborn & Martins, 1996; Heywood & Parker, 1997; Iding, 1997; Nashon, 2000, 2003; Baker & Lawson, 2001;

Yanowitz, 2001; Heywood, 2002; Snir et al., 2003; Besson & Viennot, 2004; Blake, 2004; Paris & Glynn, 2004; Chiu & Lin, 2005; Harrison & de Jong, 2005; Justi & Gilbert, 2006; Oliva et al., 2007; Kim, 2009), and (iii) their value in raising students' levels of motivation for learning (Thiele & Treagust, 1994a; Glynn & Duit, 1995; Glynn et al., 1995; Jarman, 1996; Venville & Treagust, 1996; Gay, 2002; Paris & Glynn, 2004; Harrison, 2006). Little is to be gained by rehearsing the various arguments here or cataloguing the research findings, save to note that careful teacher exposition is essential if students are to distinguish clearly between the analogue and the target concept and conceptual relationships it is meant to represent and make effective use of the analogy/metaphor in concept acquisition and development (Thiele & Treagust, 1991, 1994a,b; Dyche et al., 1993; Dagher, 1995b; Newton & Newton, 1995; Smit & Finegold, 1995; Solomon, 1995; Saari & Viiri, 2003; Crawford & Cullin, 2004). Interestingly, when teachers use analogies in their teaching it seems to have beneficial impact on their own conceptual understanding as well as on the understanding of their students. In addition, because analogies focus attention very directly on conceptual relationships, their use seems to direct much closer attention to the conceptual understanding of individual students, leading to more frequent teacher-student and student-student interactions, a wider range of explanations, examples and applications, and the use of less formal language (James & Scharmann, 2007).

It should be noted that considerable care is necessary in interpreting research findings because different researchers focus on different aspects of learning: recall (immediate or delayed), comprehension, deployment in problem solving, and so on. A number of authors (Jarman, 1996; Harrison & Treagust, 2000; Nashon, 2003; Harrison & de Jong, 2005) have pointed out the importance of discussing the limitations of analogies with students, a stage that is given special status (as step 6) in Glynn's (1991) *Teaching with Analogies* model (TWA).

1. Introduce the target concept.
2. Remind students of the analogue concept.
3. Identify relevant/similar features of the target and analogue.
4. Map similarities.
5. Draw conclusions about the target.
6. Indicate where the analogy breaks down.

Elegant and appealing as this model is, it is often difficult for teachers faced with constant demands, interruptions and diversions to attend to each of the six steps as carefully and systematically as the model specifies. In response, Treagust et al. (1998) propose the Focus-Action-Reflection (FAR) guide: (i) a pre-lesson *focus* on concepts (are they difficult, unfamiliar or particularly abstract?), students (what ideas do the students already have about the concept?) and experiences (what relevant experiences do students have?); (ii) in-lesson *action*, such as checking students' familiarity with the analogue, discussing ways in which the analogue is like the target and ways in which it is unlike the target; (iii) post-lesson *reflection* on the clarity and usefulness (or not) of the analogy and any changes that should be made for next time. As Harrison and Treagust (2006) note, the three stages of the FAR guide strongly resemble the planning phases of expert teaching and action research.

Some analogies are used across widely different social and cultural contexts, and are commonly used in textbooks – for example, the factory-biological cell and the water flow-electricity flow analogies. Nashon (2003) refers to these as *universal* analogies. In contrast, some analogies are drawn directly from students' immediate sociocultural environment. Lagoke et al. (1997) refer to these as *environmental* analogies (a term with obvious scope for misunderstanding). Lubben et al. (1999) have demonstrated how culturally-located metaphors interact in complex ways with student learning, sometimes positively, sometimes negatively. Harrison and Treagust (2000a) note that some teachers shy away from using purely verbal analogies: "Perhaps teachers are conscious of the unreliable way that students interpret spoken analogies" (p. 1014). In consequence, teachers often supplement analogies with visual representations such as diagrams, pictures and demonstrations: "Although a word mapping of features is sometimes sufficient for an analogy to be drawn, additional graphic or pictorial mapping are desired because ... students can form better representations of the analogy" (Glynn et al., 1995, p. 252). For example, Bean et al. (1990) observed that a group of students taught by analogies with accompanying pictures scored significantly higher marks on the post-test than a group taught with analogies alone. The term *elaborate* analogy is used by Glynn and Takahashi (1998) to describe the utilization of a combination of verbal and visual components to carefully and systematically compare and contrast the analogue and the target learning – in this case, using the TWA guidelines. There is good evidence that elaborate analogies presented by the teacher are more effective than textbook-based versions, probably because of the flexibility of teacher talk and the capacity to adjust the overall complexity and the specific details of the analogy to the capabilities, experiences and interests of the learner (Hogan et al., 2002). Visual aids help students to see the analogue as intended by the teacher; they function as a form of scaffolding. Like other forms of scaffolding a particular analogy (and its associated images) may be discarded once it has served its purpose. The notion of elaborate analogies has been extended by Kim (2009) to include a strong kinaethetic component – in this case, the use of dance in a series of analogies focused on molecular motion, changes of state and reaction kinetics. Kim's work links the formal aspects of analogy-based teaching with the motivational elements of role-play commonly used to model electric current, circulation of the blood and differences between electrovalent and covalent bonding (Cosgrove, 1995; Aubusson et al., 1997; Aubusson & Fogwill, 2006).

It has been argued that more sophisticated understanding can be generated when teachers deploy multiple analogies (Spiro et al., 1989) and when they encourage students to generate their own metaphors, analogies and models, and then compare them carefully and systematically with those developed by scientists, or the school version of them (Flick, 1991; Bloom, 1992a,b; Abell & Roth, 1995; Jarman, 1996; Christidou et al., 1997; Pittman, 1999; BouJaoude & Tamim, 2000; Cameron, 2002; Jakobson & Wickman, 2007). As noted earlier in the chapter, scientists frequently generate analogies for use in problem-solving and theory building (Hesse, 1966; Leatherdale, 1974; Clement, 1988, 1998; Nersessian, 1992, 1995; Dunbar & Blanchette, 2001), so encouraging students to generate their own analogies can be an important aspect of enculturation into the practices of science. So, too, can the criticism of analogies, ascertaining their limitations, refining them, and using them to communicate ideas to others (Clement, 1989; Duit, 1991; Wong, 1993a,b). A number of researchers have noted that students often spontaneously generate analogies when trying to make sense of new ideas, solve problems, report observations and explain phenomena and events to others (Shapiro, 1994; Gallas, 1995; Eilam, 2004; May et al., 2006). According to Gallas (1994), children often make sense of the ideas and experiences they encounter in science lessons by telling stories, in which metaphors and analogies frequently have a prominent role (see discussion in Chapter 9). Spier-Dance et al. (2005) also report that using dramatic role-play to enact student-generated analogies (in their case, focused on electron transfer in redox reactions) can be particularly effective in enhancing learning among lower-achieving students.

The effectiveness of analogies and metaphors in bringing about the desired learning is impacted by a complex of interacting factors: the difficulty of the target scientific concept(s); the complexity of the language deployed; students' existing scientific knowledge; students' familiarity with the chosen analogue/metaphor, their level of interest in it, and their ability to think in abstract terms (clearly the analogue should be cognitively more accessible than the target learning); the teacher's background scientific knowledge and ability to deploy it effectively and appropriately; the teacher's experience and skill in using analogies and metaphors (Duit et al., 2001; Kurtz et al., 2001). This latter factor comprises the following components: (i) awareness of the need to ascertain students' prior understanding of the analogue and to explain the analogue carefully (pointing out the relevant and irrelevant features and where the analogy breaks down); (ii) capacity to provide appropriate guidance and support using visual representation and, when possible and appropriate, physical manipulation of materials; and (iii) awareness of the dangers of over-extending the analogy.

BY WAY OF SUMMARY

To reiterate a point made earlier, students need to acquire a robust understanding of a number of important scientific ideas, principles, models and theories – the *learning science* component of science education. They need to know how those important theories were developed and validated, and appreciate the role and status of models and theories in scientific practice - the learning *about* science component. They also need experience of theory building (part of the *doing* science component), although it would be absurd to suggest that all curriculum content should be approached in this way. First, the concepts involved are such that students could not arrive at such understanding unaided. Second, even if it were possible, it would be a very time consuming and inefficient way of learning content. Third, students need to experience a variety of learning approaches;

any one method becomes stale through overuse. Of course, proper understanding of a scientific model, including the ability to evaluate its effectiveness in a particular context and use it appropriately, is intimately bound up with taking a critical stance towards the means by which it was produced and the argument used to substantiate it. In other words, understanding of the substantive structure of science, including the second order senses articulated here, is inseparable from understanding the syntactical structure of science (item 5 in King and Brownell's list of disciplinary characteristics) and the language of science (item 7 in the list). Indeed, it could be argued that a full understanding also requires an appreciation of the cultural and historical context of its generation (item 4 on the list), an understanding of the ways in which the community of scientists functions to monitor the work of practitioners and validate new knowledge claims (item 1), and an awareness of the valuative and affective stance that underpins it (item 8). It might also include some awareness of the ways in which the scientific knowledge students learn in school is impregnated with culturally determined metaphor and analogy. For example, Robert Hooke used the word "cell" to describe what he observed in magnified plant tissue because these features reminded him of the tiny rooms in which monks live; Kepler used metaphors from music and geometry to formulate views about planetary orbits; Rutherford and Bohr used the analogy of planetary motion to think about atomic structure. Dunbar (1995), Gentner et al. (2000), Paton (2002) and Darden and Craver (2002) provide many examples of the ways in which scientists use everyday analogies and metaphors as frameworks for understanding phenomena and as epistemological tools in theory building and clarification. Students should also be made more conscious of how the everyday language of science is rich in embedded metaphor, simile and analogy: electrical *resistance*, genetic *code*, *messenger* RNA, *killer* T-cells, computer *virus*, and even hormonal *permissiveness* and *antagonism*. Because the effectiveness of such imagery is rooted in its familiarity, metaphors and analogies do not always cross cultural boundaries successfully.

As an aside, it is also worth considering the metaphors used by teachers and teacher educators to describe or model their actual or ideal practice (see examples in Chapter 4). Tobin (1990, 1993) and Tobin and Tippens (1996) show how a change in metaphor can sometimes be a stimulus to initiating a shift in classroom practice. Thomas and McRobbie (1999, 2001) have adapted these ideas and strategies for understanding, challenging and changing students' views of learning and developing their metacognitive awareness.

NOTES

[1] Hacking (1999) uses the term "ontologically subjective" for real objects that owe their existence to human activity.

[2] According to Pepper (1942), "a world hypothesis is determined by its root metaphor" (p. 96), of which there are six: animism, mysticism, formism, mechanism, contextualism and organicism. Cobern (1991) provides a helpful introduction to world view theory in relation to science education.

[3] The term antirealist refers to a number of philosophical positions opposing realism: instrumentalism, idealism, phenomenalism, empiricism, conventionalism, constructivism and pragmatism

(Fine, 1991). Instrumentalism and constructivism are the two antirealist positions most relevant to the discussion in this chapter.

[4] Although instrumentalists regard these entities as fictions, they are purposefully created fictions, not mere flights of fancy. Using invented concepts like gene and magnetic field, and building them into complex explanatory structures, is not just academic game playing or an engaging scientific fantasy.

[5] Nersessian (1992) suggests that this is usually the case.

[6] Zone of proximal development (ZPD) is a term used by Vygotsky (1978) to indicate the level of cognitive demand within which learning can occur with appropriate teacher assistance: "it is the distance between the actual developmental level as determined by independent problem solving and the level of potential development as determined through problem solving under adult guidance or in collaboration with more capable peers" (Vygotsky, 1978, p. 86). In other words, ZPD extends from the most difficult task a learner can accomplish independently to the most difficult task she/he can accomplish with assistance from a more knowledgeable person. In Vygotsky's view, teachers should concentrate their efforts in the ZPD. Instruction should lead development by providing opportunities for students to engage in and acquire competence in intellectual functions that are on the verge of development. The problem with learning geared only to the child's level of independent problem-solving (as in the Piagetian approach) is that it lags permanently behind the child's developing mental processes. "The only good learning", says Vygotsky (1978), is "that which is in advance of development" because it "awakens and rouses to life functions which are in a stage of maturing" (p. 89). Scott (1998) provides a helpful overview of Vygotskian ideas in the context of science teaching and learning.

[7] My definition of "practical work" includes any classroom, fieldwork or laboratory activity involving the use of scientific apparatus, chemicals or biological specimens (living or dead), either by students or by teachers. It includes computer-simulated experiments and computer supported experiments; it does not include "cut and paste" activities, work that is solely text-based, drama or dance.

[8] Interestingly, many science teachers who oppose conjuring on ethical grounds see no problem with using computer simulations, in which the "right answer" is also guaranteed by the way the simulation is designed (Hodson, 1993 studies).

[9] Lynch and Macbeth (1998) use a related metaphor in which they see students as witnesses of the scientific phenomena presented or organized by the teacher, an idea that has real potential for development into a view of scientific practice as an adversarial/judicial system in which evidence for and against a particular scientific model/theory is presented, argued and judged.

[10] Empiricist oriented students incline towards the following views: (i) scientific knowledge is unproblematic and constitutes the "right answers" about the universe, (ii) scientific knowledge is discovered by means of objective data-gathering via a universal scientific method, and (iii) scientific knowledge is additive and "bottom up", and carefully accumulated evidence will establish infallible knowledge. Those with constructivist oriented views tend to believe that: (i) observation and experiment are theory-laden, (ii) scientific knowledge is cooperatively constructed through negotiation among scientists, and (iii) science proceeds via a series of paradigm shifts.

[11] My interpretation is slightly different from Tsai's. I would regard "understanding" how scientists do/did science as part of learning about science. Doing science, by contrast, involves students in designing, conducting, interpreting and communicating their own investigations or in model building activities.

[12] Amati (2000) provides some striking evidence of the progress middle school students can make in their understanding of models, and in building their own models, by using a computer program called ModelIt.

[13] For Jonassen and Carr (2000), "mindtools" are computer software applications such as databases, spreadsheets, semantic networking programs, expert systems, systems modelling tools, microworlds, hypermedia authoring tools and computer conferencing, that enable learners to represent knowledge/information using different representational formalisms.

[14] Gilbert and Reiner (2000) have devised a typology of thought experiments (destructive, constructive and Platonic) for use in building, refining and criticizing models. Further discussion of the characteristics of thought experiments and their development in both science and science education

can be found in Brown (1986, 1991, 2006), Winchester (1990), Sorensen (1992), Norton (1996), Reiner (1998, 2006), Park et al. (2001) and Klassen (2006b).

[15] These student teachers regarded theories as hypothetical, abstract and more fundamental than models, which they considered to be illustrations of theories and, therefore, always visual in nature.

[16] Mercer (1995) provides the following guidelines for effective group discussion: all relevant information is shared; all members of the group are invited to contribute to discussion; all opinions and ideas are respected and considered; everyone is asked to make their reasoning clear; challenges and alternatives are made explicit and are negotiated; the group seeks to reach agreement before taking a decision or engaging in action.

[17] Analogies make explicit use of familiar ideas to explain unfamiliar ones (e.g., a factory to explain a biological cell). The familiar domain is known as the analogue, base or source; the unfamiliar one is the target. Metaphors operate at a more implicit level, making sense of the unfamiliar by using descriptive terms or phrases in situations where they are not literally applicable – for example, coal as "bottled sunshine", genetic mutation as "spelling errors" and the familiar environmentalists' metaphor of "Spaceship Earth".

CHAPTER 7
THE SYNTACTICAL STRUCTURE OF SCIENCE

Chapter 6 emphasized the importance of being clear about the distinction between practical work intended to bring about particular conceptual understanding (activities that I have previously referred to as *learning science*) and practical work intended to provide students with first hand experience of doing science (see also, Roth, 1994; Woolnough, 2000; Chinn & Malhotra, 2002; Hmelo-Silver et al., 2002; Rahm et al., 2003; Kang & Wallace, 2005; Park et al., 2009). Between these two extremes is practical work designed to teach students *about* science, essential components of which include building an awareness of the investigative methods and data gathering techniques available to scientists, acquisition of basic laboratory skills to a level of competence that permits students to collect reliable data, and first hand experience of what is involved in designing a scientific inquiry. It is important that the latter extends beyond learning about "fair tests" by systematically controlling variables, an activity that has come to dominate practical work in many schools, especially in the United Kingdom. For the purposes of learning *about* science there may be some value in regarding scientific investigation as comprising four major elements, and in designing learning experiences that systematically focus on the particular characteristics and distinctive features of each phase.

— A *design and planning phase*, during which overall research aims are formulated, specific questions are asked, problems identified, investigative procedures devised and data gathering techniques chosen. In some circumstances, predictions may be made and/or hypotheses formulated.
— A *performance phase*, during which the various operations are carried out and data are collected and checked for accuracy and reliability.
— A *reflection phase*, during which the findings are considered and interpreted in terms of preferred theoretical frameworks, conclusions are reached, possible explanations considered and chains of argument established.
— A *recording and reporting phase*, during which the procedure, its rationale and the various findings, interpretations and conclusions are rationalized, justified and recorded for personal use and/or communication to others.

Useful as this analysis is for the purpose of teaching and learning *about* science, it can be very misleading when it comes to *doing* science. Although there are these four phases of activity, they are not entirely separate. Scientists refine their approach to an investigation, develop greater understanding of it and devise more appropriate and productive ways of proceeding *all at the same time*. As soon as an idea is developed and an investigation begun, they are subjected to evaluation. Sometimes that evaluation leads to new ideas, to further and different investigative methods, or even to a complete re-casting of the original idea and re-formulation of the underlying problem. Thus, almost every move that a scientist makes during an inquiry changes the situation in some way, so that the next decision is made and the next action is taken in an altered context. Consequently, *doing* science is an holistic and fluid activity, not a matter of following a strictly defined set of actions formulated well in advance. Science is an organic, dynamic, interactive activity, a constant interplay of thought and action. As it proceeds, the whole is continuously evaluated, re-planned and re-directed. The path from initial idea to final conclusions may involve many backtracks, re-starts, short cuts and dead ends. In other words, *doing* science is an untidy, unpredictable activity that requires each scientist to devise her or his own course of action. In that sense, science has no one method, no set of rules or sequence of steps that can, and should, be applied in all situations – as Paul Feyeraband (1975) so eloquently argued. His case against methodologists like Karl Popper (1959) and Imre Lakatos (1978b) is that their arguments for a fixed method of science rest on too naïve a view of what is involved in conducting investigations and building theories, and that the only principle that is always applicable in conducting the complex and sometimes chaotic business of science is the principle *Anything Goes*. In similar vein, Percy Bridgman (1950) remarked: "scientific method, as far as it is a method, is nothing more than doing one's damnedest with one's mind, no holds barred" (p. 278). Lewis Wolpert's (1997) advice to young scientists is to "try many things, do what makes your heart leap, think big, challenge expectations, *cherchez le Paradox*, be sloppy so that something unexpected happens, but not so sloppy that you cannot tell what happened, seek simplicity, seek beauty ... and so on: perhaps one should try them all" (p. 15).

However, too literal an interpretation of Feyerabend's dictum "anything goes" is unhelpful. Implying that science has no methods at all constitutes a gross disservice to students. The crux of Feyerabend's argument is that in order to make progress, scientists should do whatever suits the particular circumstances in which they find themselves: the nature of the problem, the phenomenon or event under scrutiny, the range of theoretical perspectives available to the inquirer, the opportunities and facilities for observation and experiment, particularly the technology and instrumental techniques available, and so on. In approaching a particular situation, scientists choose a method they consider to be appropriate to the task by making a selection of processes and procedures from the range of those available and approved by the community of practitioners. Sometimes one kind of move assists progress, sometimes another. Most importantly, when the community comes to appraise a piece of scientific research, it uses as one of its criteria of judgement a careful consideration of the methods employed. Were they well chosen? Were they satisfactorily performed? Could/should the investigation have been conducted differently? Consensus about the appropriateness and reasonableness of the chosen methods is essential if the knowledge generated is to be accepted as part of the scientific community's body of knowledge. In that sense, it is not true that "anything goes". While Feyerabend's free-wheeling approach certainly applies to the creative phase of scientific inquiry (what some have called the context of discovery); it is less applicable to the context of justification, for which there are strict procedures relating to judgements about reliability, validity and appropriateness; it is certainly not applicable to the ethical dimensions of scientific investigations. This more circumspect position might also incline us to consider that Feyerabend's anarchistic view is fine (maybe even necessary) for what Kuhn (1970) calls *revolutionary* science, but is entirely inappropriate for the effective conduct of *normal* science, where a more dispassionate and systematic approach is essential. Revolutionary science is about the overthrow of existing science and its replacement by new ways of explaining phenomena and events; normal science (what we might call the "stock-in-trade" of practising scientists) is about the development, refinement and extension of existing theory. In normal science, scientists work within the paradigm, assuming that its basic premisses are valid; they follow its rules and procedures; they use its concepts and theories to explore, develop and extend the paradigm, widen its scope, solve its internal puzzles and problems (both theoretical and procedural), formulate quantitative relationships among concepts, and so on. In Kuhn's (1977) words, normal science "aims to elucidate the scientific tradition in which (the scientist) was raised rather than to change it ... The puzzles on which he concentrates are just those which he believes can be both stated and solved within the existing scientific tradition" (p. 234). Careful attention to detail and painstaking accumulation of data are crucial to many aspects of good science, to the effective conduct of laboratory testing, to the maintenance of safety and to the sound operation of science-based industries. But, of course, theoretical breakthroughs are not made this way. Science needs both technicians and creative theory-builders, and we need to provide students with knowledge, understanding and experiences that acknowledge this dual need and lay the groundwork for all manner of future careers. Kuhn himself wonders whether "flexibility and open-mindedness have not been too exclusively emphasized as the characteristic requisites for basic research" (Kuhn, 1963, p. 342). He continues, "almost none of the research undertaken by even the greatest scientists is designed to be revolutionary, and very little of it has such effect" (p. 343).

Decisions about how to conduct an inquiry are also a consequence, in part, of the social context in which the work is located. What equipment, materials and facilities are available? What level of technician support is available? Are other scientists with expertise in related or complementary fields able and willing to participate in the project, or at least to discuss strategies and interpretations? The support of funding agencies (or lack of it) and political pressures within the institution will also play a prominent role in decision-making about how (or whether) to proceed. In addressing the highly personalized, idiosyncratic and innovative ways in which gifted scientists behave in these complex circumstances, McGinn and Roth (1999) employ the notion of scientists as *bricoleurs*,[1] "who improvise to make the most out of each situation as they pursue their interests and goals; they construct local theories by arranging and rearranging, negotiating and renegotiating with materials; and they draw on a variety of social, cultural, material, political, and economic resources to proceed" (p. 15). In making their choices and in implementing their chosen strategy scientists make use of a kind of expertise that has been variously labeled: *tacit knowledge, scientific intuition* and *scientific flair*. It is the kind of knowledge, often not well articulated or even consciously applied, that can be acquired only through the experience of doing science; knowledge that links what we know, in a conscious sense, with what we feel and experience. It is the kind of knowledge that a centipede uses all the time when it walks, though it might be unable to explain, if it could talk, how it manages to control all those legs! It is the kind of knowledge that good games players, chess masters, expert musicians and gifted teachers have, and use all the time. Often, they cannot analyze, describe and explain what they do; they just do it, intuitively. In the same way, scientists are often not consciously aware of what they are doing in confronting and solving conceptual and methodological problems. They also just do it! What I am attempting to describe is the kind of knowledge responsible for the (sometimes) seemingly non-logical, intuitive, idiosyncratic, inspired moves of the gifted scientist, summed up in Henri Poincaré's (1952) famous remark that "logic remains barren unless it is fortified by intuition" (p. 193).

This amalgam of experiential knowledge, creativity and intuition constitutes the central core of the art and craft of the scientist. It is not separate from laboratory expertise, on the one hand, and deep conceptual understanding, on the other. Rather, it is the capacity to use both in a purposeful way in order to achieve particular goals. It combines deep conceptual understanding and well-developed bench skills with elements of creativity, experimental or investigative flair (the scientific equivalent of the gardener's "green fingers") and a complex of affective attributes that provides the necessary impetus for the determination and commitment essential to making progress in addressing complex and difficult problems. With experience, it develops into what Polanyi (1958) and Oakeshott (1962) call *connoisseurship*, though connoisseurship also includes a strong metacognitive dimension (scientists reflect on what they do and use that reflection to inform and modify future actions, though they may do so unconsciously). Thus, scientists proceed partly by rationalization (based on their theoretical understanding) and partly by intuition rooted in their tacit knowledge of how to do science (their connoisseurship): "a practicing scientist is continually making judgements for which he can provide no justification beyond saying that that is how things strike him" (Newton-Smith, 1981, p. 232). Holton and Brush (1973) comment: "some of the greatest theoretical scientists have stated that during those early stages of their work they do not even think in terms of conventional communicable symbols and words" (p. 188). Rather, they rely on tacit understanding of the phenomena under investigation. Albert Einstein is reputed to have said: "The supreme task of the physicist is to arrive at those universal elementary laws from which the cosmos can be built up by pure deduction. There is no logical path to these laws;

only intuition, resting on sympathetic understanding, can lead to them" (cited by Wertheim, 1995, p. 187). When children reply to a teacher's or a researcher's query about their choice of investigative procedure with comments such as "I had a hunch", "I felt that it was the best thing to do", "I just knew to do it that way" or "It seemed right", they are exhibiting an embryonic form of tacit scientific understanding. We need to recognize it not only as a legitimate part of doing science, but as a *crucial* part – one that we should strive to develop and to celebrate (Hodson, 1992c). Stephen Jay Gould (1995) captures this view particularly well.

> Scientists reach their conclusions for the damnedest of reasons: intuitions, guesses, redirections after wild-goose chases, all combined with a dollop of rigorous observation and logical reasoning to be sure ... This messy and personal side of science should not be disparaged, or covered up, by scientists for two major reasons. First, scientists should proudly show this human face to display their kinship with all other modes of creative human thought ... Second, while biases and preferences often impede understanding, these mental idiosyncrasies may also serve as powerful, if quirky and personal, guides to solutions. (p. 94)

Of course, a scientific investigation is not finished when data has been collected and interpreted, and conclusions drawn. It is still necessary for scientists to make sense of the whole undertaking, construct a convincing rationale for the study, justify the methods, construct an argument that leads persuasively to the conclusions, and present that argument to the scientific community for critical appraisal. In Galison's (1987) words, "Experiments are about the assembly of persuasive arguments, ones that will stand up in court ... The task at hand is to capture the building-up of a persuasive argument about the world even in the absence of the logician's certainty" (p. 277). The meaning of an experiment or investigation is not self-evident; rather, it is constructed and negotiated. If we are to engage students in *doing* science in any meaningful sense then engagement in the construction of explanations and evaluation of evidence is crucial. They need to justify the design and data collection methods, interpret the results and reach conclusions; they need to build an argument that shows how the conclusion is supported by the evidence, and whether other plausible conclusions might be reached; they need to present their work in a form that is intelligible to others and to practice the conventions of scientific communication. It is through these activities that students recognize that scientific knowledge is constructed and justified through argument and debate. Explanations do not emerge unaided, straightforwardly and unerringly from data, though the way science is often presented in school might lead students to believe this to be the case.

> The image is one of action, an individual engaged in manipulating the material world, exploring and exposing its inner secrets, but not one of discourse, i.e., reading, writing and communicating science. For the school pupil such images are substantiated by the fact that science, in common with design and technology, music and drama, has a specialised location for its teaching, one that supports and enables practical manipulation of the material world. (Osborne, 2002, p. 203)

Over the past decade or so, the emphasis on "fair testing" and process skills acquisition, compounded by the current obsession with analytical assessment schemes, has resulted in practical work that is unrelentingly "hands-on" and only minimally "minds-on". In consequence, many students fail to recognize that the practical work in which they engage is intended to generate evidence for or against knowledge claims, fail to make links between theory, design, evidence and conclusions, and fail to appreciate that a substantial part of *doing* science is argumentation (Millar, 1998; Robertson, 1999; Blank, 2000; Kim & Song, 2006; Zion et al., 2004; Haigh et al., 2005; Gott & Duggan, 2007; Grandy & Duschl, 2008; Hume & Coll, 2008; Park et al., 2009). In the provocative words of Martins et al. (2001): "observation and experiment are not the bedrock on which science is built. Rather they are the handmaidens to the rational activity of generating arguments in support of knowledge claims" (p. 191).[2] The language of science and the structure of scientific argumentation will be addressed in some detail in Chapter 8.

AUTHENTIC INQUIRY AND SCHOOL SCIENCE

Presenting an authentic view of science[3] in the curriculum requires that students be given the opportunity to frame their own questions and problems, devise their own investigative strategies and data collection methods, interpret and make sense of their findings, share their experiences with others, and generally move in the direction of connoisseurship. In traditional laboratory work all theorizing (identifying the area of concern, stating the problem to be investigated, designing the inquiry strategies and techniques, interpreting the data, designing the report or communication) is the responsibility of the teacher, with periodic but rather cursory consultation with the more vocal students. Teachers decide what to study, how to study it and how to explain (or explain away) unanticipated results; students act largely as technicians, carrying out the laboratory operations specified by someone else. Gough (1993) is particularly scathing in his criticism of the kind of activities that take place in most school science laboratories, their organization, and the kind of equipment made available.

> [They] neither resemble the sites in which most scientists work nor are they used for the kind of experimentation and data generation that characterizes much professional science. Rather, they are places where students follow recipes, perform routine procedures, rehearse technical skills (e.g., manipulating apparatus, monitoring instruments, measuring and recording), demonstrate the reliability of selected ("well-accepted") scientific "laws" or phenomena – and falsify their data when the procedures and the demonstrations produce inconclusive or "unexpected" results. (p. 20)

There is little or no room for messy data and creative imagination, no acknowledgement of the theory ladenness of inquiry methods, no suggestion that there may be alternative interpretations of data. There is seldom any opportunity for students to pose their own question(s) for investigation, formulate hypotheses to be tested, predict experimental outcomes, design investigative procedures, and so on. And in many schools the deference afforded to inflexible assessment requirements has reduced scientific inquiry to a set of formulaic procedural steps that must be rigidly followed. Donnelly et al. (1996) give a detailed account of the pernicious effect on practical work of the imposed restrictive assessment schemes of the National Curriculum for England and Wales. Abrahams and Millar (2008) also note that the overwhelming emphasis of practical work in most textbooks concerns the way data should be collected, analyzed and interpreted.

What is needed is a shift from "student as technician" to "student as creative scientist", with sole responsibility for the inquiry. While students will necessarily be constrained in their approach by what they currently know and the laboratory skills they currently possess, it is ownership of the inquiry, and sole responsibility for the design of the investigation that provide the essential stimulus for the development of both their conceptual and procedural knowledge. Further learning occurs when students are encouraged to write about their inquiries in less formal ways. An investigator's logbook is an ideal vehicle for encouraging students to reflect on the progress of their investigation: What am I going to do? Why am I doing it? What difficulties and frustrations can I anticipate? Is the investigation working out as planned? What data has been collected? What data is significant and what can be ignored? Do I need to re-think anything or re-plan anything? Could anything be improved? Logbooks capture and hold ideas as they arise, record thinking processes and document developing ideas and tentative conclusions; they reveal the intuitive base of research, the unpredictable timeline of creative thought and the constant recycling of concepts, ideas and perspectives. Their use encourages a less prescriptive approach to scientific investigation. What is being acknowledged here is that one of the key elements in doing science, and a consequence of the dynamism of scientific inquiry discussed above, is *changing one's mind*, going back to an earlier stage in order to try something different. Working in this relatively unstructured fashion is the only way for students to experience what it is like to conduct a scientific inquiry. Of course, poor design, unexpected difficulties and just plain bad luck will sometimes intervene and render the inquiry inconclusive. But much will have been learned about the vagaries and uncertainties of scientific inquiry, and much will have been done to combat the myths that currently surround the curriculum portrayal of scientific practice.

Part of that "myth reduction" exercise focuses on a shift away from the belief that experiments are the *sine qua non* of scientific investigation. Many fields of scientific endeavour deal with events that are remote and inaccessible in time and space, and so make little or no use of experiments. Indeed, if experiments are considered essential to scientific progress, then Nicolaus Copernicus and Charles Darwin, two prominent "scientific revolutionaries", should not be regarded as scientists. In some fields, experimentation may be possible in principle but is ruled inadmissible for ethical reasons. In some fields, experimentation is too difficult, too dangerous or too expensive. This is often the case in school. It is here that correlational studies play a crucial role.[4] In a sense, studies seeking correlations among variables are more faithful than experimental inquiry to the supposed open-mindedness of science because they make fewer assumptions about the nature of the interactions and provide opportunities for studying phenomena and events in natural settings rather than the contrived situation of the laboratory. It is also the case that many field settings are simply unsuitable for sustained experimental inquiry because they are too complex or too fragile and uncertain. An experimental approach might so distort the natural setting that it no longer represents natural behaviour, as Bowen and Roth (2002) illustrate in their discussion of observational work with lizards. Making science experiences in school more authentic includes a shift away from the current preoccupation with experimentation, especially the so-called fair test and the "illusion of certainty" about scientific knowledge that it promotes (Hodson & Bencze, 1998). The view I wish to promote is that the relationship between experiment and theory is complex and varies with circumstances, much like the relationship between science and technology discussed in Chapter 5. Ian Hacking (1983) expresses this position particularly well.

> Some profound experimental work is generated entirely by theory. Some great theories spring from pretheoretical experiment. Some theories languish for lack of mesh with the real world, while some experimental phenomena sit idle for lack of theory. There are also happy families, in which theory and experiment coming from different directions meet. (p. 159)

The data does not speak for itself, as some textbooks suggest when they liken scientific investigation to detective work. Rather, the data means what the theory says it means. An important part of the misrepresentation of experiment in school science is that scientific claims are tested against reality. Not so! They are tested against our *interpretation* of reality, often using evidence collected by instruments that encode in material form a great deal of assumed theory that is subsequently implicit in the experimental conclusions. It is not uncommon for science textbooks to assert that a hypothesis can be rejected or a theory falsified on the evidence provided by a single experimental test. Indeed, many textbooks suggest that this is the *only* role for experiments. This kind of naïve interpretation of Popperian falsificationism carries with it the assumption that theory-independent evidence is obtainable and that unambiguous testing is possible. In practice, scientific experimentation is far from being a simple, straightforward matter, and science education that portrays it as such is grossly misleading. It is important for students to realize that every experiment is set within a theoretical matrix (particular conceptual schemes and theories), a procedural matrix (a community approved "method" or practice underpinned by theories and conventions about how to conduct, record and report experiments) and an instrumental matrix (theories underpinning the design and construction of all scientific instruments employed in the experiment, together with the theories of perception that underpin all observations). It is this complex of theoretical understanding and assumptions that gives both form and purpose to experiments. What counts as good research design, what kind of observations are sought, what measurements are regarded as legitimate, what instruments may be utilized, and what sort of evidence is seen as crucial, are all determined by the theoretical matrix within which the scientist is working. Experimental data are not collected by passive observation; rather, they are purposefully sought using theory-laden instruments designed specifically to modify and simplify complex phenomena so that particular, measurable quantities can be recorded. Through technology, scientists gain access to aspects of the natural world that would otherwise remain hidden. On occasions, scientists may even design experimental set-ups and use technological devices to produce interesting phenomena and entities that do not otherwise exist. In other words, the technologies we utilize in scientific investigations impact very significantly on the knowledge we create through their use.

Because experiments form a critically important part of the scientist's repertoire, it is crucial that the school science curriculum does not misrepresent their role. It is misleading to present students with the idea that theories are discarded because of a few negative results. In practice, all theories have to live with anomalous data; it is a natural feature of science. We seriously mislead students when we pretend that the kinds of experiments they perform in class constitute a straightforward and reliable means of choosing between rival theories. Moreover, because experiments are conceived, designed and executed with a particular complex of theoretical understanding, considerable judgement is involved in appraising the significance of the evidence they furnish. Whether to accept the evidence, and in consequence to accept or reject the theory, reject the evidence, or conclude that some matters are still problematic and the experiment should be re-planned, is a decision that is not easily made.

Before leaving this discussion it is worth mentioning that a great deal of the power of experimentation lies in the capacity to control and contrive situations so that scientists do not need to deal with all the complexities of studying an object or phenomenon *as it is* (it can be simplified, idealized and stylized) or *where it is* in the real world (by re-locating it to the laboratory, all kinds of technological resources can be brought to bear). Nor do scientists need to accommodate to an event *when it happens to occur* (natural cycles can be replaced by more convenient ones – more frequent, more predictable). Thus, experiments are *representations* of reality – processed, manipulated and partial versions of it, rather than the "real thing". This line of argument can be extended further. Everything in the laboratory is, in some sense, a human construct: the knowledge from which scientists begin the investigation, the materials, the measuring devices and other technical tools are all "created entities"; laboratory animals are selectively bred; plants are specially grown; pure chemicals are manufactured; and even the water supply is sterilized and deionized. Because of the complexity, fluidity and context-specific character of experimentation, scientists themselves could be regarded as a kind of "technical device": repositories of experiential knowledge, tacit understanding, scientific intuition and capacity for critical judgement. Technicians, through their capacity to optimize experimental conditions, to build, maintain and use apparatus specially suited to the circumstances, and to "feel" what is reasonable and acceptable as experimental results, can also be regarded as "instruments" for experimentation (Knorr-Cetina, 1992). It is important to recognize the "constructed" quality of experiments and the knowledge they produce, though that should not lead us to lose faith in the theories that are constructed from the data they generate. That faith, as argued previously, is rooted in the highly public forms of critique within the scientific community. It is also important to note that experiments, for all their "created" qualities, cannot create *anything* and *everything* a scientist can imagine. Nor can scientists make the entities they study in the laboratory behave according to their own will. The real world exerts some constraints on scientific activity; in a host of ways, the natural world is independent of human perception, imagination, inquiry and action (Hacking, 1983; Fine, 1986). There is also a wide range of technological, ethical, social, political and economic constraints on scientific activity – that is, on "the science that gets done", to use Young's (1987) phrase. This aspect of scientific inquiry will be discussed further in Chapter 10.

Because the conduct of scientific investigations is neither fixed nor entirely predictable in advance, and because it involves a component that is experience-driven in a very personal sense, it is not teachable by conventional methods. That is, one cannot learn to do science by learning a prescription or set of procedures to be applied in each and every situation. The only effective way to learn how to do science is by doing science alongside a skilled and experienced practitioner who can provide on-the-job support, criticism and advice. In Polanyi's (1958) words, "connoisseurship" can only be communicated by example, not by precept. To become an expert wine-taster ... or a medical diagnostician, you must go through a long course of experience under the guidance of a master" (p. 54). It is here that the notion of learning by apprenticeship or internship is useful. Apprentices learn to think, reflect, argue, act and interact in increasingly knowledgeable and skilful ways by engaging in "legitimate, peripheral participation" (Lave, 1988, 1991; Lave & Wenger, 1991) with people who already have the appropriate knowledge and skills. Apprenticeship is not just a process of internalizing knowledge and skills, it is the process of becoming a member of a community of practice (Wenger, 1998). Thus, it fits very well into the King and Brownell framework of disciplinary characteristics. Developing an identity as a member of the community and becoming more knowledgeable and skilful are part of the same process, with the former motivating, shaping and giving meaning to the latter: "Newcomers become oldtimers through a social process of increasingly centripetal participation, which depends on legitimate access to ongoing community practice" (Lave, 1991, p. 68).

When they are given opportunities to participate peripherally in the activities of the community, newcomers pick up the relevant social language, imitate the behaviour of skilled and knowledgeable members, internalize their values, recognize what excites and motivates them, and gradually start to act in accordance with the community norms. Clear and skilful demonstration of expert practice and the provision of opportunities for critical questioning, interspersed with opportunities for guided participation by the "novice", provided they are informed by critical feedback from the "expert", comprise the stock-in-trade of the apprenticeship approach long used for teaching and learning in the trades and crafts. For centuries, bread, cheese and wine makers, blacksmiths, carpenters and shoemakers have all acquired their expertise this way. Thus, many science educators argue that students can best learn to do science by learning directly from scientists through internship programmes (Ritchie & Rigano, 1996; Smith et al., 2000; Barab & Hay, 2001; Schenkel, 2002; Knox et al., 2003). As Barab and Hay (2001) note, there are two key elements to this kind of experience: one is the opportunity to design, conduct and report one's own research, with access to tools and facilities unavailable in schools; the other is the mentorship provided by participating scientists. The first element is essential for building a sense of ownership, enhancing confidence and stimulating creativity; the second element helps students to acquire the language, understand the norms and appreciate the values associated with scientific practice, and build a sense of identity as a scientist. It is through this combination of experiences that future scientists themselves learn when they embark on doctoral studies. They certainly do not learn to do science by taking courses in the history and philosophy of science, as Peter Medawar so pungently states.

> Most scientists receive no tuition in scientific method, but those who have been instructed perform no better as scientists than those who have not. Of what other branch of learning can it be said that it gives its proficients no advantage; that it need not be taught or, if taught, need not be learned? (Medawar, 1969, p. 8)

> Science, broadly considered, is incomparably the most successful enterprise human beings have ever engaged upon; yet the methodology that has presumably made it so, when propounded by learned laymen, is not attended to by scientists, and when propounded by scientists is a misrepresentation of what they do. Only a minority of scientists have received instruction in scientific methodology, and those that have done so seem no better off. (Medawar, 1982, p. 80)

Richmond and Kurth (1999) report that the main benefits of a 7-week summer research experience programme for grade 11 and grade 12 students were: (i) increased facility in using the language of science to construct, refine and communicate meaning; (ii) greater awareness of the uncertainty and tentativeness of scientific knowledge; and (iii) recognition that science is both cumulative and collaborative, in the sense that the development of a particular idea or theory extends over long periods of time and involves contributions from many researchers. Following a similar 8-week programme for students in grades 10 and 11, Bell et al. (2003) comment that most of the scientist mentors thought that the participants had learned a great deal about the scientific enterprise. However, data collected by means of a variant of the VNOS questionnaire, together with interviews involving student participants, revealed that views on key NOS items were unchanged by the experience. Charney et al. (2007) report that a 4-week placement in a research facility (including daily seminars), together with follow-up research work in schools, assisted the 30 high school participants in developing what the authors call "more sophisticated ways of thinking". In particular, increased ability to generate hypotheses and consider alternative hypotheses, utilize models and carefully constructed arguments in explanations, connect ideas, extend conceptual understanding and ask pertinent questions. In a study by Painter et al. (2006), students' perceptions of scientists were radically changed by as little as one week's experience of interacting with scientists and using cutting edge technology – in this case, nanotechnology. The majority of the 54 grade 7 and grade 10 students no longer saw scientists as "nerdy" or "weird", and no longer saw science as "boring". Instead, they recognized that scientists are like "regular people"; they enjoy the same kinds of leisure activities as other members of the community; they can be women and people of colour. In addition, they saw science as something they themselves might contemplate as a career. Importantly, these views were still in place one year later. Schwartz et al. (2004) designed an internship programme for 13 preservice secondary school teachers involving 5 hours per week (for ten weeks) working in a research laboratory. The authors conclude that it is not the doing science activity *per se* that is key to enhancing student teachers' NOS understanding, but reflection on their experiences through journal writing and seminars designed to encourage participants to discuss research experiences with other interns. Samarapungavan et al. (2006) arrive at essentially the same conclusion following their study of a research apprenticeship programme for undergraduates: the success of the experience seemed to depend on "the degree of research autonomy given to students, and on opportunities to engage with expert researchers in conversations and reflections" (p. 493). Finally, it is worth noting the tentative finding by Roberts and Wassersug (2009) that those who participate in hands-on summer science research programmes for high school students are significantly more likely to enter and maintain a career in science than those whose experience of first hand research is delayed until the university level. Similarly, Fields (2009) notes that several of the 14–18 year olds who participated in a summer astronomy camp expressed a very strong intention to pursue a career in science, with a significant number commenting on their commitment to a research-oriented career.

LEARNING BY APPRENTICESHIP IN SCHOOL

While apprenticeship/internship experiences offer enormous potential for enhancing NOS understanding, they create major logistical problems. Few teachers are able to provide opportunities for all students to work with real scientists. However, they can create opportunities for students to engage in a form of apprenticeship in school via a three-phase approach.

— *Modelling*, where the teacher demonstrates and explains the desired behaviour.
— *Guided practice*, where students perform specified tasks with help and support from the teacher.
— *Application*, where students perform independently of the teacher.

Teacher modelling of scientific practice is predicated on the assumption that observation of a skilled performer facilitates the learning of skills, in this case, the strategic skills of planning, conducting, interpreting and reporting scientific investigations. When teachers model scientific inquiry it is important that they choose an authentic question: one that addresses a real issue and one for which they do not already know the answer. Too often, laboratory work in schools creates the impression that scientists spend their time confirming knowledge they already possess. Too often, also, it creates the impression that science is unrelated to everyday life. Thus, it is important, especially with young children, to investigate something of interest and concern in the immediate environment of the students.

A number of writers have noted some of the ways in which the practices of scientists differ from the kind of activities commonly provided in schools (Chinn & Malhotra, 2002; Hmelo-Silver et al., 2002; Schwartz & Crawford, 2004; Abd-El-Khalick, 2008; Duschl & Grandy, 2008; Ford, 2008b; Park et al., 2009; Wong & Hodson, 2009). It is important that teachers take whatever steps they can to represent authentic practice, including the varying motives for engaging in research and the differing ways in which a successful outcome is defined. Research is not always directed towards the production of new knowledge or the testing and refinement of existing theories. Some efforts are directed towards the development of new inquiry techniques, the accumulation of new data or the resolution of inconsistencies in previous investigations. I share the views of Elby and Hammer (2001) that the notion of scientific practice presented in most school science curricula is too broad and genneralized, and that we should be mindful of the need to show students that investigative methods vary (to an extent) from sub-discipline to sub-discipline. However, while teachers may find much of interest in Knorr-Cetina's (1999) detailed account of differences between the sub-disciplines of high energy physics and molecular biology, for example, they need to be circumspect in balancing curriculum representation of differences with similarities. Following their questionnaire and interview-based study of 24 scientists, Schwartz and Lederman (2008) report that scientists' NOS views and what they say about their own investigative approaches seem to differ in relation to specific individual contexts and experiences rather than in terms of the broader sub-disciplines of biology, chemistry, earth sciences and physics. They argue that focusing on subtle nuances and differences might confuse students and draw attention away from features that are common, concluding that a generalized treatment of NOS across science disciplines is appropriate for K-12 science. My own view is that in the early stages of a student's experience of scientific inquiry, especially in primary schools, it is important to ensure that modelled investigations involve as many as possible of the individual sub-processes in which children are expected to develop proficiency, and might expect to employ subsequently in their own investigations. Later on, modelled investigations and case studies can begin to show students just how different investigations in the different disciplines can be, and can show that the business of designing and conducting experiments in the real world is rather different from the way school textbooks tend to describe it. This point is given graphic illustration in the comparison by Hmelo-Silver et al. (2002) of the strategies adopted by experts and novices in designing a clinical drug trial. This discussion is further developed later in the chapter.

Predicting, observing, measuring, identifying and manipulating variables, recognizing trends in data, using suitable scientific concepts to hypothesize and model, and describing, recording and reporting in appropriate scientific language can all be modelled by the teacher, with attention focused on their essential features. It is important, too, despite my earlier criticisms, to explore the notion of a "fair test" and to discuss the importance of standardization of technique and accuracy of measurement in ensuring reproducibility of data. At all levels, attention should be directed towards the need to record procedures and data fully, clearly, carefully and accurately, using lists, charts, graphs and so on, as appropriate. Young students are often very careless of such matters, erroneously believing that they will remember all the important details of the day's activities. Having reached this point in the investigation, students then need to make sense of the whole enterprise: first, they need to reflect on and justify the design of the inquiry and the choice of data collection methods; second, they need to interpret the data, using whatever theory is most appropriate, and draw conclusions; third, they need to construct an argument that justifies those particular conclusions in terms of pre-existing evidence and further evidence collected in the inquiry, having also considered alternative conclusions; fourth, they need to present their findings in an appropriate form for communication to others. To achieve all this, students need ample time for guided reflection and critical discussion on the role of theory and its relationship to design and evidence. It almost goes without saying that (i) the form and structure of scientific argumentation will need to be taught (see Chapter 8), (ii) students will need to be provided with opportunities to practice their argumentation skills and to give and receive criticism, and (iii) the interdependence of argument and experiment (i.e., experimental findings form the basis of many arguments; arguments provide guidance for further experiments) will need to be made explicit – a matter to be revisited in Chapter 8. Furthermore, criteria for good recording and reporting will need to be discussed, with opportunities provided for writing up findings in the manner required by scientific journals or scientific conferences.

It is anticipated that students will become more expert in each of the four phases of investigation (design, performance, reflection, reporting) as a consequence of observation, practice and experience, through evaluative feedback provided by the teacher and generated in inter-group criticism and discussion, and through intra-group reflection on the activity, both as it progresses and on completion. In several respects, then, the teacher's role shifts from instructor to research director. Crucial to the notion of apprenticeship is dialogue about the way the inquiry is progressing, including frank discussion of problems encountered, lines of inquiry that prove fruitless, and barriers to progress that prove insurmountable. Students' suggestions and advice should be sought, and some acted upon. This is the stage during which teacher and students are *co-investigators*, with both parties asking questions, contributing ideas, making criticisms and lending support (see Crawford et al., 2000; Kelly et al., 2000). When teachers accept students' ideas and contribute ideas to students' own investigations, those students are encouraged to help each other, thus helping to build a climate conducive to cooperative learning. Interestingly, Rogoff (1990) points out that working with a partner can sometimes serve as a distraction because it requires attention to be focused on negotiating the division of labour and attending to social issues instead of attending to cognitive issues. Also, she argues, some tasks may simply be too difficult to coordinate with someone else, especially for young children. Crucial also, if the goal is for students to gain understanding of authentic scientific practice, is constant comparison between what students are doing in their inquiries and what scientists do in their day-to-day activities. Such reflections are essential if the gap between what Hogan (2000) calls *proximal* and *distal* knowledge of NOS (see Chapter 2) is to be avoided or minimized (see also Vhurumuku et al., 2006).

It is, perhaps, not surprising that Berry et al. (2001) found that hands-on investigations designed to raise students' awareness of the role of experiments in "establishing science facts" were more successful than investigations intended to raise awareness of the socially constructed nature of scientific knowledge. As discussed in Chapter 2, learning gains associated with sociocultural aspects of scientific practice are generally less substantial than gains relating to other NOS elements. The authors also note that learning may have been compromised, to an extent, by the simultaneous emphasis on learning content. Earlier work by Hart et al. (2000) had shown that students were more successful in learning about scientific inquiry (especially in terms of developing, using and communicating scientitifc procedures) when they were assured that content learning was not a focus of attention and results obtained would not form the basis of further learning. In other words, overall learning goals need to be clearly stated. It is also clear, as with other aspects of NOS learning, that the curriculum needs to focus carefully and explicitly on key elements of scientific inquiry. Too often, teachers expect this understanding to arise from implicit messages about scientific inquiry embedded in lab activities. Toh et al. (1997) have demonstrated that students make progress in their ability to cope with open-ended inquiry more quickly when they are given what the authors call "pre-training" in some key elements of scientific inquiry relating to tolerance of ambiguity and uncertainty. It is also the case that younger students are much more accommodating of uncertainty than older students because they have *not* had years of experience of conventional, rigidly structured school science practical work (Metz, 1995, 2004). They are also more accommodating of plurality of methods. Indeed, young children have no expectations of a *particular* method for doing science; they readily appreciate that differences in concerns and starting points for scientific investigations, and differences in scientists' previous experiences and access to technological resources, will necessarily lead them to proceed differently. Teachers and science textbooks create the expectation of a single method through their continual reference to *the* scientific method. There is good evidence that carefully designed and sensitively implemented strategies for guided practice (level 2 in my proposed scheme) can assist students at all levels of education, including kindergarten, in developing the capacity to ask pertinent, researchable questions and design appropriate strategies for answering them (Hofstein et al., 2004, 2005; Apedoe, 2008; Park Rogers & Abell, 2008; Samarapungavan et al., 2008).

To achieve intellectual independence, students must eventually take responsibility for their own learning and for planning, executing and reporting their own inquiries. In other words, learning as assisted performance must enable students, in time, to go beyond what they have learned and to use knowledge in creative ways, for solving novel problems and building new understanding. Consequently, alongside the modelled investigations, students should work through a carefully sequenced programme of investigative exercises, during which the teacher's role is to act as learning resource, facilitator, consultant and critic, and simple holistic investigations in which careful planning by the teacher can almost guarantee that students will succeed, while creating the impression that students are acting independently of the teacher. This latter point derives from my observation that students focus almost exclusively on procedural matters when addressing teacher-directed problems, with attention focusing increasingly on conceptual and design issues as students assume more control of problems. Investigative exercises provide opportunities for students to learn through a cycle of practice and reflection, and to achieve, with the careful assistance and support of the teacher, a level of performance they could not achieve unaided.

There is some evidence that it may be more productive to begin involvement in holistic investigations with engineering-type problems, where the goal is to optimize desirable or interesting outcomes, because they more closely resemble children's intuitive problem-solving strategies and their everyday ways of thinking (Schauble et al., 1991b). Many teachers, especially at elementary school level, also find this kind of activity easier to organize. Alternatively, Lock (1987) and Fensham (1990) both advocate the use of consumer-type problems, which have the added appeal of familiarity and immediacy. A particularly intriguing suggestion is to engage students in a systematic study of so-called "urban myths", such as the Murphy's Law variant that buttered toast always lands butter-side down, soaking conkers in vinegar makes them harder[5] and dangling a chain from the back of a car reduces travel sickness (Matthews, 2001). Interestingly, when Chin and Chia (2004) allowed grade 9 students in a PBL-oriented biology course to choose their own questions for investigation, all suggestions ran along similar lines: Does MSG in food cause hair loss? Does strenuous exercise following a heavy meal result in appendicitis? These are certainly not questions that are subject to the kinds of investigative approach I have in mind in this current discussion, though they may well support other kinds of inquiry-based learning. Perhaps it is inevitable that young students' first attempts at posing questions for investigation will be geared towards testing their own, often idiosyncratic ideas about the immediate world around them, as in the study by Keys (1998) at the grade 6 level. They may have little connection with the conceptual knowledge being acquired in other areas of the science curriculum. Personally, I do not see this as a problem in the early stages. However, effecting the transition to more strictly science problems associated with the search for explanations and causes is a key step.

Eventually, as students gain experience and take on increasing control of decision-making, they can proceed independently: choosing their own topics and problem situations, and approaching them in their own way. By this stage the work in which students engage should increasingly resemble the work of scientists. For Roth (1995), this means ill-defined problems, preferably of an ambiguous and social nature, driven by students' current knowledge. In tackling the investigation students should engage in shared discourse and have access to the knowledge of others (teachers, scientists, the Internet, etc.). Lottero-Perdue and Brickhouse (2002) have some useful suggestions for how science teachers can provide opportunities for "tinkering" and can begin to create the kind of cooperative investigative climate that fosters problem-based learning in the real world sense – that is, how scientists (and particularly engineers) approach the ill-defined problems they encounter in their day-to-day work. From this point on, students are responsible for the whole process, from initial problem identification to final evaluation. As a consequence, they experience both "the excitement of successes and the agony that arises from inadequate planning and bad decisions" (Brusic, 1992, p. 49). Because students are given the opportunity to experience failure it is imperative that the class atmosphere is both forgiving and supportive. Indeed, throughout these activities the teacher's role is crucial: model scientist, advisor, learning resource, facilitator, consultant, emotional support and critic. As Ravetz (1971) has commented, learning to do science occurs "almost entirely within the interpersonal channel, requiring personal contact and a measure of personal sympathy between the parties" (p. 177). By engaging in holistic scientific investigations, alongside a trusted and skilled critic, students increase both their understanding of what constitutes *doing* science and their capacity to do it successfully. In other words, *doing* science is a reflexive activity: current knowledge and expertise informs and determines the conduct of the inquiry and, simultaneously, involvement in inquiry (and critical reflection on it) refines knowledge and sharpens expertise.

CREATING A SUITABLE ENVIRONMENT

In an apprenticeship model of learning both the nature and the timing of teacher intervention are crucial: deciding how to attend to each learner in a way that is appropriate to them, taking into account their unique personal framework of understanding (Hodson, 1998a), including its affective and social components; and deciding when to encourage and support, when to direct and instruct, and when to involve others. Knowing when, where, how much and what type of guidance, critical feedback and support are needed to facilitate effective learning and the development of good learning behaviours and inquiry skills is a matter of professional judgement deriving from experience and thoughtful reflection on it. Too much guidance can interfere with students' thought processes, act to frustrate problem solving and lead to premature closure; too little guidance can leave students unable to make satisfactory progress and can lead to feelings of frustration, and even alienation. To be effective, teacher guidance and assistance needs to be pitched slightly beyond the current level of unaided performance – that is, in each student's *zone of proximal development*.

> A novice works closely with an expert in joint problem solving in the zone of proximal development ... Development builds on the internalization by the novice of the shared cognitive processes, appropriating what was carried out in collaboration to extend existing knowledge and skills. (Rogoff, 1990, p. 141)

As discussed in Chapter 6, *scaffolding* is the term commonly used to describe the ways in which teachers can (i) reduce the complexity of tasks (delaying more complex and difficult aspects until students have mastered simpler aspects), (ii) sequence a complex task into a series of simple ones, (iii) make suggestions, ask questions and provide cues to enable students to focus on particular elements and clarify their understanding, and (iv) provide constructive feedback and support to help students diagnose their own problems and difficulties, and build their self-confidence (see Hodson & Hodson, 1998a,b for a fuller discussion). Constant dialogue between teachers and students is essential if good intervention decisions are to be made. It is the nature of the language used by the teacher during these exchanges that establishes the interpretive framework within which students are able to make scientific sense of whatever is being studied. Again, there is an issue of fine professional judgement: neither imposing meaning nor permitting students to construct whatever meaning happens to suit them, for whatever reasons. Rather, teachers should seek to create a dialogic context in which meaning is *co-constructed*.

> The teacher remains in charge, but his or her exercise of control is manifested not in a once-and-for-all choice between intervening with the "correct" answer or standing back and leaving students to find their own solution, but in the making of moment-by-moment decisions about how to proceed, based on knowledge of the topic, understanding of the dynamics of classroom interaction, intentions with respect to the task, and a continuous monitoring of the ongoing talk. (Wells & Chang-Wells, 1992, p. 46)

Teachers create a space in which students express their thoughts, ideas and tentative understanding, make comments, ask questions and explore their feelings; teachers' comments and responses seek to stimulate, sustain and develop students' efforts (Wells, 1999a; Pappas & Zecker, 2001a,b).

Communication needs to be in a form that makes sense to the learner, and little is to be gained by too early an insistence of formal write-ups, though transition to such write-ups is a long-term goal. Much can be achieved by creating opportunities for students to justify their choice of topic, design, interpretation and conclusions to an audience of peers and experts. Conferencing and other forms of public presentation embody three important learning outcomes: they assist the development of conceptual and procedural understanding, mirror authentic scientific practice, and contribute to the establishment of a learning community. They also give students a valuable opportunity to practice the language of science and to explore the complexity of relationships among theory, design of scientific inquiries, data interpretation and formulation of conclusions. Through such experiences, students learn to formulate a convincing warrant for belief and refine their capacity for engaging in scientific argumentation (see discussion in Chapter 8).

214

Paradoxically, experiencing the idiosyncratic and personal nature of scientific investigation enables students to recognize the mutual dependence of scientific practice and scientific theory. It enables teachers to appreciate the interrelatedness of learning science, learning *about* science and *doing* science, and to understand why *doing* science, in itself, cannot meet all the goals of science education. In any scientific inquiry, students achieve three kinds of learning. First, enhanced conceptual understanding of whatever is being studied or investigated. Second, enhanced procedural knowledge, that is, learning more about experiments and correlational studies, and acquiring more sophisticated understanding of relationships among observation, inquiry design and theory. Third, enhanced investigative expertise, which may eventually develop into scientific connoisseurship. Providing opportunities for students to report and debate their findings, and supporting them in reflecting critically on personal progress made during the inquiry, are key elements in achieving this integrative understanding. Sandoval (2005) makes the point that if students do not have to decide what kind of data to collect they are unlikely to engage in epistemological consideration of what kind of data are appropriate in other circumstances, and if they are not responsible for using data to justify their own knowledge claims they are unlikely to consider the basis on which other claims might be warranted. In other words, doing science activities make important contributions to the kind of understanding we seek, and may well be essential to the development of the sophisticated epistemological understanding that constitutes critical scientific literacy. Kelly et al. (2000) show how extensive and carefully orchestrated discussion of experimental design and possibilities for interpretation of generated data (including anomalous data) can lead students to a much more sophisticated view of scientific inquiry and recognition that experiments can be "surprising, inconclusive, sometimes in contradiction to theory, and subject to multiple interpretations" (p. 650). In similar vein, Windschitl et al. (2008) describe an approach that they call "model-based inquiry" (a term also used by Khan, 2007), in which students are consistently and repeatedly reminded that the primary purpose of scientific inquiry is to build theoretical models that are *testable* (i.e., state a relationship with observable events, processes and phenomena), revisable (i.e., may be changed in the light of new evidence or a new way of interpreting evidence), *explanatory* (i.e., causal rather than just descriptive), *conjectural* (in the sense that they involve theoretical entities that are not directly observable) and *generative* (i.e., lead to new insights, hypotheses and predictions). Students proceed via a series of carefully structured and guided conversations that enable them to make decisions about what they know and what they want to know, generate testable hypotheses, collect evidence, reach a conclusion and construct an argument. Systematic scaffolding through teacher questioning, constant monitoring of progress and provision of appropriate feedback are also the elements emphasized in Haigh's (2007) successful approach to fostering three types of creativity in science (as defined by Boden, 2001): *combinational creativity* – that is, combining old ideas in new ways; *exploratory creativity* – that is, using existing ideas and procedures to generate new data or

215

new insight; and *transformational creativity* – that is, challenging and changing existing ideas, conventions and procedures.

However, because of the idiosyncratic nature of scientific investigation, and the highly specialized but necessarily limited range of conceptual issues involved in any particular inquiry, *doing* science is insufficient in itself to bring about the breadth of conceptual development that a curriculum seeks or is necessary for critical scientific literacy. Students cannot learn sufficient science by restricting activities to *doing* science. It takes too long and is too uncertain. Moreover, not all topics lend themselves to a *doing* science approach. It is also the case that a *doing* science approach is insufficient, in itself, to bring about the breadth of knowledge and critical understanding of the syntactical structure of science encapsulated in the notion of critical scientific literacy, as I elaborate later in this chapter. Hapgood et al. (2004) attempt to address this problem through an approach they call "guided inquiry supporting multiple literacies" (GisML). Students and teachers participate in two forms of investigation: (i) "first hand investigations", in which students are guided to collect data and draw conclusions about phenomena, and (ii) "second hand investigations", in which they use text-based resources to gather information about scientists' investigations and explorations of the same phenomena. Second hand investigations are guided by "notebook texts", which the authors describe as "a hybrid of exposition, narration, description, and argumentation" (p. 460). Used cautiously and sparingly, and bearing in mind the danger of compromising good NOS understanding by simultaneous emphasis on content (as discussed in previous chapters), this approach could prove useful in highlighting some key NOS issues.

It should also be apparent that a curriculum restricted to *learning science* and learning *about* science will almost certainly leave most students unable to *do* science for themselves (something that I consider to be an essential part of scientific literacy). Though necessary, conceptual knowledge and knowledge about procedures that scientists can adopt, and may have adopted in particular situations in the past, are insufficient in themselves to enable students to engage successfully in scientific inquiry. That ability is only developed through hands-on experience of *doing* science in the kind of critical and supportive atmosphere described above. Moreover, conducting whole investigations and engaging in practical problem solving in authentic scientific contexts are valuable because the tension that arises when individuals confront obstacles that seem to prevent them achieving their goals is a powerful incentive, forcing them to "proceduralize" knowledge that might otherwise remain inert (Prawat, 1993).

> Struggling (or even failing) to formulate and carry out empirical investigations in science may teach students more meaningful lessons about how science is accomplished than flawlessly executing cookbook labs or solving carefully formulated problems. (O'Neill & Polman, 2004, p. 262)

Through such activity, procedural and propositional knowledge become fused into *strategic knowledge* – knowledge that can be used in real contexts. It has long been a legitimate criticism of school science education that many students learn a great deal of "textbook knowledge" but are unable to deploy it appropriately and successfully in real contexts. Because the notion of critical scientific literacy developed in chapter 1 incorporates deployment of understanding about science in real contexts, doing science experiences would seem to be a vital element of the curriculum, a view endorsed a decade and a half ago by the American Association for the Advancement of Science (1993).

> Before graduating from high school, students working individually or in teams should design and carry out at least one major investigation. They should frame the question, design the approach, estimate the time and costs involved, calibrate the instruments, conduct trial runs, write a report and finally, respond to criticism. (p. 9)

RESEARCH EXPERIENCE, COMMITMENT AND LABORATORY SKILLS

Of course, the success of an apprenticeship approach rests on *teachers* having some experience and expertise in conducting their own scientific investigations, something that cannot always be assumed, especially for teachers in elementary schools. So adopting this approach as a key curriculum component has major implications for teacher education, both at preservice and inservice levels. At the very least, we need to design science methods units in preservice teacher education programmes to sensitize future teachers to the importance of student-led inquiry and to provide guidance on how to implement such an approach in school (Bencze, 2000). Research internships for student teachers might well have a crucial role to play, a view endorsed by the scientists interviewed by Taylor et al. (2008). Although Schwartz et al. (2004) report that such experiences can have a favourable impact on teachers' NOS understanding, they do not comment on whether this improved understanding encompasses enhanced investigative capability or increased willingness to involve students in open-ended inquiry. These authors also note that student teachers learned more from discussions and carefully designed reflections than they did from the experiences themselves. In Brown and Melear's (2006) study, teachers who had experienced authentic scientific inquiry as part of their preservice programme reported that it had provided valuable insight into the frustrations their students would be likely to encounter. However, there was little evidence that they were implementing open-ended inquiry in their classrooms, even after two or three years teaching experience. In contrast, Jeanpierre et al. (2005) report that many of the teachers who participated alongside a pair of their students in a two-week residential research programme on monarch butterflies did subsequently provide more opportunities for open-ended inquiry in class.

Bencze and Elshof (2004) created an opportunity for three secondary school science teachers to spend one week in a field ecological research centre in a remote part of Northern Yukon, participating in the day-to-day work of a resident scientist and his graduate students as they studied local flora and fauna. The experience had noticeable impact on the teachers' NOS views and their enthusiasm for engaging students in *doing* science activities, though whether these teachers did subsequently engage students in open-ended project work is not known. Windschitl's (2003, 2004) account of a self-directed inquiry project for preservice secondary school teachers reveals that, predictably, teachers' prior views of scientific inquiry had a profound influence on the way they designed and conducted their investigations. Although all participants were science graduates, many held mistaken, overly simplistic or confused views about scientific inquiry, especially with regard to the proper role of theory in framing research questions, designing inquiries and interpreting findings. Often they would select research questions that they believed would result in a "do-able investigation" (i.e., easy to complete) rather than questions that addressed important theoretical issues. There was rarely any reference to claims and arguments, model/theory building or the possibility of alternative explanations of phenomena and events (Windschitl, 2008). In other words, many students utilized what Driver et al. (1996) call "relation-based reasoning", where inquiry is focused on correlating variables or finding a linear causal sequence, rather than "model-based reasoning", where the goal of inquiry is to check, test, challenge or develop a theoretical model. Significantly, engagement in personal inquiry did sharpen their views of scientific inquiry, most notably for those who already had more sophisticated procedural understanding. In subsequent classroom practice, the participants who used student-directed, open-ended inquiry most extensively were not always those with the most sophisticated NOS views. Rather, they were those with undergraduate, postgraduate or professional experience of authentic scientific research. It seems that there is a confidence and familiarity issue at work here, which further exposure to personal inqury at the preservice level may be able to address. Windschitl and Thompson (2006) report that an intensive course on models and modelling did help preservice teachers to develop more sophisticated understanding of scientific models and encouraged them to implement model-based teaching in their classrooms. However, they report that even with scaffolding, the majority still found it difficult to use theoretical models to design their own empirical investigations. The authors identify two factors shaping participants' thinking: first, previous school-related experience, "which influenced not only what they recognized as models but also the way they believed models could be incorporated into inquiry"; second, "a widely held simplistic view of "the scientific method" that constrained the procedures and epistemic frameworks they used for investigations" (p. 783). Akerson and Hanuscin (2007) report that a 3-year long action research approach to NOS and inquiry-based learning did have beneficial impact on the NOS views of the six elementary teachers involved and did result in more frequent engagement of their students in inquiry-based learning, resulting in substantial improvement in students' NOS views. Crucial factors seem to have been the frequency and regularity of meetings, the extended time of the project, and the sense of ownership of the curriculum materials generated by the action research approach.

What is clear is that teachers will not incorporate *doing* science experiences into the curriculum unless they regard them as important and relatively easy to organize. In terms of the former, it seems that the significance teachers attach to engaging in extended student-driven project work is related to their NOS views (Bencze et al., 2006). For example, teachers who hold social constructivist views (defined by the authors as anti-realist and naturalist[6]) are much more likely to promote student-driven open-ended inquiry than teachers who have rationalist-realist views of science, a finding that has much in common with Kang and Wallace's (2005) conclusion from their study of teachers' use of technological design projects. It is also the case that teachers who are curious, skeptical and open-minded are more likely to encourage these same attributes in students. But even when teachers' NOS views might favour adoption of project work, it is not all "plain sailing". First, some teachers (especially at the elementary school level) may feel that their subject matter knowledge is insufficient to support this kind of work. Assisting students to formulate good research questions, design inquiries and interpret data demands a substantial level of familiarity with scientific knowledge and confidence in using it, and requires a substantial level of familiarity with the pedagogical strategies and techniques of inquiry-based learning. Both these areas of concern can be readily addressed through inservice and preservice teacher education programmes. As research by Carlsen (1991a,b, 1992, 1997) has shown, teachers tend to speak less and to encourage much greater student participation, especially student-initiated discussion and student-student interaction, when they are teaching familiar subject matter. When teaching unfamiliar material they tend to ask more questions, but the questions are of a much lower cognitive level (fact-oriented) or serve a class control function. Carlsen reports that, far from opening up discussion, such questions tend to put students on the defensive.

There are many other factors that can impact adversely on teachers' decision-making regarding *doing* science, including insufficient time, excessive cost, shortage of apparatus, poor laboratory facilities, absence of technicians, concerns about safety, and the demands of an overcrowded curriculum. Many teachers are concerned about the problems of class organization and management during open-ended work, especially with large class sizes; many are unwilling to cede control to students; some are anxious about the uncertainty and unpredictability of open-ended work and fear that they will be unable to respond adequately and quickly to unexpected events or unanticipated questions; some believe that there is pressure from parents, senior teachers and even students to concentrate on "proper science" (i.e., learning the material specified in the official curriculum document). As Trumbull et al. (2006) point out, an overall school climate that puts great value on orderly classrooms, tightly controlled teaching/learning activities and efficient coverage of a content-heavy curriculum is likely to discourage even the most enthusiastic teacher from undertaking time-consuming and "messy" project work (see also, McGinnis et al., 2004). Similarly, in Crawford's (2007) study of five student teachers engaged in an inquiry-oriented teacher education programme, involving a year long school placement, the views and usual teaching practices of the school-based mentor and other senior staff had a very profound impact on the student teachers' deployment (or not) of open-ended inquiry. As the author comments, "A novice teacher lacking confidence to stand in front of a classroom of students may certainly be cautious in using an inquiry-based approach, when that approach is in direct contrast to one used by the Mentor" (Crawford, 2007, p. 637). Rudolph (2008) also emphasizes the importance of ensuring that "the expectations, power relations, and reward structure of the existing classroom culture provide a coherent set of incentives for students to play the inquiry game" (p. 120). Writing from a New Zealand perspective, Hume and Coll (2008) conclude that teachers' reluctance to engage students in open-ended scientific inquiry, especially at higher grade levels, is largely a consequence of the narrowly defined assessment requirements of a high stakes national qualification examination system.

Bencze and DiGiuseppe (2006) describe a paradoxical situation in which teachers in a school belonging to the Canadian Coalition for Self-Directed Learning (CCSDL)[7] still engaged students in practical activities that were predominantly teacher-directed and closed-ended. There seem to be two sets of assumptions in play. First, teachers' views about the routines, roles and responsibilities of classroom teachers; second their views about the purpose of practical work and the extent to which they draw a distinction between *learning science* and *doing* science. With regard to the first cluster of concerns, it seems that many teachers (perhaps most teachers) do not consider the fluidity, uncertainty and sometimes frustrating nature of authentic scientific inquiry, and especially the ceding of control over time and procedures to students, as compatible with their views of what constitutes a well-ordered and well-managed classroom activity. Encouragingly, Ford and Wargo (2007) report that involvement of student teachers in activities designed to give them first hand experience of authentic inquiry *and* stimulate discussion of how such experiences can be provided in a school setting, can begin to shift their views. Crawford's (2000) detailed study of one highly experienced high school biology teacher (Jake), with a long history of successful inquiry-oriented teaching, identified ten key roles that he adopted towards his students: *motivator* (including encouragement of students to take responsibility for their own learning), *diagnostician* (monitoring learning), *guide* (helping students to develop good learning and investigative strategies), *innovator* (constantly seeking to use new ideas, information and methods), *experimenter* (ensuring a strong focus on experiments), *researcher* (researching his own teaching and seeking feedback from students), *modeller* (exhibiting the attitudes, attributes and values of scientists), *mentor* (supporting and guiding students in NOS issues), *collaborator* (being willing to exchange ideas and allow students to take on the role of teacher) and *learner* (always curious and eager to learn). Crawford also noted that Jake's students took on roles generally not taken by students in traditional science classrooms – most notably, active collaborator, leader, apprentice, planner and teacher. Herein lies a potential agenda for professional development programmes. To paraphrase Lawrence Stenhouse (1975), there is no curriculum development (and its associated student learning) without teacher development. Interestingly, Blanchard et al. (2009) report that a 6-week marine ecology programme for teachers, designed to provide research experience in tandem with reflection on the teaching of scientific inquiry, was much more successful in fostering good understanding of scientific research and stimulating subsequent *doing* science activities in the classroom when teachers entered the programme with more sophisticated theory-based understanding of teaching and learning gleaned from their studies in graduate level courses. Following a 2-week intensive inservice programme for secondary school teachers on inquiry-based learning, Lotter et al. (2007) conclude that subsequent implementation of open-ended inquiry is determined by a cluster of four "core teaching conceptions", each of which needs to be systematically and critically addressed: (i) conceptions of science (teachers' NOS beliefs); (ii) beliefs about students (in particular, regarding their capacity to cope with open-ended work); (iii) views about what counts as good teaching; and (iv) understanding of the purpose(s) of education. However, as Songer et al. (2003) remind us, "best practice in one setting may not look like best practice in another" (p. 514). Consequently, teachers need exemplars, guidance, support and opportunity to reflect. In similar vein, Wong et al. (2006), Yung et al. (2007) and Bencze et al. (2009) advocate the use of videos of exemplary practice, together with resource packages in which experienced and successful practitioners discuss critical aspects of inquiry-based teaching. In an ideal world, these practitioners would be available online as consultants. Perhaps the best way of approaching the complex task of teacher professional development is via the kind of action research strategies described by Hodson et al. (2002) and Zion et al. (2004).

With regard to views about practical work (the second issue raised by Bencze and DiGiuseppe, 2006, above) it is evident that some teachers see it as a means of bringing about conceptual understanding, or even as a way of "learning facts", while others see it as a means of engaging students in the activities of scientific inquiry, and may even subscribe to the view that learning science is best achieved by doing science, as the three case studies presented by Kang and Wallace (2005) illustrate. In many schools, practical work is both over-used and under-used. It is over-used in the sense that teachers engage in practical work as matter of course, expecting it to assist the attainment of all learning goals. It is under-used in the sense that its real potential is only rarely fully exploited. Indeed, much that we provide is ill-conceived, muddled and lacking in real educational value. The first step in planning a more effective curriculum is to be clear about the purpose of a particular lesson. The second step is to choose a learning activity that suits it. A learning experience intended to assist concept acquisition or concept development will almost certainly need to be very different in design from one that aims to help students develop an understanding of particular elements of scientific inquiry, generate interest and excitement about science, or provide personal experience of experimental design. My own view, as I have tried to make clear on several occasions (Hodson, 1990b, 1992b, 1993b, 1996, 2005, and several locations in this book), is that teachers should design practical work and other teaching and learning activities in relation to three distinct but inter-dependent activities: learning science, learning *about* science and *doing* science.

It is also likely that students' NOS views will influence the ways in which they engage in practical work, influencing their decisions about how to conduct an investigation, what data to collect, how to make sense of the data, and so on (see discussion in Chapter 2). For example, those who see knowledge claims as emerging directly and unambiguously from data collection and analysis are likely to concentrate on collecting additional data and repeating experiments as a way of addressing problems. One of the purposes of *doing* science activities is to invert this influence, so that what students do in the lab and in the field impacts favourably on their NOS views. As with all NOS-oriented teaching and learning activities, an explicit and reflective approach is more likely to be successful than an implicit or indirect approach, although Roth and Roychoudhury (1993) claim that when grade 8 and grade 12 students are given the freedom to perform "experiments of personal relevance in authentic contexts", their ability to identify and define variables, interpret, transform and analyse data, plan and design experiments and formulate hypotheses "seem to develop holistically without being taught explicitly" (p. 148).

In addition to influencing their own classroom behaviour, student expectations can be a major factor on teachers' decision-making. It is often the case that students expect teachers to "teach", in the sense of dispensing knowledge and paying close attention to the curriculum specification or textbook, and to prepare them carefully and systematically for examinations (Yung et al., 2008). Oddly, high ability students can often be the most conservative in this respect, preferring the teacher to continue using the kind of class activities through which they have been successful in the past. Perhaps it is inevitable that both teachers and students will put most value on that which is examined and for which academic credit can be gained. Thus, it seems sensible to follow Kempa's (1986) advice to channel this influence in such a way that it becomes a positive factor, rather than a negative one. In other words, we should use assessment procedures to bring about desirable curriculum development. Practical work assessment that focuses on students' ability to design, conduct, interpret and report their own scientific inquiries will certainly encourage, and may even *require* teachers to provide students with opportunities to *do* science for themselves, and will ensure that teachers consider very carefully what constitutes rigorous and authentic scientific investigation.[8] It would also be hopelessly naïve to assume that all students are both interested in science (and scientific inquiry) and highly motivated. Some are focused solely on gaining marks and achieving grades, and will pay little heed to wider learning goals; some are reluctant or unwilling participants; some are openly hostile. The complex and difficult task of engaging these diverse students in meaningful, purposeful activity through a *personalized* approach to learning is discussed at length in Hodson (1998a).

It may be that the most powerful factor militating against adoption of *doing* science approaches is teacher anxiety about the ability of students to cope with student-directed, open-ended investigations and the epistemological position they entail (see, for example, Roehrig & Luft, 2004). Hodson (1993a) notes that teachers tend to adopt more conservative approaches with students they consider to have lower academic ability. However, Bencze (2000) provides good evidence that when teachers create the kind of classroom environment that is conducive to *doing* science, and when institutional support is forthcoming, most students are able to carry out scientific investigations and technological design projects with minimal teacher direction, though not always with successful outcomes. There is no doubt that some students find open-ended inquiry frustrating: they may have insufficient conceptual knowledge, procedural "know-how" or experience to tackle problems with confidence, and may feel that teachers provide insufficient guidance. Apedoe (2008) reports that, even at undergraduate level, many students fail to track down the knowledge they lack, despite ready access to libraries and computers, prompting speculation about their sense of commitment, interest and self-confidence. Some students are unable or feel that they are unable to express ideas, plans, interpretations and conclusions in writing, even though they are able to do so verbally. It is here that the use of a writing heuristic can play a valuable scaffolding function (see Chapter 9).

My experience is that in their first experiences of self-directed inquiry students need considerable help in formulating proposals. They have a tendency to concentrate on the minutiae of specific *procedures* for collecting data rather than the wider *purposes* of data collection. They are often guilty of collecting insufficient data, and sometimes collect inappropriate or inadequate data. It is not uncommon for students to collect irrelevant data, such as the name or sex of the person who made the observation or took the measurement, or to prioritize the most recently acquired data over older and possibly more secure or useful data. They often reach conclusions that are inconsistent with or unwarranted by their data. Many students have a preference for numeric data rather than descriptive data, even when its use is unhelpful or inappropriate. When students have a reason to disbelieve the evidence (for example, because it is inconsistent with their existing beliefs) they will often make strenuous efforts to find flaws in the data or to reject it entirely. When they cannot discern a pattern in the data and reach a quick conclusion some students are tempted to *impose* a pattern. While this does not amount to fabrication of data, it could lead to it if teachers do not reinforce the message that accumulation of good data can be a lengthy, complex, difficult, uncertain and sometimes very frustrating business, and that what students are currently experiencing is exactly what real scientists experience. One of the problems, as Millar and Lubben (1996) comment, is that previous experience of school science has led students to believe that they *ought* to be able to state a firm conclusion. The reality of scientific inquiry is that data are sometimes inconclusive and the investigation procedure or data collection methods may need to be modified if good data are to be acquired.

Before students can evaluate data in terms of its support (or not) for theoretical propositions, they need the skills to collect reliable data and the ability to convert raw data into meaningful "results" (see Duschl et al., 2007, for a detailed review of the relevant research literature). With regard to the latter ability, Toth et al. (2000) report that many students can design experimental inquiries and collect data but cannot put these data into a form that supports meaningful conclusions. Kanari and Millar (2004) observe that students often have greater difficulty making sense of their own data (for example, whether a variable has increased, decreased or remained the same) than evaluating the significance of pre-existing data as a warrant for belief. Koslowski (1996) argues that the interpretation students place on data depends on whether they can imagine an underlying mechanism to account for patterns in the data.[9] If students can imagine *why* two variables might co-vary, data showing no covariance is likely to be rejected, unless there is a great deal of it; conversely, if a plausible mechanism cannot be imagined, the same amount of data showing covariance is unlikely to persuade students that the variables are related. In other words, students use theory as a way of judging acceptability of data. A more cynical interpretation is that students see what they expect to see, and do not see what they do not anticipate. From a longitudinal study of 14 children, Tytler and Peterson (2005) conclude that although children as young as ten can seek relevant evidence to confirm a particular idea or proposition, their understanding of the formal process of arguing back from evidence to clearly establish an idea takes much longer to establish.

Of course, a robust understanding of data includes the capacity to recognize and deal with errors. A study by Masnick and Klahr (2003) examined students' understanding of four types of experimental error: design errors, measurement errors, interpretation errors and what the authors call "execution errors" (those arising from something entirely unanticipated, such as imperfections in the manufacture of objects used in the experiment, uncontrollable changes in local conditions such as the weather, and so on). They conclude that even very young children (grades 2 to 4, for example) have some understanding of experimental error, especially measurement and execution errors, though their understanding is not coherent and systematic. In an effort to make sense of findings such as these, Jeong et al. (2007) introduce the notion of *evidentiary competence*: the concepts and reasoning skills involved in the collection, organization and interpretation of data, and awareness that empirical data is often central to theory appraisal. The 13 components of evidentiary competence postulated by the authors relate principally to experimental investigations. At the planning stage, they include identification of data relevant to the investigation, understanding of dependent and independent variables, choice of appropriate sample size and design of fair tests. At the data collection stage, they include the need for objectivity and accuracy in data collection and establishment of reliability through successive replication. At the interpretation stage, students need to be able to interpret graphs and tables of data (and know how to code their own data in these ways) and deal appropriately with anomalous data. In addition to developing their proto-taxonomy, the authors provide an insightful review of published research focusing on a number of elements of evidentiary competence, and an account of their own findings on grade 6 students' competence related to objectivity, replicability and interpretation of data. My concern is that incautious researchers might use this framework as though the components are theory-free, generalizable and transferable, that is, they may make the same fundamental errors as the process-skills enthusiasts made in the 1970s and 1980s (see Hodson, 1996). These same concerns inform my response to the notion of "concepts of evidence", which Gott and Duggan (1996) and Gott et al. (2003) see as comprising three broad categories: concepts associated with design, including variable identification, fair test, sample size and variable type; concepts associated with measurement, including relative scale, range of interval, choice of instrument, repeatability and accuracy; concepts associated with handling data, such as use of tables and graphs. The capacity to carry out any of the sub-skills in the Jeong et al. (2007) taxonomy appropriately and competently (or any of the skills listed in other taxonomies) depends on conceptual understanding of the phenomenon or event under investigation. Clearly, also, the capacity to collect valid and reliable data depends on students' laboratory skills.[10]

If students are to engage successfully in holistic scientific investigations they need robust laboratory skills that enable them to collect reliable data. The level of expertise necessary for students to carry out procedures and make measurements quickly, accurately, and as a matter of routine, is rarely attained through conventional laboratory classes. Indeed, I have argued elsewhere that the acquisition of laboratory skills has little, if any, value in itself and should not be regarded as a major goal of most practical work (Hodson, 1992b, 1993b). Rather, lab skills are a means to an end – that end being further learning. Attempting to justify practical work in school in terms of skill development is to be guilty of putting the cart before the horse. It is not that practical work should be geared to the acquisition of particular laboratory skills; rather, it is that particular skills are necessary if students are to engage successfully in practical work. Two points follow: we should teach only those skills that are of value in the pursuit of other learning and, when such is the case, we should ensure that those skills are developed to a satisfactory level of competence. If we try to teach too many skills the outcome may be that students acquire none of them to the level of unconscious expertise necessary for doing science successfully. The level of expertise we should seek to develop (in the chosen skills) is similar to that shown when we drive a car successfully through dense traffic without being aware of shifting gears, checking mirrors and adjusting pressure on the brake pedal or accelerator. The skills have become automatic. Recognizing that this kind of expertise is only acquired through repeated use in varying contexts has clear implications for the way we conduct practical work in school, as Masters and Nott (1998) make clear. There is much to be said for "skills clinics" to equip students with the lab skills they need for a particular investigation, as the need becomes apparent.

It should also be noted that computers have a major role in enhancing hands-on investigations, and more extensive use of computer-supported labwork and fieldwork might render some of my concerns about inadequate skill levels redundant. By displaying data in a variety of ways, plotting graphs, performing calculations, analyzing data, collecting data and even monitoring and controlling experiments, computers can eliminate much of the "noise" associated with complex apparatus and procedures, the boredom of long waiting times and the mathematical problems caused by lengthy and sometimes difficult calculations, as well as by-passing the need for some hands-on lab skills. Computer-enhanced experimentation enables students' attention to be focused on the creative phases of scientific inquiry, leaving the technician's role of carrying out the investigative procedures to the computer. In addition, because of the "stamina" and versatility of the computer, students can investigate a much wider range of phenomena and

225

events, and can utilize techniques that would be otherwise unavailable. In many ways, computer-assisted laboratory work (and computer-assisted fieldwork using portable computerized probes) is more like real scientific research than most conventional school practical activities. Advances in computer technology have even made it possible to conduct investigations at remote sites. For example, Harvard University's *MicroObservatory* enables students to make telescopic observations that would be impossible using school equipment. An additional advantage is that students can observe and record images of a distant night sky during local daylight hours (Sadler et al., 2001). Scanlon et al. (2002) provide details of several well-established websites that allow remote access to findings from on-going research projects. Some organizations enable students to be active researchers, requesting data, making interpretations and sharing conclusions with other site users. Recent developments in ICT have made it possible for students living in widely separated locations to be involved in collaborative research activities, either through informal arrangements between interested teachers or via organizations such as Science Across the World (http://scienceacross.org), NASA (http://quest.nasa/govt), the Pathfinder Science Community (http://pathfinderscience.net) and the Learning through Collaborative Visualization (CoVis) project (Edelson et al., 1996).

GOING BEYOND BENCHWORK

Laboratory benchwork is the stock-in-trade of most science teachers, and very few venture beyond it. Interestingly, although research evidence suggests that laboratory bench work is only partially successful in achieving its learning goals, at least as planned and implemented in most schools (Hodson, 1993b, 2005), teachers have confidence in it and invest considerable time and effort in it. By contrast, they do not invest in fieldwork, even though the research evidence attesting to its educational value is very encouraging (Gerber et al., 2001; Braund & Reiss, 2004; Hamilton-Ekeke, 2007). For example, Killerman's (1996) study shows that well-planned fieldwork not only raises students' attainment levels in tests of conceptual understanding but also has a beneficial impact on their environmental awareness and attitudes towards conservation – further key elements in critical scientific literacy. Because they do not disturb, violate or destroy what is being studied, as experimental investigations often do, field-based correlational studies are more in keeping with non-exploitative environmental values, and so might be regarded as education *for* the environment as well as education *about* and *in* the environment. While fieldwork conjures up images of groups of students involved in time consuming and expensive visits to distant field centres, with consequent large-scale disruption of the regular curriculum and potential alienation of other teachers, it does not have to be that way. As Lock (1998) points out, quality fieldwork is possible at low cost and minimal disruption by imaginative use of sites within or adjacent to the school grounds and by the construction and utilization of "model environments", often using cheap and readily available items such as old baths and drainpipes. Barton and Tan's (2009) case study of how four 11-year olds investigated the phenomenon of "urban heat islands" is a splendid example of fieldwork in an urban setting. Local markets, parks, and even the schoolyard, provide rich sites for simple but rewarding investigations. In the context of the foregoing discussion of learning through apprenticeship, it is worth noting that fieldwork sometimes provides opportunities for students to work alongside practising scientists, as in the Wetlands Project described by Helms (1998a,b) and the extensive study of the impact of water pollution and farming methods on insect-eating birds outlined by Cunniff and McMillen (1996).

Albone et al. (1995) list 28 projects, half of which are field-based activities, involving cooperative research between practising scientists and school students, while Robinson (2004) provides information on 16 field research opportunities open to students in a wide variety of fields, including marine biology, rain forest ecology, geology and archaeology. Posch (1993) describes some fascinating work conducted under the auspices of the Environment and Schools Initiative Project, including a collaborative project involving five Italian secondary schools in which students examined the quality of ground and surface water in their immediate localities by means of chemical, bacteriological and micro-plankton analysis. Similar opportunities may arise in zoos, aquaria, botanical gardens, nature trails, local conservation areas, theme parks, amusement parks, outdoor pursuits centres, science centres, museums and interpretive centres (Rennie & McClafferty, 1996; Griffin, 1998; Krapfel, 1999; Ash & Klein, 2000; Falk & Dierking, 2000; Bencze & Lemelin, 2001; Heering & Muller, 2002; Pedretti, 2002, 2004; Anderson et al., 2003; Braund & Reiss, 2004; Black, 2005; Kisiel, 2006; Rahm, 2006, 2008; Tal & Steiner, 2006; Rennie, 2007a; Tal & Morag, 2007; Tran, 2007; Nielsen et al., 2009).[11] In discussing the potential of such sites for providing *doing* science experiences, Braund and Reiss (2006) draw useful distinctions among three learning contexts: the *actual* world (e.g., field trips and visits to industrial installations and research facilities), the *presented* world (e.g., science centres and zoos) and the *virtual* worlds available through information and communications technology. With regard to the third category, Robinson (2004) provides valuable information on a number of online simulated research activities.

There is no doubt that inquiry-based learning and problem-based learning (whether in the lab or in the field) can be enormously enhanced by providing opportunities for students to consult practising scientists via contacts with a local university, college, community science centre or weekend science camp (Lawless & Rock, 1998; Bencze & Hodson, 1999; Caton et al., 2000; Chowning, 2002; Schaillies & Lembens, 2002; King & Bruce, 2003; Collins, 2004; MacDonald & Sherman, 2007) or through telemonitoring and related ICT technologies (Roup et al., 1993; Crawford et al., 1999; Comeaux & Huber, 2001; O'Neill, 2001; Novak & Krajcik, 2004; Songer, 2007). One intriguing way of putting students in contact with practising scientists is by means of a Researchers-in-Residence programme. For example, Earland (2004) describes a scheme funded by the Engineering and Physical Sciences Research Council (UK) in which PhD students are seconded to schools to assist teachers, motivate students, introduce students to contemporary scientific practice and act as role models of scientists. A similar scheme operates in the United States under the auspices of the National Science Foundation (2007). The young researchers also gain valuable experience of school science education and improve their teaching and communication skills. Closely related is the Teacher Scientists Network, which maintains a register of scientists willing to give advice to teachers and to involve students in small-scale research projects (Pringle, 1997). Evans et al. (2001) provide similar advice and guidance for teachers working in a US context.

Among a range of specific collaborative research ventures, Paris et al. (1998) describe a museum-school-university partnership established to deliver a 6-week biology programme in which students in grades 3–5 attended classes in the university's laboratory and used living exhibits and materials from the museum to carry out investigations, with university students assisting them as part of their own course requirement. The activities culminated in a "family biology night" at which the students gave an account of their experiences and presented their findings. Buntting and Jones (2009) report on a partnership between a research organization (The Cawthron Institute) and a New Zealand high school, through which students collected data on the impact of a range of environmental factors on the size and weight of green-lipped mussels. An interesting variant described by Schaillies and Lembens (2002) puts students in the dual role of researcher and researched. In a study of the impact of diet on immune systems, students monitored various indicators of a healthy immune system after being assigned to one of three experimental diet groups (defined as "healthy", "unhealthy" and "normal"). Lee and Songer (2003) describe the *Kids as Global Scientists* project: an 8-week long, inquiry-based programme designed to give middle school students some of the experiences of being a meteorologist. Students collect local weather data, compare and contrast it with data collected by other groups of students, interpret weather maps, make weather forecasts, and consult with a weather specialist and with students in other locations through the project Website (http://www.onesky.umich.edu) (Songer et al., 1999). This project, and the *Students as Scientists: Pollution Prevention through Education* initiative described by Comeaux and Huber (2001), provide a glimpse of authentic practice in a way that raises awareness of the potential social and political impact of scientific inquiry. Similar motives underpin the various citizen science initiatives,[12] a typical example of which is the investigation of the feeding habits of birds organized by Cornell University, which involved some 17,000 participants (Trumbull et al., 2000). Through these kinds of activities students gain experience of working on real research projects, have opportunities to engage in creative problem-solving and have the satisfaction of having their ideas taken seriously by real scientists. When their work is subsequently presented at conferences, which is not uncommon, students receive an enormous boost to their confidence and gain further valuable insight into the workings of the scientific community. However, something as simple to organize as a series of visits to laboratories and industrial sites can also have substantial impact on students' views of science, scientists and scientific practice, as in Scherz and Oren's (2006) work with grade 9 and grade 10 students at three Israeli high schools. The key, of course, is to provide students with a working brief that requires them to scrutinize personnel, processes and products carefully and systematically, and to present their findings in written, oral or multimedia format.

COMPUTERS AND MULTIMEDIA EXPERIENCES

Computer simulations, modelling software and visualization tools are especially powerful techniques for enabling students to engage in the more creative aspects of scientific inquiry that help to consolidate their understanding of the syntactical structure of science (Edelson, 2001; Winn et al., 2006; Kim et al., 2007; Winberg & Berg, 2007; Songer, 2007). As discussed earlier, many students do not have sufficient opportunities to engage in hypothesis generation and experimental design because teachers are unwilling to provide the time, meet the cost or run the risk of students adopting inappropriate, inefficient or potentially hazardous investigative strategies. Consequently, teachers tend to design all the inquiries, usually in advance of the lesson, and students merely follow their instructions. With a computer simulation, poor designs can go ahead and any problems can be discovered by the students themselves and modified or eliminated quickly and safely. Thus, students learn from their mistakes and are led to investigate more thoroughly and more thoughtfully. More importantly, they learn that designing investigations is something that they can do; it is not the exclusive preserve of white-coated scientists in laboratories full of expensive equipment. Too often, students come to believe that the way in which an inquiry is conducted in class is the only way it can be done. Computer simulations enable different groups of students to formulate and implement different procedures, some of which will work well, some less well, some not at all. This is more like real science. There are at least three learning goals embedded in such experiences. First, students learn much more about the phenomena under investigation and the concepts that can be used in accounting for them, because they have more time and opportunity to manipulate concepts and ideas. Second, they acquire some of the thinking and strategic planning skills of the creative scientist. Third, they learn that science is about people thinking, speculating and trying things that sometimes work and sometimes fail.

The use of computer simulations, as opposed to real experiments, enables the teacher to tailor the learning experience precisely to the teaching/learning goals, instead of the more usual situation of having to fit the learning goals to the complexities of reality. Teachers can increase or decrease the level of complexity, include or exclude certain features, adopt idealized conditions and generally create an experimental situation that enables learners to concentrate on the central concepts without the distractions, waywardness of materials and pedagogic noise that is so much a feature of experiments with real things. In many cases, the real learning gains are associated with thinking about ideas, working out a means of conducting the inquiry and interpreting the results. The actual performance of the experiment contributes very little. In some cases, because of excessive pedagogic

229

noise, it serves merely to distract; in others, because the procedures are lengthy, tedious and uninteresting, it serves to alienate learners. Computer simulations enable teachers to place the emphasis on the principal learning goals of the lesson and to "freeze", re-run or modify an experiment quickly and easily in order to clarify or develop a teaching point. However, the capacity to reduce complexity is a two-edged sword: the simplicity and orderliness that enables learners to engage so productively with the ideas that underpin the simulation can distort their views of reality. Omissions and errors can cause serious misconceptions. Clarity and accessibility without trivialization or distortion is sometimes difficult to achieve. A major problem is that simulations may give rise to a belief that the variables that determine a process are both easily controlled and independent of each other. A skilled teacher would seek to make students aware, on the one hand, that nature is not so simple, and on the other hand, that part of the scientific approach is the deliberate attempt to contrive artificial situations in which variables are controlled and manipulated as a way of gaining understanding. Thus, there are two learning messages here. First, a message about the nature of scientific investigations. Second, a message that science is very powerful in situations that can be controlled and controllable (i.e., when experiments can be used) but is less powerful, and therefore less reliable, in situations that are not closely controlled. Thus, computer simulations provide significant opportunities for countering the all-too-prevalent myths that science is universally applicable, certain in the knowledge it generates and capable of solving all problems (see discussion of curriculum myths about science in chapter 5). Of course, any conclusions drawn during exploration of a computer simulation are only as valid as the theory on which the simulation is based. The validity and status of such theories need to be evaluated before utilizing any results derived from the simulation. A real danger of simulations is that they can be used by unthinking teachers and students as evidence for the very theory assumed in their design. Evidential support for a theory must be obtained in other ways. One of these ways, of course, is through laboratory and fieldwork – the very activities that simulations are designed to replace.

Interrogation of interactive multimedia packages and Internet websites incorporating databases enable students to explore ideas, make predictions and speculate about relationships, and to check them against "the facts" (pre-assembled data) quickly and reliably without the restrictions of teacher directions and without the attendant noise and frustrations of benchwork.[13] Similarly, the "evidence evaluation tasks" described by Bell and Linn (2000) and Davis and Linn (2000), in which students are presented with written reports of evidence to evaluate, provide opportunities for students to coordinate and judge data from multiple sources. WISE (Web-based Inquiry Science Environment) supports students in revising misconceptions by providing scaffolding and metacognitive supports such as inquiry maps for guided inquiry activities, hints on inquiry questions and "evidence pages" with relevant scientific ideas and examples (Linn et al., 2003). These methods have made it possible for learners to investigate thoroughly their own questions about a whole range of things without the constraints imposed by inadequate laboratory facilities, under-developed practical skills, insufficient time, lack of materials or considerations of personal safety. Students can make predictions or speculate about relationships, and check them against the "facts" quickly, reliably and in a secure and private learning environment. Students are enabled to explore ideas, construct and reconstruct knowledge, hypothesize and test their ideas for themselves and by themselves, free from the restrictions of worksheets and teacher directions. Although there is no benchwork involved, these activities constitute *doing* science in a very real sense. By using the computer as a tool to find answers to their own questions, students develop real problem-solving and inquiry skills. Problem situations have to be analysed, questions formulated and searches planned. Through such activities students learn to identify problems that are significant, worth investigating and susceptible to systematic inquiry. At the same time, they learn that not all questions and problems have a unique solution or a right answer, and that many solutions are tentative and in need of refinement through further inquiry. In other words, *doing* science often generates as many questions as it answers. Interestingly, when Pedretti et al. (1998) interviewed students in grades 10 and 11 about their views of the computer-based approaches employed in a technology-enhanced science programme (use of computer simulations, MBL activities and multimedia materials), emphasis predominantly concerned learning issues rather than science or technology. Andrew commented, "We're learning physics, of course, but the main objective of the class is to try and teach ourselves independently" (p. 585), while Justine said, "You're learning how to teach yourself something with the information you're given ... when you get older and have your own career, you're not always going to have someone there to teach you everything. You're going to have to figure it out yourself" (p. 586).

CASE STUDIES

As discussed in Chapter 6, much can be learned about the nature of science and scientific inquiry through well-chosen case studies in the history of science. For example, Galili and Hazan (2001) report that a series of case studies in the history of optics was successful in shifting the views of grade 10 students from "science is dogmatic truth" towards "science is dynamic, uncertain, tentative and controversial". A similar argument for the value of historical case studies is made at some length by Kipnis (2007). Likewise, Irwin (2000) showed that a case study of atomic theory led grade 9 students to appreciate the tentative nature of scientific knowledge, while Lin and Chen (2002) describe how a history of science approach had a favourable impact on preservice chemistry teachers' understanding of the creative aspects of scientific investigation, the theory-dependence of observation and the role and status of scientific theory. However, findings from a study of undergraduates, graduate students and preservice secondary school teachers following three different history of science courses were disappointing. The authors (Abd-El-Khalick & Lederman, 2000b) report few and somewhat limited changes in participants' NOS views. Notably, however, those students

231

who entered the course with "more adequate" (i.e., more sophisticated) NOS views showed greater improvement, which may suggest that history of science is best deployed in an enrichment or consolidation role. A case study approach may have particular value in providing richly detailed examples of how argument is deployed in science to establish knowledge claims and resolve theoretical or interpretive disputes. The overthrow of phlogiston theory by the modern theory of combustion in oxygen or the triumph of the heliocentric view of the solar system are particularly well-suited for this kind of emphasis.

A perennial difficulty for approaches using the history of science to teach science content or NOS knowledge is the difficulty students experience in putting themselves into the appropriate conceptual and sociocultural framework necessary to fully comprehend the nature of the problems, the resources available, and the motivating influences on the key players in the events. Many are guilty of adopting a "Whiggish" approach[14] – that is, interpreting everything from a 21st century perspective. Many students seem unable to see the ideas of previous generations as anything other than wrong, or even absurd – a perspective that is too often reinforced by school science textbooks. Students' scientific understanding is often insufficiently secure and sophisticated for them to understand the relationships between previous scientific ideas and current science. Many experience difficulty in immersing themselves in the sociocultural context and social conventions of the time, and so fail to understand the needs and interests that drove the research. In addition, there is often a welter of historical detail concerning political intrigue and commercial interests that can divert attention from whatever NOS understanding the study was designed to promote. In consequence, many students' retain (or even reinforce) the NOS misconceptions they held at the outset. These problems remind us just how important it is that a history of science approach directs explicit attention to key NOS items, lest they be submerged in what we might call "historical noise".

Case studies in contemporary science or recent scientific history may be more effective than historical studies in overcoming these problems. Their immediacy, personal relevance and (sometimes) controversial nature can be particularly engaging for students – see, for example, the case study of sickle cell anaemia developed by Howe and Rudge (2005) and the work on SARS research by Lee (2008) and Wong et al. (2008, 2009). Buxton (2001, 2006) suggests that teachers draw on the science studies literature to assist construction of such case studies – in particular, on ethnographic studies of laboratory practice. This approach has much in common with that recommended by McGinn and Roth (1999) and deployed in recent work by Ford and Wargo (2007). Wong et al. (2009) took the radically different step of seeking the views and firsthand accounts of scientists involved directly in the events described in the case study, on the grounds that scientists working at the frontiers of science are particularly well-positioned to comment on the practices of the community of scientists, the nature of scientific work, the aims behind it, and the inter-relationships with the society in which it is embedded. The outcome was a multimedia teaching and learning package on the SARS epidemic that combined a wide range of television news bulletins and documentaries, newspaper reports and video-interviews with nine Chinese and Canadian scientists who had been prominent in the research efforts leading to the identification of coronavirus as the pathogen of SARS, the decoding of the genomic sequencing of the SARS-coronavirus, the development of new diagnostic tests, and elucidation of the transmission modes of SARS. The authors report significant enhancement of NOS understanding in the particular aspects of scientific inquiry featured in the case study when used with preservice and inservice science teachers (Wong et al., 2008). Arguing along similar lines, Abrams and Wandersee (1995) advocate the use of "research-based interactive vignettes"[15] compiled from original papers and interviews with scientists as an effective means of teaching students about the day-to-day intricacies of cutting edge scientific research.

Of course, a case study of any one series of events cannot achieve all the NOS understanding we consider desirable, nor serve to dispel all stereotyped, mistaken or confused views of science acquired through formal and informal educational processes. Case studies are necessarily limited in scope and geared closely to particular contexts and situations. Nevertheless, a series of detailed and richly textured case studies (both contemporary and historical) can make a substantial contribution to enhanced NOS understanding, especially with regard to what we might call "cutting edge science". As noted earlier, Elby and Hammer (2001) assert that the NOS items usually included in school science curricula can sometimes present a view of science that is too generalized and too broad. To paraphrase Sandra Harding (1986), too much of what is said to characterize science is really about physics, and physics is an atypical science. The syntactical structure of each discipline is deeply grounded in the discipline's substantive aspects and the specific purposes of the inquiry. As the important research questions vary in form and substance, so too do the methods and technologies employed to answer them, the kind of evidence sought, the standards by which they are judged, the kinds of arguments deployed and, perhaps, even the underlying values (Cartwright, 1999). And the differences in approach that arise are just too extensive and too significant to be properly accounted for by generic models of inquiry. There are few similarities between the day-to-day activities of palaeontologists and epidemiologists, for example, or between scientists researching in high energy physics and those engaged in molecular biology. While physicists may spend time designing critical experiments to test daring hypotheses, as Popper (1959) states, most chemists are intent on synthesizing new compounds.

Chemists make molecules. They do other things, to be sure ... they study the properties of these molecules; they analyze ... they form theories as to why molecules are stable, why they have the shapes or colors that they do; they study mechanisms, trying to find out how molecules react. But at the heart of their science is the molecule that is made, either by a natural process or by a human being. (Hoffmann, 1995, p. 95)

Mayr (1988, 1997, 2004) has also criticized the "standard" NOS views promoted in curriculum documents on grounds that they are nearly always derived from physics. Biology, he argues, is markedly different in many respects, not the least significant of which is that many biological ideas are not subject to the kind of falsificationist scrutiny advocated by Karl Popper and given such prominence in school science textbooks: "It is particularly ill-suited for the testing of probabilistic theories, which include most theories in biology ... And in fields such as evolutionary biology ... it is often very difficult, if not impossible, to decisively falsify an individual theory" (Mayr, 1997, p. 49). Sciences differ substantially in the ways they use observation and experiment, and in the extent to which they use mathematics (Knorr-Cetina, 1999). In addition, as Jenkins (2007) notes, "the criteria for deciding what counts as evidence, and thus the nature of an explanation that relies upon that evidence, may also be different" (p. 225). It is more than 70 years since Bachelard (1934/1985) expressed broadly similar views about diversity among the sciences.

> When one looks at science what is immediately striking is that its oft-alleged unity has never been a stable condition, so that it is quite dangerous to assume a unitary epistemology. Not only does the history of science reveal a regular alternation between atomism and energetics, realism and positivism, continuity and discontinuity, rationalism and empiricism; and not only is the psychology of the scientist engaged in active research dominated one day by the unity of scientific laws and the next by the diversity of things; but even more, science is divided, in actuality as well as in principle, in all of its aspects. (p. 14)

As a particular science progresses, and new theories and procedures are developed, the nature of scientific reasoning may change. Indeed, Mayr (1988, 2004) has distinguished two different fields even within biology: *functional* or mechanistic biology and *evolutionary* biology, distinguished by the type of causation addressed. Functional biology addresses questions of proximate causation; evolutionary biology addresses questions of ultimate causation.

> The functional biologist is vitally concerned with the operation and interaction of structural elements, from molecules up to organs and whole individuals. His ever-repeated question is "How?" ... The evolutionary biologist differs in his method and in the problems in which he is interested. His basic question is "Why?" (Mayr, 1988, p. 25)

In similar vein, Ault (1998) argues that the geosciences are fundamentally historical and interpretive, rather than experimental. The goal of geological inquiry, he argues, is interpretation of geologic phenomena based on observations, carefully warranted inferences and integration or reconciliation of independent lines of inquiry, often conducted in diverse locations. These interpretations result in a description of historical sequences of events, *sometimes* accompanied by a causal model.

Thus, any consideration of NOS must be "contextual, conditional, with an eye to open horizons: "closed" answers must, for that reason alone, be suspect, indeed rejected" (Suchting, 1995, p. 20). In short, we should seriously question whether views in the philosophy of science that were arrived at some years ago can any longer reflect the nature of 21st century science, especially in rapidly developing fields in the biological sciences. Notwithstanding the cautions urged earlier regarding differences in investigative strategies and the need to simplify matters for younger students, scientific inquiry is perhaps better understood as *situated practice*, in which the particular conceptual schemes deployed and the specific procedures and instrumental techniques utilized play a crucial role. As Bowen and Roth (2007) point out, situatedness is the principal feature of field ecology. For example, research design is driven by local conditions to such an extent that little in the way of observational techniques and field approaches is generalizable or transferable. Equipment is rarely standardized, robust and utterly reliable; rather it is often purpose-made, idiosyncratic in design, intended for short-term use, and made from locally available materials. Replication, one of the classic hallmarks of good science, is virtually impossible because of population variation and inevitable changes in the local ecological conditions. Even when the same animals are identifiable and available for study, their behaviour may be substantially different because of ageing, sickness, changes in the food supply, presence of youngsters, different weather conditions, and any of the myriad factors that might impact the population group and the individuals that comprise it, including previous experience of field researchers. Bowen and Roth's remarks make good sense in light of Baker and Dunbar's (1996) argument that the general characteristics of experimental design and the conditional rules for "proper conduct" (what they call the "experiment schema") will vary quite extensively from domain to domain.

All those who try to cross borders between the different disciplines of science will quickly learn that each has its distinctive vocabulary and style of communication. Although knowledge building is the major concern of all science disciplines, it is not too much of an exaggeration to state that the different sub-disciplines of science also employ distinctive styles of argumentation, rules of evidence and criteria of judgement. For example, Ruse (1988, 1998) and Ruse and Hull (1998) present a strongly argued case for a distinctive philosophy of biology. In other words, the specifics of scientific rationality vary between sub-disciplines, with each sub-discipline playing the game of science according to its own rules. It is not surprising, therefore, that students often experience great difficulty in generalizing even the simplest process skills from one context to another, an observation that has been given theoretical legitimacy by the situated cognition movement (Lave, 1988, 1991; Brown et al., 1989; Lave & Wenger, 1991; Hennessy, 1993; McLellan, 1996). This argument has been eloquently and forcefully made by Rudolph (2000).

Understanding of the operations of science has changed dramatically over the past 40 years. Educators need to begin to exploit the vast literature of the science studies community, not to develop some universalist picture of science, the value of which is questionable, but to begin to understand what the various practices of science look like in all their myriad forms, in order to provide some reasonably authentic context in which to situate the scientific knowledge claims of the curriculum. (p. 409)

235

Elby and Hammer (2001) also note that "a sophisticated epistemology does not consist of blanket generalizations that apply to all knowledge in all disciplines and contexts; it incorporates contextual dependencies and judgments" (p. 565). Essentially the same point is made by Clough (2006) when he says that "while some characteristics [of NOS] are, to an acceptable degree, uncontroversial ... most are contextual, with important and complex exceptions" (p. 463). Thus, instead of trying to find and promote broad generalizations about the nature of science, scientific inquiry and scientific knowledge, teachers should be building an understanding of NOS from examples of the daily practice of diverse groups of scientists engaged in diverse practices, and should be creating opportunities for students to experience, explore and discuss differences in the kind of knowledge and the methods of generation across multiple contexts – again, being mindful of the need to avoid confusing younger students. It is for this reason that NOS-oriented research needs to study the work of scientists active at the frontier of knowledge generation.

A series of case studies can provide the rich context-specific detail necessary for emphasizing important differences in practice between the disciplines of science. Situating NOS elements in contexts of authentic, contemporary scientific practice may help to dispel some persistent stereotypes about the uniformity of scientific inquiry. Moreover, case studies of contemporary science can make students more cognizant of the rapid advances in scientific knowledge and investigative procedures that have taken place in recent years, particularly in the biological sciences. Teachers need to recognize that NOS understanding is as tentative and subject to change as scientific knowledge itself (Losee, 1993; Heidelberger & Stadler, 2002), and serious questions should be asked about whether the selection of NOS components for most school science curricula fully and adequately reflects 21st century scientific practice. Case studies of contemporary scientific practice, especially in the newer scientific domains, can address this concern. Furthermore, by using the comments of practising scientists, as in the case study of SARS described by Wong et al. (2008, 2009), we can make the scientific enterprise appear exciting, interesting and accessible to students, even though we may reveal one or two less-than-savoury aspects of the day-to-day lives of scientists (see Wong & Hodson, in press).

NO ONE METHOD

What is abundantly clear is that there is no one method for bringing about a satisfactory level of understanding of the substantive structure and syntactical structure of science (items 6 and 5 in the King and Brownell list). Lab work, fieldwork, computer simulations, case studies, biography and autobiography, and even creative writing, dance and drama (see Chapter 9), can all play a part, but none is sufficient in itself. We need to exploit the particular strengths of each approach and to stimulate student interest through variety. In essence, we need to adopt what Leach et al. (2003) call the "drip feed" approach, that is, lots of opportunities to focus on NOS knowledge, using a variety of methods, spread throughout a students' science education experience. Also, we need to take account of research findings that NOS instruction is more successful when it is explicit, rather than implicit, and when it provides frequent opportunities for reflection. It is also very important that we include NOS knowledge in the assessment and evaluation schemes, and find ways to assess what it is we really want students to know. It is always worth making the point that assessment determines the priorities of both students and teachers; it is always worth reminding teachers that assessment schemes too often fail to focus clearly on the important learning objectives.

Research findings concerning the relative effectiveness of integrating NOS learning with content learning, embedding it in consideration of socioscientific issues, or approaching it via decontextualized, generic NOS teaching units, are less clear. The research may be unclear because researchers fail to recognize that these three approaches serve fundamentally different purposes relating to introduction, exploration, consolidation and application of knowledge, and meet fundamentally different student needs relating to immediacy, familiarity, curiosity and emotional involvement. While it is tempting to think that a content-rich approach will invariably be preferable, simply because it gives "life" to NOS concepts, a decontextualized approach may be helpful in the early years of a student's science education (activities relating to theory-laden observation, for example) and may be essential for addressing long-standing and firmly held misconceptions about science that can prejudice further learning, simply because students do not have to contend with potentially distracting content issues. In such a situation, there is much of value in adopting a conceptual change strategy (Posner et al., 1982): first, elicit students' NOS views; second, explore these views in the light of specific examples from the history of science, as a way of creating dissatisfaction; third, present "more adequate" conceptions of key NOS elements; finally, provide opportunities for students to explore the intelligibility, plausibility and fruitfulness of these ideas in making sense of scientific knowledge and practice in a wide variety of disciplinary and sociohistorical contexts. At all times, teachers need to model behaviour and language that reflect authentic scientific activity, address explicit and implicit messages about science embedded in classroom tasks, make frequent use of examples of authentic practice (both historical and contemporary), and provide ample opportunities for students to reflect on the NOS issues relevant to their current activities. Essentially the same suggestion for embedding NOS teaching in a concept-rich environment is captured by the Monk and Osborne (1997) 6-step model combining historical material, detailed consideration of issues in philosophy of science and constructivist pedagogy.

- Presentation of phenomena or events – carefully chosen for their historical significance, inherent interest and capacity to generate questions, problems and predictions.
- Eliciting students' views, ideas and explanations – using the usual range of constructivist strategies: skilful teacher questioning, use of concept maps, word association, art work, group discussion, writing activities of various kinds, etc.

237

- Historical studies – presented orally by the teacher or via text, multimedia, student research, museum visits, and the like.
- Devising suitable tests – designing experiments and other kinds of investigations to evaluate and judge the range of ideas collected.
- Presenting the current scientific view – if it has not already arisen, the current view (or the school science curriculum version of it) should be introduced by whatever means is appropriate, with particular emphasis on the reasons why this view has been accepted.
- Review and evaluation – reviewing learning progress (always an essential element of constructivist pedagogy) and reiterating the basic thrust of the approach, that is, how to distinguish between justified and unjustified beliefs.

The value of this approach is that it shifts emphasis from the mere acquisition of theoretical knowledge towards a concern with the warrant for belief: How do we know? What are the reasons for holding this belief? What is the evidence for this view? How was that evidence acquired? Interpreting theoretical disputes from the history of science requires students to adopt rival perspectives on the interpretation of data, and to enter discussion about what counts as relevant data and appropriate ways of collecting it. Thus, it is excellent preparation for appraising new theoretical propositions in relation to existing ideas. In addition, by studying the ways in which scientists construct knowledge, students are assisted in building metacognitive insight into their own knowledge construction. By utilizing historical material and emphasizing epistemological concerns, the approach meets several of the learning *about* science goals embedded in the notion of critical scientific literacy. It also simultaneously addresses one of problematic issues in constructivist pedagogy: the reluctance of some students to contribute their own ideas. It is sometimes comforting for students to know that others, including some eminent scientists, have held erroneous beliefs. Hence contributing ideas to class discussion can be seen not so much as a potential threat to self-esteem as an opportunity to enrich discussion. It may be that successful experiences of this kind can lead to increased tolerance for and appreciation of the views of others. Support for these speculations can be found in Crawford's (1993) description of how she has used Michael Faraday's experimental diaries to create a classroom climate that encourages students to (i) articulate their anxieties and uncertainties, (ii) develop the emotional commitment to struggle with theoretical and practical problems, and (iii) learn to think independently. Strategies appropriate to NOS in relation to language use will be discussed in Chapters 8 and 9.

In short, we need a wide range of approaches, with the choice at any one time being determined by the students' understanding of the particular NOS item, their familiarity with the science content, their interest in the context, and their level of emotional maturity. I incline towards Clough's (2006) view that NOS-oriented activities can be described as lying along a continuum from decontextualized to highly contextualized, and that NOS instruction would be more effective if teachers "deliberately scaffolded classroom experiences and students' developing NOS understanding back and forth along the continuum" (p. 463).

NOTES

[1] Bricolage is a term used by Levi-Strauss (1968) to contrast the intuitive, unplanned ways in which "savages" (his term) solve everyday problems without formal problem-solving strategies. Parker and McDaniel (1992) note that: "bricoleurs tackle a problem not by reading a manual or taking a course of study, but by using a personal bag of tricks. They are masters of improvisation, using whatever tools and devices are on hand or can be invented" (p. 99).

[2] See also Wellington and Osborne (2001) for a very similar statement: "Observation and experiment are not the bedrock on which science is built, they are the handmaidens to the rational activity of generating arguments in support of new ideas about the way the world behaves" (p. 140).

[3] Authenticity is a tricky and elusive concept. Wallace (2004) discusses three possible meanings: (a) child-centred authenticity, where teaching is personally meaningful and based on students' experiences; (b) subject matter authenticity, where teaching aligns with the work of scientists; and (c) situated authenticity, where learning focuses on real world situations. Meaning (b) is closest to the ideas discussed and issues raised in this chapter, allied to a sense of personal ownership of the investigation rooted in meaning (a). Of course, truly authentic scientific inquiry in the science classroom is impossible, simply because science communities (at all levels, from pairs of students working on a problem to whole school groups attending a lecture or a science fair) are vastly different from the community (or communities) of professional science in all manner of ways: purposes, expectations, values, organization, norms of behaviour, resources (both cognitive and material), incentives and rewards. The best we can do is to simulate, reproduce or represent those aspects of the complex business of doing science that are central to the notion of critical scientific literacy as faithfully as we can.

[4] Bencze (1996) provides a detailed argument for the adoption of correlational studies in science education and a critical discussion of their distinctive features, including systematic inquiry via statistical control.

[5] Conkers are the dark brown nuts of the horse chestnut tree. Generations of British children have played a game in which they take turns to use a conker threaded onto a string to break the opponent's conker. Avid players use all manner of treatments, including soaking them in vinegar or baking them in an oven, in an effort to produce a champion conker.

[6] In accordance with Loving's (1991, 1998) Scientific Theory Profile.

[7] One of the goals of the CCSDL is to enable students "to construct their own knowledge, in ways and directions suiting their individual needs, interests, perspectives, and abilities" (Bencze & DiGiuseppe, 2006, p. 334).

[8] See Hodson (1992c, 1993c), Champagne et al. (2000) and Yung (2001) for an extended discussion of some of the key issues relating to assessment of open-ended inquiry.

[9] See Russ et al. (2008) for an extended discussion of students' abilities to think mechanistically.

[10] Millar et al. (1996) acknowledge these points when they say that a student's capacity to carry out experimental investigations requires four types of understanding: understanding of the purpose of the task, declarative understanding, skills in carrying out manipulations, and understanding of evidence. Ryder and Leach (2000) report some interesting findings on interpretation of experimental data by students at upper secondary and undergraduate level.

[11] It is a matter of some concern that adult visitors to two informal education sites in Western Australia (one traditional museum and one interactive science centre) seemed to become less scientific in their thinking as a consequence of the visit, becoming more likely to think that science has an answer for all questions, scientific explanations are definite, and scientists always agree among themselves (Rennie & Williams, 2006, 2008). I am assuming that carefully planned and properly followed-up visits can move the thinking of school-age students in the opposite direction.

[12] Examples include the At-Bristol project on biomedical issues run by the University of Bristol and the Wellcome Trust (www.at-bristol.org.uk), Galaxy Zoo (www.galaxyzoo.org), Pathfinder Science (http://pathfinderscience.net) and Be a Citizen Scientist! (http://home.twcny.rr.com/allenz/citizen_scientist.htm). See www.citizensci.com for a list of citizen science projects and Robinson (2004) for details of a range of what he calls "backyard research activities", largely long-term butterfly and amphibian monitoring projects.

[13] The presence of material on Internet websites that is not subject to the "quality control" of editors and peer review raises major problems concerning accuracy, credibility, bias and so on. These matters will be addressed, albeit briefly, in Chapters 8 and 9.

[14] The term "Whig history" was coined by Herbert Butterfeld (1931). It may be defined as an attempt to explain the past in terms of the present, to investigate the past in order to support conclusions in the present and to fit the past into an explanatory scheme applicable to the present, as distinct from the attempt to explain the past on its own merits. It is not too much of an exaggeration to say that Whiggish history sees the beliefs, practices and institutions of the present as the goals of previous efforts (see discussion in Chapter 10).

[15] "Interactive vignettes" or "interrupted stories" are based on Kieran Egan's (1986) notion of teaching as story telling. At regular intervals, as the story unfolds, readers are presented with questions to discuss.

CHAPTER 8
THE LANGUAGE OF SCIENCE AND SCIENCE EDUCATION

Many would argue that the language of science (item 7 in King and Brownell's list of disciplinary characteristics) is the feature that most clearly distinguishes science from other forms of knowledge, at least in the public domain. In addition to addressing the distinctive features of scientific language (and how students can best understand it and use it effectively), this chapter looks at language as a potential barrier to effective science learning and a factor controlling access to science. Chapter 9 considers language use in the science classroom, that is, reading, writing and talking activities as powerful pedagogical tools in learning science and learning *about* science.

In his influential essay titled *Realms of Meaning and Forms of Knowledge*, Paul Hirst (1974) argued that the disciplines of knowledge[1] can be distinguished in terms of three major demarcation criteria: (i) their distinctive concepts and conceptual structures; (ii) the methods practitioners use to develop new knowledge; (iii) the ways in which knowledge claims are judged and decisions made concerning admission to the community's corpus of knowledge. Chapters 6 and 7 were devoted to the first two criteria, and attention can now turn to the third criterion, which Hirst maintains is the most important and the most distinctive feature of a discipline. A key element in this third criterion is the quality of the argument constructed by scientists to establish the warrant for belief in a particular knowledge claim. Discussion of scientific argumentation,[2] and how students can gain familiarity with it, and competence and confidence in using it, is the second major component of this chapter.

Few would quibble with the assertion that one cannot fully know and understand French culture without a substantial working knowledge of the French language. So many cultural assumptions, values and attitudes are implicit in the way the language is structured and used that ignorance of the language, its subtleties and idiomatic deployment, renders French culture unknowable beyond the merely superficial level. So, too, with the culture of science: to be considered scientifically literate, an individual must be able to read, write and talk the language of science appropriately, comfortably and effectively. Facility with the language is necessary in order to share scientific experiences and insights with others, address problems, formulate and evaluate solutions, give and receive criticism, and make important decisions on socioscientific issues.

241

All of what we customarily call "knowledge" is language. Which means that the key to understanding a "subject" is to understand its language ... what we call a subject is its language. A "discipline" is a way of knowing, and whatever is known is inseparable from the symbols (mostly words) in which the knowing is codified. (Postman & Weingartner, 1971, p. 102)

Social and functional linguistics regards our use of language as socially and culturally contextualized meaning-making, with language functioning as a system of resources for engaging in meaningful action. Thus, particular language *genres* have developed in particular social circumstances in order to meet the particular needs and interests of the social group. As Bazerman (1988) puts it:

A genre is a socially recognized, repeated strategy for achieving similar goals in situations socially perceived as being similar. A genre provides a writer with a way of formulating responses in certain circumstances and a reader a way of recognizing the kind of message being transmitted ... Thus the formal features that are shared by the corpus of texts in a genre and by which we usually recognize a text's inclusion in a genre, are the linguistic/symbolic solutions to a problem in social interaction. (p. 62)

As Bakhtin (1981, 1986) points out, each of us regularly communicates in a range of language genres – the characteristic modes of expression within the particular social groups in society to which we belong. For example, any one individual may engage on a regular basis in everyday greetings and dinner table conversations with friends, verbal exchanges involved in buying and selling goods and services, communications with colleagues or friends engaged in some specialized tasks, intimate conversations with lovers, and parent-infant talk. Our area of employment and our leisure pursuits will bring further exposure to distinctive language genres: for example, barristers will engage in cross examination of witnesses, soldiers will give and receive military commands, and teachers will engage in classroom dialogue, while mountaineers, photographers and stamp collectors will use languages specifically designed to convey ideas and feelings about their particular interests and passions quickly and reliably. Because these language genres are socioculturally constituted, each carries with it the common assumptions, beliefs, interpretations, norms, expectations and values of the group whose genre or social language it is. As Agar (1994) remarked, "You can't use a new language unless you change the consciousness that is tied to the old one" (p. 22); as a new language is acquired, "old unconscious ways of doing things are dusted off and new ways are built up" (p. 143). Use of a particular language signals one's membership (or aspiration to achieve membership) of the community that developed it. Thus, Gee (1996, 1999) likens our acquisition of a language genre to the adoption of an "identity kit", complete with "costume" and "set of instructions" on how to act so that others will recognize us as someone already enculturated into the community's ways of thinking, acting, valuing and making sense of things. It follows that an individual's level of success within the community of scientists is determined, in part, by her or his capacity to use the language of science appropriately and effectively for presenting ideas in a clear and persuasive way to other scientists, including research colleagues and students, journal reviewers and editors, and evaluation panels for research grant awarding bodies. However, as Gee (2004) observes, "in much of science education, language is pushed into the background or ignored, while thinking or doing is brought into the foreground as if these tools had little to do with language" (p. 13).

The argument I wish to develop here is that scientific literacy entails familiarity with the logical form and structure of the language of science, the cultural traditions, intellectual and behavioural norms and inherent values it encapsulates, its underlying worldview and epistemological assumptions, and the specific rhetorical devices and tools of persuasion it deploys. The language of science is a particular way of conceptualizing, representing and talking about phenomena and events; it embodies distinctive scientific modes of thinking, involving such things as conjecture, hypothesis, speculation, inference, prediction, generalization and evidence, and a commitment to discussion, questioning, criticism and persuasion; it includes approved ways of assembling and interpreting data, articulating explanations, and making and justifying knowledge claims using particular data tabulation and graphing techniques, chemical formulae and equations, mathematical symbolism and visual representation. Students will not just "pick up" this complex language unaided. It has to be taught, practiced, deployed in authentic contexts and evaluated in action, such that students come to see themselves as members of the scientific community, or the school version of it. Effecting a smooth transition from everyday language to scientific language is a matter of *personalization*, that is, of "making the language one's own" and using it comfortably and willingly, as well as appropriately.

[It] becomes "one's own" only when the speaker populates it with his own intention, his own accent, when he appropriates the word, adapting it to his own semantic and expressive intention. Prior to this moment of appropriation, [it] ... exists in other people's mouths, in other people's contexts, serving other people's intentions. (Bakhtin, 1981, p. 293).

Moje (1995) points out some of the ways in which a teacher's use of language can promote a sense of belonging within the classroom that may extend to a sense of belonging within the scientific community. Of course, if particular forms of language and language use can influence the thinking and attitudes of students in a positive way, they can also act to exclude or alienate students, or contribute to their feelings of frustration and inadequacy. Thus, adoption of particular classroom language is just as much a part of curriculum decision-making as selecting curriculum content and choosing among theories of learning and the teaching methods they inform. Language barriers can be considerable for those who do not normally hear or read scientific language in their everyday lives. Arons (1973) describes what for many students is a common experience: "an incomprehensible stream of technical jargon, not rooted in any experience accessible to the student himself, and presented much too rapidly and in far too high a volume for the assimilation of any significant understanding of ideas, concepts, or theories" (p. 772). Significantly, in Brown's (2006) study of socially disadvantaged and under-represented urban students participating in a grades 9 to 10 life sciences class, few of the students had problems with the conceptual content of the course or with the hands-on learning methods adopted by the teacher, but almost all reported difficulty with the language of science and the language of school. These difficulties arise, in part, because acquisition of a particular language entails the assumption of a particular identity. As Bakhtin (1981) points out, language is not neutral; it does not pass freely and easily between individuals; its acquisition changes the learner in several important respects and may impact on her/his sense of self, sometimes positively and sometimes negatively. Our language use also impacts on how others perceive *us*, because it signals our sociocultural affiliations. In the words of Nancy Brickhouse (2001), "learning is not merely a matter of acquiring knowledge, it is a matter of deciding what kind of person you are and want to be and engaging in those activities that make one a part of the relevant communities" (p. 286). Hymes (1972) uses the term "communicative competence" to denote an appreciation of the rules, conventions and values that accompany particular language use and an ability to switch language codes as the social situation changes. The ability to switch codes appropriately and comfortably can be seen as part of what Aikenhead and Jegede (1999) call "secured collateral learning" and also as a key element of critical scientific literacy.

SOME CHARACTERISTICS OF SCIENTIFIC LANGUAGE

The following is typical of the kind of knowledge students are expected to acquire in the later years of a school chemistry course.

> The temperature falls when ammonium nitrate dissolves in water because the combined enthalpies of hydration of the ammonium and nitrate ions is less than the lattice energy of ammonium nitrate.

This statement has a number of features that can be taken as characteristic of scientific knowledge in general. First, it is expressed in very specialized language, rather than in colloquial terms. The language of science includes words purpose-built for particular contexts, sometimes using Latin and Greek roots. In general, these words should present few difficulties for students because teachers can introduce them carefully and pay particular attention to clarifying their meaning. Much more problematic are dozens of common words (like *contract*, *efficient* and *abundant*) that are either not understood by students or understood as having the opposite meaning. About 50% of 12 to 14-year-olds think that *contracts* means "gets bigger", *initial* means "final", *negligible* means "a lot" and *abundant* means "scarce", while *efficient* is often taken to mean "sufficient" (Cassels & Johnstone, 1985). Table 8.1 lists more than 50 words commonly used in science lessons and science textbooks that may cause problems for some students (Pickersgill & Lock, 1991; Farrell & Ventura, 1998; Prophet & Towse, 1999).

Research by Gardner (1975) identified logical connectives such as *conversely*, *essentially*, *moreover*, *in practice*, *respectively* and *hence* as a particular source of difficulty for students. As Wellington and Osborne (2001) point out, many of

TABLE 8.1

Words that commonly cause difficulty for students.

---

abundant adjacent assumption average calibrate characteristic complex component composition concept consecutive constituent contract contrast converse criterion devise disintegrate diverge diversity effect efficient emit equilibrium estimate exert factor ideal illustrate initial incident liberate limit linear maximum miscible negligible random rate regular relative relevant retard saturated secrete sequence source specific spontaneous stimulate tabulate uniform valid

---

these logical connectives are widely used in comparing and contrasting, formulating hypotheses, making inferences and attributing cause and effect. It is also the case that scientific language has a much higher "informational density" than everyday language, that is, scientific language contains more essential content words per sentence than everyday speech, which is frequently littered with non-content words (Fang, 2005). Problems for students can be considerable when faced with statements such as "Our eyes are sensitive to electromagnetic waves in the frequency range 4×10^{14} to 7.5×10^{14} Hz (wavelength between 4×10^{-7} and 7×10^{-7})" and "Once transcription has been successfully initiated, the RNA polymerase continues along the DNA molecule until it encounters terminator sequences on the non-transcribed DNA strand".

Science also makes use of common everyday words, but uses them in a restricted or very specialized way. Words like *force*, *energy* and *work* (in physics), *element*, *conductor* and *compound* (in chemistry), and *plant*, *animal* and *cell* (in biology) spring to mind. Often these words create significant learning problems because students think that they understand them, but do not always appreciate the particular specialized scientific meaning, and when its use is appropriate or necessary. Everyday meanings can interfere with scientific ones. One wonders, for example, what images the term "North pole" conjures up for young children using magnets for the first time. As if all this was not confusing enough for students, scientists in different disciplines sometimes use the same word to mean entirely different things: consider, for example, *cell*, *nucleus* and *molar* as used by biologists and chemists. Sometimes the precise meaning of a word is only apparent in its context of use. Just as words like *surgery* and *wicket* have multiple meanings in medical practice and the game of cricket, so science teachers might use the word *electricity* to refer to current, voltage or power (see Wellington and Osborne, 2001, for an extended discussion of these issues).

One of the most pernicious myths perpetrated by science teachers, and constantly reinforced by textbook emphasis on definitions and assessment schemes that reward verbatim reproduction of such definitions, is that the meaning of words used in science is immutably fixed and certain. In reality, meaning changes over time as theories are developed. For example, an *acid* may be defined in terms of its behaviour (for example, a substance with a sour taste that reacts with certain metals to produce hydrogen) or in theoretical terms – that is, as a proton donor or a molecule that can accept a lone pair of electrons for covalent bond formation, depending on the preferred theory. Sometimes the changes in meaning are quite spectacular. Compare, for example, the concepts of *mass*, *time*, *space* and *energy* in the Newtonian and Einsteinian paradigms. In Einstein's special theory of relativity, the mass of a body depends on the observer's frame of reference and, moreover, mass can be converted into energy; in Newtonian physics, the mass of a body is fixed and is constant for all observers, regardless of their frame of reference. Although the term *mass* is used in both paradigms, the respective concepts are incommensurable. Indeed, in a Newtonian world, the notions of relative mass and time dilation would be absurd.

> Since new paradigms are born from old ones, they ordinarily incorporate much of the vocabulary and apparatus, both conceptual and manipulative, that the traditional paradigm had previously employed. But they seldom employ these borrowed elements in quite the traditional way. Within the new paradigm, old terms, concepts, and experiments fall into new relationships one with the other. (Kuhn, 1970, p. 149)

Proper use of the language of science involves a considerable measure of conceptual understanding. Indeed, the language of science is heavily impregnated with theoretical assumptions. Understanding the earlier statement about ammonium nitrate requires an understanding of solubility, ionic bonding, hydration and so on. Scientific knowledge is predominantly abstract knowledge, and becomes increasingly abstract as learners progress. We may initially engage students in interaction with familiar real objects through laboratory activities, and so on, but our intention is to ascend a hierarchy of abstraction. According to Ausubel (1968), concepts with high levels of abstraction are "more stable" (that is, more resistant to forgetting) and more useful in dealing with new situations than are concepts with lower levels of abstraction. Consequently, their acquisition should be a major goal of teaching and learning science. The development of increasing abstraction and the search for relationships between these abstractions (scientific theory-building) is also, of course, a major goal of science itself, and it is this high level of abstraction that gives scientific knowledge its particular explanatory power. It is the abstract concepts of a theoretical system that enable natural events and phenomena to become susceptible to scientific study. As argued several times in earlier chapters and at length in Hodson (2008), theory does not create the objects being studied, nor does it derive from the naïve observation of them. Rather, it is an enabling device by means of which we idealize, model and interpret. Recognizing this is central to understanding the distinctive nature of scientific knowledge and a vital part of learning *about* science. "The concepts and techniques of a theoretical system enable natural events to become *scientific* events ... It is as a scientific event that they are stripped of their everyday guises and become data or evidence in theoretical debate" (Matthews. 1994, p. 130).

In scientific theory-building, the complexity and diversity of the world is reduced and made more manageable, understandable and susceptible to control. Through idealization and abstraction, phenomena and events can be subjected to more systematic study. The power of concepts to provide understanding is located in their connections with other concepts via interconnected conceptual systems, in which, of course, meaning may shift somewhat from context to context, or problem to problem. The task for teachers is to assist students in making the crucial move from a commonsense, empirically based set of descriptions of phenomena and events to the abstract, idealized (and sometimes mathematical) descriptions characteristic of theoretical science. Idealization and abstraction are achieved through *nominalization* – that is, the replacement of active verbs and adjectives by nouns. Thus, simple observations such as "magnets repel" and "magnets attract" are replaced by abstract concepts such as *magnetic repulsion* and *magnetic attraction*, "diverge" becomes *divergence*, "evaporate" becomes *evaporation* and "emits energy" becomes *energy emission*. By this means, particular events are also made universal. Nominalization creates new terms and new concepts to systematize knowledge, synthesize seemingly disparate ideas and impute cause and effect relationships (Veel, 1997). The significant details of complex multi-step processes may even be encapsulated in a single word-concept, such as *photosynthesis*. Difficulties for learners are compounded when clusters of nouns are used to express an abstract idea underlying an observed phenomenon, as in Halliday's (1996) example of *glass crack growth rate* to replace "the speed with which glass cracks under certain conditions of stress".

Moreover, as discussed in earlier chapters, the concepts and ideas of science are related within a broad network of interdependent meanings. As Halliday (1996) notes, science makes extensive use of *interlocking definitions*, where the meaning of any one item is dependent on the meaning of others in the same sentence, and of *technical taxonomies* or highly ordered (usually hierarchical) constructions in which every term has a definite functional value, as in biological classifications. Most significantly, through its adoption of a third person, past tense and passive voice, science is seemingly presented as an account of the natural world that is independent of people, as a simple and straightforward presentation of "the facts", with little or no scope for challenge, argument or alternative interpretation. As Pinch and Collins (1984) remark, the language of science is *self-effacing* and implies that "the experimenter played only the role of facilitator, or "amplifier" of Nature's voice" (p. 522). However, a sense of authority is created through emphasis on accurate statements of data, clarity of expression and objective presentation of conclusions and explanations. In Lemke's (1990) words, this formalized and potentially alienating language "sets up a pervasive and false opposition between a world of objective, authoritative, impersonal, humorless, scientific fact and the ordinary, personal world of human uncertainties, judgments, values and interests" (p. 120). Even in the context of scientific controversy or dispute, scientists maintain the use of this austere and impersonal language of formal communication, systematically understating their claims and respectfully dealing with counter claims. Accounts of scientific investigations are distinctive in terms of what they do *not* include. Usually there are no references to failures, preliminary experimental runs, false starts, confusions, indecisions, reformulations, accidents, chance occurrences and frustrations. There is even less likelihood of any reference to the scientist's interests, emotions, commitments and values, or to

247

the difficulties under which the work was conducted (time constraints, financial restrictions, illness, family pressures, and the like). And there is certainly no reference to the author's sex, race, religion and ethnicity, though there may be some clues present in the author's name and institutional affiliation.

The sometimes counter-intuitive nature of scientific explanations, the high level of abstraction of scientific knowledge, its divorce from ordinary daily experience, and its presentation via unfamiliar linguistic conventions are among the factors that make science so difficult to learn. Unless teachers take explicit steps to present a counter view, this cluster of characteristics can also play a key role in positioning the teacher and the textbook as *authorities*, not to be challenged by mere students. Too often, teachers provide the visually rich and personally engaging concrete experiences of laboratory work and fieldwork and then demand a description and explanation for the phenomena or events observed in an abstract and impersonal language that is both unfamiliar and difficult to use. Instead of using descriptions such as "light is split into many diverging rays when it passes through a glass prism", students are expected to talk about *refraction*. Even greater problems can be created by too early an insistence on representing observational data in symbol form, such as chemical or algebraic equations. When students think in terms of concrete experiences and teachers think in terms of abstract theoretical concepts, they are, in effect, "speaking different languages". As a consequence, much of the laboratory work in school science does little to build students' understanding (Hodson 1993b). Similar problems may arise from the use of analogies, similes, metaphors and over-simplified versions of theoretical models to "get the point across" more easily and more quickly (see discussion in Chapter 6). Osborne (1996, 1998) has argued that Harre's notion of "three levels of theory" might be helpful in assisting students to ascend the level of abstraction towards sophisticated theoretical understanding and language use. Level 1 theories are concerned with describing and classifying tangible, directly observable entities. Level 2 theories concern entities that are only accessible to our senses through instrumentation. Level 3 theories deal with theoretical entities for which there is no direct evidence. Osborne's suggestion is not for an invariable teaching progression in which students are first introduced to a simple conceptual language to describe events and phenomena, proceed to theoretical structures that relate these observations in cause and effect relationships, and finally learn to use abstract systems. Rather, he emphasizes the need to pitch the level of abstraction at a point that is appropriate to the particular learners and the particular learning context, and to impress on students that the sophisticated theories of science ultimately rest on directly and indirectly observable events. In doing so, he is making a case for a much more overt and systematic teaching *about* science and its epistemology, a view that I fully endorse.

Because meaning in science is also conveyed through symbols, graphs, diagrams, tables, charts, chemical formulae and equations, 3-D models, mathematical expressions, computer-generated images, and so on, Lemke (1998) refers to the language of science as "multi-modal communication".[3] Any one scientific text might contain an array of such modes of communication, such that it may be more appropriate to refer to the *languages* of science.

> Science does not speak of the world in the language of words alone, and in many cases it simply cannot do so. The natural language of science is a synergistic integration of words, diagrams, pictures, graphs, maps, equations, tables, charts, and other forms of visual mathematical expression. (Lemke, 1998, p. 3)

Thus, the overall meaning of a scientific text or a science lesson is built by combining a partial meaning from the words with a partial meaning from the diagrams, equations and other "inscriptional devices" (as Latour, 1990, calls them) and a partial meaning from the mathematics. In Lemke's (1998) words, it is as if "we said the first words of each sentence in Chinese, then the next few in Swahili, and then the last few in Hindi, and in the next sentence we started with Swahili ... And so on" (p. 6). The key to effective communication in science, and to understanding the communications of others, resides in appreciation of how these different forms of representation interact and support each other (Sherin, 2001; Ainsworth, 2006). Research by Peacock and Weedon (2002) has identified the visual elements of scientific text as a major source of comprehension difficulties for students, confirming early research findings on the problematic aspects of pictures, diagrams and maps, especially for lower achieving students (Reid & Beveridge, 1986; Constable et al., 1988; Reid, 1990; Liben & Downs, 1993; Mayer, 1993). However, as Ford (2004) reports, teachers are often seduced into buying books for classroom use that have lots of large colour photographs, diagrams and artistic illustrations. It is a matter of some concern that the preservice elementary school teachers interviewed in Ford's research tended to choose "easy-to-read" books rather than "information-rich" books that demand some effort on the part of the reader. Why I regard this as a matter of concern will become apparent in Chapter 9. It is also worth noting the advice of Hand et al. (2009) to ensure that students fully understand the link between mathematical and textual representation of knowledge before proceeding to graphical representation. Combining all three modes too early can lead to confusion.

Scientific reporting is also distinctive in being expository, analytical and impersonal, and making little or no use of metaphoric or figurative language. As noted earlier, it is frequently expressed in the past tense, passive voice. Enculturation entails being able to use this distinctive language appropriately and being able to present ideas and findings in the various sub-genres of science, including the scientific paper and the laboratory or fieldwork report. Practising scientists work within a number of sub-communities. Within their own research group and with other close colleagues they engage in informal discussion of plans, ideas, inquiries and findings as a means of exploring their own thinking and gaining critical feedback. Within the wider community, they participate in conferences and publish in scientific journals, with the express purpose of persuading others of the validity of knowledge claims. These public forms of communication set and apply rigorous standards for presentation of claims and arguments. Some scientists may have a teaching role and/or a role in supervising, guiding and monitoring the progress of novice scientists and graduate students. In addition, they may be required to address grant awarding bodies, television audiences, groups of "concerned citizens" or courts of law. Each of these activities requires scientists to frame information, explanations and arguments in different ways. Each form of communication has its ground rules, assumptions, conventions and underlying values. There is a strong case to be made for providing students with a similarly wide range of language-based activities and to foster awareness of, and ability to use the various sub-genres of science. However, as Lemke (1990) argues at some length, many science teachers seem to expect all this to occur unaided:

> Students are not taught how to talk science: how to put together workable science sentences and paragraphs, how to combine terms and meanings, how to speak, argue, analyze, or write science. It seems to be taken for granted that they will just "catch on" to how to do so ... When they don't catch on, we conclude that they weren't bright enough or did not try hard enough. But we don't directly teach them how to. (p. 22)

Indeed, many aspects of learning science and learning *about* science are simply taken for granted; they are rarely, if ever, questioned by teachers or students. As Cobern (1995) remarks, "science education appears to assume that students share with scientists particular worldview presuppositions about what the world is like, what questions are important, and what methods ought to be used in pursuing answers" (p. 289). The conduct of science lessons and students' roles within them are also regarded as unproblematic and taken-for-granted.

> In extreme cases, students are expected simply to know the cultural norm of scientific appropriateness as embodied in the teacher's mind and mind set, how they are expected to think in science, the routines of activity that they are to observe, and how they are to behave. (Shapiro, 1998, p. 229)

In consequence, for many students, the language and cultural norms of science remain for ever a barrier to learning and a major hurdle in gaining access to science and feeling comfortable within the scientific community.

LEARNING THE LANGUAGE OF SCIENCE

The notion of learning through apprenticeship (as utilized in Chapters 6 and 7) assumes that students can and will learn the language of science by interaction with someone who is already an expert, and by using it themselves in carrying out authentic tasks. Thus, teachers should model appropriate language use, make explicit reference to its distinctive features, provide language-based activities that focus on these features, create opportunities for students to act as autonomous users of the language and provide critical feedback on their success in doing so. There needs to be much more *metatalk* (talk about language), with teachers explaining why they are adopting a particular linguistic form. Students need to know that while everyday language will suffice on some occasions, a specialized language of science is necessary on others. They need to know the circumstances in which different codes are applicable and they need lots of practice in switching between them.

Necessarily, teachers organize learning experiences and manage the activities of the classroom through linguistic exchanges with students. They help students to understand, they guide and support their efforts, monitor their progress, and provide feedback on their learning progress through dialogue. Thus, talk is one of the principal means by which students move from everyday commonsense understanding, and the personal language of everyday discussion in which it is usually expressed, to scientific understanding and the formal technical language in which it is expressed. It is teacher talk that scaffolds this transition. However, teacher talk can also serve to maintain and reinforce certain myths, stereotypes and falsehoods about science, and for some students can contribute to the problems of border crossing into the subculture of science. Given the sociocultural location of language and its accompanying sociopolitical cargo of meaning, important questions of authority, culture and power are raised. Whose view of reality is being promoted? Whose voices are heard? And why? When used carelessly, classroom language projects a set of explicit and implicit messages that privilege some views, interests and values, and discount or reject others. In many classrooms there is a conscious or unconscious reflection of middle-class values and aspirations that serves to promote opportunity for middle-class children and to exclude children of ethnic minority background and low SES status, who quickly learn that their voices and cultures are not valued. When the speech genres and interpretive frameworks of some children are disregarded or specifically rejected as inferior, school science becomes implicated in a continuing suppression of opportunity and perpetuation of privilege. As O'Loughlin (1992) says:

> To the extent that schooling negates the subjective, socioculturally constituted voices that students develop from their lived experience ... and the extent that teachers insist that dialogue can only occur on their terms, schooling becomes an instrument of power that serves to perpetuate the social class and racial inequalities that are already inherent in society. (p. 816)

Many students already believe that school is substantially worse than a waste of time: it confines them against their will in physically unattractive surroundings, imposes on them a code of conduct that is unfamiliar and unwelcome, and often denies them any measure of choice and self-determination about what and how they study. To compound matters, these already disenchanted students are presented with a science curriculum that is remote from real life and couched in an alien language. Even if they make the effort to learn science, they are presented almost daily with unappealing messages about the nature of science and scientific practice. Science is presented as complex and difficult, and so only accessible to "experts" who have subjected themselves to a long and arduous training. Scientists are portrayed as dispassionate and disinterested experimenters, who painstakingly reveal the truth about the world. Scientific knowledge is delivered as established and certain knowledge that is not to be challenged or doubted by students. Groisman et al. (1991) identify some other features of science classrooms, including the size and arrangement of the teacher's and the students' desks and benches, use of time constraints and emphasis on individual written work as the (only) basis for student assessment, that contribute to the distinctive culture of the science classroom and to the establishment of its language, values, expectations and code of behaviour. For most high school students this is a startling change from what they were used to in primary (elementary) school; for many, it may constitute a formidable barrier to entry into science. One wonders how much has changed in the 30 or more years since Michael F.D. Young (1976) remarked that "school science separates science from pupils' everyday lives, and in particular their non-school knowledge of the natural world. It is learnt primarily as a laboratory activity, in a room full of special rules, many of which have no real necessity except in terms of the social organization of the school" (p. 53). Taken together, the rules about the conduct of lessons, the conventions concerning who can speak and who must listen, what can be spoken about (including what can be challenged), and how school talk and science talk are to be constructed, impose a set of conventions and restrictions that can be so formidable that many youngsters are prevented from gaining access to science education or dissuaded even from attempting it. In other words, science discourse has the potential to reflect, and even reinforce, sociocultural inequalities.

> It is not surprising that those who succeed in science tend to be like those who define the 'appropriate' way to talk science: male rather than female, white rather than black, middle- and upper-middle class, native English-speakers, standard dialect speakers, committed to the values of North-European middle-class culture (emotional control, orderliness, rationalism, achievement punctuality, social hierarchy, etc.). (Lemke 1990, p. 138)

For many students, learning to participate in the cultural practices of the science classroom is not a simple and straightforward business. Teaching and learning practices in many schools favour students from particular backgrounds and those familiar with particular ways of speaking, and can seriously disadvantage those from different backgrounds, who may be unfamiliar with interaction styles used in science classrooms. Unfamiliarity with the imposed discourse patterns, including turn taking, listening, use of certain grammatical conventions, awareness of implied meaning and sensitivity to certain non-verbal elements can create major problems for students whose culturally determined ways of interacting are substantially different. These students may have difficulty knowing when to talk, how to present their ideas, how to acknowledge and support the views of others and how to express criticism and difference of opinion (Lee & Fradd, 1996, 1998). Facial expressions, gestures, maintenance of eye contact (or not), and even the distance apart at which we can comfortably stand, are all culturally determined. Moreover, discourse practices also embody and presuppose particular beliefs, values and sociopolitical identity. In consequence, engaging in scientific discourse in school can create a cultural conflict for some students, forcing them to choose between the dominant school culture (in this case, school *science* culture) and their own sub-culture, which the school often seeks, consciously or unconsciously, to marginalize. There can be substantial cultural costs, with considerable intrapersonal tension for some minority students who must try to balance or resolve tensions between their academic/school identity and their personal/community identity. Some choose to reject mainstream cultural endeavours and find common ground in estrangement from or indifference to academic achievement. Those who choose to participate in school science risk losing membership of their home sub-culture and may even be accused of "trying to be another race" (Gilbert & Yerrick, 2001, p. 584). In consequence, some students will avoid certain linguistic forms and certain codes of behaviour, and adopt others, in an effort to maintain their personal identities and group affiliations and to signal or "symbolically cue" their preferred identity to others (Brown et al., 2005). To ignore these matters and to act as if power relationships do not impregnate classroom events is to ensure the perpetuation of the political *status* quo and the continued exclusion of significant numbers of students from a satisfactory science education. To recognize the sociopolitical context of the classroom and to acknowledge the ways in which values and culturally determined meanings permeate all aspects of the language used in classrooms is to take a major step in facilitating border crossings and rebuilding society along more socially just lines.[4]

Lee and Fradd (1996) point out just how complex and pedagogically problematic these differences in language use and language familiarity can be. Not only are there sometimes substantial differences between the language regularly used by teachers and that used by students, and between the language preferences of different sub-groups of students, but also differences between the "official" language of science and the traditional language of the science classroom.

> In teacher-student interactions, both teachers and students from diverse backgrounds bring with them ways of looking at the world representative of the environment in which they have been reared. These habits of mind or ways of knowing may or may not be compatible with scientific habits of mind or ways of knowing typically associated with scientific discourse ... Interactional patterns within a language and culture group may be incompatible with scientific practices as taught in the mainstream. For instance, the nature and practice of science involves the use of empirical standards, logical arguments, skepticism, questioning, criticism, and rules of evidence. These practices may be incongruent with cultural interactions that favor cooperation, social and emotional support, consensus building, and respect for authority. Scientific practices to encourage critical, creative, and independent thinking may also be incongruent with cultural interactions in which information is given to learners by authority, be it a teacher or a textbook. (p. 274)

The science classroom is a nexus and contested space for three distinct languages: (i) the students' everyday, home and community-based language(s); (ii) the language of instruction and school; and (iii) the language of science. For students, acquisition of the language of science and the language of school is a matter of crossing borders from the familiar to the less familiar. As with other enterprises of border crossing, the more closely a student's usual language resembles the language of instruction and the language of science, the easier the transition will be, and the smoother the appropriation of scientific language for authentic communication. Sensitive teachers are aware of the problems and barriers; they deliberately take steps to help students overcome them and establish a level of comfort in these different sociolinguistic settings. In an approach reminiscent of Aikenhead's (1996, 1997) idea of maintaining a "dichotomized notebook" (with "my ideas" or commonsense understanding on one page and scientific ideas on the facing page), Moje et al. (2001, 2004) argue that teachers should explicitly direct students' attention to these language issues, develop their awareness of the similarities and differences among the three languages, evaluate their effectiveness in relation to purpose and audience, and teach students how to negotiate and transfer meaning between them, as the situation demands (see also Rosebery et al., 1992; Lee & Fradd, 1998; Reveles et al., 2002; Lee, 2004; Brandt, 2008; Reveles & Brown, 2008). Put simply, students need lots of practice in moving backwards and forwards among these genres and lots of teacher-supported opportunities to reflect on their experiences in doing so. Lisa Delpit (1988, 1995) and Gale Seiler (2001) go further, and advocate that the rules and conventions of these language genres are taught explicitly in the context of power issues relating to race-ethnicity, gender and class.

WHERE TO START?

A number of writers have described meaning as comprising a central core of *denotative* meaning and a wide-ranging periphery of *connotative* aspects (see Sutton, 1992, for an extended discussion). Thus, our understanding of a term such as water, for example, comprises denotative elements such as covalent molecule with the formula H_2O, intermolecular hydrogen bonding, H-O-H bond angle = 105° and boiling point of 100°C, together with all the other non-scientific associations it has for us, such as: it is runny, wet, cold, used for making tea and washing the car. Often this framework of associations and connotations will include attitudinal and emotional elements deriving from and located in previous experience. Thus, *water* may conjure up happy memories of windsurfing or distressing ones of nearly drowning; *force* may arouse feelings of anxiety, fear or anger; *spider* may trigger feelings of revulsion. For any individual, the meaning of a word or phrase is its current array of denotative and connotative aspects. It will necessarily vary from individual to individual; it will change over time in response to experience; and it will be strongly influenced by the sociocultural contexts in which the individual moves. Learning is a matter of adding to, modifying and sometimes deleting elements from this complex of meanings and understandings. In teaching science, it is the teacher's task to assist students in modifying and developing *their personal framework of understanding* in order to incorporate the desired scientific aspects of meaning and to foster an appreciation of when their use is appropriate. Of course, these "approved meanings" jostle alongside a wide range of personal, idiosyncratic meanings and associations. It is the capacity to select and use appropriate aspects of our personal framework of understanding in response to different circumstances that is at the core of science learning: knowing, for example, when to use a model and when to use a theory, or which particular aspects of a concept's meaning to use for different tasks. It is often the array of personal, idiosyncratic and emotional connotations that render the scientific aspects more meaningful to us. Sadly, it has been common practice in science education to ignore these "other" aspects of meaning, even to attempt to suppress or eliminate them in favour of specialized scientific terms and insistence on a formalized linguistic code. Students are instructed not to use "fizzing", much less to use "a seething frothing turmoil" as I once saw in a 12 year-old's notes on the reaction of dilute hydrochloric acid with zinc. *Effervescence* is the approved term, precisely because it is not an everyday term. Similarly, Greek and Latin terms are often employed in science with the specific intent of eliminating "unwanted" associations. Although specialized terms like *photosynthesis* and *effervescence* bring increased explanatory power, it is likely that these "other", personal aspects of meaning, with their everyday associations, can provide the key anchoring points for new learning, and so render it more meaningful. We should be encouraging rather than discouraging the connotative aspects of understanding, just as we should be encouraging the use of familiar analogies and striking metaphors as a way of relating the new terminology to familiar knowledge and experience (see Chapter 6). Similar arguments extend to the formalized writing style often demanded of students, where they are dissuaded from saying that "Julie and I did such and such" in favour of "Procedure x was performed on ...".

It seems almost self-evident that the most sensible starting point for effective student participation in linguistic exchanges is the everyday language that students bring to school, a discourse that has little in common with scientific language. Through increasing levels of participation, managed and supported by the teacher, everyday discourse can be brought to resemble more closely the discourse adopted by the community of scientists. It is worth noting Lemke's (1990) observation that students pay much more attention to what is being said when teachers shift from the formal language of science to the colloquialisms of everyday speech, adopt more humanized ways of talking about science, and lace their explanations with humour and references to personal experience. This should not be interpreted as a case for abandoning scientific language in class in favour of the colloquial. It is a case, however, for a more thoughtful use of familiar language to assist learning, and it is a case for helping students comprehend the distinctive features of scientific language and why the language is structured that way. It is a case for alerting teachers to the ways in which the language of science is sometimes used, consciously and unconsciously, to disadvantage, exclude, alienate and disempower, and to assist students in recognizing it, too. It is also a case for students to acquire a self-conscious and critical understanding of the language of science and the ability to use it in pursuit of their own and the local community's interests.

Lemke (1987, 1990) criticizes teachers for emphasizing the formal language of science to the exclusion of everyday ways of speaking and writing, arguing that too great an insistence on careful and precise language may help to promote an ideology of authority concerning science and lead students to believe that

255

scientific knowledge is fixed and certain. By contrast, he says, more familiar vocabulary and language forms help students to see the relationship between science and the real world and to appreciate how scientific knowledge derives from everyday commonsense knowledge: "students will begin to grasp semantic and conceptual relationships in colloquial language first. Then they will substitute scientific, technical terms for colloquial words. Only much later will they be able to speak "pure science" (Lemke, 1990, p. 172). There are two key points here: first, students will understand the concepts better when they are expressed in familiar language; second, having secure conceptual understanding will support students' acquisition of the formal language of science. Claxton (1997) also argues that too early an insistence on the "correct form" of scientific reports and accounts of investigations can be counter-productive: "Insist on the proper names, measurements and ways of communicating and you will, for many children, kill the spirit of scientific thinking" (p. 18). Ballenger (1997) reports that when Haitian-Creole students were encouraged to use the vernacular to discuss topics in science, their conceptual understanding was enhanced, richer discussion ensued, and students' capacity to recognize and establish relationships between claims and evidence improved (see also Warren et al., 2001). Webb and Treagust (2006) report similar learning gains in a South African context when teachers engage in code switching between the language of science and the language of home, and explicitly signal how much they value "mother tongue language". Varelas et al. (2002) provide striking evidence of the effectiveness of plays and rap songs in the vernacular in assisting a group of grade 6 African American students in an urban school to develop a robust understanding of some basic concepts in physics, and gain confidence in their ability to use the language of science to express their understanding. In the authors' words, "Black students who mostly feel alienated by science, and have not imagined how their own cultural identity allows them entrance to the mostly foreign culture of science, may be helped to construct their identity as science learners through using tools that allow them to express who they are and how they come to develop their science understandings" (p. 601). Brown and Spang (2008) note that a teacher's skilful use of what they call "double talk" (using vernacular and scientific descriptions and explanations side by side) is often mirrored by students as they engage in the long and often complex business of acquiring the language of science. Ash et al. (2007) describe a potentially powerful research tool for observing such hybrid language and the ways in which students shift to and fro between science talk and everyday talk. Finally, Brown and Ryoo (2008) provide some striking evidence, from their work with Latino and African American students at the grade 5 level, in favour of what they call a "content first" approach: using everyday language to ensure basic conceptual understanding, followed by explicit instruction in the language of science.

It should be noted that Gee (2004), like Halliday (1993, 1996), argues that classroom use of everyday language severely limits students' access to conceptual understanding in science: "It can create a symmetry that is misleading and obscures important underlying differences" (p. 27). My own view, as the subtext of the preceding two paragraphs implies, is that teachers should strenuously resist being pushed into an "either/or" position; rather, they should use everyday language as a way of personalizing and contextualizing science, establishing links with the world outside school and providing staged access to the formal language of science. In that sense, everyday language is an asset rather than an impediment. Through such "hybrid language practices", as Gutierrez et al. (1999) call them, students are enabled to make direct use of the fund of knowledge, beliefs, values, attitudes and aspirations they have acquired from their home culture, families and peer groups, and from their everyday experiences in general, in the context of *learning science*, learning *about* science and *doing* science. This juxtaposition of ideas and language enables students to build bridges between their own world and the world of science. As Moje et al. (2004) note: "It helps learners see connections, as well as contradictions, between the ways they know the world and the ways others know the world" (p. 44). It is also worth noting the valuable role that can be played by television programmes, in the form of news, documentaries, dramas and more overtly educational materials, in juxtaposing and moving between scientific language and everyday language (see Dhingra, 2006). Popular science magazines such as *New Scientist* and *Scientific American* can fulfil a similar role. Of course, the popular media are also implicated in the perpetuation of stereotyped, distorted and inaccurate views about science, scientists and scientific practice (Nelkin, 1995; Lewenstein, 2001; Reis & Galvão, 2004). Ways in which teachers might seek to address these media images are considered in Chapter 9.

Chapter 9 discusses the importance of familiarizing students with a wide range of scientific text and encouraging them to "translate" text from one form to another, including varying the purpose of the writing, the intended audience and, of course, the language in which it is expressed. Once again, it should be noted that Michael Halliday (1993) claims that it is not possible to "represent scientific knowledge entirely in common sense wordings" because "the conceptual structures and reasoning processes of physics and biology are highly complex and often far removed, by many levels of abstraction, from everyday experiences" (p. 70). Jay Lemke (2004) also makes the point that in some sub-disciplines of science, such as theoretical physics, "the running verbal text would make no sense without the integrated mathematical equations" and the verbal text "could not, in most cases, be effectively paraphrased in natural language" (p. 39). Of course there can never be absolute equivalence of meaning. All exercises in paraphrasing and re-wording raise important questions about the extent to which "true meaning" can be captured. It may well be the case that some meaning is lost when text is translated from one genre to another. It is also the case that some meaning is added and other meaning is modified. This is the very reason why "translation activities" are so valuable to learning. Major problems will, of course, arise if the meaning of the scientific text becomes confused or obscured, but this is no more than a warning to the teacher to be careful in the examples of scientific text they select for classroom use, the "translation activities" in which they engage students, and the level of advice and support they provide. Used appropriately, these activities can be very effective in enhancing students' comprehension of conceptual and procedural knowledge and in developing their critical reading skills.

Wallace (2004) reports that many students in grades 7 to 11 have recognized that writing for the teacher in "the official language" of science enables them to use scientific terms, report scientific data and state conclusions in a rote and unthinking manner, and still be confident that it will make sense to the teacher and gain them the appropriate mark or grade. When they are required to write for a different audience, in a different language (for example, using everyday language to describe something to a friend or relative), they need to think much more carefully about what to say and how to say it. As Wallace (2004) comments, "Children in school science are fully capable of understanding the occurrence and features of different discourse styles, but ... they need experience in transforming language from one discourse style to another ... unpacking the various discourse genres used in science class will contribute to the authentic use and understanding of academic science language by students" (p. 907). Similarly, Sutton (1998) comments: "Custom and form in scientific writing should be *understood*, rather than followed in a ritual manner" (p. 33, emphasis added). A related point is that we need to draw a distinction between language as a means of exploring understanding and language as a means of transmitting knowledge to others, another matter to be discussed in Chapter 9. Indeed, Sutton (1996, 1998) argues that teachers need to make students aware of two major categories of language use.[5] First, language as a kind of coding or labelling system for reporting, receiving, transmitting and storing information and knowledge as concisely and effectively as possible. Second, language as an interpretive tool for making sense of our experiences, and those of others, building explanations of phenomena and events, persuading others of our particular point of view, and building up a body of shared knowledge and thought. It is important that students do not see scientific language solely as a means for "experts" to dispense information to non-experts, but also as a way of thinking and arguing that they can use to establish their own views, evaluate the views of others and reach important decisions on socioscientific issues. Indeed, Deanna Kuhn (1992) makes the point that thinking and arguing are two sides of the same coin.

> It is in argument that we are likely to find the most significant way in which higher order thinking and reasoning figure in the lives of most people. Thinking as argument is implicated in all of the beliefs people hold, the judgments they make, and the conclusions they come to; it arises every time a significant decision must be made. Hence, argumentative thinking lies at the heart of what we should be concerned about in examining how, and how well, people think. (p. 156)

Carlsen (2007) has added a third category to Sutton's framework (in effect, the extension of one element in Sutton's second category, above): language as a *tool for participation in communities of practice*, including participating in the solution of common problems and in public discourse concerning the acceptability and further deployment of scientific knowledge. These categories are beautifully illustrated in a study by Jiménez-Aleixandre and Reigosa (2006) of how grade 10 students effected a shift from using the terms *concentration* and *neutralization* as part of a simple coding/labelling system to using them as intellectual resources to plan, perform and interpret practical actions in a chemistry lab. It is to these interpretive, persuasive and evaluative aspects of the language of science that I now turn.

SCIENTIFIC ARGUMENTATION

As noted in Chapter 6, Giere (1997) provides a useful diagrammatic overview of the relationship between a theoretical model and the real world (Figure 6.1). Reaching consensus about the most acceptable model involves a cluster of interacting, overlapping and recursive steps: (i) collection of data via observation and/or experiment, (ii) reasoning, conjecture and argument, (iii) calculation and prediction, and (iv) critical scrutiny of all these matters by the community of practitioners. Language plays a key role in all these steps as scientists establish research priorities, write grant proposals, plan investigations, give instructions to technicians, collect and interpret data, reach conclusions, monitor the activities of other research groups, and prepare their work for critical scrutiny by the community of scientists. As an integral part of these activities, arguments are constructed and evaluated at a number of different levels:

> First, within the mind of the individual scientist when struggling to design an experiment or to interpret data; second, within research groups where alternative directions for the research program are considered in light of the group's theoretical commitments and empirical base; third, within the scientific community at large, through interactions between competing positions at conferences or through journals; and fourth, in the public domain where scientists in a contested field expose their competing theories through the media. (Driver et al., 2000, p. 297)

Scientists devote considerable time, energy and ingenuity to convincing others that the work they have done is important, the data are reliable, the interpretations are valid, and the conclusions are both trustworthy and scientifically significant. In science, judgements about the acceptability of knowledge claims are made on the strength and quality of the supporting argument. If, as Hirst (1974) maintains, this is the most important criterion, it should be afforded a high priority in the science curriculum. Until comparatively recently this has not been the case. As Osborne et al. (2004) comment, "it is ironic that science, which presents itself as the epitome of rationality, so singularly fails to educate its students about the epistemic basis of belief, relying instead on authoritative modes of discourse ... that leave students with naïve images of science ... and little justification for the knowledge they have acquired" (p. 996). Even when students are given first-hand experience of the methods by which scientists collect data, the ways in which scientists use that data to construct scientific arguments usually receives scant attention. In Ratcliffe's (2007) words, "depite science being an evidence-based discipline, at the frontiers of controversy about competing theories and models, many students in science classrooms do not normally engage in discussion and argumentation, either of scientific controversies or of socio-scientific issues" (p. 123). Accepting Hirst's case for priority to be afforded to criteria of judgement regarding knowledge claims demands a curriculum focus on argumentation, and a number of science educators have recently turned their attention to what had previously been a shamefully neglected area of research and curriculum development. The research agenda set out by Newton et al. (1999), Driver et al. (2000), Osborne (2001), Duschl and Osborne (2002), Erduran et al. (2004), Osborne et al. (2004), Simon et al. (2006), Duschl (2008), and Jiménez-Aleixandre and Erduran (2008) is roughly as follows.

— Why is argumentation important?
— What are the distinctive features of scientific argumentation?
— How can it be taught? What strategies are available?
— To what extent and in what ways are the strategies successful?
— What problems arise and how can the difficulties be overcome?

The simplest answer to question 1 is located in the concern to present students with an *authentic* view of science. Real science is impregnated with claims, counter claims, argument and dispute. Arguments concerning the appropriateness of experimental design, the interpretation of evidence and the validity of knowledge claims are located at the core of scientific practice. Arguments are used to answer questions, resolve issues and settle disputes. Given that observational data cannot be used unproblematically to provide "proof" of a proposition, scientists have to argue for the relevance of particular data, justify the methods by which it was collected, and persuade other scientists that it supports a particular conclusion. In other words, the scientific knowledge admitted to the corpus of knowledge is socially constructed by critical argument within the scientific community. This is not to revisit earlier discussion on the extent of that social construction and whether it constitutes social *determination*, but to urge teachers to look at how arguments for scientific propositions are assembled and presented. As argued several times already, acknowledging that the knowledge generated by scientists is necessarily influenced, to some extent, by the sociocultural context in which the scientists are located and the values they adopt, does not entail the adoption of a relativist stance in relation to all knowledge claims. Scientists cannot claim whatever they like, for whatever reasons happen to suit them. They have to construct arguments for public scrutiny in accordance with community-designated standards, norms and conventions. Students need to know about these standards, norms and conventions if they are to judge the rival merits of competing arguments.

A second answer to question 1 is that citizens need to understand scientific argumentation in order to engage meaningfully in debate on socioscientific issues. In everyday life, decision-making on SSI is based largely on evaluation of information, views and reports made available via newspapers, magazines, television, radio and the Internet. The ability to judge the nature of the evidence presented and its validity, reliability and appropriateness, the interpretation and utilization of that data, and the chain of argument substantiating the claims, are crucial to good decision-making. Embedded in this rationale for engaging students in argumentation is the assertion that it enhances their critical thinking skills, communicative competency and capacity for careful and systematic reflection. A third answer to question 1 is that engaging in argumentation enhances conceptual understanding and assists students in developing more sophisticated understanding. Building a convincing argument for a particular knowledge claim demands clear understanding of all relevant concepts and how they are or can be related to the available evidence. Conceptual relationships have to be explored and clarified; existing knowledge may need to be discarded, modified or developed; new conceptual knowledge may be necessary before progress can be made. In other words, arguing in order to learn may have just as important a role in the curriculum as learning to argue (see discussion in Chapter 9), though von Aufschnaiter et al., (2008) note that argumentation may be far more effective in consolidating and extending existing knowledge than in acquiring new knowledge.

Chapter 6 discussed the role, status and structure of scientific knowledge, and advocated an approach in which students are engaged (from time to time) in the processes of building, criticizing and developing scientific models and theories. Chapter 7 looked at ways in which students can be familiarized with, and gain personal experience of, the procedures scientists use to investigate phenomena and events and generate data to inform theory-building. Understanding and using scientific argumentation, one of the focuses of this chapter, constitutes the third essential component in a curriculum oriented towards providing students with *authentic* experience of science. Scientific investigation, theory-building and argumentation are complementary and reflexive activities. Each informs and is informed by the others. When students conduct a scientific investigation, construct an explanation of the phenomena and events being studied through social discourse, justify their constructed model in terms of the data they have collected, and defend it against the counter proposals of others, they can be said to be engaging in authentic scientific activity. Kim and Song (2006) describe precisely this kind of approach with a group of grade 8 students: in groups of two or three, students chose their own problem or topic for investigation, designed an appropriate investigative strategy, collected data (both first hand and by library or Internet search) and wrote a report of their findings and conclusions. In the final activity, each group presented its report to the whole group and responded to the criticisms and suggestions of their peers in a format designed to mimic a scientific conference. This is not to say that *all* learning in school science should be approached in this way. It is to say that students need to have this holistic experience of doing science from time to time throughout their science education. As I have argued numerous times in this book, we sometimes need to focus on the components of an activity before we can attend to the ways in which those components interact and support each other. Students can learn about argumentation, and gain experience of constructing arguments, without necessarily using data they have collected by means of an investigative procedure they have designed.

In the early stages, lessons that focus on a limited number of learning goals are often more successful than those that aim "to do everything".

Helen Longino (1990) makes the point that "it is the social character of scientific knowledge that both protects it from and renders it vulnerable to social and political interests and values " (p. 12) (see also Chapter 4). First, she says, there are public forums for the presentation and criticism of evidence, methods, assumptions and reasoning – in particular, conferences, academic journals and the system of peer review. Second, there are shared and publicly available standards that critics must invoke in appraising work, including but not restricted to empirical adequacy (as discussed earlier in this book).[6] Third, the scientific community makes changes and adjustments in response to critical debate and is clearly seen by practitioners and members of the public to do so. Fourth, the right to submit work for peer appraisal and criticism is open to all practitioners; so, too, the right to publicly criticize the work of others. Of course, the way in which the community of scientists exercises this public scrutiny at any one time is subject to a whole range of social, cultural, political and economic factors, and the inclination of individual scientists to be persuaded by a particular argument or to be swayed by particular evidence depends, in part, on their background knowledge, assumptions and values. In these respects, as Longino states, the procedures of scientific appraisal are vulnerable to charges of social relativism. The resilience of science in the face of such charges is a consequence of its openness to criticism by individuals of diverse backgrounds, experiences, interests and underlying values. Questions will be asked about the appropriateness, extent and accuracy of the data, how it was collected and interpreted, and whether the conclusions follow directly from the data, and so on. The explanation will be scrutinized for internal consistency and for consistency with other accepted theories. Particular attention will be directed to the background theory and assumptions underpinning the research design, and to the deployment of auxiliary theories and choice of instrumentation and measurement methods. The possibility that these questions may be answered differently by different appraisers, and that different perspectives will be brought to bear in the appraisal process, is the reason for upholding the principle of academic equality and is one of the guarantees of scientific objectivity.[7] In short, Longino argues that the critical scrutiny exerted on scientific ideas by peer review and public critique via conferences and journals is the centerpiece of scientific rationality and a guarantee of the objectivity and robustness of the knowledge developed.

> The formal requirement of demonstrable evidential relevance constitutes a standard of rationality and acceptability independent of and external to any particular research program or scientific theory. The satisfaction of this standard by any program or theory, secured, as has been argued, by intersubjective criticism, is what constitutes its objectivity. (Longino, 1990, p. 75)

What is at issue in this chapter is the nature of scientific community's forums of criticism, the procedures and criteria employed, and the extent to which students can understand and participate successfully in these community-validated procedures. Students need to know the kinds of knowledge claims that scientists make and how they advance them. In particular, the form, structure and language of scientific arguments, the kind of evidence invoked and how it is organized and deployed, and the ways in which theory is used and the work of other scientists cited to strengthen the case. Neglect of scientific argumentation in the school science curriculum gives the impression that science is the unproblematic accumulation of data and theory. In consequence, students are puzzled and may even be alarmed by reports of disagreements among scientists on matters of contemporary importance. They are also unable to address in a critical and confident way the claims and counter claims impregnating the socioscientific issues with which they are confronted in daily life. Being able to assemble coherent arguments and evaluate the arguments of others, especially those appearing in the media, is crucial if students are to understand the basis of knowledge claims they encounter and make decisions about where they stand on important issues. It almost goes without saying that opportunities to scrutinize the arguments used by scientists in constructing and validating theories contributes substantially to students' understanding of those theories, and their ability to use them appropriately. Indeed, constructing and reconstructing arguments, and addressing the rival merits of arguments through debate, has long been an integral part of constructivist pedagogy and a key tool for identifying and eliminating misconceptions, clarifying existing knowledge, introducing alternative views and consolidating new knowledge. One could also argue that important learning goals associated with the affective and social dimensions of learning are promoted by a curriculum emphasis on argumentation.

Although her work has been severely criticized by Koslowski (1996) for neglecting the context dependence of argumentation in favour of generic reasoning skills, Deanna Kuhn's (1991, 1992, 1993) conclusion that the construction and deployment of valid scientific arguments "does not come naturally" to students is important. Just like other aspects of learning about science, argumentation has to be taught. And if it is to be taught effectively, teachers must know something about the structure and characteristics of scientific arguments, something that cannot necessarily be assumed to be the case, given the neglect of argumentation in the science curriculum over the years (see Driver et al. (2000) and Ratcliffe (2007), cited earlier). Stephen Toulmin (1958) describes the structure of an argument in terms of six components (Figure 8.1).

- *Claim* – makes an assertion or states a conclusion.
- *Data* – states the evidence used to provide support for the claim.
- *Warrant* – explains or justifies the relationship between the evidence and the claim.
- *Qualifier* – indicates the degree of reliance to be placed on the conclusions and/or the conditions under which the claim is to be taken as "true".
- *Backing* – states the additional evidential, theoretical and methodological assumptions underlying the warrant and establishing the validity of the argument.

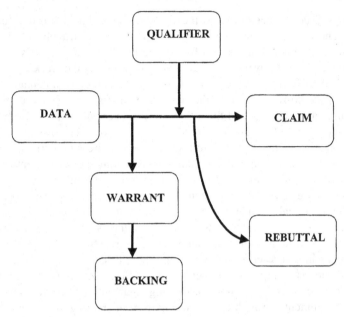

Figure 8.1. Toulmin's model of scientific argumentation (from Simon et al., 2006).

— *Rebuttal* – identifies circumstances in which the claim can no longer be sustained or introduces reservations that question the data, warrant, backing or qualifier of an argument.

The six components comprise two *levels* of argumentation: first, the construction of a basic argument establishes the relation between a claim and the evidence in support of it, and states the justification for this relationship; second, backing, rebutting and qualifying the justification complements and extends the basic argument. Each component in the model is, in effect, an answer to a question: What is being asserted? (*claim*) What evidence supports the claim? (*data*) What reasons, principles, rules or values justify the conclusion? (*warrant*) How likely is it that the conclusion is true? (*qualifier*) What theoretical assumptions justify the line of reasoning and establish the trustworthiness of the claim? (*backing*). Under what circumstance(s) would the argument break down? (*rebuttal*).

In constructing a 5-level analytical framework for assessing the quality of an argument, Zeidler et al. (2003), Erduran et al. (2004) and Osborne et al. (2004) place considerable emphasis on the systematic consideration of rebuttals.

— Level 1 – arguments comprising a simple claim versus a counter-claim, or a claim versus a claim.
— Level 2 – arguments consisting of claims with data, warrants or backings, but no rebuttals.
— Level 3 – arguments with a series of claims or counter-claims with data, warrants or backings and the occasional weak rebuttal.

- Level 4 – arguments with a clearly identifiable rebuttal. Such an argument may have several claims and counter-claims as well, though this is not necessary.

- Level 5 – extended arguments, with more than one rebuttal.

Schwarz et al. (2003) advance a case for evaluating arguments in terms of argument type, soundness of argument, overall number of reasons, number of reasons supporting counter-arguments, and types of reasons (including logical, concept-rich and theory-based reasons, appeals to authority, everyday common sense and personal experiences, and attempts to tease out the consequences of holding a particular view or engaging in a particular kind of action). Their notion of a hierarchy of arguments ranges from *simple assertions* (a conclusion unsupported by any kind of justification) through *one-sided arguments* (for which one or two reasons may be advanced) and *two-sided arguments* (including reasons that both support and challenge the conclusion, but do not weigh their rival merits) to *compound arguments* (replete with multiple reasons and critically evaluated counter-arguments). Kelly and Takao (2002) and Takao and Kelly (2003) describe how they evaluated first-year undergraduates' written arguments (in oceanography) in terms of the "relative epistemic status' of the propositions, arranged into six hierarchical levels – in essence, looking at the extent to which students go beyond data analysis and interpretation to deploy specific and general geological theory and incorporate geological knowledge acquired in other course units into the arguments they build. In a subsequent article, Kelly and Bazerman (2003) show that successful authors (i.e., those judged by the course instructors to have produced the most persuasive and convincing texts) are able to "adjust the epistemic level of their claims to accomplish different rhetorical goals, build theoretical arguments upon ... specific data ... introduce key concepts that served as anchors' and tie multiple strands of empirical data to central constructs" (p. 28). What is at issue here, in addition to sound knowledge and analytical skills, is a sense of audience and a feeling for the rhetorical purposes of different elements of the text. Finally, as Sandoval and Millwood (2005) and Sampson and Clark (2008) remind us, criteria for judging scientific arguments are not solely a matter of structure and the nature of justification. The specific scientific content of the argument is also important. For example, Sandoval and Millwood's approach focuses attention on two components of the argument: (i) the epistemological quality of the argument – that is, the extent to which the student has cited sufficient data in warranting a claim, written a coherent explanation and made appropriate use of inscriptions (graphs, tables, equations, etc.); and (ii) its conceptual quality and appropriateness – that is, how well the student has articulated the claims within an appropriate theoretical framework and warranted those claims using appropriate data.

Sampson and Clark's (2006) detailed review of the field expresses dissatisfaction with all schemes so far used by science educators to evaluate the quality of scientific arguments. They suggest that rather than focusing on technical issues relating to the precise structure of an argument, teachers should pay attention to such matters as: the kinds of claims advanced and whether they are well supported by the evidence; whether all the evidence has been utilized and discrepancies accounted for; whether the sources of data and the methods by which data were accumulated have been critically examined; how (or if) alternative claims are acknowledged, their weaknesses pinpointed and conclusions rejected. Both Duschl (2008) and Erduran (2008) provide detailed commentaries on the key issues relating to the systematic study of argumentation in the science classroom. In his critique, Duschl (2008) advocates an approach based on Walton's (1996) notion of "presumptive reasoning" for analyzing students' patterns of reasoning in terms of elements such as statements of belief and commitment, reference to an expert, acknowledgement of the need for further information, conjectures and hypotheses, inference of causal relationships, reference to likely consequences and whether these are judged to be "good" or "bad", and similarity to other cases, events and situations. Before leaving this discussion it is worth noting the report by Naylor et al. (2007) that very few of the grade 3 and grade 4 students in their study supported their claims (concerning two topics in physics) with warrants, and even fewer used backings and theoretical justifications. The authors' response was to shift the focus of analysis from the structure, content and language of the argument to the nature of the interactions within the discussion groups. At level 1, students are unable or unwilling to enter into discussion; at level 2, they make a claim to knowledge; and at level 3, they begin to provide justifications for their claims. At level 4, students offer further evidence to support their claims; at level 5, they respond to ideas from others in the group; and at level 6, they are able to sustain an argument in a variety of ways. Finally, at level 7, students evaluate the evidence and make judgements. Interestingly, the authors found that students were more productive when the teacher was not involved and they were able to work out their own ground rules for presenting ideas, criticizing views and reaching consensus – a finding at variance with advice on group organization offered by Mercer et al. (1999), as discussed in Chapter 9.

STUDENTS' ARGUMENTATION SKILLS

It may seem odd to look at research on the development of students' argumentation skills *before* engaging in a detailed discussion of appropriate teaching and learning methods. However, the methods best suited to developing argumentation skills, namely talking, reading and writing activities, are also methods well-suited to further development of conceptual and procedural understanding, enhancing students' understanding of the history of science and achieving several other important learning goals. Justice can only be done to the important pedagogical issues by allocation of a whole chapter (chapter 9) to these matters. Chapter 9 will also look at ways to develop students' ability to write appropriate scientific reports.

Research findings from studies using the Toulmin framework (Figure 8.1) are somewhat mixed. Bell and Linn (2000) conclude that students tend to rely on data to support their claims, but rarely use warrants or backings; Jiménez-Aleixandre et al. (2000) found that many students do not even use data to support their claims. Simple and straightforward as the Toulmin model seems, its deployment in science education research to monitor the development of students' argumentation skills is beset with difficulties – most notably, the difficulty of determining exactly what counts as *claim, data, warrant* and *backing* in a particular set of circumstances (Erduran et al., 2004). Thus, ascertaining the extent to which students have made use of data, warrants, backings and qualifiers to support arguments, and the extent to which they use rebuttals to elaborate and extend or oppose an argument, can be problematic. Even with written arguments, the deployment of Toulmin's model as a research tool can sometimes be difficult; with verbal arguments it is even more problematic. The natural flow of conversation can serve to disrupt the logical structure of the argument. Moreover, students often use language that is vague or ambiguous, and they frequently contradict themselves as they struggle to sort out their ideas. Boundaries between the categories of argument become blurred and fluid, with key elements in the argument being implied rather than explicitly stated. They may even be conveyed by gesture. In addition, elements of an argument may be omitted because the arguer simply assumes that it is already well known and does not need to be re-stated, and may have picked up unspoken signals to that effect from listeners. In short, real face-to-face argument is dynamic and interactive. In trying to ascertain students' capabilities it is essential to take account of the context in which the argument is located, and its familiarity and interest for the student. Cross et al. (2008) report that four additional codes were necessary in order to do justice to the argumentative structures produced by a group of grade 10 and grade 11 students following a technology-mediated biology course: (i) prompt for claim – an argument type in which a participant directs one of the other participants to state their claim about a particular topic; (ii) prompt for warrant – an argument type in which a participant directs one of the other participants to warrant their claim about a particular topic; (iii) prompted claim – an argument type in which a participant provides a claim after a prompt by another participant; (iv) prompted warrant – an argument type in which a participant warrants a claim after a prompt by another participant. McNeill et al. (2006) move in the opposing direction, reducing Toulmin's model to just three components: claim, evidence (or data) and reasoning (a combination of warrant and backing, deployed as considered appropriate).

Common sense would seem to indicate that content knowledge is crucial, and that those who know more about the topic/issue under consideration will be better positioned to construct arguments, criticize arguments and engage in argumentation. For example, Sadler and Zeidler (2005) note that students with deeper understanding of genetics made fewer errors of reasoning and made more frequent and explicit reference to content knowledge in their reasoning about gene therapy and cloning than students with more naïve understanding. Studies by Patronis et al. (1999), Hogan (2002), Dawson and Schibeci (2003), Sadler (2004), Keselman et al. (2004) and Lewis and Leach (2006) provide further confirmation. A study by Sadler and Donnelly (2006) in the context of gene therapy and cloning suggests that a basic level of knowledge is essential in order to understand the nature of the problem and to know what might constitute appropriate evidence. Beyond that, the authors say, there is little evidence that background knowledge in genetics impacts significantly on argumentation skills. What may be much more important is *context* knowledge – in this case, specific knowledge of gene therapy and cloning. In a closely related study, Sadler and Fowler (2006) show that at the College level, science majors with advanced biological knowledge (that may have included knowledge of genetic technologies) significantly outperformed non-science majors and high school students in the sense of making repeated, explicit and appropriate reference to scientific knowledge in construction and criticism of arguments. In other words, when students' science content knowledge is extensive in depth, breadth and organization it does make a difference to their ability to deploy it effectively. It should be noted that students' familiarity with and facility in using devices such as diagrams, charts, tables and equations, especially when deployed using sophisticated technology, can also play a key role.

As noted in earlier discussion, social relations and affective factors are crucial determinants of success for any group-based activity. So, too, the overall classroom climate and the unspoken "rules of the science education game". For example, students' attempts to construct and evaluate scientific arguments can sometimes be disrupted by the intrusion of other classroom priorities – principally, "getting the job done" as quickly as possible in order to gain the mark or credit. When students perceive that the social climate of the classroom is such that the teacher's knowledge and style of discourse is highly valued and their own is not, they may become reluctant to proffer alternative views, however sensible those views may be; they may be content simply to repeat verbatim what the teacher says. In too many schools, this suits the teacher very well because it gives him or her a quiet life.[8] Indeed, Aikenhead and Jegede (1999) and Aikenhead (2000) illustrate some of the ways in which teachers and students conspire to reduce the cognitive demand of the science curriculum in favour of playing a version of "Fatima's rules", that is, going through the motions of meaningful learning but actually practising rote memorization and algorithm use.[9]

Sadly, many students are socialized to see their task as memorizing a series of definitions and reproducing them on demand, mastering a set of algorithms for solving standard problems, and carrying out the teacher's instructions for pseudo-experimental inquiries intended to obtain a particular set of results. They do not see their role as one that requires or even allows them to think, or to question the source, relevance, validity and reliability of the views and ideas presented to them. Nor are they given opportunities to design, conduct and interpret scientific inquiries for themselves and by themselves, or to build their own arguments. Therefore, although these students may "succeed" in school, in the sense of being able to say and do the "right things" and gain the marks that are made available for such conformity, they also "fail", in the sense that they do not acquire a robust and usable set of meanings to incorporate into a personal framework of meaning and understanding (Costa, 1997; Loughran & Derry, 1997). What many students learn is how to do classroom tasks, how to be neat, how to finish on time, how to look busy and to fill up the available time, how to avoid attracting the teacher's attention and, in practical lessons, how to tidy away, wash up and write things up in the approved form. Drawing on the work of Bloome et al. (1989) and their notion of "procedural display",[10] Jiménez-Alexandre et al. (2000) identify many occasions during science lessons, particularly during hands-on activities, when student priorities are oriented towards "doing the lesson" (engaging in activities that "get the job done") rather than "doing science". Because they are not intellectually engaged, some students may develop passive resistance techniques such as "silence, accommodation, ingratiation, evasiveness, and manipulation" (Atwater, 1996, p. 823). What they do *not* learn is how to employ their knowledge in novel situations and how to use it to develop a deeper and richer understanding. Such students have not been enculturated into science; rather, they have been assimilated into school. In Kelly and Green's (1998) words, they have achieved "communicative competence" rather than meaningful learning. Perhaps they have learned their own version of Fatima's rules.

Taking account of all these factors in designing an approach suitable for the school science classroom, and evaluating its effectiveness in developing students' capacity to engage in argumentation, is no easy matter. Research shows that students do improve their capacity for constructing and presenting effective arguments through practice, though not always as rapidly and predictably as we might wish (Jiménez-Aleixandre et al., 2000; Zoller et al., 2000; Osborne et al., 2004; Garcia-Mila & Andersen, 2008). What is clear is that development of those skills is a long-term undertaking, and one or two brief experiences will not suffice. As with other aspects of NOS, teaching of argumentation needs to be explicit and systematically planned. For example, students in the study described by Duschl and Ellenbogen (2002) often failed to assemble the data they had collected in a lab activity (and had successfully represented in a range of graphs) into a coherent argument to justify their conclusions. The authors recommend the modification of lab instructions to include scaffolding prompts to guide students in construction of arguments and selection of evidence. Several other researchers have shown how the quality of students' arguments can be considerably enhanced by judicious scaffolding, encouraging student reflection, fostering metacognition and providing timely and constructive feedback (Kuhn et al., 2000; Bell, 2002; Cho & Jonassen, 2002; Engle & Conant, 2002; Zembal-Saul et al., 2002; Nussbaum & Sinatra, 2003; Felton, 2004; Nussbaum & Kardash, 2005; Sandoval & Millwood, 2005; Erduran, 2006; Chinn, 2006; Kenyon et al., 2006; McNeill et al., 2006; Andriessen, 2007; McNeill & Krajcik, 2007, 2008; Reigosa & Jiménez-Aleixandre, 2007; Chinn & Samarapungavan, 2008; Jiménez-Aleixandre, 2008; Varelas et al., 2008; Berland & Reiser, 2009; McNeill, 2009).

In discussing the syntactical structure of science (Chapter 7) I suggested that teachers can create opportunities for students to engage in a form of apprenticeship via a 3-phase approach: *modelling*, *guided practice* (using carefully structured tasks) and *application*. A similar approach seems entirely appropriate for teaching and learning scientific argumentation.[11] Through hearing, reading, speaking and writing scientific arguments, students acquire familiarity with their form and structure (including presentation of data in the form of tables, graphs, photographs, X-ray pictures and supportive deployment of chemical and mathematical equations) and develop confidence in using them, both in school and beyond. Initial understanding is established when teachers direct students' attention to the essential features of a scientific argument and explain why argument is a central feature of scientific practice, when they use evidence to justify particular theoretical positions, construct and evaluate arguments, engage in counter argument and debate, and help students to reflect on the nature of argumentation and the criteria that distinguish between good and poor arguments. Developing an understanding of what counts as a *good* argument (i.e., that it is adequately warranted, coherent, comprehensive and, perhaps, parsimonious) requires teachers to create opportunities for students to consider alternative theoretical accounts of a phenomenon or event, and the arguments used to support or justify them, under the expert guidance of the teacher. It is important to provide lots of opportunities for students to collaborate with others in: (i) discussing rival theoretical ideas and the evidence that does or does not support them; (ii) constructing and defending arguments in favour of their preferred theoretical position, and (iii) rebutting challenges and counter arguments. In addition to working with case studies of important episodes in the history of science and in contemporary practice, students need opportunities to construct and argue for their own scientific ideas, models and theories (see discussion in Chapter 6). Engaging in argumentation, in both oral and written form, is a powerful way of rationalizing and developing one's own thinking. As Billig (1996) comments: "humans do not converse because they have inner thoughts to express, but they have thoughts because they are able to converse" (p. 14). Thus, learning to think is learning to argue, and learning to argue is learning to think. It is also a powerful way of teaching NOS understanding. As noted earlier in this chapter, and in Chapter 4, Helen Longino (1990, 2002) identifies four key aspects of the social construction of scientific knowledge: recognized and public venues for presentation and criticism of ideas; publicly recognized standards governing appropriate forms of criticism; tolerance of dissent and response to criticism; equality of intellectual authority, albeit a tempered equality that respects differing levels of experience and expertise. By creating similar conditions in the science classroom, students' learning of science and their learning about science can both be enhanced.

Nussbaum et al. (2008) have shown that providing a brief set of guidelines relating to criteria for a good scientific argument (relates two or more variables, describes a causal mechanism, supports claims with facts, accounts for all facts, searches for counter examples and considers alternative theories or explanations), together with examples of arguments displaying these characteristics, resulted in preservice teachers using more of the criteria in forming their own arguments, conducting more thought experiments, and considering more alternatives. Felton (2004) noted some substantial improvements in the quality of arguments when students were provided with a checklist to assist critical evaluation of arguments generated by peers and guided reflection on their own arguments. Palincsar and Magnusson (2001) and Hapgood et al. (2004) report equally substantial improvements when students are provided with "notebook texts" that remind them about such aspects of scientific inquiry as using standardized investigative procedures, controlling for experimental error, documenting all procedures, using diagrams, data tables and graphs as rhetorical devices, and so on. Similarly, Herrenkohl et al. (1999) report enhanced performance in constructing arguments when students are provided with "Thinking Like a Scientist" guidelines. Herrenkohl and Guerra (1998) noted marked improvement in the argumentation skills of grade 4 students when they were assigned specific critical roles during group presentations, including asking questions, challenging claims and expressing alternative viewpoints. Pilkington and Walker (2003) describe a similar approach in which students were instructed to adopt one of three "argumentation roles': role 1 was to challenge others to provide evidence, identify contradictions and point out alternatives; role 2 was to ask for explanations and clarifications; role 3 was to provide information, either spontaneously or in answer to a question. The authors report that the role play experience had considerable beneficial impact, particularly in terms of students taking on much greater responsibility for sustaining the debate. Cross et al. (2008) report on the use of a novel way of presenting guidelines: an animated cartoon video in which the cartoon characters engage in one argumentation routine considered "good" and one considered "poor". The criteria employed focused on the quality of the argument produced and the nature of the engagement, including participation in discussion, listening to others and maintaining a sense of purpose. It is also worth noting the report by Toth et al. (2002) that the capacity to organize data and theoretical hypotheses diagrammatically through evidence mapping software known as BELVEDERE[12] can substantially enhance grade 9 students' capacity to evaluate possible relationships between multiple explanations and complex, wide-ranging data (in this case, data relating to the causes of dinosaur extinction). In their work with grade 7 students, McNeill and Krajcik (2008) used a combination of four pedagogical strategies: (i) defining scientific explanation; (ii) modelling scientific explanation; (iii) making the rationale for a particular scientific explanation explicit; and (iv) connecting scientific explanations to everyday explanations. In a later study, McNeill (2009) discusses three additional pedagogical strategies: providing timely and appropriate feedback, connecting the discussion to students' prior knowledge, and ensuring a robust understanding of the relevant science concepts. In both studies, strategy (iii) had the most significant impact on students' conceptual understanding and their ability to construct their own explanations. Helping students to understand the rationale behind a particular scientific explanation seems to help them see why they need good evidence and sound reasoning to support their own claims. McNeill and Krajcik (2008) found that strategy (iv) had a negative impact on student understanding, though the authors point out that because only three of the 13 teachers used such examples it is unwise to draw a generalized conclusion from this finding. Much depends, of course, on teachers' own conceptual understanding and on how skillfully they can juxtapose scientific language and everyday language (see earlier discussion). McNeill and Krajcik also warn about the dangers of providing too much guidance on the construction of arguments and explanations.

271

When teachers focus on close definition of the essential elements of an experiment or argument, without adequate discussion of particular cases, students' attention can be directed towards algorithmic or formulaic constructions, and productive discussion is neglected.

It is clear from research reported earlier in the chapter that familiarity with the conceptual content of the argument and interest in the context in which it is located can have substantial impact on students' ability to construct and evaluate arguments. Interestingly, improvements seem to be more marked when argumentation is deployed in the context of socioscientific issues, especially those with a moral-ethical dimension, presumably because students can draw on ideas and knowledge accumulated outside formal educational settings and can deploy their own ethical values in a context of personal interest (Patronis et al., 1999; Simonneaux, 2001; Zohar & Nemet, 2002; Osborne et al., 2004; Kolstø, 2006). An SSI context also puts emphasis much more firmly on the need to *persuade* others of one's point of view, a key element in building a robust argument. As noted earlier, students often see their efforts in class as being directed towards showing the teacher that they know or can locate the "right answer". A focus on argumentation, especially in the context of SSI, shifts emphasis from the task of reporting "correct answers" to careful and systematic appraisal of the reasons why a particular individual or group might accept an idea or be persuaded to give it serious consideration. As Berland and Reiser (2009) note, "the act of persuading carries with it the social expectations that individuals have an audience for their arguments, that this audience will interact with the presenters to determine whether they are persuaded of one another's ideas and that the audience may not have had the same experiences and beliefs as the authors of the argument" (p. 49). The relationship between audience and form of presentation will be discussed at some length in Chapter 9. The use of argumentation in the context of SSI will be explored in a book currently in preparation. As an aside, it should be noted that Chang and Chiu (2008) propose that the Lakatosian model of scientific research programmes (especially the notion of "hard core" and "protective belt", see Lakatos, 1978b) is more appropriate than the Toulmin model for representing the controversial and sometimes highly emotional arguments relating to socioscientific issues, where views are often well-entrenched and constantly reinforced outside school.

ARGUMENTATION, EPISTEMIC BELIEFS AND SOCIAL DYNAMICS

There is some evidence that engaging students in the construction and presentation of arguments can improve both their conceptual understanding and their understanding of some key aspects of scientific investigation, such as the distinction between observation and inference, the role of prediction and hypothesis, and the need for controls in experimental investigations (Richmond & Striley, 1996; Kawasaki et al., 2004; Sandoval, 2005). These experiences can also impact positively on attitude towards science (Lazarowitz & Hertz-Lazarowitz, 1998). In addition, there is strong evidence that students who hold the view that scientific knowledge is created by people, rather than uncovered or revealed from nature by scientific method, are better able to use laboratory work and talking, reading and writing activities to build strategies and attitudes that may eventually lead to good argumentation skills, presumably because they see themselves as having a mandate to question, criticize, establish connections, construct and reconstruct ideas (Tsai, 1999; Wallace et al., 2003; Stathopoulou & Vosniadou, 2007). Those students who regard science as a body of proven knowledge have no such mandate. In other words, a good understanding of other NOS issues is key to developing the knowledge and skills of scientific argumentation. Kuhn and Franklin (2006) claim that students' epistemological understanding develops through four distinct stages. At the *realist* stage (pre-school years), children regard knowledge statements as copies of an external reality. At the *absolutist* level, they have recognized that knowledge is a product of the mind and is either "right" or "wrong". Scientific knowledge located in textbooks is, of course, "right"; it is fixed and certain. At the *multiplist* level, students have recognized that people can disagree, and so they regard knowledge as freely chosen opinion rather than facts. Knowledge is "true only in the eye of the beholder", that is, all views are legitimate. To reach the *evaluatist* stage, students need to recognize that "some opinions are more right than others to the extent that they are better supported by argument and evidence" (Kuhn, 2005, p. 32). In other words, knowledge is "verified true belief", that is, knowledge based on critical evaluation of evidence. Kuhn and Franklin (2006) postulate a close connection between level of epistemological understanding, willingness to engage in argumentation and success in learning science.

> If facts can be ascertained with certainty and are readily available to anyone who seeks them, as the absolutist understands, or, alternatively, if any claim is as valid as any other, as the multiplist understands, there is little reason to expend the intellectual effort that argumentation entails. One must see the point of arguing to engage in it. This connection extends well beyond but certainly includes science, and in the field of science education a number of authors have made a case for the connection between productive science learning and a more mature epistemological understanding of science as more than the empirical verification of facts. (p. 985)

In confirmation, Nussbaum et al. (2008) report that among 88 undergraduates categorized as absolutists, multiplists or evaluatists, evaluatists interacted more critically, constructed arguments of higher quality, and raised many more issues. Multiplists interacted less with their discussion partners, were more accepting of inconsistencies and possible misconceptions (and certainly acknowledged them less often), and were less critical of arguments. These authors also speculate that multiplists probably did not see the point of arguing because any one explanation is as good as any other. There are some interesting dynamics relating engagement in argumentation with NOS views, general epistemic beliefs and some affective and social dimensions of learning. For example, Nussbaum and Bendixen (2003) have shown that less assertive students tend to avoid engagement. Indeed, many students are made uncomfortable because they regard argument as discord and conflict. Anxiety is especially likely among those students who regard science as simple, certain and fixed. Clearly there is an important role here for the teacher in creating an emotionally secure and non-threatening environment, fostering more sophisticated NOS views and giving guidance on the qualities of a good argument.

Chapter 4 noted, albeit briefly, some of the myths, distortions, stereotypes and falsehoods about science, scientists and scientific practice that continue to be promoted by the science curriculum. We should be very careful not to promote similar myths about scientific argumentation. Just as it is mistaken to believe that there is one all-purpose method of scientific inquiry, so it is mistaken to believe that there is a generic, lock-step approach to the construction and presentation of scientific arguments. Although the models advanced by Toulmin and others are enormously useful for teaching about the basic structure of scientific arguments, we need to recognize that scientists are likely to modify their approach to suit particular contexts and particular audiences. *What* is being argued, *why* it is being argued, *where* and through what medium it is being argued and *to whom* it is being argued, will all impact on the structure of the argument and the language in which it is expressed. While it is important to study the style of arguments used in scientific articles, research reports and conference presentations, and to practice writing them, including whatever differences we detect among and between the sub-disciplines of science, we also need to be cognizant that scientists use other media for communicating arguments. As Bricker and Bell (2008) remind us, "scientists also attempt to convince each other through more informal communication channels, such as phone calls, emails, and face-to-face informal interactions (e.g., shared dinners at conferences or over drinks)" (p. 483). The form of the argument will change as the purpose and medium of communication change. Scientists will also argue differently with immediate colleagues, research competitors, students, television audiences and newspaper readers. Chapter 9 includes further discussion of the importance of sensitizing students to the ways in which the form and substance of scientific communication is impacted by purpose, medium and audience, and the importance of providing students with opportunities to apply this understanding in the construction and critical reading of scientific text.

Of course, real explanations, especially those constructed by students, are full of "detours, dead ends and fallacious arguments" (Duschl & Osborne, 2002). For example, Zeidler (1997) notes that many students exhibit a confirmation bias and pay insufficient attention to potentially disconfirming data. They are more easily convinced by arguments for propositions that are consistent with their own beliefs than by those counter to their beliefs. The more strongly a belief is held, the more difficulty students have in accepting counter evidence and counter arguments. In many situations, students are unsure what counts as sufficient evidence and are likely to jump to generalizations and conclusions on grounds of very little evidence; they may be unable to distinguish significant from less significant evidence; they may incorporate indirect or even unrelated data in an effort to strengthen an argument; they may over-generalize or rely on personal or anecdotal experience and intuition to "fill the gaps". On other occasions, they may fall into the trap of giving equal weight to all data regardless of its merit. Other examples of fallacious reasoning include circular reasoning, simple appeals to authority figures and sources, and *ad hominen* attacks (attacking the speaker/author rather than the legitimacy of the argument). Making students aware of these faults, and noting that scientists are also sometimes guilty of these same errors (see Chapters 4 and 5), reminds them that although norms and standards exist they are not always easy to follow in the hurly-burly of real situations, and that the development of expertise is the outcome of lengthy experience and opportunities for critical reflection. Research shows, as might be predicted, that students and non-scientists do not place as much emphasis as scientists and technicians on empirical consistency and plausibility of conclusions (Hogan & Maglienti, 2001). It is also worth noting Sandoval and Millwood's (2008) point that having a preference for arguments supported by good quality and consistent evidence does not necessarily enable students to locate that evidence, though it does make it more likely that they will use it if they find it.

The main problem, of course, is that many teachers do not provide sufficient opportunities for students to construct, use and evaluate arguments, question and criticize each other's ideas, and challenge the views expressed in textbooks or by teachers (Newton et al., 1999; Driver et al., 2000). The dominant form of interaction in many classrooms, even these days, is teacher talk. Whole class discussion is often dominated by a teacher-led pattern of teacher initiation, student response, teacher evaluation (I-R-E),[13] which Lemke (1990) calls "triadic dialogue". The teacher asks a question (to which she/he already knows the answer), calls on a student to respond, and passes some evaluative comment on the answer. Such methods have been strongly criticized by researchers on the grounds that they inhibit students working out meanings for themselves, reduce their levels of language competence, limit possibilities for discussion, treat students as subordinate, reinforce the teacher as authority on matters of truth, and sometimes promote a "cat and mouse" guessing game through which students frequently end up frustrated and unsuccessful (Wells, 1986; Fisher, 1992; Dillon, 1994; Edwards & Westgate, 1994). When teachers keep strictly to the IRE strategy there are few, if any, opportunities for students to respond to other students' ideas, either to support or contradict them, and little encouragement to build the persuasive arguments characteristic of the discourse of professional science. As Cornelius and Herrenkohl (2004) comment: "What would be the point of trying to convince your classmates that your idea has merit if the teacher would step in and solve the controversy with a simple yes or no?" (p. 485). Even when opportunities are provided for students to work in groups during practical work, discussion tends to focus on procedural aspects and deciding what data to include in the lab report rather than how the data inform the theoretical issues under consideration. Ritchie and Tobin (2001) found that student discussions tend to be dominated by statements that simply reproduce information previously given by the teacher or located in textbooks, often verbatim. Moreover, teachers tend to reward such "reproductive talk" and to use cues and hints that lead students towards textbook knowledge expressed in approved "official" language. In consequence, the teacher's role as the authoritative dispenser of scientific knowledge is rarely challenged. In many classrooms, little has changed since Schwab (1962) pointed out, more than 40 years ago, that teachers present science as a "rhetoric of conclusions", emphasizing *what* we know rather than *how* we know.[14] Little wonder, then, that students are generally poor at recognizing links between theoretical ideas and the evidence that supports them. In reading scientific text, most students can recognize observations, descriptions of methods and conclusions, but are less adept at recognizing justifications, causal statements and statements of evidence (Norris & Phillips, 1994; Ratcliffe, 1999; Naylor et al., 2007).

OVERCOMING PROBLEMS

There are many reasons why this approach continues to be the dominant form of science education. First, it is often what official science curricula and their associated assessment schemes tend to prioritize. Second, teachers claim that there is insufficient time for a curriculum programme oriented towards the rhetoric of argument. Moreover, many teachers lack confidence in their ability to engage students productively in group discussions, fear a loss of class control, have little faith in the efficacy of the approach, have low expectations of students' capacity to achieve the learning goals associated with argumentation, and lack confidence in their ability to assess student competence in this area. Some believe that presenting students with alternative explanations of phenomena and events, and with differing arguments based on observational data, might serve to confuse them or even reinforce whatever misconceptions they already hold; many teachers believe that only high achieving students are capable of engaging in argumentation. As discussed in Chapter 3, in the context of NOS-oriented teaching in general, progress will only be made when teachers accept argumentation as an essential component of a curriculum for critical scientific literacy and become committed to including it in their science programmes. Teachers also need a range of pedagogical strategies to initiate and support argumentation, and ready access to suitable curriculum materials. Working with other teachers in mutually supportive groups can provide valuable opportunities for the critical reflection essential to accepting and developing new classroom practices. Videos and interactive multimedia materials featuring exemplary practice can be enormously helpful. Teachers also need a sufficiently robust understanding of science content to support rival arguments and a willingness to abandon their traditional position of ultimate classroom authority. Emphasis needs to shift from judging student explanations right or wrong to evaluating the quality of the argument they use in justifying the explanation. In addition, as noted by Duschl et al. (2007), "skill and persistence are required to help students grasp the difference between scientific argument, which rests on plausibility and evidence and has the goal of shared understanding, and everyday argument, which relies on power and persuasiveness and assumes that the goal is winning. It is not straightforward to get a middle schooler to see the distinction between disagreeing with an idea and disagreeing with a person" (p. 187). These problems are, of course, reinforced in many schools by an unrelenting emphasis on competition among students.

Clark and Sampson (2007) summarize the classroom conditions for support-ing argumentation as follows: "(a) students must engage with plural accounts of phenomena and evidence to support multiple points of view ... (b) the learning environment must provide a context that fosters dialogic discourse, (c) tasks and activities given to groups must require collaboration in order to promote discourse between students ... (d) students must have enough time to understand the central concepts and underlying principles ... (e) the teacher or learning environment must facilitate student-to-student talk without the limitations and rigidities char-acteristic of most teacher-student interactions" (p. 257). Clark and Sampson also describe an online discourse system that enables students to characterize the prin-ciple(s) underlying the data they have assembled, sorts students into discussion groups with those who have built different principles, facilitates discussion and mutual criticism within the groups, and assists in preparing arguments for presen-tation. Lynch (2002) provides a particularly helpful discussion of synchronous, asynchromous and simulated face-to-face communication tools. Other important initiatives include the use of the SenseMaker software to help students connect theory and evidence (Bell, 1997, 2004; Bell & Linn, 2000) and the web-based science environment (WISE) to present conflicting views of scientific phenomena (Linn et al., 1999, 2003). The Convince Me software has been used to foster stu-dents' metacognitive skills by creating representations of their reasoning patterns (Adams, 2002, 2003); the ProgressPortfolio tool provides specific prompts to en-able students to monitor the progress of an investigation, organize and represent information in a variety of ways, and reflect on conclusions and claims (Loh et al., 2001). Clearly there is much scope for research here. However, discussion of technology-supported discussion and communication tools is well outside the scope of this book; for discussion of such matters see Hill and Hannafin (2001), Reiser (2002, Andriessen et al. (2003), Veerman (2003), Bell (2004), Nussbaum et al. (2004), Andriessen (2007), Clark et al. (2007, 2008) and Clark and Sampson (2008).

Sadler's (2006) study of preservice middle and secondary school teachers demonstrates that a purpose-built course on argumentation and its importance in contemporary science education can have beneficial impact in terms of enhanced understanding of the nature of scientific argumentation and improved ability to construct convincing arguments. However, this enhanced understanding did not translate into extensive use of argumentation in these teachers' classrooms, largely because the existing school climate was unsupportive, or even hostile. It also became clear that although the student teachers developed a positive stance to-wards the use of argumentation strategies, few (if any) saw it as a major goal of science education. Simon et al. (2006) report encouraging results from a 2-year long professional development programme aiming to support the teaching of argumentation. Twelve teachers attended a series of workshops in which they collaborated to develop pedagogical strategies for subsequent trialling in their schools, thereby ensuring a feeling of ownership, engendering confidence in their appropriateness, and developing a commitment to use them. Subsequently, all the teachers attempted to implement these strategies. Many were successful in incorporating "higher order" elements of argumentation such as backings, warrants and rebuttals. As might be predicted, the teachers met with varying amounts of success with respect to students' development of argumentation skills. Much can be learned from the situations in which success was limited. For example, as Simon et al. (2006) conclude: "students need to learn how to listen and talk, justify claims, and so on, before they can debate; likewise, teachers need to value and learn how to implement group discussion and prompt justification before they can orchestrate effective counter-argument within their teaching" (p. 256). There is a key role here for a mode intermediate between spoken and written language: the lecture or teacher exposition in which there is an extended oral presentation of pre-prepared material.

Although it has been fashionable in recent years to be dismissive of teacher presentation, it does enable the language of science and the structure of scientific arguments to be modelled. Because teachers can use formal aspects of the language of science alongside more familiar everyday language, teacher presentation can be a powerful scaffolding tool in the enculturation process, introducing students to the specialized terminology of science, its distinctive mode of representing experience, and the form and style of argumentation. By shifting backwards and forwards between scientific language and everyday language, and by using everyday exemplars, similes and analogies, graphic metaphors, personal reminiscences, stories, anecdotes and jokes, teachers can make science more meaningful, exciting, humanized and accessible. Teacher talk is a way of introducing new ideas and new terminology carefully, systematically and sensitively. Skilful teacher talk helps to focus attention, picks out and emphasizes key ideas, clarifies meaning, points out similarities and differences, and so assists students in sorting out their own ideas. As discussed earlier in this chapter, skilful teachers are able to use a mix of scientific language and everyday language, including the students' vernacular, to assist students in acquiring conceptual understanding and gaining familiarity with and confidence in using the language of science. Barnett (1992) coined the term *interlanguage* to describe how teachers can scaffold learning by pitching their language somewhere between students' everyday language and more formal scientific language, and encouraging students to do so.

> As students work through the secondary school, they develop for science a personal and dynamic interlanguage which progressively approximates to the technical language of science used by scientists. Gradually, students adopt the features of science language which they find useful, and gradually their skills in using them increase (p. 8).

As noted earlier, Gutierrez et al. (1999) refer to it as "hybrid language", while Brown and Spang (2008) call it "double talk". Dawes (2004) describes how the much-derided IRE or IRF approach (see above) can be rehabilitated as an effective scaffolding method: "Teacher-led whole class IRF conversations ensure that children can explain their ideas, hear the explanations of others, use new words, refer to previous contributions of other children, share comments and information, and generally talk and think together" (p. 682). Arguing along similar lines, Gordon

Wells (1999b) makes the point that the third move from the teacher (the F in IRF or E in IRE) can serve different functions. It can serve to close down discussion and direct students' attention to a particular approved idea or it can clarify, paraphrase, confirm and extend the student's answer, signal its significance, show its relationship and relevance to other ideas, summarize, recontextualize or modify a succession of responses, request additional information or encourage further comment, exploration of ideas and development of meaning. There are also occasions when it is judicious to omit the third move altogether, thereby signalling that students do not necessarily need teacher confirmation. In this way, students are encouraged to be more autonomous – in a sense, "experts" who can work things out for themselves through further articulation and critique of ideas, additional research and vigorous debate. Of course, students have learned over many years the roles they are expected to play in classrooms. In consequence, some students may resist this more open-ended and less teacher-controlled discourse pattern, and may attempt to elicit a response from the teacher. If they are to overcome student reluctance/resistance, teachers need to spell out the "new rules" (Oliveira et al., 2007).

Mortimer and Scott (2003) and Scott et al. (2006) have extended these ideas into an analytical framework that describes teacher talk in terms of two dimensions, *dialogic-authoritative* and *interactive-non-interactive*, resulting in four categories of teacher-student communication.

- Interactive/dialogic – teacher and students collaboratively explore a range of ideas.
- Non-interactive/dialogic – teacher addresses different points of view, exploring similarities and differences.
- Interactive/authoritative – teacher leads students through a question and answer routine with the aim of establishing and consolidating a particular point of view.
- Non-interactive/authoritative – teacher presents a specific point of view.

In authoritative discourse, teacher talk aims to convey specific information; it makes few invitations to genuine dialogue and often involves instructional questions intended to guide students through a particular argument or explanation. In consequence, student contributions are usually direct responses to the teacher's question(s); they are usually short, often single word answers, or assertions of factual knowledge. In dialogic discourse, teacher talk aims to encourage thought and debate. In consequence, students often spontaneously offer their own ideas (i.e., ideas not elicited by teachers), sometimes in direct response to other students' comments, frequently in whole phrases and sentences, and expressed in ways that invite comment from others. Students and teacher approach learning tasks together through talk that is interactive, mutually supportive and purposeful, creating what Mercer et al. (1999) call "learning conversations". Clearly, different approaches are appropriate to different educational purposes and on different occasions. The expertise of the teacher resides in knowledge of these different modes, when they are appropriate, how to deploy them effectively and how to move seamlessly among them as a particular lesson unfolds. Kelly (2007) advocates the creation of comprehensive databases of classroom interactions as a resource for preservice and inservice teacher education, while Wong et al. (2006) include aspects of teacher-student talk in their multimedia packages of exemplary science teaching.

NOTES

[1] Hirst (1974) identifies seven distinct Forms of Knowledge: mathematics and formal logic, the physical sciences, the human sciences, philosophy, aesthetics, religion and moral-ethical understanding.

[2] Discussion in this chapter adopts the distinction between argument and argumentation drawn by Osborne et al. (2004, p.998): *argument* refers to the claim, data, warrant and backing that form the substance or content of a particular argument; *argumentation* refers to the processes involved in constructing and presenting arguments.

[3] Lemke (1998) reports that the journal *Physical Review Letters* includes an average of at least one, and often two to three graphical figures (tables, charts, graphs, photographs, drawings, maps, etc.) per page, together with at least two to three (and sometimes as many as six or seven) equations per page. In *Science*, which generally publishes more experiment-based articles and fewer theoretical articles, there tend to be fewer equations.

[4] Moore (2007) reminds us that teachers can also be seriously disadvantaged by the gatekeeping nature of language, especially in the context of writing academic papers and taking examinations for advanced qualifications, or being interviewed for jobs and promotion.

[5] Vygotsky (1962) refers to *three* functions: (i) language as a cognitive tool for processing knowledge; (ii) language as a social or cultural tool for sharing and shaping knowledge amongst people; (iii) language as a pedagogical tool for providing intellectual guidance to another. All three dimensions are considered in this chapter.

[6] Of course, scientists not only argue *from* these standards, they also argue *about* them, and from time to time will *change* them.

[7] Among several others, Harding (1998) has developed this into a generalized argument for the urgency of involving more women, more members of racial/ethnic minority groups and more people from non-industrialized cultures in the appraisal of scientific knowledge and scientific practice.

[8] Several years ago, Sharp and Green (1975) made the provocative statement that selection of teaching and learning methods is driven by the self-interest of teachers in maintaining classroom discipline, and that "the radicalism of the "progressive educator" may well be a modern form of conservatism, and an effective form of social control in both the narrow sense of achieving discipline in the classroom and the wider sense of contributing to the promotion of a stable social order" (p. vii).

[9] In a particularly insightful study, Larson (1995) describes a student (Fatima) who consistently and successfully "plays the game of school" according to a set of invented rules and conventions (e.g., "Don't read the textbook, just memorize the bold-faced words and phrases'). Although they may not understand the material they reproduce in tests, students like Fatima often achieve acceptable grades and can make instrumental use of science education in pursuit of other goals.

[10] "Procedural display" is the term used by Bloome et al. (1989) for the unquestioned and often purposeless habits and forms of interaction in which students and teachers engage every day in order to get them through the lesson with minimum fuss and maximum efficiency.

[11] Boulter and Gilbert (1995) identify three forms of (verbal) argumentation that can be used at different stages of this kind of apprenticeship: in *didactic* argumentation, the purpose is for the teacher to convey an understanding of the argument underpinning a consensus model as quickly and effectively as possible; in *Socratic* argumentation, the teacher asks specific questions and selects particular student responses in order to build an argument for a consensus model; in *dialogic* argumentation, control of the discussion is shared between teacher and students and ultimately among students, with appropriate teacher support and advice.

[12] BELVEDERE was developed by Alan Lesgold and his colleagues (see Cavalli-Sforza et al., 1992).

[13] Mehan (1979) refers to these exchanges as I-R-E; Sinclair and Coulthard (1975) refer to them as I-R-F (teacher initiation-student response-teacher follow-up).

[14] Schwab's accusation was that science is "taught as a nearly unmitigated rhetoric of conclusions in which the current and temporary constructions of scientific knowledge are conveyed as empirical, literal and irrevocable truths" (Schwab, 1962, p. 24).

CHAPTER 9
READING, WRITING AND TALKING FOR LEARNING

As all teachers and all aspiring writers will readily attest, struggling to convert partly formed ideas into articulated speech or into coherent written text helps to clarify, develop and elaborate those ideas. It is reasonable to assume, then, that language-based activities can be utilized in science classrooms to enable students to explore, develop, extend, enrich and reorganize their *personal framework of understanding* and their capacity to construct scientific arguments. Hodson (1998a) comments as follows:

> If learning science is about constructing a complex web of concepts and conceptual relationships into which "official" scientific knowledge is woven, if knowledge becomes meaningful once it is integrated with what is already known in ways that are personal to the learner and if understanding is extended and developed when learners reflect on the relationships between their existing understanding and new knowledge items in the light of current and previous experience, then language can play a clear and important role. What is at issue here is the shifting of emphasis from language as an instrument of teaching to language as a means of learning and a tool for thinking. This shift of emphasis entails a much more active use of talking, listening, reading and writing activities than has been usual in science teaching, especially at secondary school level. (p. 154)

It is almost 40 years since Ned Flanders (1970), a pioneer of classroom observation studies, showed that, on average, two-thirds of each lesson comprises talk, and two-thirds of this talk is teacher talk. Thus, a 45-minute lesson provides for ten minutes of student talk, and in a classroom of 30 children each student will have about 20 seconds for an activity that is arguably the most effective way of building a rich and robust personal framework of understanding. Some 30 years later, Newton et al. (1999) showed that less than 5% of class time is devoted to group discussion and less than 2% of teacher-student interactions involve genuine discussion of ideas. Clearly, we need a shift of emphasis. It is also the case that in question-answer sessions sufficient time is rarely allowed for students to assemble a response before the teacher rephrases the question, asks a different question or asks a different student. More significantly, in the context of the current discussion, no time is allowed for students to elaborate their answers. Another paradox is that as the cognitive demand of the matter under consideration increases, teacher questions often become longer and more complex and student answers get shorter, frequently descending to the level of "Yes", "No" or "I don't know". In other words, students are doing less talking and teachers are doing more in the very situations where student talk would be of most value. In short, traditional question-answer sessions seem to provide the teacher with an opportunity to use language to organize thought, but don't afford the same opportunity to students. Significantly, students working in self-directed groups often allow each other much more time to formulate a response, and to elaborate it, than teachers allow students. Moreover, they also create more opportunities for comment on each other's responses.

Major criticisms can also be levelled at the ways in which many science teachers use reading and writing activities (Glynn & Muth, 1994; Holliday et al., 1994; Rivard, 1994; Osborne & Collins, 2000; Wellington & Osborne, 2001; Bangert-Drowns et al., 2004). In general, it seems that secondary school science teachers do not place much value on reading, devoting no more than 10% of class time to it, and often restricting it to short bursts of activity pitched at what *Learning for Life*, better known as the Bullock Report (Department of Education and Science, 1975), called the *literal* comprehension level. There are two issues to consider: first, texts are often too difficult for students; second, through unfamiliarity with language-based learning methods, teachers do not create enough opportunities for students to proceed to the higher levels of *inferential* comprehension (where readers can appreciate some of the subtleties of the text and can "read between the lines" for implicit messages) and *evaluative* comprehension (where readers are able to judge for themselves the value and quality of the material, and compare its merits with those of other texts).[1] Many of the writing activities that teachers provide are similarly impoverished: students spend large amounts of time slavishly copying notes from the blackboard or from teacher dictation, transferring material verbatim from textbooks and Internet Websites to notebooks, and constructing reports of laboratory activities to a rigid specification that leaves no scope for the expression of personal understanding and feelings.

There are, of course, some very significant differences between spoken and written language. Written discourse is more abstract; often it is more complex and difficult, requiring greater intellectual effort to comprehend it. Because the author is not present, meaning cannot be supported by facial expression, gesture or intonation; nor can the reader seek further information or ask for clarification. With written discourse there are none of the feedback mechanisms through which listeners can signal active listening, comprehension, agreement, appreciation and support, disapproval or incomprehension. On the other hand, an absent author cannot be diverted from a line of argument in pursuit of "red herrings", although *readers* can be so diverted. Written text has more permanence and so has an archival function, though audiotapes and multimedia material containing spoken language are increasingly being used for such purposes. Further, because written material is produced over a longer time period than speech there are more opportunities available to the author for reconsideration of key ideas, error detection and elimination, and other forms of editing that presumably ensure a more detailed and considered product. Vagueness, uncertainty and inconsistency are replaced by clarity, accuracy, coherence and seeming authority. However, although the text is fixed, its interpretation is not. Readers can re-visit it, scrutinize it in relation to new knowledge they have acquired or new experiences they have had. They may, in consequence, re-interpret the text quite substantially, and may even reject what they had previously accepted (or accept what they previously rejected). While spoken language is well-suited to the negotiation of collaborative *action*, written language is more supportive of individual or group-based *reflection*. A dialogue can be established between the reader's thoughts and the writer's words. Text can be used as a "thinking device" for exploring, testing, reinforcing and refining existing knowledge, as well as for developing new understanding. It should also be noted that writers frequently use their own text in this way; thoughts are often clarified and elaborated through writing activities. There is also a very significant difference, says Wells (1993), between the way experience is represented in written and spoken language. Oral language is a more dynamic mode, in which reality is expressed in terms of processes, actions and events; written language takes a synoptic perspective, with reality and experience being viewed in terms of objects, definitions and explanations. The former is a language of negotiation and action, the latter is a language of symbolic representations and relationships, which makes it an ideal medium for providing detailed instructions and information. Rivard and Straw (2000) make some similar points:

> Talk is important for sharing, clarifying, and distributing scientific ideas amongst peers ... the use of writing appears to be important for refining and consolidating new ideas with prior knowledge. These two modalities appear to be dialectical: talk is social, divergent, and generative, whereas writing is personal, convergent, and reflective. (p. 588)

Given these differences, and the fact that spoken language is employed in situations involving more than one person, while text can be used by individuals working alone, text and talk are likely to be used differently by teachers. Talking is commonly used for negotiating, planning, monitoring and evaluating actions, while pre-prepared text is used for providing detailed information and instructions, and student writing is used for recording data and reporting experiences and results. However, the most valuable and productive learning may occur when talk and text are used to complement and enrich each other.

> For it is when participants move back and forth between text and talk, using each mode to contextualize the other, and both modes as tools to make sense of the activity in which they are engaged, that we see the most important form of complementarity between them. And it is here, in this interpenetration of talk, text and action in relation to particular activities, that, I want to suggest, students are best able to undertake what I have called the semiotic apprenticeship into the various ways of knowing. (Wells, 1993, p. 10)

There are three points being argued here. First, students learn science and learn *about* science by talking, reading and writing. Second, talking about text is especially productive. Third, in the same way that students learn to do science by

285

doing science alongside a skilled practitioner, students learn to read, write and talk science by doing so in the company of a more skilled practitioner, who models, guides, criticizes and supports. As Gee (1996) states, it is through reading, talking and writing that learners make the key transition from knowing what is technically correct to having the competence and understanding to say the "right" thing in the "right" way, at the "right" time and in the "right" place. While it is sometimes useful to look separately at activities concerned with talking, listening, reading and writing, as in this chapter, it is clear that they are complementary, interactive and reflexive. Thus, learning is enhanced when students talk and write about what they read, talk about what they write, and read what their peers write. And when a reader not only responds to a text but also provides spoken or written comments, possibly as a stimulus for editing and rewriting, the reader-writer relationships become even more intertwined and productive of learning. Significant learning also occurs at the intersection of language-based activities and hands-on inquiry. When students talk and write about their investigations, observations and findings, they scrutinize what they have done in greater detail, organize their thoughts better, sharpen their interpretations and arguments, identify gaps in their knowledge and understanding, consider alternative conclusions, interpretations and ways of proceeding, and speculate on the significance of what they have done and what they have learned. It is through this combination of talking, reading, writing and *doing* science, and their multiple interactions, that students are stimulated to reflect on these processes, on their learning and its development, and on the nature of science and its wider implications for society and the environment.

LEARNING THROUGH TALKING

If talking is as productive of learning as I am claiming, we need to give students the time and space to talk, and we need to resist the temptation to correct their poorly formed views too early. If students are given time to talk, they may "talk themselves into a better understanding" and, coincidentally, begin to shift the power asymmetries of the typical classroom (Gallas, 1995). However, productive talk doesn't just happen; it has to be carefully planned and sustained by judicious teacher interventions. Ultimately, the success of talk-based classroom activities depends on establishing a classroom environment in which student-student interaction is encouraged and supported. Students need to feel comfortable to listen to the ideas of others, question them, introduce their own ideas, accept criticism, work with others to articulate, modify and develop ideas, and build towards shared understanding. My own research with Canadian students in grades 6 to 12 shows that without considerable groundwork by the teacher, free exchange of ideas and criticism is quite rare and meaning is established only tentatively and hesitantly. On occasions, meaning is imposed by bold assertions from the more confident group members, rather than negotiated within the group. Often, genuine understanding does not develop because the group feels the need to reach early and easy consensus. In other words, their task orientation (commitment to "getting the job done") curtails the time needed for matters to be thoroughly discussed. Depending on students' previous experience of group learning methods, it may be necessary to lay down robust procedural guidelines. Too often, this is not done. Indeed, teachers rarely provide students with explicit guidance. In consequence, students commonly lack clear, shared understanding of the purpose of many of the activities in which they are engaged, and are unaware of the criteria by which their performance will be judged by the teacher. They are often confused, unfocused, unproductive and apathetic. When explicit guidance is provided, students can be enthusiastic and effective in sharing, criticizing and re-constructing their views and ideas (Barnes & Todd, 1995; Mercer, 1995, 1996). Although Naylor et al. (2007) note some situations in which teachers are better advised to maintain an arms-length approach with regard to organization of discussion procedures, it is beyond dispute that students need to know the appropriate way to *present* an argument, *listen* to an argument and *respond* to an argument; they need to know how to criticize, accept criticism, argue and reach consensus. This form of learning can be catastrophically undermined if students do not have the necessary language and social skills to participate appropriately or if the appropriate affective climate has not been established (Hodson, 1998a). In short, not only must students be sufficiently interested to participate, but there also needs to be a safe, trusting, non-threatening and supportive environment in which all students feel confident to contribute and allow each other sufficient "space" in which to talk, especially when their ideas are tentative and hesitantly expressed. It is important that any feelings of injustice or hurt are dealt with promptly and effectively. Both Solomon (1998) and Boulter and Gilbert (1995) warn teachers against using a confrontational or oppositional approach, urging the adoption of "inclusive language" that (i) encourages sharing of personal thoughts and experiences, rather than demonstrating one's knowledge and skill, and (ii) emphasizes listening, "creating space" for others and acknowledging all contributions.

Barnes (1988) distinguishes two kinds of talk, located at opposite ends of a continuum: (i) *exploratory* talk, through which students articulate, consider and reorganize their ideas (in effect, "listening to themselves thinking", as Thier and Daviss (2002) put it); and (ii) *presentational* talk, through which they report to others on what they currently understand or have recently learned in a more formal way.[2] Exploratory talk occurs during laboratory activities, the confrontation of puzzling phenomena and events, and particularly in activities devised specifically to encourage talking, such as formulating definitions of key concepts, constructing concept maps and other graphic organizers, responding to concept cartoons, discussing poems and stories, comparing rival explanations, addressing common misconceptions and myths, formulating a point of view, building a case for a particular position, using agree/disagree and true/false cards, constructing consequence maps, playing science word games, matching text to diagrams, devising a set of questions to aid other students in gathering information from the Internet, selecting text and visual aids to illustrate an idea, creating a timeline for important scientific discoveries and using predict-observe-explain activities. Wellington and Osborne (2001), Newton (2002) and Staples and Heselden (2002) engage in extended discussions of such methods. More elaborate activities such as jigsawing and snowballing are particularly well-suited to the development of listening skills. There is also considerable exploratory talk associated with preparing for group presentations, During inquiry-based activities, exploratory talk is also used to create a sense of group cohesion and purpose, and to manage and organize the work. It shapes the nature of the inquiry, enables consensus to be reached, establishes the limits of the group's current understanding and identifies those areas in which the teacher's help needs to be sought. Presentational talk can range from simple "show and tell" activities, in which students describe an object or recount an investigation or event to other students, to elaborate multimedia group presentations to the rest of the class, other students (both within the school and outside), parents or invited members of the local community. Learning to speak clearly and concisely in order to convey information, ideas and opinions to others in a comprehensible way is an important aspect of enculturation into science and a key component of education for responsible citizenship. Moreover, all kinds of productive talking are involved in preparing group presentations, especially if there are opportunities to edit, re-work and refine the presentation. Post-presentation evaluation can be a productive time for reflection and consolidation of learning. By reviewing each presentation from the listener's perspective, teachers and students can co-construct an evaluation checklist of points to keep in mind during the preparation of oral presentations, These kinds of activities are also invaluable in providing teachers with insight into the understanding of individual students.

Scientists themselves are involved in a wide range of talking activities: conversations with colleagues, technicians and other scientists, research group meetings, giving lectures, running seminars, talking to journalists, speaking to lay audiences at public meetings, council meetings and schools, debating ideas on radio or television, arguing a case for research funding, making oral presentations at conferences or in job interviews, and so on. Skilled communicators adapt the form and style of their language, and their use of visual images and other non-verbal content (including gesture and body language), to the purpose of the activity and to their perception of the needs, experience and interests of the audience. There are many ways in which presentational talk in the science classroom can be designed to mimic the seminars, debates and conferences of the scientific community, thus playing a significant role in teaching students *about* science and the ways in which scientific arguments are constructed and scientific knowledge is negotiated by practitioners. Other related techniques with huge potential for NOS teaching/learning and assessment include "mantle of the expert", hot-seating, debates, making multimedia presentations, role play and drama (Staples & Heselden, 2002). The immediacy of such activities can be a major stimulus to thought: feedback is immediate, critical questions are asked, ideas from others can be utilized at once as the basis for modifying one's views. When students have to explain, defend their views and answer questions, they develop a deeper understanding and are led to explore and develop their ideas. Explaining something to another person requires the explainer to think carefully about an idea, seek to present it in novel ways, relate it to the listener's prior knowledge or experience, translate it into terms familiar to them, and generate appropriate examples. When students have to convince others of the intelligibility, plausibility and fruitfulness of their ideas, they necessarily evaluate, elaborate and synthesize aspects of their knowledge and understanding, make explicit what otherwise might remain implicit, examine critically their own thinking, clarify concepts and conceptual relationships, identify vagueness, discrepancies, contradictions, gaps in understanding, inconsistencies and so on. They have to paraphrase and find alternative forms of expression, making more extensive use of everyday language; they cannot just use the jargon phrases of the textbook. A key pedagogical issue is whether to provide explicit instruction on these matters prior to the need to use them in specific contexts of scientific inquiry and theory-building, or whether to embed them in those other activities, as the need arises. My own feeling is that the ideal is a judicious blend of both approaches, as discussed in earlier chapters in relation to other NOS issues.

The emphasis in Chapter 8 was, in part, on *learning to argue* – that is, understanding the nature of scientific argumentation and learning how to construct, criticize and evaluate arguments. In this chapter (also, in part), the emphasis is on *arguing to learn* – that is, participating in argumentation activities as a means of developing conceptual and procedural understanding and/or confronting socioscientific issues. In arguing to learn, students' primary purpose shifts from attempting to persuade others of the legitimacy of one's point of view or the adequacy of one's claim to knowledge (the focus in learning to argue) to cooperative exploration of an idea or issue, articulating individual understanding, challenging and being challenged, co-constructing new understanding, and so on. As Andriessen (2007) comments, "arguing contributes more effectively to learning when it is not competitive. If we want to use argumentation for learning, students need to balance an assertiveness in advancing their claims with a sensitivity to the social effects of their argument on their opponents" (p. 446). The key pedagogical question for the teacher is how to foster the negotiation and collaboration that leads to interpersonal and group achievement rather than personal success at the expense of others (Stein & Albro, 2001).

As noted earlier, most teachers make little or no use of talking as a means of bringing about learning. Presentational talk is rare; exploratory talk is rarer still, at least as a systematic and carefully planned learning strategy. Many teachers claim that the pressure of an overcrowded curriculum leaves no room for such time-intensive methods; many more are unconvinced of its educational value, unsure how to stimulate productive talk, uncertain how to manage classroom discussions, unclear about their own role during small group work or unwilling to cede control to students (Yip, 2001; Schwartz & Lederman, 2002; Polman, 2004; Treagust, 2007). If teachers are to make more extensive use of talk-based activities they need evidence of its success and access to tried and trusted strategies and materials. Dawes (2004) describes a project called *Thinking Together*, which sought (with marked success) to provide elementary school students with the talking, listening, questioning and consensus-reaching and decision-making skills necessary for effective group discussion. The project derived from an earlier one, described by Mercer et al. (1999), in which students were issued with a set of ground rules for discussion: (i) all relevant information is shared; (ii) the group seeks to reach agreement; (iii) the group takes responsibility for decisions; (iv) reasons are expected; (v) challenges are accepted; (vi) alternatives are discussed before a decision is taken; (vii) everyone in the group is encouraged to speak by other group members. When the ground rules were followed, the quality of discussion was substantially enhanced, with beneficial impact on students' individual reasoning skills. In contrast, as discussed in Chapter 8, Naylor et al. (2007) state that discussion was more productive when ground rules were *not* issued by teachers. Rivard's (2004) research with low-achieving grade 8 students showed that activities involving peer talk as an adjunct to writing enhanced their comprehension and understanding of ideas in ecology more than writing activities alone; among high achievers, activities focused on writing (without attendant discussion) were more effective. However, it would be premature to attribute these differences to ability level rather than different patterns of social interaction within the groups. As Kutnick and Rogers (1994) note, there can be all manner of social tensions that inhibit effective group work. Groups can become polarized, cliques can develop, individuals can be marginalized or even harassed by others, and others can become "freeloaders". Interactions are profoundly influenced by the status and power of the participants and their conceptions of themselves, the teacher and other students. Those of high academic or social status tend to speak more frequently and more confidently and assertively, they command more attention, and their ideas are more frequently utilized, both by the students and by the teacher (Bianchini, 1997; Moje & Shepardson, 1998). Hatano and Inagaki (1991) note that effective discussion can sometimes be derailed because some students tend to show solidarity with their friends, supporting their ideas regardless of merit. Varelas et al. (2008) comment as follows: "While some children are seen as 'smart', others are seen as 'popular', and their voice is heard in different ways. At times children are not willing to contradict each other, their ideas are weakly held or the bonds of their friendship are stronger than the power of their ideas" (p. 92). However, they also note that some students may act as mediators, making spaces for those with less power. Kempa and Ayob (1995) report the somewhat counter-intuitive finding that overt involvement in discussion is not always essential for enhancement of understanding. In their post-discussion writing, those who participated as "attentive listeners" often incorporated as many ideas generated during the discussion as did the more active group members. Even group size can be critical: the group has to be large enough to generate ideas and criticism, yet small enough for everyone to feel valued and to have an opportunity to participate without feeling rushed. My own research indicates that a group size of four to six is about right.[3]

According to Kim and Song (2006), group discussion generally proceeds through four stages, regardless of the topic: focusing, exchanging, debating and closing. The first priority is to establish the focus for the discussion and decide on procedures to be adopted. Next, students exchange information, identify gaps in their knowledge and seek to establish appropriate frames of reference. The all-important third phase, in which students criticize the views of others and respond to criticisms of their own ideas, can sometimes be curtailed by the desire to effect closure and signal to the teacher that the task has been completed. It almost goes without saying that the purpose of the activity should be made explicit, groups should be encouraged to remain "on task" (while recognizing the importance of "social talk" in preparing the ground for effective discussion) and time limits should be made clear well in advance. It is important, too, to ensure that groups reach an effective and explicit closure to the discussion, though not a premature one. Too often, discussion tails off, becomes fragmented and unclear, or is abruptly stopped by the bell or by teacher intervention. Above all, there must be *authentic* discussion in which the content of the discussion is problematized in some way, students express their views in their own way, diversity of views is encouraged and the strength of feelings is acknowledged by the teacher. It is important, too, that students are encouraged to see themselves as "authors and producers of knowledge, with ownership over it, rather than mere consumers of it" (Engle & Conant, 2002, p.404), and that they are required to meet Longino's (1990, 2002) directive to uphold certain disciplinary norms in presenting their own views and considering the views of others. However, it is crucial that teachers also provide guidance, point students in the direction of additional data or alternative ideas, introduce new ways of thinking and give guidance on appropriate use of specialized language. In the context of learning *about* science, it is not just *what* students say that it is important, but also *how they say it*. For example, while students' use of familiar everyday language may facilitate a willingness to explore ideas, its continued use can disguise a lack of understanding. Notwithstanding the points made in discussion of the valuable role played by the vernacular, in Chapter 8, it is important to emphasize that teachers need to introduce the distinctive features of scientific discourse in all its diversity of purpose and audience. Appropriate vocabulary is important. As Vygotsky (1962) points out, acquisition of new concepts is acquisition of new words. It is a crucial part of the enculturation process. However, there is also a danger that students may pick up the teacher's vocabulary and linguistic patterns and use them "mechanically" to hide their lack of understanding. A useful tactic is to require at least two very different ways of expressing the same view. As always, it is a matter of striking a balance between too much teacher direction and too little.

Clearly the nature of the stimulus for discussion is key. In a fascinating study with elementary school children, Naylor et al. (2007) found that using concept cartoons as a stimulus for discussion encouraged greater involvement and prompted students to explore the merits of alternative explanations as a way of building consensus rather than seeking to defend a personal position. Even more intriguing is the use of puppets by Simon et al. (2008) to model particular kinds of talk (including scientific argumentation), encourage particular types of verbal interaction, enhance student involvement and motivation, and raise student participation levels. Puppets can take on the role of an expert, giving information and advice, answering questions and modelling "proper" scientific behaviour, or they can adopt the role of someone who needs help with solving a problem, which the students can then provide. Many teachers are reluctant to voice uncertainty, confusion or ignorance; puppets can easily do so, thus enabling students to take on the role of teacher, advisor or scientific expert. Also, as the authors note, "since puppets do not have the status or authority of the teacher they could be particularly useful in situations where children lack confidence in putting forward or defending their ideas" (p. 1231).

READING ACTIVITIES

There are four major categories of learning goals deriving from reading activities: first, students acquire and develop conceptual and procedural understanding; second, they learn about scientific communication; third, they gain insight into their own understanding and how and when it should be deployed; fourth, they are encouraged to think about scientific knowledge and scientific practice more thoroughly and more critically, and thereby are better positioned to participate fully and effectively in societal decision-making. It is with respect to the second point that it is important for students to be given the opportunity to work with a broad range of text types: textbooks, magazines, newspapers and Internet postings, academic papers and reports, publicity leaflets and advertisements, biographical and historical material, works of fiction, and so on.

Textbooks are still widely used in science education, particularly in North America, so teachers need to be cognizant of the difficulties that young readers may face in using such materials. Because texts in science are often very rich in information, make use of unfamiliar terminology, use lots of logical connectives, adopt an impersonal and abstract style, deal with matters that are remote from the everyday experience of most children, and contain many counter-intuitive ideas, they are often more difficult to understand than texts used in other areas of the curriculum. Sutton (1992) wryly observes that many school science textbooks give the impression that they are designed to be read by teachers, rather than by students. In recent years, writers and publishers have made strenuous efforts to produce more attractive, engaging and effective school science textbooks, though the use of motivational devices has sometimes created other problems, as Sutton (1989) remarks.

> Some current books for school science are in danger of losing their clarity as texts in a welter of glossy illustration and other material such as cartoon strips ... What exactly are these books: part worksheet, part inspiration, part wallpaper, trying to cater for several ability levels, but not really meeting any one level in a sustained way? ... Their heterogeneity seems to make them increasingly ephemeral. They invite scanning ... but not engagement. (p. 158)

Presented with a text whose content and knowledge base is unfamiliar, and possibly lacking in motivational appeal, students may: (i) ignore the text altogether and continue to rely on their existing knowledge; (ii) use a "surface processing" approach to extract key words and phrases; or (iii) distort or misrepresent the text to make it compatible with their existing understanding (Roth & Anderson, 1988).

There are three ways of addressing these problems. First, teachers can model the processes of good reading, including the use of the Contents, Index and chapter sub-headings to gain an overview. They can provide advice on how to deal with vocabulary problems, how to read diagrams and charts in conjunction with the text, how to make good quality notes, and so on (Heselden & Staples, 2002). A second approach is to encourage students to learn and adopt one of the several generic reading strategies, such as SQ4R (Thomas & Robinson, 1972), MURDER (Danserau, 1985), KWL (Neate, 1992) and QUADS (Wray & Lewis, 1997). A third approach is to provide many more reading tasks that require students to interrogate the text. By engaging with the text in more active ways, students not only explore and develop understanding about the matter under consideration, they also develop the skills and supportive attitudes that enable them to use text more successfully in the future. In other words, when students are encouraged to regard text as a resource to support discussion, argument and the co-construction of understanding, they are more inclined to develop good reading habits like searching for more subtle levels of meaning in texts and evaluating texts more critically. In reciprocal reading (Palincsar & Brown, 1984), teachers model a particular technique that students can subsequently use on their own to improve their comprehension level. The method is reciprocal because teacher and students exchange roles during their discussions about the meaning of the text. First, the teacher explains that the group will concentrate on particular aspects of the reading task – for example, (i) formulating questions to be answered by the reading, (ii) clarifying parts of the reading that are hard to understand, (iii) summarizing what has been read, and (iv) predicting what the next section of the text will address. The teacher introduces the text by verbalizing the thoughts she/he has on first approaching a new text: What is it about? Why does the author consider it important? What is the main idea? As the students gain experience with the approach and confidence in using it, they can contribute to these framing questions. After both teacher and students have silently read the first section of the text, the teacher talks through the four reading tasks specified at the outset, in effect "thinking aloud" by framing questions ("What did the author mean by ..."), looking for ways to clarify difficult or confusing passages ("You can see from the diagram on page 3 that the author is saying ..."), summarizing what has been read so far ("So what the author is saying is ...") and predicting what the next section will be about ("From what we have just read, what do you think the author will deal with next ..."). The cycle is then repeated for the next section of text. Once they gain confidence, students will begin to make contributions, assist in answering questions, contribute to clarifying meanings, help to reconstruct or revise the summary, and make additional or more detailed predictions. After several such activities, the students take over as teacher and the teacher becomes part of the learning group, making contributions and assisting the "student teachers" when necessary.

Expert readers are distinguished from novice readers by the extent to which they use existing knowledge to make sense of the text, monitor their comprehension as they proceed (through self-questioning), deal promptly with any failure to understand, identify key ideas and evaluate their significance as they encounter them. They also actively search for consistency, coherence and discrepancy, attend to misunderstandings and misconceptions as they become aware of them, reformulate and synthesize knowledge as they read (Pearson et al., 1992). These attributes can be fostered by engaging students in more authentic, active reading: reading for specific purposes, made clear to the learner in advance, and requiring active engagement with the text. Davies and Greene (1984) urge teachers to replace the often rather vague and unhelpful instructions they give students in connection with reading tasks (such as, "Read pages 20–24, there will be a test next week") with much more purposeful *directed activities related to text* (DARTs) that periodically require students to stop, reflect on what they have read so far and attempt to make sense of it. DARTs can be pitched at almost any level of conceptual sophistication, but at all levels it is the specific nature of the instructions and the opportunities to get feedback from other students that is the key to ensuring active, reflective and purposeful reading. The simplest examples include: reading a section of text and discussing with others what is likely to follow it; completing sentences, tables, charts and diagrams from which items have been omitted; identifying important vocabulary and significant points by underlining and highlighting; paraphrasing key arguments; writing definitions; deciding on appropriate titles and sub-headings for lengthy extracts of text; re-ordering scrambled text; "translating" text into diagrams, data tables, flow charts and concept maps; designing posters or logos to represent something described in the text. Wray and Lewis (1995, 1997, 2000) have developed a 10-stage model called *Extending Interactions with Text* (EXIT), in which each stage has a key question: What do I already know about this subject? What do I need to find out? Where can I obtain this information? How can I remember the important points? And so on. Each question is linked with particular teaching and learning strategies (brainstorming, concept mapping, use of writing frames, etc.). Thier and Daviss (2002) describe a strategy they call *Write as You Read*, in which students use coloured markers or sticky notes to identify key words, concepts and passages, and indicate portions of text that they want to remember or do not fully understand. In addition, they write notes to themselves, including questions they wish to raise about the text and definitions of new words. Guidelines to students might include such things as: underline the main ideas, underline the parts you want to remember, make a mark next to parts you do not understand, highlight the parts you find interesting, circle the parts you agree with and (in a different colour) those you disagree with, write notes about what you need to find out, compose questions about material you do not understand, write notes about your thoughts and feelings, write a short summary of the passage, find definitions for words that are new to you and write sentences that include them. Initially, teachers can require that each student makes a certain minimum number of comments or highlights per page and shares their notes, comments and highlighted sections with others. By seeing how others use the method, students can become more comfortable and skilled with the approach. Eventually it becomes routine.

Kalman (2009) reports on a similar approach with undergraduates, which he calls "reflective reading".

Of course, it is important not to over-use DARTs, EXIT techniques and other such strategies; much of their impact lies in their novelty. They can irritate and bore students if used too frequently. As quickly as possible, and wherever possible, teachers should engage students in *authentic* reading tasks that require close attention to text and make more extensive use of alternative text types. Comprehension at the inferential and evaluative levels necessitates an understanding of the ways in which changes in the writer's purpose and intended audience lead to the adoption of different writing styles and the use of different structural conventions. Critical scientific literacy includes the ability to move freely between these different modes, as both reader and writer (see later discussion). Indeed, as argued in Chapter 1, being able to read and listen with understanding, evaluate the nature and quality of an argument, and express one's views clearly and persuasively, are perhaps the most important aspects of science education. Moreover, acquiring the ability to read effectively for further learning is essential to the development of intellectual independence and lays the necessary base for lifelong learning. The ultimate goal, of course, is that students adopt a critical stance towards the material they read, whatever its form and style – a goal rendered even more urgent by the widespread availability of Internet-based information on all manner of topics.

> In various forms, the mass media, teachers, and peers inundate students with assertions and arguments, some of them in the realm of science, mathematics, and technology. Education should prepare people to read or listen to such assertions critically, deciding what evidence to pay attention to and what to dismiss, and distinguishing careful arguments from shoddy ones. (AAAS, 1989, p. 139)

When scientists read articles in their particular specialist fields they analyze new information, evaluate it, and if deemed trustworthy, they incorporate it into their personal knowledge store, and may subsequently use it in their own work. New knowledge claims are related to previous knowledge claims, taking into account the previous work of the researcher(s), their biases and the circumstances under which the work was conducted and various claims established. Mallow (1991) observes that such reading is done with "pencil in hand", as the scientist jots down interesting ideas, records promising new techniques, checks calculations, and notes personal responses and feelings. Norris et al. (2008) make essentially the same point: "They puzzle over the meanings of what other scientists have written; question their own and other scientists' interpretations of text, sometimes challenging and other times endorsing what is written" (p. 770). Interestingly, Tenopir and King (2004) report evidence to suggest that many scientists regard reading as the primary source of stimulation for their research. When scientists read outside their specialist field (for general scientific interest) their needs are, of course, very different, and so their pattern of reading will be different. So, too, when they read in their role of reviewer of research articles for an academic journal, appraiser of a research proposal for a grant awarding body or examiner of a PhD thesis. But in all cases, information in one science communication has to be related to information in other communications, and any inconsistencies among ideas and knowledge claims detected and addressed. What we should be seeking through the school science curriculum is similar capabilities for *all* students. In essence, an ability to apply what they know about scientific argumentation to all three of the genres of scientific communication identified by Goldman and Bisanz (2002): genres for communication among scientists (refereed journals, handbook chapters and conference presentations), genres for popularizing science information (press releases, news briefs, science-oriented magazines and informational Websites) and genres of formal education and instruction (textbooks, lab workbooks, special educational Websites). In reality, most non-scientists (both students and adults) are not skilled in reading these genres. Many are unaware of inconsistencies in what they read, unable to assess reliability of data and detect bias, and only moderately capable of relating what they read to what they already know. In general, they are poor at distinguishing claims from evidence, evidence from conclusions, and beliefs from inferences made from data (Goldman & Bisanz, 2002). Like everything else in science education, critical reading skills need to be modelled and taught, carefully and systematically. Specifically, students need advice, criticism and support in their efforts to distinguish inferences from observations, appreciate differences between justifications and explanations, connect items of information within and across texts, evaluate the validity and reliability of all information used, weigh the rival merits of alternatives, assess consistency and inconsistency, and seek to resolve inconsistencies by gathering further information.

Norris and Phillips (2008) emphasize the iterative and interactive nature of reading. It is iterative in the sense that readers proceed through a number of steps (not necessarily sequentially) as they struggle to deepen and refine their understanding of the text: "Lack of understanding is recognized; alternative interpretations are created; judgement is suspended until sufficient evidence is available for choosing among the alternatives; available information is used as evidence; new information is sought as further evidence; judgements are made of the quality of interpretations, given the evidence; and interpretations are modified and discarded based upon these judgements and, possibly, alternative interpretations are proposed, sending the process back to the third step" (p. 256). It is interactive in the sense that readers use their existing knowledge, attitudes, values, aspirations and experiences to interpret and evaluate the text: "They make progress by judging whether what they know fits the current situation; by conjecturing what interpretation would or might fit the situation; and by suspending judgement on the conjectured interpretation until sufficient evidence is available for refuting or accepting it. The reader actively imagines, and negotiates between what is imagined and available textual information and background knowledge" (p. 256). When these activities are conducted in a group situation, with opportunities to share, compare and contrast ideas and interpretations with others, students extend and deepen their understanding of the relevant science, enhance their NOS understanding and acquire more sophisticated appreciation of the criteria for judging the quality of written texts. Teachers can pave the way for effective group work through a preliminary stage in which the class addresses a common text, or texts, and the teacher creates opportunities and provides emotional support for students to make comments, ask questions, state their ideas and interpretations, express feelings, concerns and problems, identify confusions, note associations and relationships to other ideas and experiences, and so on. Collaboratively, teacher and students engage with text, explore and critique ideas, develop and extend their understanding. As students struggle to express their ideas clearly and coherently, teachers assist by rephrasing, employing technical language, introducing additional ideas, re-directing attention, identifying alternative textual resources and stimulating further reflection. This kind of classroom discourse simultaneously promotes and develops students' thinking, understanding, language use and critical reading skills.

Using multiple texts, including material displayed on blackboards, overhead projectors, posters and computer screens, is especially valuable because it enables connections between the ideas and experiences of different authors to be established. As Lemke (1992) notes: "We can make meanings through the relations between two texts; meanings that cannot be made within any single text" (p. 257). There is considerable advantage in following the lead of Gordon Wells (1990) and extending the notion of "text" to include "any artifact that is constructed as a representation of meaning using a conventional symbolic system" (p. 378). For Wells, this expanded definition includes drawings, diagrams, charts, graphs, algebraic equations and chemical symbols, formulae and equations. It also includes a range of "oral texts" such as poems, rhymes, proverbs, everyday sayings, songs and speakers' recounts of previous events and experiences. Varelas et al. (2006) have expanded the definition further to include orally shared recollections of radio and television programmes, movies, previous classroom discussions and hands-on activities in the lab or in the field, while Howes et al. (2009) provide some useful insight into ways of combining hands-on student-driven inquiry with text-based inquiry. In the sense that it provides a powerful means by which teachers can scaffold students' learning, the creation of learning opportunities by juxtaposition of texts and/or text types, which some researchers refer to as "intertextuality", has many similarities with the notions of hybrid language/discourse and "double talk" discussed in Chapter 8. It is particularly effective in promoting exploratory talk as students struggle to make coherent sense of the phenomenon, event or issue under consideration through the perspectives presented by the various "texts", and in relation to current classroom activities (especially hands-on work), recollections of past conversations, events and experiences, and even "predicted future experiences" and "imagined experiences" in the typology of intertextuality developed by Varelas et al. (2006). Palincsar and Magnusson (2001) report favourably on the role of *notebook texts*, described by the authors as "a hybrid of exposition, narration, description and argumentation" (p. 174), in enhancing conceptual understanding, building on students' hands-on experiences, and better preparing students for conducting scientific investigations and building explanations. Notebook texts include "think aloud" accounts of a fictional scientist's investigation of the topic currently being addressed or shortly to be addressed by students in the lab or in the field, including reflections on its purpose, guiding questions, investigative procedures, methods of collecting, manipulating and representing data, interpretations made, conclusions drawn, arguments constructed, research questions that could not be answered and any new questions that arise.

As they gain experience and competence, students might work in groups to formulate a set of questions focused on a specific text for use by other students, or even read and criticize written material produced by other students. They might prepare a "position statement" or a set of debating points, subject texts to a rigorous appraisal for evidence of sociopolitical bias, sexism and racism, assess text readability for other students or assemble an annotated bibliography for use by other students or parents. Following explicit teaching relating to text organization, voice, use of metaphor and techniques of argumentation, students might be asked to identify the structural features employed by various authors in pursuit of their purposes (see, for example, the insightful writing of Sharon Dunwoody (1993) and Bruce Lewenstein (1995) on scientific journalism). Mallow (1991) provides some valuable insights into these matters and presents useful advice on how to read different text types. He notes, for example, that authors of popular science articles typically emphasize potential practical applications of the research and include information about who conducted the study, but omit information about methods and procedures, the very information that is essential for judging the credibility of the work. So-called "refutational texts" (Chambers & Andre, 1997; Guzzetti et al., 1992, 1993, 1997), which present standard scientific explanations and detailed arguments to refute common misconceptions, can play an important role in sensitizing students to key differences between "authoritative texts" (most textbooks) and "persuasive texts" (academic papers). Awareness of how form, structure and content are related to purpose underpins two crucial elements in the development of critical scientific literacy: the ability to communicate one's ideas concisely and effectively, and the ability to respond critically to the ideas of others.

Many significant aspects of learning science and learning *about* science could be located in an activity in which students collect, display, criticize, compare and contrast samples of writing on science and technology from newspapers and popular magazines, textbooks of various styles, science-oriented magazines, academic journals, Websites, interactive media, cartoons, advertisements and product labels (Wellington, 1991, 1993; Morvillo & Brooks, 1995; Ford, 1998; Heseldon & Staples, 2002; Ratcliffe, 1999; McClune & Jarman, 2000; Wellington & Osborne, 2001; Ebbers, 2002; Dimopoulos & Koulaidis, 2003; Jarman & McClune, 2004; Glaser & Carson, 2005; Elliott, 2006). Such activities might usefully be extended to include television programmes, museum exhibitions, movies and works of fiction (Gough, 1993; Long & Steinke, 1996; McSharry, 2002; Rose, 2001; Dhingra, 2003, 2006; Pedretti, 2004). It is also through activities such as these that students can be exposed to socioscientific issues. Of course, reading or watching media reports of science may be a new experience for many students, and so provision of guidelines may be essential in ensuring that they direct their attention to key aspects of the report, such as use of provocative headlines and illustrations, editorializing, identification (or not) of information sources, omission of alternative views, portrayal of scientists in favourable or unfavourable light, balance, bias, thoroughness of content coverage, and so on. Even a simple checklist of questions can be enormously helpful. For example, who conducted the research and where was it conducted? How was the research funded? What is being claimed? What evidence supports the claim? How was the evidence collected? How was the evidence interpreted? What assumptions are made and what theories are used in arguing from evidence to conclusion? Do the authors use well-established theory or do they challenge such theories? Are alternative interpretations and conclusions possible? What additional evidence would help to clarify or resolve issues? Have other studies been conducted, either by these scientists or by others?

An extensive survey of secondary schools in Northern Ireland by Jarman and McClune (2002) showed that the majority of science teachers who use newspapers do so within an STS or STSE orientation to highlight the link between school science and everyday life. Very few (2 out of 50) reported using newspapers to develop students' critical literacy skills, and most teachers reported incidental and opportunistic use rather than systematic deployment. The most widely cited reasons for not using newspapers in class related to the often inaccurate, misleading, superficial, sensationalist or biased nature of newspaper reports, which I am arguing here is precisely the reason why we *should* read and critique them in class. All but one of the 21 teachers involved in a similar survey in Alberta, Canada, 10 teaching grade 10 science and 14 teaching grade 12 biology, reported that they used newspapers and other media reports in class (Bisanz et al., 2002). Again, the most widely cited reason (62%) was to emphasize STS connections (a prominent feature of the Alberta science curriculum), together with reinforcing curriculum content (50%) and generating student interest (33%). Encouragingly, eight of the 14 grade 12 teachers mentioned that they use them as a means of developing students' critical reading skills, though none of the grade 10 teachers cited this reason. Levinson and Turner (2001) report that teachers in their study found newspaper articles useful when teaching about socioscientific issues, but would welcome more advice on how best to use such materials. Similarly, Gayford (2002) states that teachers regard newspapers as an appropriate source of information, critical comment and opinion on environmental matters, but feel illprepared to exploit them. Like teachers in the survey by Kachan et al. (2006), they are concerned about their own capacity to judge the accuracy and reliability of media reports. Conversely, Halkia (2003) reports that most of the teachers in her sample of 82 primary school and 70 secondary school teachers in Greece put considerable trust in the information found in newspaper reports, and use them to enhance their own understanding. Moreover, they believe that the narrative style, emotional content and rhetorical devices deployed by journalists, especially metaphors and analogies, appeal strongly to students and help to make difficult concepts accessible.

Overall, research paints a pretty depressing picture of the ability of students, at both school and university level, to read media reports with the kind of un-derstanding encapsulated in the notion of critical scientific literacy (Norris & Phillips, 1994; Korpan et al., 1997; Phillips & Norris, 1999; Norris, et al., 2003; Penney et al., 2003). Phillips and Norris (1999) suggest that this is an unfortunate legacy of the over-use of textbooks in school science, most of which adopt an authoritative style. In their desire to illustrate, define, categorize and summa-rize scientific knowledge, textbooks rarely invite readers to debate the evidence, confront the argument, speculate about alternatives or challenge the conclusions (Myers, 1992, 1997). Norris and Phillips (2008) make the following observation: "In contrast to journal articles, in which statements frequently are qualified with hedges, textbooks tend to present statements as accredited facts that require no hedging. In journal articles, photographs, diagrams, graphs, and other illustra-tions serve the function of proof by providing experimental results. In contrast, illustrations in textbooks generally are used to picture and have no argumentative function" (p. 238). Those brought up solely on a diet of school textbooks are ill-prepared to read scientific papers and media reports in a critical way. The analysis conducted by Phillips and Norris (1999) identifies three major student "stances" towards reports of science in newspapers and magazines. In the critical stance, readers attempt to reach an interpretation that takes account of the text infor-mation and how it is presented in relation to their own prior beliefs, sometimes producing a new mental model or representation of the phenomenon or events under consideration. Those readers adopting the *domination* stance allow prior beliefs to overwhelm text information, reinterpreting it (sometimes implausibly) to make it consistent with their existing frameworks and beliefs. In stark con-trast, the *deferential* stance allows the text to overwhelm prior beliefs, resulting in blind (though perhaps only temporary) acceptance of views expressed in the text and implicit trust in the author. Disturbingly, this latter position seems to be the most common among high school students. My own, admittedly informal, research with students and student teachers suggests that adoption of a critical stance towards information accessed via the Internet is even less common than in regard to information gleaned from newspapers and journals. Most readers subject such text to only a shallow and rudimentary analysis, accepting as true or dependable anything that "seems plausible" (a brief summary of the findings of three relevant research articles is included later in this chapter). The situation with respect to school science textbooks is just as depressing. As Goldman and Bisanz (2002) report, many students simply extract and memorize information perceived as important for tests and exams. However, Ratcliffe (1999) shows that with appropriate preparation, students as young as 11 years old can, in her words, "unpick an evidence chain" (p. 1097).

Shibley (2003) describes a philosophy of science course aimed largely at sci-ence, engineering and computer science majors that supplemented readings in philosophy of science with weekly reading of the *Science Times* section of *The*

New York Times (published every Tuesday), the upkeep of a personal journal in which to record their responses to the articles, and weekly discussion groups for free exchange of ideas and argument. The author reports that the stimulus of using the materials, many of which were controversial, enhanced most participants' ability to utilize ideas from the formal readings in philosophy of science in a critical way. Glaser and Carson (2005) report on an interesting project called *Chemistry is in the News*, in which student portfolios of critical readings of media-derived materials are shared within and between classes, and subsequently peer reviewed on-line by students in other countries. In the study by Craven et al. (2002), cited in Chapter 5, preservice elementary school teachers read articles from a wide variety of self-selected publications (ranging from *The Globe* and *The Sun* to *The New York Times* and *Scientific American*) in order to explore their changing views of the distinctions between scientific and pseudoscientific writing. Over the 15-week period, the student teachers did become considerably more critical with respect to the validity and reliability of the various claims made in the articles, with increasing use of notions such as evidence-based conclusions, repeatability, bias, clarity and coherence of argument. Similar enhancement of critical reading skills was evident in a British study of preservice science teachers conducted by Elliott (2006). Disturbingly, many of the student teachers involved in the study reported that reading newspaper articles on science, as required by the preservice course, had been a novel experience for them. Elliott asks: "If even people such as these do not regularly follow scientific developments and debate in newspapers, what hope is there that their less scientifically minded, scientifically educated, or academically successful peers do?" (p. 1260). If we are serious about the goal of universal scientific literacy, and the fostering of responsible citizenship, we have to change current attitudes and practices. The science classroom is probably the best place (perhaps the *only* place) where a firm foundation can be laid for a practice of regular engagement with science in the media that lasts a lifetime.

In recent years, the Internet has become the dominant medium through which the public (including students) access knowledge and information, in all areas and disciplines. Brem et al. (2001) have studied the ability of students in grades 9, 11 and 12 to evaluate information located on Websites of varying quality, including some hoax sites. Despite lots of preparatory work and continuing support by the teachers, students frequently failed to differentiate between the quality of the science and the nature of the reporting and presentation, often equating amount of detail with quality. Students were often unable to assess the accuracy, judge the credibility and evaluate the site's use of evidence to substantiate knowledge claims. Because students tended to rely on common sense as their principal guide, rather than careful analysis and critical reflection, they were too easily seduced by whatever attractive surface features the authors deployed. In a similar study at the grade 6 level, Wallace et al. (2000) found that students usually concentrated on the search aspects of the task and their ability to navigate a range of sites, and neglected to evaluate the quality of the science they located. Often they searched for key words and then slavishly copied the chunk of text in which they had located them into their notebooks. Interestingly, given the context of this book, Lin and Tsai (2008) show that those with more sophisticated NOS views are more thorough and more discriminating in accessing knowledge from the Internet. It is clear that much more research and curriculum development work is essential with respect to this crucial element of scientific literacy.

As a brief aside, it should be noted that young children may find it easier to assimilate new ideas if they are presented in a narrative style - see Bruner (1986, 1996), Egan (1986, 1988a,b, 1989a,b), Strube (1990), Alexander and Jetton (2000), Negrete and Lartige (2004) and Norris et al. (2005) for an extensive discussion. Thus, there is a strong case for utilizing stories from the history of science (Solomon, 2002), biography, autobiography and narrative accounts from the popular science literature (Myers, 1990a), together with a wide range of fiction written for children and young adults (Butzow & Butzow, 1989, 1998; Gertz et al., 1996; Cerullo, 1997; Stannard, 2001; El-Hindi, 2003), science fiction (Gough, 1993; Freudenrich, 2000; Raham, 2004; Weinstein, 2006), comic strips and cartoons (Millard & Marsh, 2001; Vilchez-Gonzalez & Perales Palacios, 2006; Weitkamp & Burnet, 2007), drama, movies and television programmes (Borgwald & Schreiner, 1994; Perkins, 2004; Dhingra, 2006). Sutton (1992) urges textbook writers to reinstate narrative accounts of scientific advances (a common feature some 50 years ago), and to focus on the lives of scientists, their struggles and their disputes with other scientists, as a way of enlivening science lessons. Stories can fascinate us because they reveal inner feelings, intentions and motivations. They are concerned with thinking, knowing, understanding and acting; the characters not only see and hear, they like, love, fear and hate. Through their capacity to stimulate the imagination and evoke an emotional response, well-written stories encourage listeners and readers to participate vicariously in the experiences and share the feelings of the protagonists. There is particular value in what Miles Barker (2002, 2004, 2006) calls "ripping yarns": stories of science and scientific achievement, together with some failures, that are designed to enthrall and motivate students by focusing on the passions and idiosyncrasies that drive scientists, the dramatic incidents that sometimes accompanied scientific breakthroughs, the disputes, petty jealousies and bitter rivalries that can arise, and the ways in which society at large responded to new ideas and discoveries.

> We learn from stories. More important, we come to understand – ourselves, others, and even the subjects we teach and learn. Stories engage us ... Stories can help us to understand by making the abstract concrete and accessible. What is only dimly perceived at the level of principle may become vivid and powerful in the concrete. Further, stories motivate us. Even that which we understand at an abstract level may not move us to action, whereas a story often does. (Noddings & Witherell, 1991, pp. 279 & 280)

In a fascinating article addressing the ways in which science and literature inform and are informed by each other, Cartwright (2007) identifies eight categories of relationship, illustrating each category with several examples suitable for class use: science as a source of images, metaphors and explanatory devices for literary purposes; science as a focus for satire and derision; science as provoking cognitive dissonance and discomfort; celebration of science and scientists; poetry as a vehicle for expressing scientific ideas; the Romantic dismissal of science as cold and inhuman; scientific irresponsibility (Faustus, Frankenstein and Dr Strangelove figures); the question of whether literature or science has ontological primacy – that is, claims by some critical theorists that science is simply another form of text, with no special claims, versus arguments by evolutionary psychologists that literature represents enduring features of a human nature shaped by natural selection. There are enormous possibilities here for addressing, among other things, both the aesthetic and affective impact of science and technology, and the wider sociocultural, economic and moral-ethical dimensions. Also of value is Schibeci and Lee's (2003) review of some recent studies of the images of science and scientists to which students are likely to be exposed outside school. For example, they note that Haynes (1994) has identified a number of common portrayals of scientists in fiction, including the obsessed or maniacal scientist, the otherwise stupid virtuoso or genius, the nerd or social misfit, the highly dedicated and idealistic scientist who sacrifices all else in pursuit of knowledge, the brave hero who struggles to overcome incredible hardships, financial constraints and/or conceptual and procedural difficulties, and the helpless scientist who has lost control over his/her discovery or the ways in which it is used. Movies, comics and cartoons project the same kinds of stereotypes; so, too, do non-fiction books and films, newspapers and television programmes (see further discussion in Chapter 10). The kind of understanding encapsulated in the notion of critical scientific literacy entails the ability to confront these stereotypes, consider the underlying motives of the author(s), and contrast them with more authentic images presented through the curriculum.

WRITING ACTIVITIES

Much of what has been said about the need for active reading can be used to make a case for active writing. Some time ago, Lunzer and Gardner (1979) found that, on average, an 11-year old science student in a UK school spent about 11% of class time engaged in writing and a 15-year old spent up to about 20%. However, at least half of that time was devoted to copying from the blackboard or from dictated notes, and another sizable chunk was spent filling in the blanks in worksheets. Sadly, in many schools, little has changed in the intervening 30 years. It is important not to confuse the physical activity involved in this kind of "closed" writing with the cognitive activity and explanation that underpins more "open" writing tasks. Indeed, active writing in the sense being explored here does not have to entail very much physical writing at all. Constructing concept maps and other kinds of graphic organizers, "instances and misconceptions" tables and glossaries/word banks, for example, involves virtually no writing, yet these methods can be particularly powerful in stimulating reflective and critical thinking, especially in group learning situations (Osborne, 1997). Similarly, in

303

"free writing" and "free association" activities (Elbow, 1973, 1981; Juell, 1985) students are able to set aside the usual concerns with spelling, grammar and sentence structure in order to concentrate on brainstorming ideas with other students. Keeping a journal or an investigator's logbook, as recommended in Chapter 7, is a personally enriching way for students to record their experiences and connect new experiences and knowledge to familiar ones. It is also a powerful stimulus to thought, largely as a consequence of its unstructured form (Shepardson & Britsch, 2001). The traditional emphasis on the technical aspects of producing a "good and clean copy" can be enormously inhibiting, and can divert students from the principal learning goals: acquiring and practising the distinctive forms of scientific discourse; exploring and developing conceptual and procedural knowledge; formulating arguments and justifications for belief; gaining insight into one's personal framework of understanding and learning how to select and utilize appropriate aspects of it for deployment in particular circumstances. While one major thrust of this chapter, as in Chapter 8, is concern that students acquire familiarity with the forms and approved conventions of scientific discourse, I wish to emphasize that too early and too rigorous a concern with these matters can be distracting, and even alienating. Students can become so concerned with the "correct" way of saying something that they do not explore their own thoughts thoroughly. In addition, without the kind of reading activities discussed earlier, students can be so intimidated by the supposed authority of the textbook that they are reduced to copying extensive passages almost verbatim. As Flick (1995) remarks: "when students research topics in the library, the result is a parody of the intended product through plagiarism and paraphrase" (p. 1068). How much more so is this apparent in material that students download from the Internet?

It is common to distinguish three main kinds of writing: *transactional, expressive* and *poetic*. In transactional writing the writer has to be logical and truthful and is required to adopt certain codes and conventions. It is the language for presenting facts and for reporting, arguing and theorizing. Whereas transactional writing is generally explicit and impersonal,[4] expressive writing assumes that the writer, and her or his experiences and feelings, are of interest to the reader. This kind of writing may have very little or no formal structure; it just follows the ebb and flow of the writer's thoughts and feelings, as in a diary or logbook. It is a style of writing that is often discouraged in science, especially in secondary school. However, if we acknowledge that science learning is profoundly impacted by students' attitudes, feelings and emotions (Hodson, 1998a; Alsop, 2005), it would make more sense to create opportunities for students to express them. If personal frameworks of understanding are highly idiosyncratic, and reflect the complexity of each student's network of sociocultural relationships, it would make sense to use a style of writing that permits their exploration and development. These points constitute a case for much more extensive use of expressive and poetic writing in science. Expressive writing may be a much better vehicle than transactional writing for expressing doubts, asking questions and speculating about ideas. It may also have value in reducing some of the barriers that can sometimes restrict access to science for girls and members of some ethnic minority groups. Expressive writing enables learners to explore a new idea or theory against and within a complex web of other ideas, memories and feelings, and to experiment with new relationships and associations. Expressive writing may be essential to the process of trying out and coming to terms with new ideas, that is, to "thinking on paper". Poetic writing, in which meaning may be deliberately ambiguous, allusive, implicit or couched in metaphor, may be a powerful way of exploring feelings and emotions. Too early an insistence on formal writing in science can hinder students' ability to personalize knowledge and communicate their ideas effectively. Learning logs and journals, especially team journals and journals that establish a dialogue between students and teacher, can be particularly effective (Hanrahan, 1999). When writing in their journals, students do not confine themselves to strictly scientific matters. They frequently incorporate other personal experiences, employ metaphor and analogy, make reference to books, movies and TV shows, and so on. This is the process by which they make personal sense of new scientific knowledge and develop strong feelings of self-worth as science learners. It is also a process through which they develop metacognitive awareness. As Richert (1992) notes, exploring personal thoughts and experience through journal writing helps you to "know what you know, know what you feel, know what you do and how you do it, and know why" (p. 192). Aikenhead's (1996, 1997) idea of maintaining a "dichotomized notebook", with "my ideas" (or commonsense understanding) on one page and "subculture of science ideas" on the facing page, may help to focus attention on the linguistic as well as the sociocultural aspects of learning science embodied in the metaphor *science teacher as anthropologist*. As Aikenhead (1996) says, "The task of border crossing is made concrete by identifying it as crossing a line on a notebook page" (p. 29). The compilation of personalized science glossaries or dictionaries has also been suggested by Feasey (1998) and by Wellington and Osborne (2001).[5]

Bruner (1990) suggests that we acquire and grow in our ability to communicate and understand our experience through language by applying narrative structures, including emphasis on human action, sequential ordering of events, taking a narrator's perspective and seeking a resolution or conclusion. Narratives are part of children's language from their very first attempts at substantial communication. Both science and science learning can be seen as narratives. In science, knowledge is accumulated by means of people systematically investigating objects, phenomena and events in the physical world, trying to make sense of their experiences, and arguing for the validity of their views. Thus, scientific discovery is a story. As we learn, our new understanding reinforces, enriches or modifies our personal framework of understanding, enabling us to see and interact with the world and with other people in new ways. Personal development of any kind, and the insight and opportunities it brings, is also a story. It follows that storytelling should be afforded a much more prominent place in the science curriculum. By encouraging students to *write* such accounts, as well as read them (see earlier discussion), we can encourage them to humanize science, express their own views, speculate, hypothesize and predict, and look for connections to other things that they know and have experienced, using a combination of familiar and authentic scientific language. Narrative may be a particularly effective means of assisting border crossing into the sub-culture of science for ethnic minority students because it enables them to highlight and confront differences in worldview and other culturally determined knowledge, experience and values (Bajracharya & Brouwer, 1997). Giving freer rein to students, and enabling them to focus on personal experiences, thoughts and feelings, may be a way of challenging, or at least providing an alternative to, the narrowly masculinist and rationalist accounts of science so commonly presented to students in school textbooks (Martin & Brouwer, 1991; Hildebrand, 1998). What teachers should keep in mind, and assist their students in appreciating, is that language not only relects our thinking (and the values and traditions that underpin it), it also *guides* our thinking. Our choice of language directs our questions, establishes our priorities, fixes the parameters of the discourse in which we engage, and so on. It follows that changing our language can play a key role in changing the image of science we hold and may impact, subsequently, on the kind of science in which we engage. There is a substantial body of literature that documents and critiques the gendered nature of scientific knowledge, scientific language and scientific practice (see Chapter 5) but, as Keller (1992) reminds us, "It is easy to say, and to show, that the language of science is riddled with patriarchal imagery, but it is far more difficult to show – or even to think about – what effect a non-patriarchal discourse would have had or would now have (supposing that we could learn to ungender our discourse)" (p. 48). The science classroom is one place where students can learn and practice the use of inclusive language. Such activities are not just an exercise on "political correctness", they can play a key role in changing students' attitudes, values and political consciousness. Indeed, Kelly et al. (2000) have shown that getting students to write *about* science, and discussing the texts they produce in class, can be an effective way of extending the scope of science education to embrace wider sociocultural, political and economic issues.

> Through discussion centered on writing in science, the course participants … considered the contextual nature of science (e.g., issues of funding, audience, economic and political ramifications), expertise (e.g., considering speakers' roles in framing arguments), evidence (e.g., supporting conclusions with an evidential base) and responsibility (e.g., citizens' role in the use and understanding of scientific knowledge). (p. 712)

Use of metaphor and other literary devices, poetic writing, art work (including cartoons and posters), music, drama and role playing may have a useful role in this context, too, and in creating opportunities for students to address some of the affective, aesthetic and moral-ethical dimensions of science (Gallas, 1994, 1995; Harms & Lettow, 2000; Monhardt & Monhardt, 2000; Watts, 2000, 2001; Feasey, 2001; Goodwin, 2001). Some years ago, Daiute (1997) identified a "form of discourse characteristic of children" that she called the *youth genre* (p. 323). It is, Daiute noted, characterized by playfulness (teasing and banter), exaggeration, emotional intensity and an argumentative disposition. We need to open up the science classroom much more readily than in the past to these youth genres

306

(there are, of course, more than one) and embrace them for their potential to assist students in making sense of the science they encounter in school in ways that are familiar and comfortable. Varelas et al. (2002), DeCoito (2008) and Barton and Tan (2009) all report studies in which non-traditional science writing, including raps, skits, plays and student-written documentaries, have played a major role in motivating groups of students who traditionally under-achieve in school science, are under-represented in science-related professions and are under-served by the education system. Especially important in the context of this book is the effectiveness of this approach in helping students gain a sense of identity as science learners and potential scientists, and assisting them in enhancing both conceptual understanding and NOS knowledge.

As noted several times in this book, an individual student can be a member of several different sociocultural groups: family, friends, school (including particular classes), sports teams and other social groups, groups based on broader defining characteristics such as gender, class, ethnicity, sexual orientation, language preferences, nationality/citizenship, political or religious affiliations, and so on. Functioning comfortably within these groups requires an individual to adopt a particular "identity" (or a series of them), characterized by a distinctive language, code of behaviour, expressed beliefs, values and attitudes. Moving comfortably between the groups entails a reasonable facility in changing identities, that is, being able to "wear different hats" as the social situation shifts. Perspectives from the literature of situated cognition suggest that learning can be thought of as a process of identity formation and identity modification (Lave & Wenger, 1991). As students decide what kind of people they are and what kind of people they aspire to be, they seek to acquire the knowledge, skills, attitudes, beliefs and values needed for effective participation in particular "communities of practice". Competence is acquired through increasingly self-directed levels of participation but with the continuing help and support of mediators whose own knowledge, experience and affiliations are deployed to help students move comfortably across the borders between communities. This view accords well with the science teacher as anthropologist metaphor and the King and Brownell model of disciplinary characteristics adopted in this book. While the processes of identity construction and development are, in one sense, an individual matter, they are also socially situated. An identity has both an "inner life" (in the individual's mind) and an outer manifestation; it comprises our conception of ourselves, the ways in which that conception is presented to the world via our utterances and actions, and the ways in which others interpret those manifestations and respond to them. Our cluster of identities is the principal factor shaping our interactions with others and, in turn, these identities are shaped by our interactions. As students experience the world and interact with other people, their conceptions of the world and other people change, the ways in which they see themselves change, and the ways in which others see them may change. Moreover, the communities to which an individual belongs, or aspires to belong, also change.

An identity, then, is a layering of events of participation and reification by which our experience and its social interpretation inform each other. As we encounter our effects on the world and develop our relations with others, these layers build upon each other to produce our identity as a very complex inter-weaving of participative experience and reification projections. Bringing the two together through the negotiation of meaning, we construct who we are. (Wenger, 1998, p. 151)

Enculturation into science involves an adjustment to one's cluster of identities to incorporate the notion of "self as scientist". This sense of identity as a scientist is constantly re-negotiated through the activities in which students engage and the roles they take on during these activities, that is, through talking, reading, writing, hands-on activities in the lab, experiences in field centres, visits to factories, hospitals, museums, zoos, aquaria, botanical gardens, and so on. As they establish and re-establish their understanding of who scientists are and what they do, students' perceptions of themselves in relation to science and scientists also changes, moving closer or further apart. All this occurs within a social environment of unequal social relations, differences in access to cultural capital and varying levels of support and opportunity, and is impacted by the biased, distorted and stereotyped images of science and scientists sometimes projected by the popular media and by school science curricula; images that, for some students, can serve to make access to the community of science (or the school version of it) seem undesirable and/or difficult or even impossible to achieve. Part of the task of enculturation is to identify and overcome these barriers and bring students' identities and their conceptions of science and scientists into closer alignment (Brickhouse et al., 2000; Brickhouse, 2001; Brickhouse & Potter, 2001; Gilbert & Yerrick, 2001; Moje et al., 2001, 2004; Reveles et al., 2002; Gilbert & Calvert, 2003; Brown et al., 2005; Ford et al., 2006; Rahm, 2007, 2008; Brandt, 2008; Brotman & Moore, 2008; Reveles & Brown, 2008).

This particularly difficult aspect of border crossing can be assisted through language-based learning experiences. It seems, for example, that students often read text as story, subconsciously regarding the author as the scientist behind the storyline (Abt-Perkins & Pagnucci, 1993). With appropriate encouragement, students can begin to see themselves in that role, identifying personally with the actions and struggles of scientists and the contexts in which they work. Re-inforcement comes, of course, through writing personal accounts of scientific investigations, both real and imagined. Personal stories provide much better opportunities than objective, third person accounts for students to explore, develop and consolidate their new understanding, make sense of new experiences, reflect on their learning progress and explore their sense of belonging within the scientific community. "It helps us to find out what we are currently thinking when we tell a new story, what we used to think when we tell an old one, and what we think of what we think when we hear what we ourselves have to say" (Schank, 1990, p. 146). Reis and Garvão (2007) describe how students' NOS views and images of science can be revealed, explored and developed by encouraging them to write science fiction stories and using the characters and events portrayed as the basis for confronting common stereotypes and misunderstandings, especially the myths about science and scientists projected by the popular media. Barton (1998, 2003) and Rahm (2007) advocate an approach in which students compile oral histories of scientists and their work, and/or engage in conversations with local residents about the impact of science and technology on their lives, as a means of stimulating reflection on one's position within the world of science.

This is not to suggest that transactional writing has no place in science education. Quite the contrary! Transactional writing promotes lucid and orderly handling of information and ideas; it permits ideas to be manipulated, juxtaposed, compared and contrasted. In the form of short expository exercises – note making, summarizing, explaining, analysing, paraphrasing, comparing and contrasting, formulating questions, elaborating, criticizing, and so on – transactional writing is an excellent way of focusing attention on discrete items of knowledge. In the form of extended essays, it is a valuable way of synthesizing ideas and reworking views. Staples and Heselden (2001) identify eight basic "non-fiction text types" with which students need to become familiar: *information, explanation, instruction, persuasion, discussion, analysis, evaluation* and *recount* (presentation in narrative or story form).[6] Scherz and Oren (2006) describe some very encouraging results with respect to increasing awareness of the nature of several of these categories as a consequence of engaging grade 8 and grade 9 students in an investigative journalism project focused on what they observed on visits to research labs, factories and workshops. In similar vein, Simpson and Parsons (2009) argue that informal science experiences can play a key role in enculturation and identity transformation because the cultural requirements and expectations of informal sessions are less demanding and less rigid than those of the classroom. In consequence, students may feel more at ease and may participate in science in ways that are more natural and more comfortable for them. Much depends, of course, on whether it is possible to overcome the fairly widespread view of museums, art galleries and science centres as institutions of social and intellectual privilege (Hein, 2000).

AUTHENTICITY, PURPOSE, AUDIENCE AND STYLE

Writing in a genre students have chosen for themselves is a powerful way for them to explore their own ideas, in the sense that they clarify, critique and extend their thinking as they write. But critical scientific literacy also demands familiarity with the community-approved writing style and with other forms of scientific writing. Further, students learn how to read written material (of all types) more critically by writing it themselves. In short, all three styles of writing (transactional, expressive and poetic) are important, but for different purposes. What students need to know, and this constitutes a crucial part of enculturation, is when a particular style is appropriate. They also need to know how to construct these different kinds of text.

Klein (1999) describes four interacting and overlapping ways in which an individual may approach a writing task.

— *Spontaneous Utterance* – writers spontaneously generate knowledge as they write. In other words, the act of writing shapes thought and makes tacit understandings explicit.
— *Forward Search* – writers elaborate and transform the ideas they initially express through on-going analysis of their texts as they carefully and systematically convert mental models and ideas into words and/or symbolic representations, note inconsistencies and contradictions, make appropriate revisions, and develop meaning.
— *Genre Orientation* – knowledge of the structure and key features of representative text types leads writers to generate, elaborate and relate ideas in a particular way, thereby identifying ideas and relationships that were not initially apparent.
— *Backward Search* – writers develop rhetorical goals (relating to audience and purpose) and generate new content to meet them, monitor the internal coherence of their writing, the strength of argument, and the relationships among claims, evidence and warrants.

He notes that expert writers use backward search to "elaborate rhetorical goals of a writing task to accommodate the interests and knowledge of their audience, the personae they wish to project and the formal characteristics of the required text" (p. 243), and further notes that novice writers who are taught to adopt these strategies produce better and more coherent texts. While use of the backward search mode is the major goal with respect to the generation of text to enhance NOS understanding, all four modes have a role in learning through writing. Teachers should also note the strong evidence that students' views of the *purposes* of writing impact substantially on the approach they adopt and the writing they produce. For example, those who regard an essay as a means of presenting facts tend to reproduce or restate ideas from textbooks and Websites, while those who regard an essay as a means of constructing an argument use the opportunity to clarify and reconstruct their understanding (Hounsell, 1997; Ellis, 2004).

Of course, students will not learn to write appropriately and effectively unless they are systematically taught how to do so, provided with opportunities to practice and given support and criticism in their efforts. Even that most ubiquitous of writing activities, taking notes, is a learned skill. Yet science teachers rarely devote any time at all to teaching students how to do it effectively. In consequence, most students, even at undergraduate level, are very poor at it, tending to amass mountains of detail and failing to recognize the one or two key salient points. In the *structured note-taking* approach devised by Thier and Daviss (2002), teachers provide students with a graphic organizer to help them "wrest order from informational chaos" (p. 62). Alternatively, teachers provide a list of questions to which students are required to respond in note form, followed by a brief summarizing paragraph. As students become familiar with the approach, they learn to design their own graphic organizers and write their own questions to fit the particular task-in-hand. In addition, teachers should describe the essential features of each text type, provide good illustrative examples, model their construction in class, and provide students with carefully designed writing tasks (perhaps making use of writing frames or templates) before expecting them to proceed unaided. The writing equivalent of DARTs, including such things as making notes to specific headings or writing frames provided by the teacher, use of spider diagrams (Buzan, 1995), rewriting and paraphrasing exercises, completing an unfinished text, and writing text from diagrams, pictures and tables of data, can play a role in encouraging students to articulate their ideas as well as in enhancing their conceptual and procedural understanding.

Developing a sense of scientific genre (one of the main scientific literacy goals) means exploring the differences between scientific forms of writing and other forms, and noting how communication style changes as audience and purpose change. In many schools, almost all writing in science is produced for the same audience and for the same purpose: the audience is the teacher and the purpose is assessment. Invariably, it entails the writer telling the teacher what she or he already knows. Moreover, the student knows that the teacher already possesses this knowledge, and is aware that the only purpose of the exercise is to enable the teacher to assign a mark or grade. In other words, it is not genuine or authentic communication. Rather, as argued earlier, it is part of the game of school, the game of "getting the right answer" and gaining an acceptable grade. By varying the audience, and by insisting that writing fulfils a genuinely communicative purpose, teachers can ensure that students are better motivated, clearer about expectations and enabled to practice and develop a wider range of writing skills. Suitable audiences include other students, parents or family members, school visitors, an Uncle George figure who invariably "gets the wrong end of the stick" and has to be "put right" (as in the Nuffield science schemes), some section of the public, government agencies, or themselves. The purpose may be to inform, describe or explain, explore ideas, interpret data and make sense of phenomena and events, simplify and clarify complex ideas, persuade, question, express uncertainty, argue a point of view, justify a position, refine and revise an explanation, predict, speculate, express emotions and feelings, apply knowledge to a new situation, address a problem, confront an issue or encourage action. Whatever the audience and whatever the purpose, it is important for the author to have a clear understanding of the likely knowledge, beliefs and values of the readers, and a good grasp of what is likely to motivate them to engage with the text at more than a superficial level. As Ferrari et al. (1998) comment, expert writers not only have a good understanding of rhetorical techniques and access to a rich vocabulary, they also have "a clear appreciation of how one's goal in writing will be received by one's audience" (p. 486). Writing may take the form of a technical report, personal journal or diary entry, field trip notes or laboratory logbook, field guide, "How to" book, TV or film script, fictional story, biography or historical case study, travelogue, letter to a friend, action or protest letter, brochure or newsletter, poster, newspaper or magazine article, Website posting, guidelines and instructions, notes for a debate, questionnaire or interview protocol, poetry, drama or role- play script. As with reading activities, it is important to begin as early as possible with *authentic* authoring tasks, in which students write for a real audience, with a real purpose in mind, and in a style that suits both audience and purpose. Email may be particularly effective in creating a sense of audience (in this case the addressee to whom the message is transmitted) and establishing a clear purpose. Also, because email messages closely resemble verbal interaction they can function as an effective bridge between everyday language and the more formal language(s) of science. Hand and Prain (2002) remind us that in addition to *audience, purpose* and *form,* we should also consider diversity in terms of *method of text production* (individual versus group, use of technology and multimedia, etc.) and *focus* (historical events, NOS, wider STSE considerations, including sociocultural, moral-ethical and economic dimensions, and so on). Klein (2006) adds a sixth consideration: diversity of *sources.* Writers necessarily utilize knowledge sources such as existing literature (including books, journals, Internet websites, multimedia materials, etc.), observational and experimental data acquired in class, personal knowledge (possibly assembled over several years of experience), and so on. Common sense would suggest that use of multiple sources creates enhanced opportunities for criticism, reflection and evaluation.

Sutton (1998) covers most of these dimensions of language use in his advocacy of a school language policy to "recover the human voice and personal expression of thought" (p. 36) by emphasizing four themes: the learner's voice, the *scientist's* voice, the *teacher's* voice and *science as story* – an effective summary of preceding discussion in this chapter. With respect to the fourth dimension, Sutton argues that presenting science as a story has many advantages over presenting it in a more traditional way: "It involves learners and invites them to think "Is this reasonable?" It corrects the idea that science deals only with certainties and admits students into a more genuine discussion of doubt" (p. 37). Its greatest strength, of course, is its familiarity. As Wellington and Osborne (2001) comment: "If the scientific genre is alienating and offputting, and, if we wish to engage children with ideas in science, we should at least offer activities that initiate writing in science in a manner which is enjoyable. Using a familiar genre at least begins the process of helping children to express their thoughts in written language through being personally engaged" (p. 76). Writing stories with a science theme is an approach that has much in common with the techniques advocated by Stinner and Williams (1993, 1998) – see Chapter 6. There is also much of value in engaging students in both the study of scientific journalism and its production (Hallowell & Holland, 1998; DeCoito, 2008). To generate effective text for a science magazine or newsletter, which might include such things as news items, opinion pieces, book reviews and reflective essays, students need to discuss complex issues as simply and concisely as possible, using a minimum of specialist vocabulary. They must engage readers' interest and attention, link abstract/theoretical knowledge to possible practical concrete applications of the science and to the real world experiences of the readers. In developing the text, the author has to organize her/his own thoughts, identify and consult reliable sources, collect suitable quotations, generate illustrative examples, and so on. Through such activities, students read, listen to, and use all four voices (learner, scientist, teacher, story); they enhance their own capacity to read articles intended for the lay public in a more critical way, and become much more aware of the role of writing in the day-to-day work of scientists. Too often, students believe that science is solely about conducting experiments in the laboratory. Through journalism-based activities they can recognize that doing science also involves writing (constructing arguments, submitting a paper, presenting at a conference) and that scientists do lots of other writing, including writing research grant proposals and research reports, composing reviews of research papers, contributing to science magazines, preparing teaching materials, writing notes from their reading of the current research literature as an *aide-memoire* and as a way of exploring and developing their own research ideas and, of course, maintaining lab and fieldwork notebooks. Rice (1998) describes a purpose-built scientific writing course in which undergraduate science majors are introduced to a range of scientific writing, including (i) scientific autobiography, (ii) description and subsequent explanation of a principle, process, phenomenon or piece of apparatus, (iii) scientific arguments, and (iv) investigative reports. His workshop-oriented approach is pitched at two levels: what he calls the "global" level (dealing with overall organization, focus, scope, emphasis and meeting expectations) and the "local" level (focusing on grammar, word choice, sentence length and structure, and the need for brevity and clarity). The purpose of the programme is to familiarize students with written scientific discourse and develop their capability for engaging in it, which entails recognizing the kinds of claims that are made, how they are advanced and substantiated, what rhetorical techniques are employed, when literature references are required, and so on. Understanding the conventions of scientific writing, and deploying them successfully, also entails gaining insight into some of the differences that exist among the sub-disciplines of science (Kelly et al., 2008) and into the effective use of graphs, tables, equations, digital pictures, maps and other inscriptional devices (Greeno & Hall, 1997; Wu & Krajcik, 2006).

Authentic authoring in a group learning setting provides rich opportunities for the co-construction of meaning. Not only are students learning how best to communicate with their audience, they are also exploring what they already know and what they need to know before they can prepare a coherent product. Hence, there are important learning experiences at all stages of the authoring process: planning, producing and evaluating. In planning, students have to assess what they already know, select what is relevant to the purpose, seek additional knowledge and understanding, organize and clarify their ideas, establish relationships among them, express them in appropriate form for the anticipated audience, ensure clarity of expression, attend to coherence and consistency of argument, and so on. These matters are in constant review throughout the production stage and, of course, are at the forefront during the evaluation stage. Thus, cooperative writing encourages students constantly to question, challenge and seek alternative perspectives. It involves them in exploration of ideas, solving problems and making decisions. In other words, it creates opportunities for what Bereiter and Scardamalia (1987) refer to as the "knowledge-transforming" mode of writing, as distinct from the "knowledge-telling" mode. The knowledge-transforming mode is essentially what Klein (1999) refers to as "backward search" and can be regarded as problem-solving within, and interaction between, a *content problem space* (including relevant facts, concepts, theories, methods, data, etc.) and a *discourse problem space* (including genre and discourse knowledge). Problem-solving in the content space involves recalling, selecting, relating, constructing and evaluating content to support rhetorical goals (persuading readers of the validity and appropriateness of a knowledge claim, for example). Problem-solving within the discourse space relates to matters of text construction, word choice and nuances of phrasing. Mature writing, say Bereiter and Scardamalia (1987), is characterized by the ease with which the writer moves between the two spaces as the act of composing the text stimulates her/him to re- evaluate items in the content space. Thus, content can be altered as particular language for composition is deployed in the discourse space and the writer reflects on what has been written. When modified or new content and new ideas are introduced into the content space they demand new forms of expression. And so on. As Galbraith (1999) notes, new knowledge results from the writing activity itself, not from some previously defined learning goal. Well-designed writing activities constitute both *learning to write* and *writing to learn*: students learn and practice the techniques and strategies of good writing while they explore, critique and develop their understanding by writing about the same topic, issue or idea for different audiences and using different genres (McGinley & Tierney, 1989; Hand et al., 1999, 2004a; Kelly & Chen, 1999; Klein, 1999, 2000).

Like most teaching and learning activities, writing tasks are too often under strict teacher control, whereas the major learning benefits accrue when students have control and make their own decisions, in collaboration with others, about *what* to write and *how* to write. Of course, students still need guidance and support from the teacher and, crucially, continuing instruction concerning the techniques of writing and the ways in which they can be deployed for particular purposes. Without an adequate and developing background knowledge of such matters, effective choices cannot be made; without regular critical feedback, significant progress will not be made. Writing does not necessarily improve with practice unless that practice is accompanied by guidance, exemplification, criticism and support. Part of this continuing instruction, guidance and support should include specific instruction in backward search and knowledge transforming strategies, together with opportunities to read and criticize both good and bad examples of text, preferably written for a variety of purposes and audiences.[7] Armstrong et al. (2008) report that students left to their own devices tend not to adopt such strategies, retreating to a knowledge-telling mode, especially when confident of the knowledge they possess. In what could be regarded as a perfect summary of the foregoing discussion, Hand et al. (2004a) state that four important conditions must be met if writing is to function effectively as a means of learning: (i) the writing task should focus on conceptual understanding and require students to articulate, elaborate and justify that understanding; (ii) the task should constitute authentic communication in the sense of having a purpose and a target readership meaningful to the students; (iii) students should be provided with support and guidance, but not to a level that preempts student initiative and sense of ownership; (iv) planning activities should engage students in purposeful backward and forward search of their developing texts. Chinn and Hilgers (2000) also note that students learn from writing when there are frequent verbal and written interactions between students and teacher, emphasis on collaborative writing, clear guidelines for writing, and opportunities to write for a range of audiences.

Rijlaarsdam et al. (2006) report very encouraging results from a series of activities in which grade 9 students wrote laboratory experiment manuals in physics for use by their peers. The manuals, which included both experimental directions and theoretical explanations for the anticipated observations and measurements, were then read, used and evaluated in action by their peers, with all lab activities and discussion videotaped. The original authors used the videotapes as feedback for revising the manuals and composing letters of advice on how to write laboratory manuals for use by other would-be authors. Couzijn and Rijlaarsdam (2004) have also successfully deployed this approach in coaching students in the construction of argumentative texts. Both studies lend support to earlier findings that *observing* other students while they are engaged in authentic reading and writing activities can sometimes be more productive than actually practising those activities oneself (Couzijn, 1999; Braaksma et al., 2001, 2002, 2004). Peer editing assignments would seem to hold considerable potential for providing constructive feedback on the quality of student-written texts. However, Chinn and Hilgers (2000) report that many of the female undergraduates with whom they worked on a language-rich programme for civil engineers disliked this activity because, they said, it placed them in a "double bind": the aggressive and competitive element of peer reviewing, seen as a necessary part of being a good scientist, conflicted with the socially constructed gender stereotype of the nurturing woman ("being a good girl", as the authors describe it). The authors report that the women did not decline to participate in the peer review activity, or protest about its inappropriateness, even though it made them uncomfortable, thereby furnishing yet another example of the ways in which the institutions of science and technology continue to oppress and discriminate against women.

In many ways, authentic writing is like authentic scientific inquiry. First, students learn to do it by doing it, alongside a skilled practitioner (the teacher) acting as guide, critic, facilitator and supporter. Second, it is a fluid, dynamic, interactive and reflexive activity. It is both untidy and unpredictable. However detailed the plan, it seems that as soon as words are put on paper they are subject to criticism and evaluation that may lead to new ways of expressing ideas, new ideas or even a reconsideration of the plan or the underlying purpose of the text. Every "writing move" (act of text creation) changes the situation in some way, so that there is a constant movement back and forth between the various components of the text, just as there is an interplay in scientific investigation between experimental design, data collection, interpretation of the data, construction of the argument and statement of conclusions. This, of course, is where the word processor is of such inestimable value, enabling even extensive changes to be made quickly and painlessly, and radically changing students' perceptions about revising, redrafting and editing, if these tasks are presented in a positive light by teachers. Computer technology also opens up the possibility of engaging in cooperative writing ventures with students in other schools, possibly in other countries, perhaps in the form of a regular exchange of newsletters on scientific matters. The *Science Across the World* project (British Council, 1997) provides just such an opportunity and one, moreover, that has enormous potential for politicizing the curriculum through a focus on socioscientific issues.

Of course, enculturation into science also involves learning to "read and write" in other modes of symbolic representation. As Lemke (1998) comments, scientists "combine, interconnect, and integrate verbal texts with mathematical expressions, quantitative graphs, information tables, abstract diagrams, maps, drawings, photographs, and a host of unique specialized visual genres seen nowhere else" (p. 89). Effective use of chemical and mathematical equations, diagrams, 3-D models, charts, graphs and pictures are the most immediately relevant, and little more needs to be said on this topic beyond expressing the hopefully uncontroversial view that students need extensive opportunities to work with and construct all styles of scientific communication, bearing in mind the warning issued by Hand et al. (2009) not to attempt too many modes of representation at the same time. Further, if the ability to read multimodal text is a key aspect of critical scientific literacy, and if we believe that engagement with multimodal text and "translating" between modes contribute substantially to both students' conceptual and procedural understanding and to their ability to construct and evaluate scientific arguments, then we should also use multimodal assessment tools to allow students to express their understanding in ways *they* consider appropriate (see Dana et al., 1992; Rennie & Jarvis, 1995a,b; Wandersee, 1999; O'Byrne, 2009).

In recent years, rapid advances in information and computer technologies have created additional opportunities (and their associated demands and problems of comprehension) in the form of organization tools such as databases and concept mapping software, dynamic modelling tools, knowledge construction tools such as hypermedia, and conversation tools (online discussion, email and the like) (Jonassen & Carr, 2000). Such technologies have radically changed the language of science and the language of science education. This more complex, multimodal, computer-based representation of science creates new problems but also new opportunities for exploration and learning (Kress et al., 2001; Kozma, 2003; Schnotz & Lowe, 2003; Alvermann, 2004). Manipulation of computer-generated text enables students to incorporate mathematical and chemical symbols, equations, diagrams, graphs, sound effects and moving images of various kinds, almost at a stroke, and to trawl the Internet for suitable supporting materials. Experience in creating multimodal text helps to build the capacity to access, read and evaluate such material, itself a crucial aspect of critical scientific literacy. Research in this area is not extensive, though what little there is suggests that students may be more adept at manipulating multimedia material than are their teachers (Buckley, 2000). It is also increasingly important for students to be literate with respect to the increasingly sophisticated techniques used by advertisers and documentary film-makers. They might learn much about these techniques and about science by designing a logo, drawing a picture, assembling a collage, writing an advertising jingle or shooting a short video. What is crucial is that students learn to move freely between these different ways of representing knowledge, are sensitive to their strengths and weaknesses, and acquire the skill and confidence to detect and evaluate whatever implicit messages they may project. Of course, if students can learn to manipulate digital information and images, then so can others - a point that serves to raise major issues of trust and highlights the fact that all text and all images seek to address the reader/viewer in a particular way. Being able to read the sub-text, identify the underlying agenda and detect the embodied values are essential elements of critical scientific literacy.

As I argued in Chapter 1, one of the principal reasons for affording learning about science a much enhanced priority in the science curriculum is its centrality to the politicization of students and their ability to address socioscientific issues in a critical and socially, ethically and environmentally responsible way. It should not be forgotten that much of the scientific knowledge and research findings informing consideration of socioscientific issues is located at or near the cutting edge of research and, therefore, is not likely to be included in traditional sources of information like textbooks and reference books. In these circumstances, media reports and Internet websites are very important, raising key issues of media literacy and Web literacy. Being "media literate" means being able to access, comprehend, analyze, evaluate, compare and contrast information from a variety of sources, including technical reports, television programmes, newspapers, websites and so on, and use them judiciously and appropriately to synthesize one's own detailed summary of the topic or issue under consideration. It means recognizing that the choice of particular language, symbols, images and sound in a multimedia presentation can all play a role in determining a message's impact, perceived value and credibility. It means being able to ascertain the writer's purpose and intent, determine any sub-text and implicit meaning, detect bias and vested interest. Students who are media literate understand that those skilled in producing printed, graphic and spoken media use particular vocabulary, grammar, syntax, metaphor and referencing to capture our attention, trigger our emotions, persuade us of a point of view and, on occasions, by-pass our critical faculties altogether. Experience in combining informational text with language specifically chosen to engage readers' attention, surprise or shock them, incite anger, generate sympathy or sway their thinking, is invaluable in understanding the ways in which the media seek to manipulate public opinion. Students can learn much about media techniques by playing the devil's advocate, for example, by writing short articles to endorse opinions they do not hold or views they actively oppose. They might learn to detect bias and distortion by engaging in it, for example, by writing text using only data, statistics, examples and "expert testimony" that are favourable to a particular viewpoint, and ignoring all other. Some kind of media checklist might be extremely helpful in helping to build up students' critical reading skills. Who is the author (or speaker) and what is the author's purpose? Is the work sponsored by anyone or by any organization? Who is the audience? How is the message tailored to that audience? Is the information complete? Are there proper citations of the sources of information? What techniques and what language are used to attract and maintain audience attention? What is assumed or left implicit? Are there any discernible underlying values and attitudes? What information and points of view are omitted? And so on.

Space precludes a more detailed consideration of the ways in which writing can be deployed to enhance students' understanding of science concepts and theories, the processes and procedures of science, and wider STSE and sociohistorical issues. Readers wishing to explore the matter further, including the tricky question of how such activities should be assessed and evaluated, will find much of value in Sutton (1992, 1996), Rivard (1994, 2004), Hand and Prain (1995, 2002, 2006), Prain and Hand (1996), Levine and Geldman-Caspar (1997), Rowell (1997), Hand et al. (1999), Hanrahan (1999), Klein (1999), McKeon (2000), Rivard and Straw (2000), Boscolo and Mason (2001), Yore et al. (2003), Prain (2006), Yore and Treagust (2006) and DeCoito (2008).

THE SCIENTIFIC PAPER

The centrality of written papers in scientific practice suggests that "official" scientific writing should play a key role in learning about science. In addition to writing standardized lab reports detailing the research question(s), procedure, results, interpretation and conclusions, students should be encouraged to keep an investigator's logbook (see Chapter 7). By comparing and contrasting the two accounts students will become sensitized to the rhetorical nature of scientific argument and the "sanitized" way in which research is presented for critical scrutiny by the community of scientists. This is the same argument used earlier in the chapter for contrasting scientific text with journalistic writing, narrative, advertising copy and other forms of communication.

The usual mechanisms for dissemination, scrutiny and evaluation of scientific research include technical reports (to research funding bodies), conference presentations and scientific papers published in refereed journals. It is at the stage of preparing these documents that the inquiry has to be fully rationalized, fuzzy thinking clarified, data cleaned up and chains of rational argument more clearly established. Scientific papers are not faithful historical records of the conduct of the inquiry. Rather, they are rational reconstructions for the purpose of persuading others of the validity of one's claims to knowledge and, sometimes, of the inadequacy of rival views. Functionally, the components of a scientific paper or technical report are intended to do the following: (i) make a case for the relevance of the investigation and its results in the light of current knowledge; (ii) provide sufficient detail about the investigative method to facilitate methodological evaluation and possible replication of the study; (iii) present the data or results of the investigation; (iv) provide an interpretation of the data that leads to specific knowledge claims; (v) justify the argument leading from data to conclusion in a way that anticipates and erases specific doubts that otherwise might be raised;

(vi) identify and acknowledge other specific and minor doubts that might be raised against the study's claims (see Suppe, 1998, for a detailed analysis of the structure of scientific papers).

It is important for students to recognize that the argumentative structure of a scientific paper is not a reconstruction of the scientist's thought processes, and certainly not a faithful recapitulation of the events of the investigation. The account is expected to be devoid of emotion; there is little if any reference to the social context in which the research was conducted. In multi-authored papers there is rarely any indication of who (among the authors) did what, and certainly no reference to technicians who may have been responsible for much of the data collection. Rather, the purpose is "to present knowledge claims and support them with sufficient explicit justification to enable discipline members and gatekeepers to evaluate whether to accept those claims and admit them into the discipline's domain of putative knowledge" (Suppe, 1998, p. 384). Severe space limitations demand that every paragraph, even every sentence, diagram and table, makes a significant contribution to establishing the author's claims. Sometimes doubts can be expressed or limitations in research design acknowledged, though the underlying purpose is to head off criticism by signalling the researcher's awareness of all problematic aspects of the research and stating the ways in which they have been resolved. Impersonality is a characteristic of modern scientific writing, although earlier publications did not eschew the voice of the author, and even nowadays not all editors insist on it. As Halliday (1996) points out, Isaac Newton's *Treatise on Opticks* (published in 1704) includes many passages such as the following: "I held the Prism in such a Posture, that its Axis might be perpendicular to that Beam. Then I looked through the Prism upon the hole." And one of the most famous papers of the past 60 years begins with the words: "We wish to propose a structure for the salt of deoxyribose nucleic acid (DNA). The structure has novel features which are of considerable biological interest" (Watson & Crick, 1953). Although editors of some contemporary academic journals allow authors to use the Introduction to describe and critique previous research in the active voice, it is still usual to insist on the passive voice for the Methods section and Results section.

Because scientists often develop very high levels of skill using particular instruments and apparatus, they tend to regard evidence generated by these means as more convincing than evidence acquired differently, no matter whether that evidence arises from their own investigation or someone else's. Clearly, then, evidence that will persuade those outside the research group may differ from that which is likely to persuade a group member. In writing a paper, all possible rebuttals and counter-arguments have to be considered and dealt with. Reference is made to the views of well-known researchers, previously published papers, established theories and accepted experimental procedures in an effort to support the case and weaken any opposing cases. Visual representations (especially tables, graphs, flow charts and diagrams) and devices such as formulae and equations are assembled as a means of illustrating and underpinning the main argument, which is "constructed like some military strategy, an ambush without an escape route.

Each time that a reading of the results different from that of the authors could possibly be made, the departure is barred by an adequate argument" (Bastide, 1990, p. 208). Many of the distinctive features and conventions of scientific communication, including the use of the past tense and passive voice, were noted in Chapter 8. Scientists write in this way because they are required to do so by journal editors, academic referees and thesis advisors, who in Ziman's (2000) words, "police the literature and censor communications that do not conform to its conventions" (p. 39). The intense pressure to publish in the premier journals has created a situation in which scientists routinely send every one of their research papers to an elite journal, even though they may be more appropriately located elsewhere. Consequently, editors of these journals report increasing numbers of rejections, appeals and complaints (McCook, 2006; Wong & Hodson, in press). Competition to establish priority in a research field increasingly drives scientists to publish "preliminary findings" in the form of letters, research notes and conference proceedings, many of which are never followed up, and to exchange "electronic preprints" with other key researchers in the field. There is increasing interest in the notion of *open publishing*: editors post papers on the Internet and issue an open invitation to other scientists to review the paper, debate the results and conclusions, contribute additional or contradictory data, and so on. Ziman (2000) expresses alarm at this prospect: "The epistemic quality of academic science depends very much on the existence of a public archive of historically dated and explicit research reports by named authors, who must take personal responsibility for their claims. This function cannot be performed satisfactorily by a collection of texts that are continually being updated and revised" (p. 259).

Much can be learned about scientific argumentation and the rhetorical features of scientific papers from a careful study of selected research publications (Janick-Buckner, 1997; Muench, 2000; Smith, 2001; Yarden et al., 2001; Kuldell, 2003; Falk et al., 2008). In addition to gaining familiarity with the language and structure of scientific communication, students are exposed to the intricacies of research design and the complexities of data interpretation, and learn something about research priorities within a discipline, competition for research funding and the lengthy timescale of most scientific investigations. They see that investigative methods vary quite substantially between the sub-disciplines of science and that each academic journal has its distinctive "house style". They also realize that research often raises as many questions as it answers. However, primary research literature can be quite intimidating, even for undergraduates, and students will need considerable guidance and support if they are to benefit from reading it. Guidelines to encourage focused, critical reading might take the form of a series of questions: Why was the study conducted? What techniques and experimental procedures were used to collect the data? What results were obtained? What conclusions did the investigators derive from these data? Were the conclusions justified? How has the research enhanced our understanding? Could the inquiry have been conducted differently, resulting (possibly) in different outcomes? What should be done next? Jigsawing, with each "expert group" taking responsibility for a particular section of the paper, is especially effective. Choice of papers is crucial. Some articles are well-suited to learning content, some are good for studying research design, and others for looking critically at argumentation. Falk et al. (2008) argue that biotechnology is a particularly appropriate field because although the research is often "cutting edge", the scientific content is reasonably understandable for students, at least in upper secondary school. Biotechnology often raises important social, cultural and moral-ethical issues and may connect to work being conducted locally, with the possibility of fieldwork, site visits, visits by scientists, and so on. As with choice of practical work, the old adage "Horses for courses" applies. Papers employing simple and direct language are ideal, especially those with good visual support. Papers should be chosen to provide variety of subject matter and research design, and simplicity of research design – ideally, a one-step, easy-to-follow investigation. It could also be argued that students should be exposed to both good and less good examples of scientific writing, especially at undergraduate level, where students might also (eventually) choose their own papers for discussion. For school students, it will almost certainly be necessary to modify the paper, reducing conceptual complexity and slimming down the data tables, while retaining the basic structure and style. Adaptation for school use may also need to involve addition of material, as well as deletion. In particular, the Introduction may need to be extended with background knowledge that the authors have omitted or merely cited by reference because they assume it is already well-known by other researchers. Details of material to which the author alludes in literature references may need to be included. The Discussion section may need to be extended to aid comprehension. Yarden et al. (2001) advise the inclusion of a Concluding Paragraph to summarize the key points of the article and point out the significance of the work in relation to our current knowledge.

Baram-Tsabari and Yarden (2005) report that when parallel groups of Israeli high school students were presented with virtually identical biological content in the form of either (adapted) primary literature *or* secondary literature (popular science writing), there were no observable differences in the students' ability to summarize the main ideas in the texts. However, as might be expected, students using the primary literature developed a better understanding of the nature of scientific inquiry, while those using the secondary literature achieved greater comprehension of the content and exhibited fewer negative attitudes towards subsequent reading tasks. This research adds further weight to my oft-repeated argument that teachers need to decide well in advance of a particular lesson whether its primary emphasis is to be on learning science (acquiring specific conceptual understanding), learning *about* science or *doing* science.

As argued already in this chapter, *writing* scientific text is arguably one of the most effective ways of acquiring the kind of understanding encapsulated in the notion of critical scientific literacy. Scientists themselves write primarily to inform and persuade other scientists, thereby establishing their position within the community and documenting their claim to specific knowledge production. However, in writing academic papers, and in attending to critical feedback from reviewers, it is reasonable to suppose that they clarify and develop their own thinking, and that these experiences will enhance the quality of both their writing and their research.

321

It is also reasonable to suppose that novice scientists acquire expertise in writing scientific text by reading lots of published and draft papers, and by working under the direction of their supervisor or research team leader on co-authoring tasks. Interestingly, from a questionnaire-based study with 17 practising scientists in an American university, Yore et al. (2002) conclude that scientists generally perceive writing as "knowledge telling" rather than "knowledge building" or "knowledge transforming" (Bereiter & Scardamalia, 1987). Their understanding of written discourse is tacit (rather than explicit), narrow in scope and strategy, structured in accordance with a very traditional view of the nature of scientific inquiry (though their own NOS views were not always traditional) and written almost exclusively for an audience of like-minded people (and certainly not for lay people). None of the scientists in Yore et al.'s (2002) study regarded writing as a means of exploring and developing their own scientific understanding. However, in a subsequent study with 19 scientists, using open-ended questionnaires and semi-structured interviews, Yore et al. (2004) found some interesting differences. Most significantly, in the context of this chapter, two thirds of the scientists expressed the view that writing, reviewing and revising a text *does* help them to clarify and enhance understanding of relevant scientific ideas: "Team research activities, coauthoring and the publication of scientific findings in refereed journals provide opportunities for convergence of diverse views, expertise and insights, intense negotiations and construction of shared understanding" (p. 344). Many described personal writing strategies characteristic of forward search, genre-oriented and backward search (Klein, 1999 – see above). The two scientists interviewed by Yore et al. (2006) stated that the processes of submission, review, criticism and revision associated with publishing in a peer-reviewed journal improved both the quality of the text and their understanding of the significance of the science. However, even though reviewers and editors often play a key role in helping authors to meet the criteria of a "good paper" and properly addressing the needs and expectations of the journal's readership, their contributions do not entitle them to co-authorship status. Indeed, it is still not universal practice for authors to acknowledge this very influential and formative work (Ford & Forman, 2006). In a detailed study of the enculturation of novice scientists (mainly graduate students) via co-authorship teams, Florence and Yore (2004) found many common procedures and activities, especially with regard to appraising data, planning and drafting, reviewing the literature and revising the text. Significantly, they found that these steps are "neither strictly sequential nor discrete" (p. 651). Writing was seen as a fluid, dynamic, interactive and recursive process. It was also regarded as flexible, in the sense that different team members assumed the lead role on different occasions (because of their particular expertise, insight or experience). All research subjects commented that the co-authoring experience had promoted their capacity for critical thinking, deepened their scientific understanding and enhanced their writing skills. There is good reason to believe that similar experiences would be of similar benefit to school-age students. Ritchie et al. (2008) describe an interesting activity in which a group of grade 4 students worked with their teacher, acting as co-author, editor, guide, learning resource and facilitator, to create what the authors call an "eco-mystery", titled *Ocean Action: An Adventure in Beachtown*, in which science content concerning marine environments was embedded in a narrative based on unravelling the complex factors that negatively impact endangered species. This is clearly an approach with enormous potential; it warrants further research and development.

It almost goes without saying that students left to their own devices are very unlikely to write an ideal persuasive text, scientific paper or lab report. They need extensive guidance, support and encouragement. Once again, I consider the 3-stage apprenticeship model to be ideal, culminating in students using the techniques of scientific argumentation to pursue their own persuasive goals relating to personal investigations and engagement in socioscientific issues. Brillant and Debs (1981) advocate an approach that starts with a critical class-based discussion of completed lab reports, both good and bad, provides a number of "model reports" for different kinds of investigations, and gradually withdraws guidelines and checklists as students take over increasing authorial responsibility. Although this approach is primarily aimed at conventional lab reports, there is no good reason to prevent it being adapted to the production of persuasive texts that mimic scientific papers for academic journals. In parallel with earlier discussion, it can be theorized that the composition of such a text requires students to move to and fro between a content problem space and a discourse problem space. What students choose to report will be the outcome of discussion within the content problem space relating to their conceptual and procedural understanding – in particular, the purpose of the investigation, its theoretical significance, the rationale for the particular techniques deployed, the significance of the data and the chain of argument leading to the conclusion. How they report it is the outcome of discussion within the discourse space concerning the structure and linguistic conventions of scientific papers and lab reports (usually regarded by students as a set of *rules* that must be followed) and the expectations of readers. The quality of the paper/report depends on the skill with which the authors marry these two areas of concern. What is reported is not what was done, but what is considered important to report. In illustrating this point in the context of undergraduate reports of designated lab activities, Campbell et al. (2000) note that when students changed the prescribed procedures, the nature of the change and reasons for it were rarely reported, indicating quite clearly the students' poor grasp of the scientific and educational purpose of the report and their inability to move comfortably between the two problem spaces. As discussed earlier, it is the ability of mature writers and skilled scientists to move between these spaces in a reflexive and informed way that results in a scientifically significant, comprehensible and linguistically appropriate text.

In the content space, the problems and beliefs are considered, while in the discourse space, the problems of how to express the content are considered. The output from each space serves as input for the other, so that questions concerning language and syntax choice reshape the meaning of the content, while efforts to express the content direct the ongoing composition. (Keys, 1999, p. 120)

As they construct, critique, modify and reconstruct the paper/report through several drafts, scientists constantly move back and forth between considerations of the purpose and design of the inquiry, data collection methods, interpretation of results, conclusions, chains of argument, justifications and warrants, and considerations of effective and appropriate ways of representing their work in the most favourable light. Keys et al. (1999) and Hand and Keys (1999) have attempted to give students some awareness of these processes in the form of a *Science Writing Heuristic* (SWH), which seeks to replace the traditional 4-part lab report (purpose, methods, results, conclusions) with a focus on knowledge claims, evidence, descriptions of data, observations and methods, and reflections on the students' own thinking – in particular, how it has changed as a consequence of the investigation. Traditional lab report headings such as Hypothesis, Procedure, Observations, Results, Discussion and Conclusion are replaced by questions such as "What is my question?" "What did I do?" "What did I see?" "What can I claim?" "What is my evidence?" "What do others say?" and "How have my ideas changed?" Research indicates that deployment of the SWH approach has been very encouraging. For example, Rudd et al. (2001) found that using SWH enhanced first-year undergraduates' understanding of certain topics in chemistry, while Keys (2000) reports that all the grade 8 students in her research group were enabled to write acceptable persuasive reports of their laboratory experiences, with nine of the 16 students showing substantial evidence of enhanced scientific understanding as a direct consequence of the writing activity. In addition, Hand et al. (2004b) have shown the positive impact of the SWH approach on grade 7 students' ability to write about the "big ideas" embedded in a topic studied via lab activities. Research by Hohenshell and Hand (2006) indicates that, in general, grade 9 and grade 10 female students using the SWH performed better on conceptual test items than males using SWH and females writing traditional lab reports; both female and male students using the SWH performed better than students writing traditional reports, regardless of whether the report was written for the teacher or for their peers. They also developed greater metacognitive awareness of the writing task. Research by Akkus et al. (2007) with students in grades 7 to 11 show that low achieving students benefit more from the implementation of an SWH approach, with regard to content learning, than high achieving students, whose performance was largely unchanged. Similar findings are reported at the teriary level by Poock et al. (2007). Grimberg and Hand (2009) rationalize these results, and their own findings with grade 7 students, by arguing that the SWH scaffolds the thinking of lower-achieving students to the cognitive level at which high-achieving students function without assistance. It should also be noted that the SWH has considerable potential as an aid to *reading* scientific articles and research reports, and can be used by teachers as a template for designing lab activities (see Hand, 2008, for an extensive discussion of such matters). Martin and Hand (2009) sound a note of caution: the shift in pedagogical orientation required by the SWH approach is not always easily achieved, and even with the support of an experienced SWH user it can take a long time for students to be comfortable with it.

Writing for a real audience and with a real purpose is the crucial transformation stage in apprenticeship, the point at which students have achieved intellectual independence. Unless the task is authentic, or is seen by students as authentic, the genre can be distorted. Perhaps the time is ripe for science teachers to give very serious consideration to the publication of a science journal that is both written *for* school-age students and written *by* them. Some years ago, the Imperial Oil Centre for Studies in Science, Mathematics and Technology Education at the Ontario Institute for Studies in Education did publish such a journal, *Eye on Science*, edited by Isha DeCoito and Maurice DiGiuseppe. Although the articles were of exceptionally high quality, and were well-received by the student audience, the economic climate was such that the venture was financially insecure and was abandoned after two years of publication. If this venture were to be resurrected it might make good sense to recruit a number of scientists to act in an advisory role to the student researcher/authors. For example, O'Neill (2001) and O'Neill and Polman (2004) describe projects in which high school students developed email-mediated mentoring relationships with volunteer scientists, who gave advice on the scope, focus and design of students' personal inquiries and acted as a critical audience as students formulated their arguments and wrote their reports.

NOTES

[1] Wood et al. (1992) make essentially the same distinctions in their categorization of levels of comprehension into *literal, interpretive* and *applied*. Similar categories underpin the four stages of literacy postulated by Kintgen (1988) (see Chapter 1).

[2] As noted in Chapter 6, Mercer (1995, 2000) identifies three distinctive categories of student talk in collaborative groups: *disputational* talk (an exchange of opposing views in which disagreements are emphasized), *cumulative* talk (students build positively but uncritically on what others say) and *exploratory* talk (students engage critically with each other and support each other in collaborative reconstruction of ideas). Mercer et al. (1999) define exploratory talk as "that in which partners engage critically but constructively with each other's ideas. Statements and suggestions are sought and offered for joint consideration. These may be challenged and counter-challenged, but challenges are justified and alternative hypotheses are offered. In exploratory talk, knowledge is made *publicly accountable* and *reasoning is visible in the talk*" (p. 97, emphasis in original).

[3] Bennett and Dunne (1995) state that the optimum group size is eight, although my experience with groups of this size is that they often fragment into smaller sub-groups or foster a very formal kind of "turn taking".

[4] When the goal is more widespread dissemination of science, Myers (1990b) comments that iconography is often more important than words in "getting the message across", citing the double helix model of DNA, the equation $E=mc^2$, and the lavish illustrations in E.O. Wilson's (1975) *Sociobiology: The New Synthesis*, as examples of particularly effective communication.

[5] As discussed in Chapter 8, Halliday (1993, 1996) and Gee (2004) have questioned not only the value but also the feasibility of attempting to "translate" understanding from the specialized language of science into everyday language, or into more expressive and poetic language, or vice versa, on the grounds that the conceptual structures of science are too complex, too abstract and too far removed from everyday experience – a view that I categorically do not share.

[6] Gallagher et al. (1993) identify five categories, though they point out that a lengthy text will be likely to involve more than one function-form relationship: *narrative* (the temporal, sequenced

discourse found in diaries, journals, learning logs and conversations), *description* (personal, common-sense and technical descriptions, informational and scientific reports, and definitions), *explanation* (sequencing events in cause-effect relationships), *instruction* (ordering a sequence of procedures to specify directions, such as a manual, experiment guide or recipe) and *argumentation* (logical ordering of propositions to persuade via an essay, discussion, debate, report or review). Ohlsson (1996) opts for seven categories: describing, explaining, predicting, arguing, critiquing, explicating and defining.

[7] Writing conferences (Graves, 1983) and "assisted monologue" (Bereiter & Scardamalia, 1987) can also be invaluable in developing the critical stance towards text that is key to scientific literacy.

CHAPTER 10
THE HISTORY, TRADITIONS AND VALUES OF SCIENCE

Adoption of the King and Brownell disciplinary characteristics approach to teaching *about* science demands the inclusion of history of science. A key element in understanding the community of scientific practice is appreciation of its history and traditions: knowing where our current body of scientific knowledge came from, who played the key roles in its development, what motivated them to pursue these particular concerns at that time, and what social, economic and moral-ethical issues were raised. Studying science as a tradition also entails consideration of the values that underpin the conduct of scientists, both as individual practitioners and as a community faced with decisions about priorities for research and development. Science is an important part of our cultural heritage and knowing something of its history and development is an important part of what it means to be a cultured and educated citizen (Arons, 1988; Jenkins, 1989; Millar, 1993, 1996; Bybee et al., 1991; Wellington, 2001).[1] As Monk and Osborne (1997) argue, teaching science and teaching about science in an ahistorical way makes as little sense as studying Shakespeare without discussing the sociocultural context of Elizabethan England. In short, science education is seriously deficient if it fails to provide satisfactory answers to the following questions in ways that are interesting and accessible to *all* students:

- Where did science come from?
- How did it get that way?
- Who were the significant people in its development?
- Why did they choose to study these issues and problems?
- In what ways did the discovery or invention impact the scientific community and society in general?

In Ernst Mach's (1960) view, those who know the history of science are much better positioned to evaluate current developments: "They that know the entire course of the development of science, will, as a matter of course, judge more freely and more correctly of the significance of any present scientific movement than they, who, limited in their views to the age in which their own lives have been spent, contemplate merely the momentary trend that the course of intellectual events takes at the present moment" (p. 8). A closely related argument is that presenting students with history of science, especially case studies of important events in 20th and 21st century science, can be an effective way to bridge the "gap" between the two cultures of arts and sciences so lamented by C.P. Snow (1962), thus simultaneously broadening the perspectives of future scientists and engineers and ensuring that future politicians and business leaders have some basic understanding of science, scientists, scientific practice and scientific developments. As Conant (1951) asserted, more than half a century ago, *all* citizens should have a robust understanding of the relationships among government, industry and commerce, education, and issues of scientific research and development, in order to develop the skills needed for addressing current policy and potential future developments in a more critical way. Studying episodes in the history of science enhances students' capacity for (i) serious and critical consideration of matters relating to environment, health, natural resources and the like, (ii) informed confrontation of ethical issues, and (iii) making an appropriate and carefully considered response to changes brought about by technological innovation.

Providing common ground for arts/humanities and science specialists to study is not without its problems. A proper appreciation of the historical development of major ideas in science requires both substantial scientific understanding and substantial historical sensitivity, and questions must be asked about whether either party would be sufficiently well-equipped to engage in serious study. Added to which is the problem that many science specialists are depressingly narrow in their interests and may be reluctant to engage seriously with material that does not contribute directly to their mastery of science content, while arts specialists may exhibit an initial aversion to studying anything relating to science because of previous unhappy and unsuccessful experiences with scientific content. In crude terms, science specialists may have little interest in history and non-science specialists may have no more interest in *history* of science than science itself. Schwartz (1995b) draws on 30 years of experience with such courses to proffer some sage advice on course structure and content, and to remind us that developing and teaching successful inter-disciplinary courses requires "imagination, self-confidence, a willingness to learn and to take risks, a fairly thick skin, and some resiliency with which to respond to the inevitable failures" (p. 1040).

Of course, approaches to the history of science can take many forms. Some are restricted to particular events and episodes, and may focus on conceptual and procedural issues, or on social, cultural, economic and political circumstances. Some take the broad historical sweep advocated by Misgeld et al. (2000) in the approach they refer to as the "historical-genetical approach", where the purpose is to use history of science to inform debate about current issues and to encourage students to consider "alternative forms of organizing production and society ... a different science in an [sic] transformed society" (p. 335). Stinner et al. (2003) have developed the notion of "units of historical presentation" (vignettes, case studies, confrontations, thematic narratives, dialogues and dramatizations) as a useful tool in discussing the appropriateness of particular approaches for addressing specific learning goals relating to science content, NOS knowledge and moral-ethical issues. Of course, the *motives* for including history of science in the curriculum will influence the kind of history studied. When the primary motive is to assist conceptual understanding, teachers may be inclined to interpret scientific history from a 21st century perspective, frequently ignoring superseded ideas or regarding them as seriously misguided – an approach that has been termed "Whiggish" history (see Chapter 7). Such accounts may seriously distort history by criticizing scientists of the past for failing to meet modern standards of data collection and experimental design, and may ridicule those scientists for being un- aware of some of our taken-for-granted modern knowledge. Scientific knowledge is portrayed as emerging in simple, straightforward and predictable fashion from scientists' struggles to solve theoretical rather than practical problems, with one scientific development leading directly and inexorably to the next until the current position was reached. When viewed from a contemporary theoretical standpoint, those who initially opposed the ideas that led directly or indirectly to current views are regarded as incompetent or perverse, while those who accepted and developed them are credited with exceptional foresight. It quickly becomes a kind of "villains and heroes" approach to scientific history. Because the curriculum fre- quently promotes the view that there has been *one* universally applicable method in use since the outset of the scientific endeavour, theories that were once accepted but were subsequently falsified are regarded as the outcome of scientists' errors. By evaluating early scientific investigations using modern criteria and standards, Whiggish historians of science ignore altogether, or put little emphasis on, any appraisals made at the time about whether an experiment was appropriate and reliable, whether a theory was intelligible and whether an argument was convinc- ing. The fact that a belief does not stand up to critical scrutiny now, in the 21st century, does not mean that it was irrational to hold such a view at the time it was proposed.

> Unencumbered by modern notions of rationality, scientists of the past had to make decisions about the acceptability of contemporary theories by *their* crite- ria rather than by *ours*. We may have the hubris to imagine that our theories of rationality are better than theirs (and they may well be), but how does it help *historical* understanding to evaluate the cogency of past theories utilizing eval- uative measures which we know were not operative (not even in approximate form) in the case at hand. (Laudan, 1977, p. 129)

We can only understand the past from the perspectives then current; the in- tellectual standards of the present sometimes have little relevance to a proper understanding of events in the distant past. Accounts that are more faithful to historical circumstances and sociocultural influences require consideration of the various by-ways, diversions, false paths and dead ends of science, recognition that science is frequently complex and uncertain, and acknowledgement that not all inquiries are fruitful. A "proper" history of science attends to both the theoretical and practical problems that motivated new ideas and new procedures, and takes cognizance of the metaphysics and worldview prevailing at the time. In these respects, "time slices" rather than "vertical history" may be more appropriate for the curriculum – that is, consideration of the range of ideas current at any one time, how these ideas were generated and how they were received, interpreted, modified and utilized in further work. However, simply noting that an idea was widely accepted at a particular time does not necessarily imply that scientists were sound in their judgement. Taking theory appraisal and acceptance at face value and invoking social influences as the major causative factor can lead to relativism, leaving no way of judging whether one theory was better than another, no way of judging whether, by the standards in force at the time, a particular judgement was well-founded. Thus, there is a problem: How can we judge episodes in the past without imposing and invoking present-day criteria and standards? Laudan's (1977) response is that scientists behave "correctly" (i.e., rationally) when they choose in favour of those ideas effective in solving the most urgent practical problems and the most significant theoretical ones. This simple criterion, he argues, is both trans-temporal and trans-cultural, and to do other than invoke it is to act irrationally.

Perhaps what suits the overall purposes of science education best is a careful delineation of the key conceptual issues set within a rich and lively account of the prevailing sociocultural context. But, as Klein (1972) comments, what the science teacher wants of historical examples is precisely what the history teacher would reject as poor history: the one delights in a "sharply defined single insight" (p. 12), the other in "the rich complexity of fact?" (p. 18). History of science in the curriculum is necessarily selective in extent, scope and style. Sometimes consciously and sometimes unconsciously, selected episodes in the history of science can easily become distorted history or distorted science. Great scientists are often portrayed as "super heroes", possessed of almost superhuman intellectual capability, extraordinary determination and startling resilience in the face adversity, while scientists who made mistakes are sometimes seen as fools. More worryingly, history of science can be used (and has been used) to promote the view that science is the exclusive province of Western males and to foster a belief that the science produced is infallible, authoritative and not to be questioned by mere students. In contrast, the focus on people, social circumstances, false trails and fierce competition between rivals can easily lead to a distorted and over-simplified version of the science, with little acknowledgement or recognition of the complexities of key conceptual, philosophical and mathematical issues.

For a number of reasons, history of technology may be more appropriate and effective than history of science in bringing about awareness of the sociocultural embeddedness of science and technology, the values that create and sustain particular priorities for research and development, the moral-ethical dilemmas arising, and the social, emotional and environmental issues and problems that may arise from deployment of new science and new technologies. The extent to which technology can have social, political, economic and environmental impact well beyond that imagined by scientists and engineers is well illustrated in D.E. Nye's (1990) book *Electrifying America*.

In the United States electrification was not a "thing" that came from outside society and had an "impact"; rather, it was an internal development shaped by its social context. Put another way, each technology is an extension of human lives: someone makes it, someone owns it, some oppose it, many use it, and all interpret it. The electric streetcar, for example, provided transportation, but there was more to it than that. Street traction companies were led into the related businesses of advertising, real estate speculation, selling surplus electrical energy, running amusement parks, and hauling light freight. Americans used the trolley to transform the urban landscape, making possible an enlarged city, reaching far out into the countryside and integrating smaller hamlets into the urban market. Riding the trolley became a new kind of tourism, and it became a subject of painting and poetry. The popular acceptance of the trolley car also raised political issues. Who should own and control it? Should its workers unionize? Did the streetcar lead to urban concentration or diffusion, and which was desirable? Like every technology, the electric streetcar implied several businesses, opened new social agendas, and raised political questions. It was not a thing in isolation, but an open-ended set of problems and possibilities. (pp. ix–x)

It has been noted by many researchers and science educators that despite its enormous educational potential, the historical development of key ideas in science is generally very poorly presented in science textbooks and other curriculum materials, rarely extending beyond a few names and dates, some pictures, and the odd colourful anecdote (Chiapatta et al., 1991; Niaz, 1998, 2000; Pessoa de Carvalho & Vannucchi, 2000; Leite, 2002; Rodriguez & Niaz, 2002; Williams, 2002). Such accounts not only trivialize the sociocultural dimensions, they also misrepresent the nature of scientific inquiry and theory building (Bauer, 1992; Hodson, 1998b; Duschl, 2004). In Schwab's (1962) words, they present science as an "unmitigated rhetoric of conclusions in which current and temporary constructions of scientific knowledge are conveyed as empirical, literal, and irrevocable truths" (p. 24). Indeed, bad history of science in the school curriculum is responsible for reinforcing many of the myths about science, scientists and scientific practice discussed in Chapter 5. It does not have to be that way! When it is done well, history of science can have enormous educational value. For example, Chapter 6 included some discussion of the value of historical case studies, including Arthur Stinner's (1989, 1994) approach via large context problems, in assisting concept acquisition and development. Chapter 7 included a similar discussion of the role of historical case studies in enhancing students' understanding of the nature of scientific inquiry. For young children, as discussed in Chapter 9, stories might prove more effective and more palatable than more formal case studies, especially in their capacity to motivate and foster careful consideration of the social, cultural and affective dimensions of science (Martin & Brouwer, 1991; Roach & Wandersee, 1995; Ellis, 2000; Barker, 2002, 2004, 2006; Solomon, 2002). Drama also has a role to play (Bentley, 2000; McSharry & Jones, 2000; Odegaard, 2003; Stinner & Teichman, 2003; de Hosson & Kaminski, 2007); so, too, simulated dialogues between important scientists (see Chapter 6). Metz et al. (2007) argue the value of "punctuated" or "interrupted" stories, an idea originally developed by Roach and Wandersee (1993) from the work of Kieran Egan (1986). The "interruption" can take the form of teacher clarification of conceptual, procedural or socioeconomic issues, class discussion of interesting or controversial aspects, lab or fieldwork activity, movies, museum visits, and so on. Tao (2002, 2003) reports some modest success in using science stories as a means of eliciting, challenging and subsequently developing students' understanding of NOS (see Chapter 3). His approach was to engage pairs of students in critical reading activities designed to help them extract and co-construct the NOS themes embedded in the stories. Although his approach was not always successful in bringing about more adequate views of NOS, and sometimes served to reinforce students' existing misconceptions, it holds out considerable promise, especially in the hands of skilled teachers who can provide appropriate criticism, guidance and support. Much depends, of course, on the nature of the stories and the implicit messages about science they carry. Klassen (2009) provides some useful advice on researching, writing, using and evaluating science stories, while Milne (1998) classifies science stories into four types, each of which promotes a particular set of assumptions about science and scientists: (i) *heroic* stories, in which a gifted and hard-working scientist single-handedly brings about a major scientific breakthrough; (ii) *discovery stories*, in which scientific knowledge is acquired accidentally, as in the familiar stories about Alexander Fleming and Wilhelm Röntgen; (iii) *declarative stories*, in which scientific ideas are presented as natural, secure and timeless, rather than the outcome of creative endeavour; and (iv) *politically correct stories*, which focus on the work of women scientists, scientists from ethnic minorities, gay and lesbian scientists, and scientists who are physically challenged (such as Stephen Hawking and Geerat Vermeij). Chapter 9 included reference to an essay by Cartwright (2007) on the various ways in which science is deployed in literature and art, and to a discussion by Schibeci and Lee (2003) focused on the images of science and scientists in contemporary fiction, movies and cartoons.

It is important that teachers are aware of the power and potential influence of the implicit messages in the stories they use in class, and their potential for promoting NOS misconceptions and falsehoods through what Allchin (2004b) calls "pseudohistory". Indeed, Allchin (2003) argues that the use of the narrative form *inevitably* leads to distortion of science because of common narrative elements: (i) monumentality – not only are scientists larger-than-life heroic figures, often working single-handedly, but the situations they face and the problems they solve are of immense significance; (ii) idealization – all nuances are laid aside in pursuit of a straightforward, "black-and-white" storyline; (iii) affective drama – stories are made more entertaining, persuasive and memorable by enhancing emotional and aesthetic elements; (iv) explanatory and justificatory narrative – the storyline is always structured to justify the authority and certainty of the scientific conclusion. Careful selection of historical events, judicious teacher intervention, and critical commentary, are essential if the all-too-common stereotypes, myths and misrepresentations of science are to be overcome. For example, attempts might be made to counteract what Stephen Brush (1985) calls "the Marie Curie Syndrome": the portrayal of Curie and other women scientists such as Marie-Anne Lavoisier and Rosalind Franklin as painstakingly accumulating mountains of data through scrupulous and tedious observations, rather than engaging in the intuitive, imaginative and creative activities commonly associated with male scientists, detaching themselves from ordinary social life in their relentless pursuit of scientific data, rather being driven by the passion of male scientists like Galileo, or acting in a subsidiary role to male scientists like Antoine Lavoisier or Francis Crick and Jim Watson.

It is important to ask whether the drive to counter common myths should necessarily lead to case studies that tell the plain unvarnished truth. "Truthful" history of science may be a Pandora's box telling students that scientists do not always behave as rational, open-minded investigators who pursue knowledge through carefully controlled experiments and dispassionate consideration of the evidence. Is a "warts and all" account of science likely to make students more enthusiastic or less enthusiastic about science? Is there some risk associated with an approach that reveals how scientists sometimes steal ideas from others, distort data and cut corners in their determination to establish the primacy of their work? Is the picture of the scientist emerging from "authentic" case studies even more harmful than the stereotyped views so frequently found in textbooks and media representations? When Stephen Brush (1974) asked this question, some 35 years ago, his answer was that an account of the ways in which scientists behave (according to historians) does not always constitute a good role model for impressionable students. Before embarking on the use of historical case studies we should ask a number of questions. Are there situations in which it may be important to project a series of stylized and mythical stories about great scientists because they make the science more memorable and more interesting? Are there other occasions when it is more sensible to "re-write" historical accounts somewhat in order to reduce conceptual complexity and focus on sociocultural issues or, conversely, to focus attention more clearly on conceptual aspects without too much distracting social, political and economic detail? Much depends, of course, on the age and stage of intellectual and emotional development of the students, and on the ways in which different versions of history contribute to the notion of critical scientific literacy. As with most things in education, what is appropriate (and when) is a matter of professional judgement by teachers about the demands and needs of particular educational situations.

In a fascinating account of popular culture in late 19th Century France, Hendrick (1992) argues that fiction (especially the novels of Jules Verne), drama (the plays of Louis Figuier featuring the lives of scientists) and popular or "laymen's terms" writing on science (notably the pamphlets on natural history written by Jules Michelet) were instrumental in building the extraordinarily high esteem in which French science and scientists were held at that time. Predictably and provocatively, Hendrick extrapolates this argument into a proposal to use such means today to raise public awareness of science and enhance general levels of scientific literacy. On a closely related matter, Stinner and Williams (1993) have convincingly argued a case for teachers and student teachers writing their own historical case studies (see also, Stinner, 1989, 1994; Stinner & Williams, 1998). Perhaps there is a case for extending this activity to school-age students – in

333

a sense, engaging them in *doing* history (in this case, history of science). With respect to a particular episode in the history of science, students would need to ask, and seek to answer through research, a number of key questions designed to tease out both science content and NOS issues, together with consideration of the social, political and economic factors prevailing at the time. What was the goal or purpose of the activity? What were the social, cultural and economic circumstances that created a particular concern or interest? What was the initial plan for investigation? What action was taken? What were the outcomes? What happened next? One notable attempt to engage students in *doing* history is described by Kafai and Gilliland-Swetland (2001). Students in grades 4 and 5 used archival and contemporary source materials (official records, manuscripts, photographs, notebooks and specimens), augmented by field trips to a local wetlands area in order to document landscape, flora and fauna through photographs and field notes, to study the work of Donald Ryder Dickey, an early 20th Century naturalist. Of course, the teacher-researchers were fortunate in having ready access to the nearby Donald Ryder Dickey Collection. Given that many archival resources (including original research articles) are now available in digital form and accessible via the Internet, this is an approach that warrants extensive further investigation as a means of acquiring deep understanding of episodes in the history of science. There is much to be gained, too, from interviewing elderly people in the community, consulting newspaper archives and studying old photographs and maps. From such activities, students quickly learn that there is no one fixed story, that all "facts" are subject to interpretation, that historical accounts and interpretation of data carry the assumptions and values of the author, and that good history involves constant citation of the evidence for assertions, all of which is also valuable learning with respect to the nature of science.

VALUES AND PRESUPPOSITIONS IN SCIENCE AND SCIENTIFIC PRACTICE

If one takes the view that scientists are members of a community of practice, and that the ideas of particular scientists will only become accepted as scientific knowledge when they achieve consensus within the community, it follows that all of the influences that impact on people in their day-to-day lives have to be considered as possible or likely influences on the conduct of the scientific community, that is, as influences on the science that is carried out and the scientific ideas that gain acceptance. Some sociologists of science have extended this acknowledgement of social *influences* on scientific practice to a position that embraces social *determination* of scientific knowledge. For simplicity, the argument can be represented as follows. If the ways in which we classify and observe are "conventional" (that is, social conventions that we choose to adopt, for whatever reasons we currently regard as important), if scientific theories are often metaphoric in origin and, therefore, socially and culturally dependent, if science is conceived, conducted and reported in a language constructed to serve the purposes of a particular group, and if knowledge is intimately bound up with human interests of various kinds, then it seems to follow that scientific knowledge is determined by the prevailing sociocultural climate (Barnes & Edge, 1982). In other words, as discussed in Chapter 6, it is a social construct and possibly *no more than* a social construct. Scientific knowledge is "what scientists say it is", for the reasons they choose. It could be otherwise. It has no special status. Science is "no better" than any other knowledge about natural phenomena and events; it is just "different".

Although I would argue strongly against this position, I would also argue strongly against the antithetical position that science proceeds undisturbed and unaffected by the values and interests of its social and cultural context, save for the impact on funding and the establishment of research priorities. What an individual scientist regards as important, puzzling or worthy of attention is a consequence of her/his *personal framework of understanding*. This unique and complex array of conceptual and procedural knowledge, ideas, beliefs, experiences, feelings, values, expectations and aspirations, what Giere (1988) refers to as the scientist's "cognitive resources", will determine the questions that are asked and the problems that are pursued, guide the way investigations are designed and conducted, and influence the way data are interpreted and conclusions drawn. There is no avoiding the fact that scientific practice is located within a cluster of social contexts at the level of research teams, the wider scientific community, and society as a whole. The ideas, beliefs and values prevailing in those social contexts will inevitably impact on scientists and influence their judgements on all manner of things. Immersed in the prevailing sociocultural milieu, scientists cannot remain immune from cultural and ideological influences; science cannot proceed undisturbed and unaffected by these influences. In other words, science is to some degree a product of its time and place, and subject to the values that pertain in the society that supports and sustains it.[2] The focus of scientific attention and, therefore, the subject matter generated is to some extent a reflection of the needs, interests, motives and aspirations of the scientists themselves and of the key decision-makers within the scientific community and the wider society. In Ziman's (2000) words, scientists would not be human if their activities were not, from time to time, "permeated with folly, incompetence, self-interest, moral myopia, bureaucracy, anarchy, and so on" (p. 5). Thus, I categorically reject Richardson's (1984) view that unearthing any evidence of contextual values influencing scientific reasoning necessarily leads to a charge of *bad* science. Values will impact on *all* scientific practice, both good and bad.

History tells us that scientific ideas come from multiple and diverse sources, many of them outside the sphere of science. The cultural context is likely to shape the questions that are asked, the topics that get pursued, the observations that are made and attended to, and what overall theoretical perspectives are likely to find favour within the community. Thus, Darwin's theorizing on evolution in terms of "survival of the fittest" was a consequence of his experiences in the ruthlessly competitive society of Victorian England, and the 19th century science of phrenology (or craniology) was promoted as a rational justification for a society that already believed women and non-Europeans to be intellectually inferior to

Caucasian men (Fee, 1979; Gould, 1981a; Hodson & Prophet, 1986). Because scientists draw ideas from their cultural location, and are strongly influenced by the values that underpin society at any particular time, there is the ever-present danger of bias and distortion (as discussed in Chapter 4). For example, Sandra Harding (1986) argues that generations of male scientists have moved whatever counts as science towards the masculine and what counts as feminine away from science: "[science] is inextricably connected with specifically masculine ... needs and desires. Objectivity vs subjectivity, the scientist as knowing object vs the objects of his inquiry, reason vs the emotions, mind vs body – in each case the former has been associated with masculinity and the latter with femininity" (p. 23). In Keller's (1985) words: "women have been the guarantors and protectors of the personal, the emotional, the particular, whereas science – the province par excellence of the impersonal, the rational and the general – has been the preserve of men" (p. 8). A language necessarily carries evidence of its history and social development. In consequence, some would argue that the very language of science is androcentric, Eurocentric and hierarchically structured, and that it reflects values relating to anthropocentrism, control, exploitation and competition, any or all of which might be a possible source of the antipathy that some women and non-Europeans feel towards science. Sociocultural pressures can function to oppose or even exclude particular lines of research and explanation, while encouraging others. The most strenuous objections to Charles Darwin's *The Origin of Species* did not concern its empirical adequacy/inadequacy but the value-laden nature of its theoretical constructs. Similar problems confronted both Newton and Copernicus.

> What chiefly troubled Copernicus' critics were doubts about how heliocentric astronomy could be integrated within a broader framework of assumptions about the natural world – a framework which had been systematically and progressively articulated since antiquity. (Laudan, 1977, p. 46)

To be admitted to the corpus of approved scientific knowledge, theories have to be socially, culturally, politically and emotionally acceptable, as well as cognitively and epistemologically acceptable. Bloor (1974) comments as follows: "The ideas that are in people's minds are in the currency of their time and place ... The terms in which they think do not emanate from their subjective psyches. They come from the public domain *into* their heads during socialization" (p. 71). Such is the power of day-to-day socialization processes that many value-laden assumptions remain unrecognized and unchallenged. As Rose (1997) points out, because modern science is hegemonic its underlying assumptions appear to be natural and universal. The great and "ultimately damaging achievement" of science, she says, "is to appear as a culture with no culture" (p. 61). Thus, unless there are substantial moral-ethical issues involved, as in recent research in the biological sciences, the priorities and practices of science usually go unchallenged. The words of Robert Young (1987), quoted at greater length in Chapter 4, provide a succinct summary of the situation: "There is no other science than the science that gets done. The science that exists is the record of the questions that it has occurred to scientists to ask, the proposals that get funded, the paths that get pursued ... matters for a given society, its educational system, its patronage system and its funding bodies" (pp.18 & 19). On occasions, sociocultural influences can be so extreme that they lead to misuse of science in pursuit of social and political goals, as discussed in Chapter 4.

By failing to address sociocultural influences, the simple-minded accounts of theory acceptance/rejection commonly promoted through the school science curriculum are insulting to students, and often flatly contradict what they read about real scientists like Galileo Galilei, Albert Einstein, Barbara McClintock, Rachel Carlson, Richard Feynman, Jane Goodall and Jim Watson. What textbooks often seem to omit from their accounts of theoretical development is the *personal* dimension, that is, the ways in which the decisions and actions of scientists are influenced by their worldviews, feelings, attitudes and prejudices. Because of the theory-laden nature of observation (indeed, the theory-impregnated nature of all the processes of science), together with the empirical under-determination of theories, what one "sees" as an observer, the proper conduct of experiments, the adequacy of a theoretical explanation, and so on, are all open to dispute, contestation and modification. Rather than attempting to reduce science to a cold, clinical, de-personalized method, and rather than presenting science as independent of the society in which it is located, we should be emphasizing the ways in which knowledge is *negotiated* within the scientific community by a complex interaction of imagination, experiment, theoretical argument and personal opinion. And we should be promoting the view that science is a theory-driven and creative endeavour, influenced throughout by social, economic, political and moral-ethical factors as they impact on the key decision-makers or what we might call the "community gatekeepers" or "the guardians of the community's store of knowledge". Not only is science a social activity in the sense that society (writ large) determines "the science that gets done", but also in the sense that the rules of scientific procedure and the legitimacy of the "product" are determined by a community of practitioners.

The question for us, as curriculum builders and teachers, is whether these social, cultural and economic factors constitute the main driving force for science or whether they are influences that simply make science a human and, therefore, imperfect but intriguing endeavour. Recognizing that science is a social activity, and that its methods and procedures were established by people and are sustained by authority and custom, is not to say that the scientific knowledge produced is empirically inadequate, socially expedient, irrationally believed or likely to be false. *Rationality* can be retained in our account of science. There is some guarantee that the methods of appraisal we choose to employ produce knowledge that is robust enough to solve empirical and conceptual problems, and has some direct relationship to the actual world. We choose particular methods because they have some objective value in helping us to reach our principal goal of "getting a handle on the nature of reality". The rationality of science is located in careful and critical experimentation, observation and argument, and in critical scrutiny of the procedures and products of the enterprise by other practitioners. It is a community-regulated and community-monitored rationality. Science is socially constructed through critical debate. And those involved in it have a commitment

337

to maintain certain rigorous debating standards. As discussed in Chapters 4 and 8, Helen Longino (1990) makes the point that "it is the social character of scientific knowledge that both protects it from and renders it vulnerable to social and political interests and values " (p. 12). First, she says, there are public forums for the presentation and criticism of evidence, methods, assumptions and reasoning – in particular, conferences, academic journals and the system of peer review. Second, there are shared and publicly available standards that critics must invoke in appraising work, including but not restricted to empirical adequacy. Third, the scientific community makes changes and adjustments in response to critical debate and is clearly seen by practitioners and members of the public to do so. Fourth, the right to submit work for peer appraisal and criticism is open to all practitioners; so, too, the right to publicly criticize the work of others. In short, as Longino (1990, 2002) argues, the critical scrutiny exerted on scientific ideas by peer review and public critique via conferences and journals is the centrepiece of scientific rationality and a guarantee of the objectivity and robustness of the knowledge developed.

> The formal requirement of demonstrable evidential relevance constitutes a standard of rationality and acceptability independent of and external to any particular research program or scientific theory. The satisfaction of this standard by any program or theory, secured, as has been argued, by intersubjective criticism, is what constitutes its objectivity. (Longino, 1990, p. 75)

This organized skepticism, which is primarily the collective responsibility of the editors and referees of reputable science journals, ensures that the accumulated store of scientific knowledge reflects rigorously scrutinized claims and supporting arguments rather than the idiosyncratic opinions and interpretations of individual scientists.

Of course, the way in which the community of scientists exercises this public scrutiny at any one time is subject to a whole range of social, cultural, political and economic factors, and the inclination of individual scientists to be persuaded by a particular argument or to be swayed by particular evidence depends, in part, on their background knowledge, assumptions and values. It is also the case that critical standards can vary substantially from journal to journal. In these respects, as Longino states, the procedures of scientific appraisal are vulnerable to charges of social relativism. The resilience of science in the face of such charges is a consequence of its openness to criticism by individuals of diverse backgrounds, experiences, interests and underlying values. Questions will be asked about the appropriateness, extent and accuracy of the data, how it was collected and interpreted, and whether the conclusions follow directly from the data, and so on. The explanation will be scrutinized for internal consistency and for consistency with other accepted theories. Particular attention will be directed to the background theory and assumptions underpinning the research design, and to the deployment of auxiliary theories and choice of instrumentation and measurement methods. The possibility that these questions may be answered differently by different appraisers, and that different perspectives will be brought to bear in the appraisal process, is the reason for upholding the principle of academic equality and is one of the guarantees of scientific objectivity, because only conclusions that are robust across varying interpretations and differing criteria will survive to become part of the corpus of accepted knowledge. Sandra Harding (1991) refers to this as a shift from *weak objectivity*, based on traditional scientific values of disinterestedness, impartiality and impersonality as guarantors of the validity and reliability of particular evidential support, to *strong objectivity*, based on diverse interpretations of what constitutes evidence, and why. Only conclusions that are robust across varying interpretations and differing criteria will survive to become part of the corpus of accepted knowledge. By seeking to exclude contextual beliefs and values from the appraisal of scientific knowledge, traditional methods fail to recognize the beneficial as well as the injurious effects they can exert on the conduct of science. In contrast, strong objectivity explicitly identifies the role that the researcher, her/his perspectives and interests, and the specific sociocultural context of the research, play in the production of scientific knowledge.

> Strong objectivity requires that the subject of knowledge be placed on the same critical, causal plane as the objects of knowledge. Thus strong objectivity requires what we can think of as "strong reflexivity". Culturewide (or nearly culturewide) beliefs function as evidence at every stage in scientific inquiry: in the selection of problems, the formation of hypotheses, the design of research (including the organization of research communities), the collection of data, the interpretation of data, decisions about when to stop research, the way results of research are reported, and so on. (Harding, 2004, p. 136)

Better scientific knowledge does not result from trying to eliminate subjectivity, and conforming to some spurious notion of objectivity, but from critical consideration of the contextual values that influence or *should* influence the scientific enterprise.

Sharon Crasnow (2008) develops a similar argument in her notion of "model-based objectivity". Modelling always involves values-based decisions and the preferences resulting from those chosen values will be reflected in the model that is built. It is crucial that we direct careful and critical attention to the values we hold, both implicitly and explicitly.

> Since models are tools, like all tools, each is designed for specific purposes. Modeling requires making choices about the world; we focus on the features that we believe are salient to what we want. As a result, what we value is an integral part of the construction of the model. In this way, values are intrinsic to science and to our negotiating our way through the world in general. Model-based objectivity directs us to examine our social values as one of a group of factors directing our choice of characteristics. (Crasnow, 2008, p. 1101)

While model-based objectivity appears to make science relative to interests and goals, charges of relativism can be avoided if we are able to make objective decisions about the values we incorporate. In other words, we can have value commitments and still have objective scientific knowledge if we are rational and

339

objective in making our decision about the values we incorporate. For Crasnow, that decision needs to be in favour of the things we should value for their contribution to human well-being and, I would add, environmental health (a discussion that will be revisited later in the chapter).

> Model-based objectivity turns the problem of science and values on its head. Instead of asking how science can manage to be objective even though values play an intrinsic role in knowledge production, I am claiming that since values do play a role, we should be asking questions about the objectivity of value claims. We should be holding and operating with values that are objectively based in projects which will be better for human beings, allowing humans to achieve goals that are more closely tied to their flourishing. (Crasnow, 2008, p. 1105)

As a set of rigorous *methods,* science embodies values such as orderliness, care and precision, accuracy, meticulous and critical attentiveness, reliability and replicability. The *knowledge* generated by scientists also has to conform to certain values – for example, clarity, coherence, universalism, stability, tentativeness, fecundity or fruitfulness, in the sense of solving problems, making predictions, etc. Elegance, simplicity and parsimony can also be significant factors in gaining support for a theory. As Martin Amis (1995) says, in his novel *The Information,* "In the mathematics of the universe beauty helps tell us whether things are false or true" (p. 8).[3] Similarly, Richard Feynman (1965) remarked: "You can recognize truth by its beauty and simplicity ... When you get it right, it is obvious that it is right. The truth always turns out to be simpler than you thought" (p. 171). In a conversation with Einstein, Werner Heisenberg (1971) argued that the simplicity of good ideas is a strong indication of their truth: "I believe, just like you, that the simplicity of natural laws has an objective character, that is not just the result of thought economy. If nature leads us to mathematic forms of great simplicity and beauty ... we cannot help thinking that they are 'true', that they reveal a genuine feature of nature" (p. 68). Max Born (1924) argues in similar vein when he says that relativity theory was accepted long before supporting experimental and observational evidence became available because it made science "more beautiful and grander". Also commenting on the elegance of Einstein's work, Paul Dirac (1980) states: "Anyone who appreciates the fundamental harmony connecting the way Nature runs and general mathematical principles must feel that a theory with the beauty and elegance of Einstein's theory *has* to be substantially correct ... One has a great confidence in the theory arising from its great beauty, quite independent of its detailed successes ... One has an overpowering belief that its foundations must be correct quite independent of its agreement with observation" (p.44). Elsewhere, Dirac (1963) says: "It is more important to have beauty in equations than to have them fit experiments" (p. 47). Miller (2006) argues that it was Dirac's insistence on beauty at the expense of "facts" that led to the discovery of antiparticles. Finally, in referring to his and Francis Crick's elucidation of the structure of DNA, Jim Watson (1980) reports that Rosalind Franklin accepted that the structure was "too pretty not to be true" (p. 124). More extensive discussion of the role of aesthetic criteria in science can be found in Yang (1982), McAllister (1996) and Girod (2007). Jakobson and Wickman (2008) point out that students (and their teachers) also use aesthetic criteria in making judgements about how best to proceed during scientific investigations, deciding what to include/exclude in their descriptions and evaluations, and how to present their findings.

A new theory is more likely to be accepted when it is consistent with other well-established theories and is less likely to be accepted when it is in conflict with them (Laudan, 1977). Thus, Copernican theory had some initial problems because it was inconsistent with Aristotelian physics, a problem that was solved by Galileo. It seems that many scientists are driven to look for common explanations or common *kinds* of explanations. Holton (1975, 1978, 1981) argues that scientific thought is governed by what he calls "themata": thematic presuppositions or predispositions to think and theorize in particular ways. He identifies 50 or so antithetical dyads, and occasional triads, that constitute the core underlying values of science and form the essential framework for building competing scientific theories. Examples include complexity-simplicity, reductionism-holism, hierarchy-unity, synthesis-analysis and constancy/equilibrium-evolution-catastrophic change. Holton (1988) also notes the conscious or unconscious preoccupation with symmetry shown by many scientists.

In Chapter 6, I used Kearney's (1984) notion of *worldview* to describe some of the fundamental presuppositions, beliefs and values that underpin the practice of science. According to Fritjof Capra (1983), there are five main ideas in the Cartesian/Newtonian worldview underpinning modern science: (i) a fundamental division between mind and matter; (ii) nature as a mechanism that works in accordance with exact and universal laws; (iii) reality as a complex system of building blocks (the whole is the sum of its parts); (iv) scientific method as the only valid way to ascertain the true nature of phenomena; (v) science as a means of understanding nature so that it can serve human wants, needs and interests. In similar vein, Smolicz and Nunan (1975) identify four "ideological pivots" inherent in scientific practice and in the image of science presented through the science curriculum. First *anthropocentrism*: the view of mankind[4] as the technologically powerful manipulator and controller of nature, with science as the means by which we control the environment and shape it to meet our interests and needs.[5] Second, *quantification*: scientists are regarded not just as observers but as measurers and quantifiers. Whatever exists in nature can (and should) be explained in mathematical terms, best of all by means of equations. Hence Lord Kelvin's dictum that we do not understand a thing until we can measure it. Third, *positivistic faith*: faith in the inevitable linear progress of science towards truth about the world, with the certainty of this knowledge being underpinned by an all-powerful and all-purpose scientific method. Fourth, the *analytical ideal*: the assumption that phenomena and events are best studied and explained via analysis; an entirely mechanistic view of the world which assumes that the whole is simply, and no more than, the sum of its parts. Such values and principles become incorporated into scientists' background assumptions. Some presuppositions about what questions are important, and how they can be answered, are simply taken for granted, and so evade critical scrutiny. Other research that shares similar presuppositions is not closely scrutinized either. Eventually, "self evident truths" are established; they form the "invisible" values underpinning scientific inquiry. Two major questions need to be considered. First, are these "ideological pivots" promoted through science education? In other words, are Smolicz and Nunan (1975) correct in their analysis? Second, are these the underlying values of science? In other words, is this a faithful representation of science and, therefore, an appropriate set of values to promote in the curriculum in pursuit of authenticity?

A decade later, as part of a major survey of Canadian science education conducted by the Science Council of Canada, Nadeau and Désautels (1984) identified what they called five "mythical values stances" suffusing science education: (i) *naïve realism* – science gives access to truth about the universe; (ii) *blissful empiricism* – science is the meticulous, orderly and exhaustive gathering of data; (iii) *credulous experimentation* – experiments can conclusively verify hypotheses; (iv) *excessive rationalism* – science proceeds solely by logic and rational appraisal; (v) *blind idealism* – scientists are completely disinterested, objective beings. As noted in Chapter 5, the cumulative message is that science has an all-purpose, straightforward and reliable method of ascertaining the truth about the universe, with the certainty of scientific knowledge being located in objective observation, extensive data collection and experimental verification. Moreover, scientists are rational, logical, open-minded and intellectually honest people who are required, by their commitment to the scientific enterprise, to adopt a disinterested, value-free and analytical stance. In quite startling contrast, Siegel (1991) states that:

> Contemporary research ... has revealed a more accurate picture of the scientist as one who is driven by prior convictions and commitments; who is guided by group loyalties and sometimes petty personal squabbles; who is frequently quite unable to recognize evidence for what it is; and whose personal career motivations give the lie to the idea that the scientist yearns only or even mainly for the truth. (p. 45)

A number of questions spring to mind. First, does the curriculum project the images identified by Nadeau and Désautels, and questioned by Siegel's remarks? Second, are these images faithful to the nature of real science and the characteristics and behaviour of real scientists engaged in doing science? In other words, do these descriptions provide an authentic view of science? Third, do students internalize these views? Fourth, does it matter what image we project or what image of science students acquire? At the time Nadeau and Désautels were writing, the answer to the first question was undoubtedly "Yes". While much has changed in the intervening years, thanks largely to the STS and STSE thrust in curriculum debate and development, many school science curricula and school textbooks continue to project these images (Lakin & Wellington, 1994; Cross, 1995; Knain, 2001; Clough, 2006). Loving (1997) laments that all too often –

(a) Science is taught totally ignoring what it took to get to the explanations we are learning – often with lectures, reading text, and memorizing for a test. In other words, it is taught free of history, free of philosophy, and in its final form. (b) Science is taught as having one method that all scientists follow step-by-step. (c) Science is taught as if explanations are the truth – with little equivocation. (d) Laboratory experiences are designed as recipes with one right answer. Finally, (e) scientists are portrayed as somehow free from human foibles, humor, or any interests other than their work. (p. 443)

Another interesting question is the extent to which these underlying assumptions and values constitute a predominantly Western way of perceiving the world. In his seminal work, *The Geography of Thought*, mentioned briefly in Chapters 5 and 6, Richard Nisbett (2003) concludes: "European thought rests on the assumption that the behavior of objects – physical, animal, and human – can be understood in terms of straightforward rules" (p. xvi). He notes that Westerners see the world in analytic, atomistic terms; they see objects as discrete and separate from their environments; they see events as progressing in linear fashion; they have a strong interest in categorization, which helps them to know what rules to apply to the objects under consideration. Moreover, because of the high priority afforded to the resolution of inconsistencies and contradictions, formal logic plays a key role in problem-solving. In contrast, East Asians "attend to objects in their broad context … Understanding events always requires consideration of a host of factors that operate in relation to one another in no simple, deterministic way" (p. xvi). They regard the world as complex and highly changeable and its components as interrelated; they see events as moving in cycles between extremes. Their lack of concern about contradiction, and emphasis on "the middle way", results in formal logic playing little role in problem-solving. When confronted with two apparently contradictory propositions, Westerners tend to polarize their beliefs, Easterners move towards equal acceptance of both propositions. In consequence, Easterners are not as surprised by unanticipated outcomes as Westerners. Nisbett's conclusion is that "Easterners are almost surely closer to the truth than Westerners in their belief that the world is a highly complicated place and Westerners are undoubtedly often far too simple-minded in their explicit models of the world. Easterners' failure to be surprised as often as they should may be a small price to pay for their greater attunement to a range of possible causal factors. On the other hand, is seems fairly clear that simple models are the most useful ones – at least in science – because they are easier to disprove and consequently to improve upon" (Nisbett, 2003, p. 134).

PERSONAL VALUES, SCIENTIFIC ATTITUDES AND THE CHANGING NATURE OF SCIENTIFIC PRACTICE

It is widely assumed, especially by the authors of school science textbooks, that scientists display and adhere to certain personal values: objectivity, rationality, intellectual integrity, accuracy, diligence, open-mindedness, self-criticism, skepticism and circumspection (in the sense of suspending judgement until all the evidence is in hand). In addition, they are expected to be dispassionate and disinterested, and to readily share their thoughts and findings with others. Like Mertonian norms (see Chapter 4), these so-called "scientific attitudes"[6] are said to guarantee proper scientific practice by ensuring that (i) all knowledge claims are treated skeptically until their validity can be judged according to the weight of evidence, (ii) all evidence is carefully considered before decisions about validity are made, and (iii) the idiosyncratic prejudices of individual scientists do not intrude into the decision-making. In choosing to become a scientist, one makes a commitment to "a set of preferences for such things as a non-dogmatic, anti-fideistic, critical attitude in which strength of belief is attuned to evidence, and for 'open horizons' over closures" (Suchting, 1995, p. 16). Because science is a communal practice each practitioner has to trust that others will conform to the community-approved standards, unless or until they learn otherwise (see discussion of error, misconduct and fraud in Chapter 4). Since no scientist can be expert in all fields it is essential to have trust in the work of scientists working in related fields and in the work of those who design the complex laboratory instruments on which so much modern science depends. Without mutual trust the scientific enterprise cannot function effectively or productively. Also at issue is the trust that non-scientists have to be able to invest in anyone regarded as an expert.

Sociologists and ethnographers of science have cast considerable doubt on the extent to which practitioners *do* exhibit these characteristics or adhere to these commitments. It is now more than 40 years since Roe (1961) suggested that scientists rarely possess these scientific attitudes, although (she says) they think that they do. They, too, subscribe to the myth of the emotionally-detached, disinterested and impartial scientist. Or they continue to promote this false image because they perceive it to be in their interests to imply a connection between a disinterested approach and the "truth" of the findings in the battle to maintain high levels of public funding for scientific research. Roe concludes: "The creative scientist, whatever his field, is very deeply involved emotionally and personally in his work" (p. 456). Mahoney's (1979) conclusions about the attitudes and characteristics of scientists, derived from writings by sociologists, historians and scientists, make particularly interesting reading.

— Superior intelligence is neither a prerequisite nor a correlate of high scientific achievement.
— Scientists are often illogical in their work, particularly when defending a preferred view or attacking a rival one.
— Scientists' perceptions of reality are dramatically influenced by their theoretical expectations.
— Scientists are often selective, expedient and not immune to perceptual bias and distortion of the data.
— Scientists are among the most passionate of professionals. Their theoretical and personal biases often colour their alleged openness to the data.

- Scientists are often dogmatically tenacious and inflexible in their opinions, even when contradictory evidence is overwhelming.
- Scientists are skilled in "expedient reasoning" – that is, bending their arguments to fit their purposes.
- Scientists are not paragons of humility and disinterest. Rather, they are often selfish, ambitious and petulant defenders of personal recognition and territoriality. Vitriolic episodes and bitter disputes over personal credit and priority are common.
- Scientists often behave in ways that are diametrically opposite to communal sharing of knowledge. They are frequently secretive and suspicious of others, and will frequently suppress data until they have established priority of discovery.
- Far from being a "suspender of judgement", the scientist is often an impetuous truth-spinner who rushes to hypotheses and theories long before the data warrants.

Following their study of scientists and engineers involved in NASA's Apollo Project, Mitroff and Mason (1974) concluded that scientists are arranged along a continuum from *extreme speculative scientists*, who "wouldn't hesitate to build a whole theory of the solar system based on no data at all" (p. 1508), to *data bound scientists*, who "wouldn't be able to save their own hide if a fire was burning next to them because they'd never have enough data to prove the fire was really there" (p. 1508). Contrary to the school textbook stereotype, the scientists who produce the most significant work are those who disregard the so-called scientific attitudes; those who conform may be excellent technicians, but they do not make theoretical breakthroughs or establish procedural innovations. This is a reminder that we need to be sure which "game" we are playing in school science education. We need to provide our students with experiences capable of developing both creative scientists and dependable technicians (see discussion in Chapter 7).

It is essential that students learn to look critically at the realities of contemporary science and the values that impregnate and underpin it. In the traditional forms of "basic" or "fundamental" research envisaged by Merton (1973), usually located in universities and/or government research institutes, "pure scientists" constitute their own audience: they determine the research goals, recognize competence, reward originality and achievement, legitimate their own conduct and discourage attempts at outside interference. In the contemporary world, universities are under increasing public pressure to deliver more obvious value for money, and to undertake research that is likely to have practical utility or direct commercial value. There are increasingly loud calls for closer links between academia and industry. Scientific research is often dependent on expensive technology and so must meet the needs and serve the interests of those sponsors whose funds provide the resources. It is often multidisciplinary and involves large groups of scientists working on problems that they have not posed, either individually or as a group. Within these teams, individual scientists may have little or no understanding of the overall thrust of the research, and no ownership of the scientific knowledge that

345

results. Indeed, there is a marked trend towards patenting and commodification of knowledge.

> Post-academic research is usually undertaken as a succession of "projects", each justified in advance to a funding body whose members are usually not scientists. As the competition for funds intensifies, project proposals are forced to become more and more specific about the expected outcomes of the research, including its wider economic and social impact. This is no longer a matter for individual researchers to determine for themselves. Universities and research institutes are no longer deemed to be devoted entirely to the pursuit of knowledge for its own sake. They are encouraged to seek industrial funding for commissioned research, and to exploit to the full any patentable discoveries made by their academic staffs, especially when there is a smell of commercial profit in the air (Ziman, 1998, p. 1813).

In consequence, scientists have lost a substantial measure of autonomy. The vested interests of the military and commercial sponsors of research, particularly tobacco companies, the petroleum industry, the food processing industry, pharmaceutical companies and the nuclear power industry, can often be detected not just in research priorities but also in research design, especially in terms of what and how data are collected, manipulated and presented. More subtly, in what data are not collected, what findings are omitted from reports and whose voices are silenced. Commercial interests may influence the way research findings are made public (e.g., press conferences rather than publication in academic journals) and the way in which the impact of adverse data is minimized, marginalized or ignored. For example, in publicizing the value of oral contraceptives and hormone replacement therapy the increased risks of cervical cancer, breast cancer and thromboembolism are often given little attention. Many researchers are so dependent on government contracts, commercial sponsorship and/or closely defined employment contracts that they become, in all but name, agents of government policy and particular business interests. In some situations they are advocates for a particular point of view rather than disinterested arbiters of the "truth". Many scientists are employed on contracts that prevent them from disclosing all their results. As Ziman (2000) comments, they are forced to trade the academic kudos of publication in refereed journals for the material benefit of a job or a share in whatever profit there might be from a patented invention. This snapshot of some of the realities of contemporary science is in direct contradiction of three, if not all four, of the "functional norms" or "institutional imperatives" said by Robert Merton (1973) to govern the practice of science and the behaviour of individual scientists, whether or not they are aware of it: *universalism, communality, disinterestedness* and *organized scepticism* (see Chapter 4). Teachers and students should ask whether Merton was right or wrong about these norms of scientific conduct. Is there, for example, a key distinction between "science as it ought to be conducted" and "science as it is currently practised"? If there is any substance to a claim of collision between ideal and actual practice, what is an appropriate view to present to students in school science classes?

In contrast to Ziman's pessimism about the impact of competition on contemporary science, Paul Kitcher (1993) and Miriam Solomon (2001) argue that intense competition and distribution of research efforts across large teams is enormously beneficial to science because it guarantees a much higher level of critical scrutiny as competitors seek to promote their views and defend them against the claims of others. Debate is both quickened and deepened when individual researchers or research teams pursue different theories and/or different research strategies. Rolin (2008) illustrates the argument for diversity of approach through the now familiar story of Francis Crick, Jim Watson and Rosalind Franklin. Whereas Franklin continued to collect "hard data" by means of X-ray crystallography, Crick and Watson opted to construct whatever models of the DNA structure were theoretically possible, and then to see if the empirical data would fit. As we know, their approach proved the more successful. Rolin (2008) comments: "The lesson to be learned from this example is that it is difficult, if not impossible to know beforehand which approach will be successful. Therefore, it is rational to distribute research efforts among different approaches" (p. 1123).

In striving for a "balanced view" for the school science curriculum it is useful to draw Longino's (1990) distinction between the *constitutive* values of science (such as the drive to meet criteria of truth, accuracy, precision, simplicity, predictive capability, breadth of scope and problem-solving capability) and the *contextual* values that impregnate the personal, social and cultural context in which science is organized, supported, financed and conducted. Allchin (1999) draws a similar distinction between the *epistemic* values of science and the *cultural* values that infuse scientific practice. With regard to the contextual or cultural values, science is no longer the disinterested search for truth and the free and open exchange of information portrayed in the school textbook versions of science. Rather, it is a highly competitive enterprise in which scientists may be driven by self-interest and career building, desire for public recognition, financial inducements provided by business and commerce, or the "political imperatives" of military interests. The cynical interpretation of this state of affairs is that the science that "gets done" (Young, 1987) is the science that is in the interests of the rich and powerful, the self-interest of particular scientists, or the interests of the companies or government agencies that provide the research funding. Chambers (1984) deals with these matters admirably and appropriately, for both school science and science teacher education, in his compelling account of the manufacture and use of pesticides, the DDT controversy and the work of Rachel Carson.

With regard to the personal values and attitudes that drive individual scientists, it is true, as Ziman (2000) remarks, "that the formal mask of disinterestedness sometimes slips, revealing power ploys, wheeling and dealing, quasi-military alliances and other sordid manoeuvres. It is a sad fact of life that academic communities are prone to clannishness, careerism, nepotism and other worldly vices ... In distributing scientific recognition and its material rewards, scientists often favour their own colleagues and fellow-countrymen. In pursuit of material resources they often behave as swinishly as any other political lobby" (p. 15). The fact is that no social institution could possibly live up to the lofty ideals encapasulated in the Mertonian norms, at least not every single day. The scientific community is inhabited by ordinary people, who sometimes meet the highest possible standards and sometimes do not. However, although the scientific community cannot regulate the behaviour of individual practitioners throughout every working day, it can regulate the overall behaviour of the community and it can ensure that the seal of community approval is only given to work that is seen to meet the highest standards of professional conduct. As discussed earlier in this chapter, and in Chapters 4 and 6, it is the social character of science that makes it vulnerable to all manner of human weaknesses but also protects it from these influences through its rigorous and highly visible systems of critical scrutiny. While we cannot design and carry out entirely value-free science, we can sharpen our ability to detect underlying values and, if necessary, challenge those values and perhaps replace them by other values. By encouraging individuals of varying backgrounds, experience and value positions to participate in the critical phases of identifying research priorities and appraising the outcomes, we can subject science to more vigorous and rigorous critical scrutiny. This assertion is based on the assumption that a change of background assumptions can sometimes lead to a revision of the value placed on particular science, a change in perception of the status of evidence, and a shift in opinion about the plausibility of explanations. In other words, a plurality of values within the community establishes a system of "epistemic checks and balances", with the ultimate justification for any knowledge claim lying in the intersubjective judgement of the community of practitioners. A process of "transformative interrogation", as Longino (1990) calls it, determines which social values are retained and which are eliminated at any one time; it seeks to minimize the domination of science by any particular set of values deemed prejudicial to the objectivity of science.

> That theory which is the product of the most *inclusive* scientific community is better, other things being equal, than that which is the product of the most *exclusive*. It is better not as measured against some independently accessible reality but better as measured against the cognitive needs of a genuinely democratic community. This suggests that the problem of developing a new science is the problem of creating a new social and political reality. (Longino, 1990, p. 214, emphases added)

As noted several times in this book, empirical data are rarely sufficient in themselves to establish the validity of a scientific theory. Other criteria have to be invoked, including simplicity, scope, internal and external consistency, fruitfulness and technological success. Kourany (2003) has argued that in choosing between empirically under-determined theories we should adopt criteria that promote egalitarian goals relating to gender, race and ethnicity, sexual orientation, age and other struggles for social justice, freedom and elimination of poverty. For example, theories or models that posit "complexity of relationship", or that treat relationships between entities and processes as mutually interactive rather than unidirectional and hierarchical, promote egalatariansm better than simpler models that only posit unidirectional causal relationships. As Intemann (2008) comments, "theories and models that treat relationships as complex are more likely to treat causal actors as having equally important and valuable roles" (p. 1071). In Kourany's view, research priorities and research policy should be determined by these same egalatarian goals. However, as noted earlier, Ziman's (2000) description and evaluation of contemporary science, what he calls "post-academic science", suggests that research priorities are largely determined by the interests of the military, by the priorities of the pharmaceutical industry, chemical industry, petroleum industry, agribusiness and biotechnology firms, or by the government on behalf of these industries. In consequence, we have "agricultural research that revolves around pesticides, herbicides and growth hormones, and other petrochemicals, of little help to smaller, poorer farmers around the world; and medical research that revolves around expensive high-tech treatments and cures rather than the less lucrative preventive knowledge that would help so many more people, especially poorer people" (Kourany, 2003, p.9). Using Sandra Harding's notion of strong objectivity and Sharon Crasnow's idea of model-based objectivity to justify replacing one set of social goals by another set that is more favourable to human well-being and environmental health would not pose insurmountable problems for the validity and reliability of scientific knowledge, though it would have enormous implications for the lives of millions of people. A book currently in preparation details my proposal for a politicized form of science education that involves students in consideration of such matters. The purpose of such an education is to enable young citizens to look critically at the society we have, and the values that sustain it, and to ask what can and should be changed in order to achieve a more socially just democracy and to bring about more environmentally sustainable lifestyles, especially in the industrialized countries of the world. This view of science education is overtly and unashamedly political. It takes the Advisory Group on Education for Citizenship and the Teaching of Democracy in Schools (QCA, 1998) at its word – not just education *about* citizenship, but education *for* citizenship.

> Citizenship education is education *for* citizenship, *behaving and acting as a citizen*. Therefore it is not just knowledge of citizenship and civic society; it also implies developing values, skills and understanding (p. 13, emphasis added).

A key part of an education *for* citizenship is critical scrutiny of the way we live and the values that underpin contemporary society, especially in the industrialized (and industrializing) countries of the world. As discussed in Chapter 4, it is our worldview that determines how we interpret reality, how we make sense of our experiences, and how we see ourselves as human beings in relation to others and the world as a whole. It determines our values, aspirations and overall sense of responsibility. It is now more than a quarter century since Capra (1982) argued that we need to shift from traditional mechanistic thinking, with its simple linear chain of cause and effect ascertained through close experimental control, to a more holistic, systems style of thinking able to deal with complex webs of relationships, multiple interdependencies, feedback systems and unpredictability. Our current social and environmental problems, he states, are rooted in the fact that "we are trying to apply the concepts of an outdated world view – the mechanistic world view of Cartesian–Newtonian science – to a reality that can no longer be understood in terms of these concepts. We live today in a globally interconnected world, in which biological, psychological, social, and environmental phenomena are all interdependent. To describe this world appropriately we need an ecological perspective which the Cartesian world view does not offer" (p. 15). In similar vein, Wilber (1995) argues that the only way to ameliorate the planetary and societal crises we currently face is "by replacing this fractured worldview with a worldview that is more holistic, more relational, more integrative, more Earth-honoring, and less arrogantly human-centred" (p. 4). When we acknowledge that our worldview influences how we live, and we decide to interpret and live in the world *differently*, we change our understanding of what is acceptable and what is possible, and new horizons for action become available. Education for citizenship is about changing values, changing the way we live, and thereby changing the future.

NOTES

[1] Brockman (1995) argues that the great intellectual achievements of science constitute a "third culture" alongside and equally as valuable as literacy and numeracy.

[2] "Society" is used here in the several senses that the term "community" was used in earlier discussions: the laboratory group in which a scientist works, the group of scientists engaged in similar and related research, the various official governing bodies (including publishing mechanisms), and society as a whole. The term "values" is taken to mean "the principles, fundamental convictions, ideals, standards, or life stances which act as general guides or as points of reference in decision-making or the evaluation of beliefs or actions, and which are closely connected to personal integrity and personal identity" (Halstead, 1996, p. 5).

[3] More famously, John Keats (in *Ode on a Grecian Urn*) stated: "Beauty is truth, truth beauty" and George Santayana (1955b) commented: "We know on excellent authority that beauty is truth, that it is the expression of the ideal, the symbol of divine perfection, and the sensible manifestation of the good" (p. 11).

[4] I am consciously using *mankind* rather than humankind to lend increased force to Smolicz and Nunan's proposition with respect to gender bias in scientific practice and theorizing. They used the term *man*, but at a time when authors (and readers) were less sensitive to gendered language.

[5] Of particular interest in this "nature in the service of man" view is the underlying assumption that we have the *right* to control and manipulate the natural environment.

[6] Some years ago, Colin Gauld (1982) wrote a particularly insightful essay on scientific attitudes and their distinction from attitudes to science. My own thoughts on these matters owe a considerable debt to Gauld's writing.

REFERENCES

Abd-El-Khalick, F. (2001). Embedding nature of science instruction in preservice elementary science courses: Abandoning scientism, but ... *Journal of Science Teacher Education, 12*(3), 215-233.

Abd-El-Khlaick, F. (2004). Over and over again: College students' views of nature of science. In L.B. Flick & N.G. Lederman (Eds.), *Scientific inquiry and nature of science: Implications for teaching, learning, and teacher education* (pp. 389-425). Dordrecht: Kluwer.

Abd-El-Khalick, F. (2005). Developing deeper understandings of nature of science: The impact of a philosophy of science course on preservice science teachers' views and instructional planning. *International Journal of Science Education, 27*(1), 15-42.

Abd-El-Khalick, F. (2008). Modeling science classrooms after scientific laboratories. In R.A. Duschl & R.E. Grandy (Eds.), *Teaching scientific inquiry: Recommendations for research and implementation* (pp. 80-85). Rotterdam: Sense Publishers.

Abd-El-Khalick, F. & Akerson, V.L. (2004). Learning as conceptual change. Factors mediating the development of preservice elementary teachers' views of the nature of science. *Science Education, 88*(5), 785-810.

Abd-El-Khalick, F., Bell, R.L. & Lederman, N.G. (1998). The nature of science and instructional practice: Making the unnatural natural. *Science Education, 82*(4), 417-437.

Abd-El-Khalick, F. & BouJaoude, S. (1997). An exploratory study of the knowledge base for science teaching. *Journal of Research in Science Teaching, 34*, 673-699.

Abd-El-Khalick, F. & Lederman, N.G. (2000a). Improving science teachers' conceptions of the nature of science: A critical review of the literature. *International Journal of Science Education, 22*(7), 665-701.

Abd-El-Khalick, F. & Lederman, N.G. (2000b). The influence of history of science courses on students' views of the nature of science. *Journal of Research in Science Teaching, 37*(10), 1057-1095.

Abd-El-Khalick, F., Waters, M. & Le, A-P. (2008). Representations of nature of science in high school chemistry textbooks over the past four decades. *Journal of Research in Science Teaching, 45*(7), 835-855.

Abell, S.K. & Roth, M. (1995). Reflections of fifth-grade life science lesson: Making sense of children's understanding of scientific models. *International Journal of Science Education, 17*(1), 59-74.

Abell, S.K. & Smith, D.C. (1994). What is science? Preservice elementary teachers' conceptions of the nature of science. *International Journal of Science Education, 16*, 475-487.

Abell, S., Martini, M. & George, M. (2001). "That's what scientists have to do": Preservice elementary teachers' conceptions of the nature of science during a moon investigation. *International Journal of Science Education, 23*(11), 1095-1109.

Abrahams, I. & Millar, R. (2008). Does practical work really work? A study of the effectiveness of practical work as a teaching and leaning method in school science. *International Journal of Science Education, 30*(14), 1945-1969.

Abrams, E. & Wandersee, J. (1995). How does biological knowledge grow? A study of life scientists' research practices. *Journal of Research in Science Teaching, 32*(6), 649-663.

Abt-Perkins, D. & Pagnucci, C. (1993). From tourist to storyteller: Reading and writing science. In S. Tchudi (Ed), *The astonishing curriculum: Integrating science and humanities through language* (pp. 99-111). Urbana, IL: National Council of Teachers of English.

Acher, A., Arcà, M. & Sanmarti, N. (2007). Modeling as a teaching learning process for understanding materials: A case study in primary education. *Science Education, 91*(3), 398-418.

REFERENCES

Adams, H.H. (1990). *African and African American contributions to science* [Baseline essay]. Portland, OR: Multnomah School District.

Adams, S.T. (2002). Use of a computer environment to analyze the coherence of argumentation about policies proposed to ameliorate global warming. Paper presented at the annual meeting of the American Educational Research Association, New Orleans, LA, April. Available as ERIC document ED464952.

Adams, S.T. (2003). Investigation of the "Convince Me" computer environment as a tool for critical argumentation about public policy issues. *Journal of Interactive Learning Research, 14*(3), 263-283.

Agar, M. (1994). *Language shock: Understanding the culture of conversation.* New York: William Morrow.

Aguirre, J., Haggerty, S. & Linder, C. (1990). Students teachers' conceptions of science, teaching and learning: A case study in preservice science education. *International Journal of Science Education, 12*, 381-390.

Ahkwesahsne Mohawk Board of Education (1994). *Lines & circles.* Cornwall, ON: Author.

Aikenhead, G.S. (1996). Science education: Border crossing into the subculture of science. *Studies in Science Education, 27*, 1-52.

Aikenhead, G.S. (1997). Towards a First Nations cross-cultural science and technology curriculum. *Science Education, 81*, 217-238.

Aikenhead, G.S. (2000). Renegotiating the culture of school science. In R. Millar, J. Leach & J. Osborne (Eds), *Improving science education: The contribution of research* (pp. 245-264). Buckingham: Open University Press.

Aikenhead, G.S. (2001a). Integrating Western and Aboriginal science: Cross-cultural science teaching. *Research in Science Education, 31*(3), 337-355.

Aikenhead, G.S. (2001b). Students' ease in crossing cultural borders into school science. *Science Education, 85*, 180-188.

Aikenhead, G.S. (2002). Cross-cultural science teaching: Rekindling traditions for Aboriginal students. *Canadian Journal of Science, Mathematics and Technology Education, 2*(3), 287-304.

Aikenhead, G.S. (2006). Towards decolonizing the Pan-Canadian science framework. *Canadian Journal of Science, Mathematics & Technology Education, 6*(4), 387-399.

Aikenhead, G.S., Fleming, R.W. & Ryan, A.G. (1987). High school graduates' beliefs about science-technology-society. I. Methods and issues in monitoring student views. *Science Education, 71*(2), 145-161.

Aikenhead, G.S. & Jegede, O. (1999). Cross-cultural science education: A cognitive explanation of a cultural phenomenon. *Journal of Research in Science Teaching, 36*, 269-287.

Aikenhead, G.S. & Ogawa, M. (2007). Indigenous knowledge and science revisited. *Cultural Studies of Science Education, 2*(3), 539-591.

Aikenhead, G.S. & Otsuji, H. (2000). Japanese and Canadian science teachers' views on science and culture. *Journal of Science Teacher Education, 11*(4), 277-299.

Aikenhead, G.S. & Ryan, A.G. (1992). The development of a new instrument: Views on Science-Technology-Society (VOSTS). *Science Education, 76*, 477-491.

Aikenhead, G.S., Ryan, A.G. & Fleming, R.W. (1989). *Views on science-technology-society.* Saskatoon: Department of Curriculum Studies (Faculty of Education), University of Saskatchewan.

Ainsworth, S. (2006). A conceptual framework for considering learning with multiple representations. *Learning and Instruction, 16*, 183-198.

Aitken, H.G.J. (1976). *Syntony and spark – The origins of radio.* New York: John Wiley.

Akerson, V.L., Abd-El-Khalick, F. & Lederman, N.G. (2000). Influence of a reflective activity-based approach on elementary teachers' conceptions of nature of science. *Journal of Research in Science Teaching, 37*(4), 295-317.

Akerson, V.L. & Abd-El-Khalick, F. (2003). Teaching elements of nature of science: A yearlong case study of a fourth-grade teacher. *Journal of Research in Science Teaching, 40*(10), 1025-1049.

Akerson, V.L. & Buzzelli, C.A. (2007). Relationships of preservice early childhood teachers' cultural values, ethical and cognitive developmental levels, and views of nature of science. *Journal of Elementary Science Education, 19*, 15-24.

Akerson, V.L., Buzzelli, C.A. & Donnelly, L.A. (2008). Early childhood teachers' views of nature of science: The influence of intellectual levels, cultural values, and explicit reflective teaching. *Journal of Research in Science Teaching, 45*(6), 748-770.

Akerson, V.L. & Hanuscin, D.L. (2007). Teaching nature of science through inquiry: Results of a 3-year professional development program. *Journal of Research in Science Teaching, 44*(5), 653-680.

Akerson, V.L., Hanson, D.L. & Cullen, T.A. (2007). The influence of guided inquiry and explicit instruction on K-6 teachers' views of nature of science. *Journal of Science Teacher Education, 18*, 751-772.

Akerson, V.L., Morrison, J.A. & McDuffie, A.R. (2006). One course is not enough: Preservice elementary teachers' retention of improved views of nature of science. *Journal of Research in Science Teaching, 43*(2), 194-213.

Akerson, V.L. & Volrich, M.L. (2006). Teaching nature of science explicitly in a first-grade internship setting. *Journal of Research in Science Teaching, 43*(4), 377-394.

Akkus, R., Gunel, M. & Hand, B. (2007). Comparing an inquiry-based approach known as the science writing heuristic to traditional science teaching practices: Are there differences? *International Journal of Science Education, 29*(14), 1745-1765.

Albone, A., Collins, N. & Hill, T. (1995). *Scientific research in schools: A compendium of practical experience*. Bristol: Clifton Scientific Trust.

Alexander, P.A. & Jetton, T.L. (2000). Learning from text: A multidimensional and developmental perspective. In M.L. Kamil, P.B. Mosenthal, P.D. Pearson & R. Barr (Eds.), *Handbook of reading research*, Vol. 3 (pp. 285-310). Mahwah, NJ: Lawrence Erlbaum.

Alic, M. (1986). *Hypatia's heritage: A history of women in science from antiquity through the 19th century*. Boston, MA: Beacon Press.

Allchin, D. (1995). Points east and west: Acupuncture and teaching the cultural contexts of science. In F. Finley, D. Allchin, D. Rhees & S. Fifield (Eds.), *Proceedings of the Third International History, Philosophy and Science Teaching Conference*, Vol. 1 (pp. 23-32). Minneapolis, MN: University of Minnesota Press.

Allchin, D. (1996). Points east and west: Acupuncture and comparative philosophy of science. *Philosophy of Science, 63*, S107-S115.

Allchin, D. (1997). Rekindling phlogiston: From classroom case study to interdisciplinary relationships. *Science & Education, 6*(5), 473-509.

Allchin, D. (1999). Values in science: An educational perspective. *Science and Education, 8*, 1-12.

Allchin, D. (2003). Scientific myth-conceptions. *Science Education, 87*(3), 329-351.

Allchin, D. (2004a). Should the sociology of science be rated X? *Science Education, 88*(6), 934-946.

Allchin, D. (2004b). Pseudohistory and pseudoscience. *Science & Education, 13*, 179-195.

Allen, L.C. (1970). A bonding model for anomalous water. *Nature, 227*(5256), 372-373.

Allen, L.C. (1971). An annotated bibliography for anaomalous water. *Journal of Colloid and Interface Science, 36*(4), 554-561.

Allen, L.C. & Kollman, P.A. (1970). A theory of anomalous water. *Science, 167*(3924), 1443-1454.

Allen, L.C. & Kollman, P.A. (1971). Theoretical evidence against the existence of polywater. *Nature, 233*(5321), 550-551.

Alsop, S. (2005). *The affective dimensions of cognition: Studies from education in the sciences*. Dordrecht: Kluwer.

Alsop, S. (2009). Not quite the revolution: Science and technology education in a world that changed. In A.T. Jones & M.J.deVries (Eds.), *International handbook of research and development in technology education* (pp. 389-401). Rotterdam/Taipei: Sense.

Alters, B.J. (1997). Whose nature of science? *Journal of Research in Science Teaching, 34*(1), 39-55.

Alvermann, D. (2004). Multiliteracies and self quesdtioning in the service of science learning. In E.W. Saul (Ed.), *Crossing borders in literacy and science instruction* (pp. 226-238). Newark, DE: International Reading association.

Amati, K. (2000). Constructing scientific models in middle school. In J. Minstrell & E.H. van Zee (Eds.), *Inquiring into inquiry learning and teaching in science* (pp. 316-330). Washington, DC: American Association for the Advancement of Science.

American Association for the Advancement of Science (AAAS) (1989). *Science for all Americans. A Project 2061 report on literacy goals in science, mathematics, and technology*. Washington, DC: AAAS.

American Association for the Advancement of Science (AAAS) (1993). *Benchmarks for scientific literacy*. Oxford: Oxford University Press.

Amis, M. (1995). *The information*. London: Flamingo.

Anderson, C.W. (1999). Inscriptions and science learning. *Journal of Research in Science Teaching, 36*, 973-974.

Anderson, D., Lucas, K.B. & Ginns, I.S. (2003). Theoretical perspectives on learning in an informal setting. *Journal of Research in Science Teaching, 40*, 177-199.

Anderson, R.D. (2007). Teaching the theory of evolution in social, intellectual, and pedagogical context. *Science Education, 91*(4), 664-677.

Anderson, W. (1984). *Between the library and the laboratory: The language of chemistry in eighteenth century France.* Baltimore, MD: Johns Hopkins University Press.

Andre, T., Whigham, M., Hendrickson, A. & Chambers, S. (1999). Competency beliefs, positive affect, and gender stereotypes of elementary students and their parents about science versus other school subjects. *Journal of Research in Science Teaching, 36*, 719-747.

Andriessen, J. (2007). *Arguing to learn. In K. Sawyer (Ed.), The Cambridge handbook of the learning sciences* (pp. 443-460). Cambridge: Cambridge University Press.

Andriessen, J., Baker, M. & Suthers, D. (Eds.) (2003). *Arguing to learn: Confronting cognitions in computer-supported collaborative learning environments.* Dordrecht: Kluwer.

Anees, M.A. (1995). Islam and scientific fundamentalism. *Technosciences, 8*(1), 21-22.

Apedoe, X.S. (2008). Engaging students in inquiry: Tales from an undergraduate geology laboratory-based course. *Science Education, 92*(4), 631-663.

Appiah, A. (1994). Identity, authenticity, survival. In A. Gutmann (Ed.), *Multiculturalism* (pp. 149-163). Princeton, NJ: Princeton University Press.

Armah, A.K. (2005). African identity (public talk). Human Science Research Council, Pretoria, Februray 25th. Cited by Keane (2008).

Armstrong, N.A., Wallace, C.S. & Chang, S-M. (2008). Learning from writing in college biology. *Research in Science Education, 38*(4), 483-499.

Arons, A.B. (1973). Toward wider public understanding of science. *American Journal of Physics, 41*, 769-782.

Arons, A.B. (1988). Historical and philosophical perspectives attainable in introductory physics courses. *Educational Philosophy & Theory, 20*(2), 13-23.

Ash, D. & Klein, C. (2000). Inquiry in the informal learning environment. In J. Minstrell & E.H. van Zee (Eds.), *Inquiring into inquiry learning and teaching in science* (pp. 216-240). Washington, DC: American Association for the Advancement of Science.

Ash, D., Crain, R., Brandt, C., Loomis, M., Wheaton, M. & Bennett, C. (2007). Talk, tools, and tensions: Observing biological talk over time. *International Journal of Science Education, 29*(12), 1581-1602.

Association for Science Education (1988). Technological education and science in schools. Report of the Science and Technology Sub-comnmittee. Hatfield: ASE.

Atkin, J.M. & Helms, J. (1992). Testing in mathematics and science – for everyone. TIMSS document reference ICC413/NPC061. Vancouver: TIMSS International Coordinating Centre.

Atran, S. (1990). *Cognitive foundations of natural history: Towards an anthropology of science.* Cambridge: Cambridge University Press.

Atwater, M.M. (1996). Social constructivism: Infusion into the miulticultural science education research agenda. *Journal of Research in Science Teaching, 33*, 821-837.

Aubusson, P., Barr, R., Perkovic, L. & Fogwill, S. (1997). What happens when students do simulation-role-play in science. *Research in Science Education, 27*, 565-579.

Aubusson, P.J. & Fogwill, S. (2006). Role play as analogical modelling in science. In P.J. Aubusson, A.G. Harrison & S.M. Ritchie (Eds.), *Metaphor and analogy in science education* (pp. 11-24). Dordrecht: Kluwer.

Ault, C.R. (1998). Criteria of excellence for geological inquiry: The necessity of ambiguity. *Journal of Research in Science Teaching, 35*(2), 189-212.

Ausubel, D.P. (1968). *Educational psychology: A cognitive view.* New York: Holt, Rinehart & Winston.

Avery, O.T., MacLeod, C.M. & McCarthy, M. (1944). Studies on the chemical nature of the substance inducing transformation of pneumococcal types: Induction of transformation by a deoxyribonucleic fraction isolated from pneumococcus type III. *Journal of Experimental Medicine, 79*, 137-158.

Ayala, F.J. (1994). On the scientific method, its practice and pitfalls. *History and Philosophy of the Life Sciences, 16*(2), 205-240.

Ayala, F.J. (2000). Arguing for evolution: Holding strong religious beliefs does not preclude intelligent scientific thinking. *The Science Teacher, 67*(2), 30-32.

Bachelard, G. (1934/1985). *The new scientific spirit* (orig pub. 1934). Trans: A. Goldhammer. Boston, MA: Beacon.

Baird, D. (2003). Thing knowledge: Outline of a mmaterialist theory for experiments. In H. Radder (Ed.), *The philosophy of scientific experimentation* (pp. 39-67). Pittsburgh, PA: University of Pittsburgh Press.

Baird, D. (2004). *Thing knowledge*. Berkeley, CA: University of California Press.

Bajracharya, H. & Brouwer, W. (1997). A narrative approach to science teaching in Nepal. *International Journal of Science Education, 19*, 429-446.

Baker, D. & Taylor, P.C.S. (1995). The effect of culture on the learning of science in non-western countries: The results of an integrated research review. *International Journal of Science Education, 17*(6), 695-704.

Baker, L.M. & Dunbar, K. (1996). Constraints of the experimental design process in real-worl science. In G. Cottrell (Ed.), *Proceedings of the Eighteenth Annual Conference of the Cognitive Science Society* (pp. 154-159). Mahwah, NJ: Erlbaum.

Baker, W.P. & Lawson, A.E. (2001). Complex instructional analogies and theoretical concept acquisition in college genetics. *Science Education, 85*, 665-683.

Bakhtin, M.M. (1981). *The dialogic imagination: Foure essays*. Austin, TX: University of Texas Press.

Bakhtin, M.M. (1986). *Speech genres and other late essays*. Austin, TX: University of Texas Press

Baldry, P.E. (1993). *Acupuncture, trigger points and musculoskeletal pain*, 2nd ed. Edinburgh: Churchill Livingstone.

Ballenger, C. (1997). Social identities, moral narratives, scientific argumentation: Science talk in a bilingual classroom. *Language and Education, 11*(1), 1-14.

Bangert-Drowns, R.L., Hurley, M.M. & Wilkinson, B. (2004). The effects of school-based writing-to-learn interventions on academic achievement: A meta-analysis. *Review of Educational Research, 74*(1), 29-58.

Barab, S.A. & Hay, K.E. (2001). Doing science at the elbows of experts: Issues related to the science apprenticeship camp. *Journal of Research in Science Teaching, 38*, 70-102.

Barab, S.A., Hay, K.E., Squire, K., Barnett, M., Schmidt, R. & Karrigan, K. (2000). Virtual solar system project: Learning through a technology-rich, inquiry-based, participatory learning environment. *Journal of Science Education and Technology, 9*(1), 7-25.

Barad, K. (2000). Reconceiving scientific literacy as agential literacy. In R. Reid & S. Traweek (Eds.), *Doing science + culture* (pp. 221-258). New York: Routledge.

Baram-Tsabari, A. & Yarden, A. (2005). Text genre as a factor in the formation of scientific literacy. *Journal of Research in Science Teaching, 42*(4), 403-428.

Barber, B. (1962). *Science and the social order*. New York: Collier.

Barbour, I. (1974). *Myths, models and paradigms*. London: SCM Press.

Barbour, I. (1990). *Religion in an age of science: The Gifford lectures, 1989–1991*. San Francisco, CA: Harper.

Barbour, I.G. (1997). *Religion and science: Historical and contemporary issues*. San Francisco, CA: Harper.

Barbour, I.G. (2000). *When science meets religion: Enemies, stramgers or partners*. New York: Harper.

Barker, M. (1995). Esteeming prior knowledge: Historical views and students' intuitive ideas about plant nutrition. In F. Finley, D. Allchin, D. Rhees & S.Fifield (Eds.), *Proceedings of the Third International History, Philosophy and Science Teaching Conference*, Vol. 1 (pp. 97-102). Minneapolis, MN: University of Minnesota Press.

Barker, M. (2002). Ripping yarns – Science stories with a point. *NZ Science Teacher, 101*, 31-36.

Barker, M. (2004). Spirals, shame and sainthood – More ripping yarns. *NZ Science Teacher, 106*, 6-14.

Barker, M. (2006). Ripping yarns: A pedagogy for learning about the nature of science. *NZ Science Teacher, 113*, 27-37.

Barman, C.R. (1997). Students' views about scientists and science: Results from a national study. *Science and Children, 35*(9), 18-24.

Barman, C.R. (1999). Students' views about scientists and school science: Engaging K-8 teachers in a national study. *Journal of Science Teacher Education, 10*(1), 43-54.

Barnea, N. (2000). Teaching and learning about chemistry and modelling with a computer-managed modelling system. In J.K. Gilbert & C. Boulter (Eds.), *Developing models in science education* (pp. 307-324). Dordrecht: Kluwer.

Barnes, B. & Edge, D. (1982). General introduction. In B. Barnes & D. Edge (Eds.), *Science in context: Readings in the sociology of science* (pp. 1-12). Milton Keynes: Open University Press.

Barnes, D. (1988). Oral language and learning. In S. Hynds & D. Rubin (Eds.), *Perspectives on talk and learning* (pp. 41-54). Urbana, IL: National Council of Teachers of English.

Barnes, D. & Todd, F. (1995). *Communication and learning revisited.* Portsmouth, NH: Heinemann.

Barnett, J. (1992). Language in the science classroom: Some issues for teachers. *Australian Science Teachers Journal, 38*(4), 8-13.

Barnett, J. & Hodson, D. (2001). Pedagogical context knowledge: Toward a fuller understanding of what good science teachers know, *Science Education, 85*(4), 426-453.

Barr, S.M. (2003). *Modern physics and ancient faith.* Notre Dame, IN: University of Notre Dame Press.

Bartholomew, H., Osborne, J. & Ratcliffe, M. (2004). Teaching students "ideas-about-science": Five dimensions of effective practice. *Science Education, 88*(5), 655-682.

Barton, A.C. (1998). *Feminist science education.* New York: Teachers College Press.

Barton, A.C. (2003). *Teaching science for social justice.* New York: Teachers College Press.

Barton, A.C. & Osborne, M.D. (1998). Marginalized discourses and pedagogies: Constructively confronting science for all. *Journal of Research in Science Teaching, 35*(4), 339-340.

Barton, A.C. & Tan, E. (2009). The evolution of da heat: Making a case for scientific and technology literacy as robust participation. In A.T. Jones & M.J. deVries (Eds.), *International handbook of research and development in technology education* (pp. 403-424). Rotterdam/Taipei: Sense.

Baskaran, A. & Boden, R. (2006). Globalization and the commodification of science. In M. Muchie & X. Li (Eds.), *Globalisation, inequality and the commodification of life and well-being* (pp. 42-72). London: Adonis & Abbey.

Bastide, F. (1990). The iconography of scientific texts: Principles of anlaysis. In M. Lynch & S. Woolgar (Eds.), *Representation in scientific practice* (pp. 187-229). Cambridge, MA: MIT Press.

Bauer, H.H. (1992). *Scientific literacy and the myth of the scientific method.* Chicago, IL: University of Illinois Press.

Bausor, J. & Poole, M. (2002). Science-and-religion in the agreed syllabuses: An investigation and some suggestions. *British Journal of Religious Education, 25*(1), 18-32.

Bazerman, C. (1988). *Shaping written knowledge: The genre and activity of the experimental article in science.* Madison, WI: University of Wisconsin Press.

Bean, T.W., Searles, D., Cowen, S., & Singer, H. (1990). Learning concepts from biology text through pictorial analogies and an analogical study guide. *Journal of Educational Research, 83*(4), 233-237.

Becker, J. (1996). *Hungry ghosts: Mao's secret famine.* New York: Henry Holt.

Behe, M.J. (1996). *Darwin's black box: The biochemical challenge to evolution.* New York: Touchstone.

Behe, M.J. (1998). Intelligent design theory as a tool for analysing biochemical systems. In W. Dembski (Ed.), *Mere creation: Science, faith and intelligent design* (pp. 177-194). Downer's Grove, IL: Intervarsity Press.

Behe, M. (2003). The modern intelligent design hypothesis: Breaking the rules. In N.A. Manson (Ed.), *God and design: the teleological argument and modern science* (pp. 277-291). London: Routledge.

Behe, M.J. (2004). Irreducible complexity: obstacle to Darwinian evolution. In W.A. Dembski & M. Ruse (Eds.), *Debating design: From Darwin to DNA* (pp. 352-370). Cambridge: Cambridge University Press.

Bell, P. (1997). Using bargument representation to make thinking visible for individuals and groups. In C. Hall, N. Miyake & N. Enyedy (Eds.), *Proceedings of CSCL'97: The Second International Conference on Computer Support for Collaborative Learning* (pp. 10-19). Toronto: University of Toronto Press.

Bell, P. (2002). Using argument map representations to make thinking visible for individuals and groups. In T. Koschmann, R. Hall & N. Miyake (Eds.), *CSCL-2: Carrying forward to conversation* (pp. 449-485). Mahwah, NJ: Erlbaum.

Bell, P. (2004). Promoting students' argument construction and collaborative debate in the science classroom. In M.C. Linn, E.A. Davis & P. Bell (Eds.), *Internet environments for science education* (pp. 115-143). Mahwah, NJ: Lawrence Erlbaum.

Bell, P. & Linn, M.C. (2000). Scientific arguments as learning artefacts: Designing for learning from the web with KIE. *International Journal of Science Education, 22*(8), 797-818.

Bell, R. (1992). *Impure science*. New York: John Wiley.

Bell, R.L. (2004). Perusing Pandora's box. In L.B. Flick & N.G. Lederman (Eds.), *Scientific inquiry and nature of science: Implications for teaching, learning, and teacher education* (pp. 427-446). Dordrecht: Kluwer.

Bell, R.L., Blair, L.M., Crawford, B.A. & Lederman, N.G. (2003). Just do it? Impact of a science apprenticeship program on high school students' understandings of the nature of science and scientific inquiry. *Journal of Research in Science Teaching, 40*(5), 487-509.

Bell, R.L., Lederman, N.G. & Abd-El-Khalick, F. (2000). Developing and acting upon one's conception of the nature of science: A follow-up study. *Journal of Research in Science Teaching, 37*(6), 563-581.

Bencze, J.L. (1996). Correlational studies in school science: Breaking the science-experiment-certainty connection. *School Science Review, 78*(282), 95-101.

Bencze, J. L. (2000). Procedural apprenticeship in school science: Constructivist enabling of connoisseurship. *Science Education, 84*(6), 727-739.

Bencze, L., Bowen, M. & Alsop, S. (2006). Teachers' tendencies to promote student-led science projects: Associations with their views about science. *Science Education, 90*(3), 400-419.

Bencze, L. & DiGiuseppe, M. (2006). Explorations of a paradox in curriculum control: Resistance to open-ended science inquiry in a school for self-directed learning. *Interchange, 37*(4), 333-361.

Bencze, L. & Elshof, L. (2004). Science teachers as metascientists: An inductive-deductive dialectic immersion in northern alpine field ecology. *International Journal of Science Education, 26*(12), 1507-1526.

Bencze, L., Hewitt, J. & Pedretti, E. (2009). Personalizing and contextualizing multimedia case methods in university-based teacher education: An important modification for promoting technological design in school science. *Research in Science Education, 39*(1), 93-109.

Bencze, L. & Hodson, D. (1999). Changing practice by changing practice: Toward more authentic science and science curriculum development. *Journal of Research in Science Teaching, 36*, 521-539.

Bencze, L. & Lemelin, N. (2001). Doing science at a science centre: Enabling independent knowledge construction in the context of schools' museum visits. *Museum Management and Curatorship, 19*(2), 139-155.

Bennett, E.L. & Calvin, M. (1964). Failure to train planarians reliably. *Neurosciences Research Program Bulletin*, July-August, 3-24.

Bennett, N. & Dunne, E. (1995). Managing groupwork. In B. Moon & A Shelton Mayes (Eds.), *Teaching and learning in the secondary school* (pp. 166-172). London: Routledge.

Benson, A. (1986). *Children's understanding of science in four comprehensive schools*. Manchester: Unpublished MEd thesis, University of Manchester.

Bentley, M.L. (2000). Improvisational drama and the nature of science. *Journal of Science Teacher Education, 11*(1), 63-75.

Bereiter, C. (1994). Implications of postmodernism for science, or, science as progressive discourse. *Educational Psychologist, 29*(1), 3-12.

Bereiter, C. & Scardamalia, M. (1987). *The psychology of written composition*. Hillsdale, NJ: Erlbaum.

Bereiter, C., Scaradamalia, M., Cassells, C. & Hewitt, J. (1997). Postmodernism, knowledge building, and elementary science. *The Elementary School Journal, 97*(4), 329-340.

Berland, L.K. & Reiser, B.J. (2009). Masking sense of argumentation and explanation. *Science Education, 93*(1), 26-55.

Berry, A., Gunstone, R., Loughran, J. & Mulhall, P. (2001). Using laboratory work for purposeful learning about the oractice of science. In H. Behrendt, H. Dahncke, R. Duit, W. Graber, M. Komorek & A. Kross (Eds.), *Research in science education – Past, present and future* (pp. 313-318). Dordrecht: Kluwer.

Besson, U. & Viennot, L.D.S.P. (2004). Using models at the mesoscopic scale in teaching physics: Two experimental interventions in solid friction and fluid statics. *International Journal of Science Education, 26*(9), 1083-1110.

Bevilacqua, F. & Giannetto, E. (1995). The history of physics and European physics education. *Science & Education, 5,* 235-246.

Bhaskar, R. (1975). *A realist theory of science.* Hemel Hempstead: Harvester Wheatsheaf.

Bhaskar, R. (1989). *Reclaiming reality: A critical introduction to contemporary philosophy.* London: Verso.

Black, E. (2003). *War against the weak: Eugenics and America's campaign to create a master race.* New York: Thunder's Mouth Press.

Borgwald, J.M. & Schreiner, S. (1994). Science and the movies: The good, the bad, and the ugly. *Journal of College Science Teaching, 33*(6), 367-371.

Bianchini, J.A. (1997). Where knowledge constrution, equity, and context intersect: Student learning of science in small groups. *Journal of Research in Science Teaching, 34,* 1039-1065.

Bianchini, J.A. (1999). From here to equity: The influence of status on student access to and understanding of science. *Science Education, 83,* 577-601.

Bianchini, J.A. & Colburn, A. (2000). Teaching the nature of science through inquiry to prospective elementary teachers: A tale of two researchers. *Journal of Research in Science Teaching, 37*(2), 177-209.

Bianchini, J.A., Whitney, D.J., Breton, T.D. & Hilton-Brown, B.A. (2002). Toward inclusive science education: University scientists' views of students, instructional practices, and the nature of science. *Science Education, 86*(1), 42-78.

Billeh, V.Y. & Hasan, O. (1975). Factors affecting teachers' gain in understanding the nature of science. *Journal of Research in Science Teaching, 12*(3), 209-219.

Billig, M. (1996). *Arguing and thinking: A rhetorical approach to social psychology.* Cambridge: Cambridge University Press.

Binnie, A. (2001). Using the history of electricity and magnetism to enhance teaching. *Science & Education, 10,* 379-389.

Birke, L. (1986). *Women, feminism and biology.* New York: Methuen.

Birke, L. (1992). In pursuit of difference: Scientific studies of women and men. In G. Kircup & L.S. Keller (Eds.), *Inventing women: Science, technology and gender* (pp. 81-102). Milton Keynes: Open University Press.

Bisanz, G., Kachan, M., Guilbert, S., Bisanz, J., Sadler-Takach, B. & Noel, E. (2002). Reading science: New challenges outside and inside classrooms highlighting the need for curricular reform. Paper presented at the Conference on Ontological, Epistemological, Linguistic, and Pedagogical Considerations of Language and Science Literacy, Victoria, BC, September.

Black, G. (2005). *The engaging museum: Developing museums for visitor involvement.* London: RoutledgeFalmer.

Blake, A. (2004). Helping young children to see what is relevant and why: supporting cognitive change in earth science using analogy. *International Journal of Science Education, 26*(15), 1855-1873.

Blanchard, M.R., Southerland, S.A. & Granger, E.M. (2009). No silver bullet for inquiry: Making sense of teacher change following an inquiry-based research experience for teachers. *Science Education, 93*(2), 322-360.

Blank, L.M. (2000). A metacognitive learning cycle: A better warranty for student understanding? *Science Education, 84,* 486-506.

Blatt, F.J. (1992). *Modern physics.* New York: McGraw Hill.

Bleier, R. (1984). *Gender and science: A critique of biology and its theories on women.* Oxford: Pergamon Press.

Bleier, R. (Ed.) (1986). *Feminist approaches to science.* Oxford: Pergamon Press.

Blinderman, C. (1986). *The Piltdown inquest.* New York: Prometheus Books.

Bloom, J.W. (1992a). The development of scientific knowledge in elementary school children: A context of meaning perspective. *Science Education, 76*(4), 339-413.

Bloom, J.W. (1992b). Contexts of meaning and conceptual integration: How children understand and learn. In R.A. Duschl & R.J. Hamilton (Eds.), *Philosophy of science, cognitive psychology, and educational theory and practice* (pp. 177-194). Albany, NY: State University of New York Press.

Bloome, D., Puro, P. & Theodorou, E. (1989). Procedural display and classroom lessons. *Curriculum Inquiry, 19*, 265-291.

Bloor, D. (1974). Popper's mystification of objective knowledge. *Science Studies, 4*, 65-76.

Boden, M. (2001). Creativity and knowledge. In A. Craft, B. Jeffrey & M. Leibling (Eds.), *Creativity in education* (pp. 95-102). London: Continuum.

Bodzin, A. & Gehringer, M. (2001). Breaking science stereotypes. *Science and Children, 39*(1), 36-41.

Bohrmann, M.L. & Akerson, V.L. (2001). A teacher's reflections on her actions to improve her females students' self-efficacy toward science. *Journal of Elementary Science Education, 13*(2), 41-55.

Borenstein, J. (2008). Textbook stickers: A reasonable response to evolution? *Science & Education, 17*(8&9), 999-1010.

Borges, A.T., Tecnico, C. & Gilbert, J.K. (1998). Models of magnetism. *International Journal of Science Education, 20*(3), 361-378.

Born, M. (1924). *Einstein's theory of relativity*. Trans. H.L. Brose. London: Methuen.

Born, M. (1934). *Experiment and theory in physics*. Cambridge: Cambridge University Press.

Boscolo, P. & Mason, L. (2001). Writing to learn, writing to transfer. In P. Tynjala, L. Mason & K. Lonka (Eds.), *Writing as a learning tool* (pp. 83-104). Dordrecht: Kluwer.

Botton, C. & Brown, C. (1998). The reliability of some VOSTS items when used with preservice secondary science teachers in England. *Journal of Research in Science Teaching, 35*(1), 53-71.

BouJaoude, S. & Tamim, R. (2000). Analogies generated by middle-school science students: Types and usefulness. *School Science Review, 82*(299), 57-63.

Boulter, C.J. & Gilbert, J.K. (1995). Argument and science education. In P.S.M. Costello & S. Mitchell (Eds.), *Competing and consensual voices: The theory and practice of argumentation* (pp. 84-98) Clevedon (UK): Multilingual Matters.

Bowen, G.M. & Roth, W.-M. (2002). Of lizards, outdoors and indoors: Translating worlds in ecological fieldwork. In D. Hodson (Ed.), *OISE Papers in STSE Education*, Vol. 3 (pp. 167-189). Toronto: Imperial Oil Centre for Studies in Science, Mathematics and Technology Education, OISE-University of Toronto.

Bowen, G.M. & Roth, W-M. (2007). The practice of field ecology: Insights for science education. *Research in Science Education, 37*(2), 171-187.

Bowler, P.J. (1985). Scientific attitudes to Darwinism in Britain and America. In D. Kohn (Ed.), *The Darwinian heritage* (pp. 641-681). Princeton, NJ: Princeton University Press.

Bowler, P. (1992). *The Fontana history of the environmental sciences*. London: HarperCollins.

Bowler, P.J. (1996). *Life's splendid drama: Evolutionary biology and the reconstruction of life's ancestry, 1860–1940*. Chicago, IL: University of Chicago Press.

Boylan, C.R., Hill, D.M., Wallace, A.R. & Wheeler, A.E. (1992). Beyond stereotypes. *Science Education, 76*(5), 465-476.

Braaksma, M.A.H., Rijlaarsdam, G. & van den Bergh, H. (2002). Observational learning and the effects of model-observer similarirty. *Journal of Educational Psychology of Education, 94*(2), 405-415.

Braaksma, M.A.H., Rijlaarsdam, G., van den Bergh, H. & van Hout-Wolters, B.H.A.M. (2004). Observational learning and its effects on the orchestration of writing processes. *Cognition and Instruction, 22*(1), 1-36.

Braaksma, M.A.H., van den Bergh, H., Rijlaarsdam, G. & Couzijn, M. (2001). Effective learning acitivities in observation tasks when learning to write and read argumentative texts. *European Journal of Psychology of Education, 1*, 33-48.

Brandt, C.B. (2008). Scientific discourse in the academy: A case study of an American Indian undergraduate. *Science Education, 92*(5), 825-847.

Braund, M. & Reiss, M. (Eds.)(2004). *Learning science outside the classroom*. London: Routledge-Falmer.

Braund, M. & Reiss, M. (2006). Towards a more authentic science curriculum: The contribution of out-of-school learning. *International Journal of Science Education, 28*(12), 1373-1388.

Brayboy, B.M.J. & Castagno, A.E. (2008). How might Native science inform "informal science learning"? *Cultural Studies of Science Education, 3*, 731-750.

Brem, S.K., Russell, J. & Weems, L. (2001). Sciencqe on the web: Student evaluations of scientific arguments. *Discourse Processes, 32*(2&3), 191-213.

Brewer, W. & Samarapungavan, A. (1991). Children's theories versus scientific theories: Differences in reasoning or differences in knowledge? In R. Hoffman & D. Palermo (Eds.), *Cognition and the symbolic processes: Applied and ecological perspectives* (pp. 209-232). Hillsdale, NJ: Erlbaum.

Bricker, L.A. & Bell, P. (2008). Conceptualizations of argumentation from science studies and the learning sciences and their implications for the practices of science education. *Science Education, 92*(3), 473-498.

Brickhouse, N. (1990). Teachers' beliefs about the nature of science and their relation to classroom practice. *Journal of Teacher Education, 41*, 53-62.

Brickhouse, N.W. (2001). Embodying science: A feminist perspective on learning. *Journal of Research in Science Teaching, 38*(3), 282-295.

Brickhouse, N.W., Dagher, Z.R., Letts IV, W.J. & Shipman, H.L. (2000). Diversity of students' views about evidence, theory, and the interface between science and religion in an astronomy course. *Journal of Research in Science Teaching, 37*(4), 340-362.

Brickhouse, N.W., Dagher, Z.R., Letts, W.J. (IV) & Shipman, H.L. (2002). Evidence and warrants for belief in a college astronomy course. *Science & Education, 11*, 573-588.

Brickhouse, N.W., Lowery, P. & Schultz, K. (2000). What kind of girl does science? The construuction of school science identities. *Journal of Research in Science Teaching, 37*(5), 441-458.

Brickhouse, N.W. & Potter, J.T. (2001). Young women's scientific identity formation in an urban context. *Journal of Research in Science Teaching, 38*(8), 965-980.

Bridgman, P.W. (1950). *The reflections of a physicist.* New York: Philosophical Library.

Brígida, E. & Falcão, M. (2008). Religious beliefs: Their dynamics in two groups of life scientists. *International Journal of Science Education, 30*(9), 1249-1264.

Brillant, L.V. & Debs, M.B. (1981). Teaching writing: A scientist's responsibility. *Journal of College Science Teaching, 10*(5), 303-304.

British Council (1997). Science across the world. *Science Education Newsletter, 134* (October), 7-8.

Broad, W. & Wade, N. (1982). *Betrayers of the truth.* New York: Simon & Schuster.

Brockman, J. (1995). *The third culture: Beyond the scientific revolution.* London: Charles Scribner.

Brooke, J.H. (1991). *Science and religion: Some historical perspectives.* Cambridge: Cambridge University Press.

Brotman, J.S. & Moore, F.M. (2008). Girls and science: A review of four themes in the science education literature. *Journal of Research in Science Teaching, 45*(9), 971-1002.

Brown, A.L. (1990). Domain specific principles affect learning andtransfer in children. *Cognitive Science, 14*, 107-133.

Brown, B.A. (2006). "It isn't no slang that can be said about this stuff": Language, identity, and appropriating science discourse. *Journal of Research in Science Teaching, 43*(1), 96-126.

Brown, B.A., Reveles, J.M. & Kelly, G.J. (2005). Scientific literacy and discursive identity: A theoretical framwework for understanding science learning. *Science Education, 89*(6), 779-802.

Brown, B.A. & Ryoo, K. (2008). Teaching science as a language: A "content-first" approach to science teaching. *Journal of Research in Science Teaching, 45*(5), 529-553.

Brown, B.A. & Spang, E. (2008). Double talk: Synthesizing everyday and science language in the classroom. *Science Education, 92*(4), 708-732.

Brown, J.R. (1986). Thought experiments sibce the scientific revolution. *International Studies in the Philosophy of Science, 1*(1), 1-15.

Brown, J.R. (1991). *The laboratory of the mind: Thought experiments in the natural sciences.* New York: Routledge.

Brown, J.R. (2006). The promise and perils of thought experiments. *Interchange, 37*(1&2), 63-75.

Brown, J.S., Collins, A. & Duguid, P. (1989). Situated cognition and the culture of learning. *Educational Researcher, 18*, 32-42.

Brown, K. (2004). *Penicillin man: Alexander Fleming and the antibiotic revolution.* Stroud: Sutton Publishing.

Brown, S.L. & Melear, C.T. (2006). Investigation of secondary science teachers' beliefs and practices after authentic inquiry-based experiences. *Journal of Research in Science Teaching, 43*(9), 938-962.

Brown, T.M. & Dronsfield, A.T. (1991). The phlogiston theory revisited. *Education in Chemistry, 28*(2), 43-45.

Bruner, J.S. (1986). *Actual minds, possible worlds.* Cambridge: Cambridge University Press.

Bruner, J.S. (1990). *Acts of meaning.* Cambridge, MA: Harvard University Press.

Bruner, J.S. (1996). *The culture of education*. Cambridge, MA: Harvard University Press.

Brush, S. (1974). Should the history of science be rated X? *Science, 183*(4130), 1164-1172.

Brush, S.G. (1985). Women in physical science: From drudges to discoverers. *The Physics Teacher, 23*, 11-19.

Brush, S.G. (1989). History of science and science education. *Interchange, 20*(2), 60-70.

Brusic, S.A. (1992). Achieving STS goals through experiential learning. *Theory into Practice, 31*(1), 44-51.

Buckley, B.C. (2000). Interactive multimedia and model-based learning in biology. *International Journal of Science Education, 22*(9), 895-935.

Bunge, M. (1966). Technology as applied science. *Technology and Culture, 7*(3), 329-347.

Buntting, C. & Jones, A. (2009). Unpacking the interface between science, technology and the environment: Biotechnology as an example. In A.T. Jones & M.J.deVries (Eds.), *International handbook of research and development in technology education* (pp. 335-348). Rotterdam/Taipei: Sense.

Burdett, P. (1989). Adventures with N-rays: An approach to teaching about scientific theory and theory evaluation. In R. Millar (Ed.), *Doing science: Images of science in science education* (pp. 180-204). Lewes: Falmer Press.

Burns, J.C., Okey, J.R. & Wise, K.C. (1985). Development of an integrated process skills test: TIPS II. *Journal of Research in Science Teaching, 22*(2), 169-177.

Burton, C. & Scott, C. (2003). Museums: Challenges for the 21st century. *International Journal of Arts Management, 5*(2), 56-67.

Butterfield, H. (1931). *The Whig interpretation of history*. London: G. Bell & Sons.

Butzow, C.M. & Butzow, J.W. (Eds.) (1989). *Science through children's literature: An integrated approach*. Englewood, CO: Teacher Ideas Press.

Butzow, C.M. & Butzow, J.W. (Eds.) (1998). *More science through children's literature: An integrated approach*. Englewood, CO: Teacher Ideas Press.

Buxton, C.A. (2001). Modeling science teaching on science practice? Painting a more accurate picture through an ethnographic lab study. *Journal of Research in Science Teaching, 38*(4), 387-407.

Buxton, C.A. (2006). Creating contextually authentic science in a "low-performing" urban elementary school. *Journal of Research in Science Teaching, 43*(7), 695-721.

Buzan, T. (1995). *Use your head*. London: BBC Books.

Bybee, R. (2001). Teaching about evolution: Old controversy, new challenges. *Bioscience, 51*, 309-313.

Bybee, R.W., Powell, J.C., Ellis, J.D., Giese, J.R., Parisi, L. & Singleton, L. (1991). Integrating the history and nature of science and technology in science and social studies curriculum. *Science Education, 75*(1), 143-155.

Cahill, T.A., Schwab, R.N., Kusko, B.H., Eldred, R.A., Moller, G., Dutschke, D. & Wick, D.L. (1987). The Vinland Map revisited: New compositional evidence on its inks and parchment. *Analytical Chemistry, 59*, 829-833.

Cameron, L. (2002). Metaphors in the learning of science: A discourse focus. *British Educational Research Journal, 28*(5), 673-688.

Campbell, B., Kaunda, L., Allie, S., Buffler, A. & Lubben, F. (2000). The communication of laboratory investigations by university entrants. *Journal of Research in Science Teaching, 37*(8), 839-853.

Capra, F. (1983). *The turning point: Science, society and the rising culture*. New York: Bantam.

Carey, R.L. & Strauss, N.G. (1968). An analysis of the understanding of the nature of science by prospective secondary science teachers. *Science Education, 58*(4), 358-363.

Carey, R.L. & Strauss, N.G. (1970). An analysis of experienced science teachers' understanding of the nature of science. *School Science & Mathematics, 70*, 366-376.

Carey, S. (1986). Cognitive science and science education. *American Psychologist, 41*, 123-130.

Carey, S., Evans, E., Honda, M., Jay, E. & Unger, C. (1989). "An experiment is when you try it and see if it works": A study of grade 7 students' understanding of the construction of scientific knowledge. *International Journal of Science Education, 11*, 514-529.

Carey, S. & Smith, C. (1993). On understanding the nature of scientific knowledge. *Educational Psychologist, 28*(3), 235-251.

Carlone, H.B. (2004). The cultural production of science in reform-based physics: Girls' access, participation, and resistance. *Journal of Research in Science Teaching, 41*, 392-414.

Carlsen, W.S. (1991a). Questioning in classrooms: A sociological perspective. *Review of Educational Research, 61*, 157-178.

Carlsen, W.S. (1991b). Subject matter knowledge and science teaching: A pragmatic approach. In J.E. Brophy (Ed.), *Advances in research on teaching* (pp. 115-143). Greenwich, CT: JAI Press.

Carlsen, W.S. (1992). Closing down the conversation: Discouraging student talk on unfamiliar science content. *Journal of Classroom Interaction, 27*, 15-21.

Carlsen, W.S. (1997). Never ask a question if you don't know the answer: The tension in teaching between modeling scientific argument and maintaining law and order. *Journal of Classroom Interaction, 32*(2), 14-23.

Carlsen, W.S. (2007). Langauge and science learning. In S.K. Abell & N.G. Lederman (Eds.), *Handbook on research in science education* (pp. 57-74). Mahwah, NJ: Lawrence Erlbaum.

Carson, R.N. (1998). Science and the ideals of liberal education. In B.J. Fraser & K.G. Tobin (Eds.), *International handbook of science education* (pp. 1001-1014). Dordrecht: Kluwer.

Cartwright, J. (2007). Science and literature: Towards a conceptual framework. *Science & Education, 16*, 115-139.

Cartwright, N. (1983). *How the laws of physics lie*. Oxford: Oxford University Press.

Cartwright, N. (1999). *The dappled world: A study of the boundaries of sciences*. Cambridge: Cambridge University Press.

Cassels, J.R.T. & Johnstone, A.H. (1985). *Words that matter in science*. London: Royal Society of Chemistry.

Castro, R.S. & Carvalho, M.P. (1995). The historic approach in teaching: Analysis of an experience. *Science & Education, 4*(1), 65-85.

Caton, E., Brewer, C. & Brown, F. (2000). Building teacher-scientist partnerships: Teaching about energy through inquiry. *School Science & Mathematics, 100*(1), 7-15.

Cavalli-Sforza, V., Lesgold, A.M. & Weiner, A.W. (1992). Strategies for contributing to collaborative arguments. In *Proceedings of the Fourteenth Annual Conference of the Cognitive Science Society* (pp. 755-760). Hillsdale, NJ: Lawrence Erlbaum Associates.

Cavallo, A.M.L., Rozman, M., Blickenstaff, J. & Walker, N. (2003). Learning, reasoning, motivation, and epistemological beliefs. *Journal of College Science Teaching, 33*, 18-23.

Cavicchi, E.M. (2008). Historical experiments in students' hands: Unfragmenting science through action and history. *Science & Education, 17*(7), 717-749.

Cawthron, E.R. & Rowell, J.A. (1978). Epistemology and science education. *Studies in Science Education, 5*, 31-59.

Celik, S. & Bayrakçeken, S. (2006). The effect of a "science, technology and society" course on prospective teachers' conceptions of the nature of science. *Research in Science & Technological Education, 24*(2), 255-273.

Cerullo, M.M. (1997). *Reading the environment: Literature in the science classroom*. Portsmouth, NH: Heinemann.

Chambers, D.W. (1983). Stereotypic images of the scientist: The draw-a-scientist test. *Science Education, 67*(2), 255-265.

Chambers, D.W. (1984). *Worm in the bud: Case study of the pesticide controversy*. Geelong: Deakin University Press.

Chambers, S.K. & Andre, T. (1997). Gender, prior knowledge, interest, and experience in electricity and conceptual change text manipulations in learning about direct current. *Journal of Research in Science Teaching, 34*, 107-123.

Champagne, A.B., Kouba, V.L. & Hurley, M. (2000). Assessing inquiry. In J. Minstrell & E.H. van Zee (Eds.), *Inquiring into inquiry learning and teaching in science* (pp. 447-470). Washington, DC: American Association for the Advancement of Science.

Chang, S-N. & Chiu, M-H. (2008). Lakatos' scientific research programmes as a framework for analysing informal argmentation about socio-scientific issues. *International Journal of Science Education, 30*(13), 1753-1773.

Charney, J., Hmelo-Silver, C.E., Sofer, W., Neigeborn, L., Coletta, S. & Nemeroff, M. (2007). Cognitive apprenticeship in science through immersion in laboratory practices. *International Journal of Science Education, 29*(2), 195-213.

Chen, D. & Novick, R. (1984). Scientific and technological education in an information society. *Science Education, 68*(4), 421-426.

Chen, S. (2006). Development of an instrument to assess views on nature of science and attitudes toward teaching science. *Science Education, 90*(5), 803-819.

Chi, M.T. (1992). Conceptual within and across ontological categories: Examples from learning and discovery in science. In R. Giere (Ed.), *Cognitive models of science: Minnesota Studies in the Philosophy of Science* (pp. 129-186). Minneapolis, MN: University of Minnesota Press.

Chi, M.T.H., Feltovich, P. & Glaser, R. (1981). Categorization and representation of physics problems by experts and novices. *Cognitive Science, 5,* 121-152.

Chi, M.T.H., Glaser, R. & Farr, M.J. (Eds.) (1988). *The nature of expertise.* Hillsdale, NJ: Lawrence Erlbaum.

Chiappetta, E.L.M., Sethna, G.H. & Dillman, D.A. (1991). A qualitative analysis of high school chemistry textbooks for scientific literacy themes and expository learning aids. *Journal of Research in Science Teaching, 28,* 939-951.

Chin, C. & Chia, L.G. (2004). Problem-based learning: Using students' questions to drive knowledge construction. *Science Education, 88*(5), 707-727.

Chinn, C.A. (2006). Learning to argue. In A.M. O'Donnell & C.E. Hmelo-Silver (Eds.), *Argumentation and technology* (pp. 355-383). Mahwah, NJ: Erlbaum.

Chinn, C.A. & Brewer, W.F. (1993). The role of anomalous data in knowledge acquisition: A theoretical framework and implications for science instruction. *Review of Educational Research, 63,* 1-49.

Chinn, C.A. & Brewer, W.F. (1998). An empirical test of a taxonomy of responses to anomalous data in science. *Journal of Research in Science Teaching, 35,* 623-654.

Chinn, C.A. & Malhotra, B.A. (2002a). Children's responses to anomalous data: How is conceptual change impeded? *Journal of Educational Psychology, 94,* 327-343.

Chinn, C.A. & Malhotra, B.A. (2002b). Epistemologically authentic inquiry in schools: A theoretical framework for evaluating inquiry tasks. *Science Education, 86*(2), 175-218.

Chinn, C.A. & Samarapungavan, A. (2008). Learning to use scientific models: Multiple dimensions of conceptual change. In R.A. Duschl & R.E. Grandy (Eds.), *Teaching scientific inquiry: Recommendations for research and implementation* (pp. 191-225). Rotterdam: Sense Publishers.

Chinn, P.W.U. (2007). Decolonizing methodologies and indigenous knowledge: The role of culture, place and personal experience in professional development. *Journal of Research in Science Teaching, 44*(9), 1247-1268.

Chinn, P.W.U. & Hilgers, T.L. (2000). From corrector to collaborator: The range of instructor roles in writing-based natural and applied science classes. *Journal of Research in Science Teaching, 37*(1), 3-25.

Chittleborough, G.D., Treagust, D.F., Mamiala, T.L. & Mocerino, M. (2005). Students' perceptions of the role of models in the process of science and in the process of learning. *Research in Science & Technological Education, 23*(2), 195-212.

Chiu, M.H. & Lin, J.W. (2005). Promoting fourth graders' conceptual change of their understanding of electric current via multiple analogies. *Journal of Research in Science Teaching, 42*(4), 429-464.

Cho, K-L. & Jonassen, D.H. (2002). The effects of argumentation scaffolds on argmentation and problem solving. *Educational Technology, Research & Development, 50*(3), 5-22.

Chomsky, N. (1969). *American power and the new mandarins.* Harmondsworth: Penguin.

Chowning, J.T. (2002). The student biotechnology expo: A new model for a science fair. *American Biology Teacher, 64*(6), 331-339.

Christidou, V., Kouladis, V. & Christidis, T. (1997). Children's use of metaphors in relation to their mental models: The case of the ozone layer and its depletion. *Research in Science Education, 27*(4), 541-552.

Christie, M. (1991). Aboriginal science for the ecologically sustainable future. *Australian Science Teachers Journal, 37,* 26-31.

Chuang, H.C. (2003). Teaching evolution: Attitudes and strategies of educators in Utah. *American Biology Teacher, 65,* 669-674.

Clark, D.B. & Sampson, V.D. (2007). Personally-seeded discussions to scaffold online argumentation. *International Journal of Science Education, 29*(3), 253-277.

Clark, D.B. & Sampson, V.D. (2008). Assessing dialogic argumentation in online environments to relate structure, grounds, and conceptual quality. *Journal of Research in Science Teaching, 45*(3), 293-321.

REFERENCES

Clark, D.B., Sampson, V.D., Weinberger, A. & Erkens, G. (2007). Analytic frameworks for assessing dialogic argumentation in online environments. *Educational Psychology Review, 19*(3), 343-374.

Clark, D.B., Stegmann, K., Weinberger, A., Menekse, M. & Erkens, G. (2008). Technology-enhanced learning environments to support students' argumentation. In S. Erduran & M.P. Jiménez-Aleixandre (Eds.), *Argumentation in science education: Perspectives from classroom-based research* (pp. 217-243). Dordrecht: Kluwer.

Claxton, G. (1990). *Teaching to learn: A direction for education.* London: Cassell.

Claxton, G. (1997). Science of the times: A 2020 vision of education. In R. Levinson & J. Thomas (Eds.), *Science today: Problem or crisis?* (pp. 71-86). London: Routledge

Clement, J. (1988). Observed methods for generating analogies in scientific problem solving. *Cognitive Science, 12,* 563-586.

Clement, J. (1989). Generation of spontaneous analogies by students solving science problems. In D. Topping, D. Crowell & V. Kobayashi (Eds.), *Thinking across cultures* (pp. 303-308). Hillsdale, NJ: Erlbaum.

Clement, J. (1993). Using bridging analogies and anchoring intuitions to deal with students' preconceptions in physics. *Journal of Research in Science Teaching, 30,* 1241-1257.

Clement, J. (1998). Expert novice similarities and instruction using analogies. *International Journal of Science Education, 20,* 1271-1286.

Clement, J.J. & Rea-Ramirez, M.A. (Eds.) (2008). *Model based learning and instruction in science.* Dordrecht: Springer.

Close, F. (1991). *Too hot to handle: The race for cold fusion.* Princeton, NJ: Princeton University Press.

Clough, M.P. (1997). Strategies and activities for initiating and maintaining presuure on students' naïve views concerning the nature of science. *Interchange, 28*(2-3), 191-204.

Clough, M.P. (2006). Learners' responses to the demands of conceptual change: Considerations for effective nature of science instruction. *Science Education, 15,* 463-494.

Clough, M.P. & Olson, J.K. (2008). Teaching and assessing the nature of science: An introduction. *Science & Education, 17*(2&3), 143-145.

Cobern, W.W. (1991). *World view theory and science education research.* NARST Monograph No. 3. Manhattan, KS: National Association for Research in Science Teaching.

Cobern, W.W. (1993). Contextual constructivism: The impact of culture on the learning and teaching of science. In K. Tobin (Ed.), *The practice of constructivism in science education* (pp. 51-69). Hillsdale, NJ: Lawrence Erlbaum.

Cobern, W. (1995). Science education as an exercise in foreign affairs. *Science & Education, 4*(3), 287-302.

Cobern, W.W. (1996). Worldview theory and conceptual change in science education. *Science Education, 80*(5), 579-610.

Cobern, W.W. & Aikenhead, G.S. (1998). Cultural aspects of learning science. In B.J. Fraser & K.G. Tobin (Eds.), *International handbook of science education* (pp. 39-52). Dordrecht: Kluwer.

Cobern, W.W. & Loving, C.C. (2001). Defining "science" in a multicultural world: Implications for science education. *Science Education, 85,* 50-67.

Cobern, W.W. & Loving, C.C. (2002). Investigation of preservice elementary teachers' thinking about science. *Journal of Research in Science Teaching, 39*(10), 1016-1031.

Cobern, W.W. & Loving, C.C. (2008). An essay for educators: Epistemological realism really is common sense. *Science & Education, 17*(4), 425-447.

Code, L. (1987). *Epistemic responsibility.* Hanover, MA: University of New England Press.

Code, L. (1991). *What can she know?* Ithaca, NY: Cornell University Press.

Colburn, A. & Henriques, L. (2006). Clergy views on evolution, creationism, science, and religion. *Journal of Research in Science Teaching, 43*(40), 419-442.

Coll, R.K. (2006). The role of models, mental models and analogies in chemistry teaching. In P.J. Aubusson, A.G. Harrison & S.M. Ritchie (Eds.), *Metaphor and analogy in science education* (pp. 65-77). Dordrecht: Kluwer.

Coll, R.K., France, B. & Taylor, I. (2005). The role of models/and analogies in science education: Implications from research. *International Journal of Science Education, 27*(2), 183-198.

Coll, R.K. & Taylor, I. (2004). Probing scientists' beliefs: How open-minded are modern scientists? *International Journal of Science Education, 26*(6), 757-778.

Coll, R.K. & Taylor, I. (2005). The role of models and analogies in science education: Implications from research. *International Journal of Science Education, 27*, 183-198.

Coll, R.K. & Treagust, D.F. (2002). Learners' mental models of covalent bonding. *Research in Science & Technological Education, 20*(2), 241-268.

Coll, R.K. & Treagust, D.F. (2003a). Learners' mental models of metallic bonding: A cross-age study. *Science Education, 87*(5), 685-707.

Coll, R.K. & Treagust, D.F. (2003b). Investigation of secondary school, undergraduate and graduate learners' mental models of ionic bonding. *Journal of Research in Science Teaching, 40*(5), 464-486.

Collins, A., Brown, J.S. & Newman, S.E. (1989). Cognitive apprenticeship: Teaching the craft of reading, writing, and mathematics. In L.B. Resnick (Ed.), *Knowing, learning, and instruction: Essays in honor of Robert Glaser* (pp. 453-494). Hillsdale, NJ: Lawrence Erlbaum.

Collins, F.S. (2006). *The language of God: A scientist presents evidence for belief.* New York: Simon & Schuster.

Collins, H.M. & Pinch, T.J. (1993). *The golem: What everyone should know about science.* Cambridge: Cambridge University Press.

Collins, N. (2004). Scientific research and school students. *School Science Review, 85*(312), 77-86.

Comeaux, P. & Huber, R. (2001). Students as scientists: Using interactive technologies and collaborative inquiry in an environmental science project for teachers and their students. *Journal of Science Teacher Education, 12*(4), 235-252.

Conant, J.B. (1951). *On understanding science: An historical approach.* New York: New American Library.

Constable, H., Campbell, B. & Brown, R. (1988). Sectional drawings from science textbooks: An experimental investigation into pupils' understanding. *British Journal of Educational Psychology, 58*, 89-102.

Cooley, W.W. & Klopfer, L.E. (1961). *Manual for the test on understanding science.* Princeton, NJ: Educational Testing Service.

Cornelius, L. & Herrenkohl, L.R. (2004). Power in the classroom: How the classroom environment shapes students' relationships with each other and with concepts. *Cognition and Instruction, 22*(4), 467-498.

Cosgrove, M. (1995). A case study of science-in-the-making as students generate an analogy for electricity. *International Journal of Science Education, 17*, 295-310.

Costa, V.B. (1995). When science is "another world": relationships between worlds of family, friends, school, and science. *Science Education, 79*, 313-333.

Costa, V.B. (1997). How teachers and students study "all that matters" in high school chemistry. *International Journal of Science Education, 19*, 1005-1023.

Cotham, J. & Smith, E. (1981). Development and validation of the conceptions of scientific theories test. *Journal of Research in Science Teaching, 18*, 387-396.

Council of Ministers of Education, Canada (1997). *Common framework of science learning outcomes.* Toronto: CMEC Secretariat.

Couzijn, M. (1999). Learning to write by observation of writing and reading processes: Effects on learning and transfer. *Learning and Instruction, 9*, 109-142.

Couzijn, M. & Rijlaarsdam, G. (2004). Learning to read and write argumentative text by observation. In G. Rijlaarsdaam, H. van den Bergh & M. Couzijn (Eds.), *Studies in writing. Vol. 14: Effective learning and teaching of writing*, 2nd ed. Part 1: Studies in learning to write (pp. 241-258). Dordrecht: Kluwer.

Crasnow, S. (2008). Feminist philosophy of science: "Standpoint" and knowledge. *Science & Education, 17*(10), 1089-1110.

Craven III, J.A., Hand, B. & Prain, V. (2002). Assessing explicit and tacit conceptions of the nature of science among preservice elementary teachers. *International Journal of Science Education, 24*(8), 785-802.

Crawford, B.A. (2000). Embracing the essence of inquiry: New roles for science teachers. *Journal of Research in Science Teaching, 37*(9), 916-937.

Crawford, B.A. (2007). Learning to teach science as inquiry in the rough and tumble of practice. *Journal of Research in Science Teaching, 44*(4), 613-642.

Crawford, B.A. & Cullin, M.J. (2004). Supporting prospective teachers' conceptions of modelling in science. *International Journal of Science Education, 26*(11), 1379-1401.

REFERENCES

Crawford, B.A., Krajcik, J.S. & Marx, R.W. (1999). Elements of a community of learners in a middle school science classroom. *Science Education, 83*(6), 701-723.

Crawford, E. (1993). A critique of curriculum reform: Using history to develop thinking. *Physics Education, 28,* 204-208.

Crawford, T., Kelly, G.J. & Brown, C. (2000). Ways of knowing beyond facts and laws of science: An ethnographic investigation of student engagement in scientific practices. *Journal of Research in Science Teaching, 37,* 237-258.

Crosland, M. (1962). *Historical studies in the language of chemistry.* Cambridge, MA: Harvard University Press.

Cross, D., Taasoobshirazi, G., Hendricks, S. & Hickey, D.T. (2008). Argumentation: A strategy for improving achievement and revealing scientific identities. *International Journal of Science Education, 30*(6), 837-861.

Cross, R. (1995). Conceptions of scientific literacy: Reactionaries in ascendancy in the state of Victoria. *Research in Science Education, 25*(2), 151-162.

Crouch, R.D., Holden, M.S. & Samet, C. (1996). CAChe molecular modeling: A visualization tool earl in the undergraduate chemistry curriculum. *Journal of Chemical Education, 73*(10), 916-917.

Culliton, B.J. (1983a). Coping with fraud: The Darsee case. *Science, 220*(4592), 31-35

Culliton, B.J. (1983b). Emory reports on Darsee's fraud. *Science, 220*(4600), 936.

Culver, R.B. & Ianna, P.A. (1984). *The gemini syndrome: A scientific evaluation of astrology.* Buffalo, NY: Prometheus Press.

Cunniff, P. & McMillen, J. (1996). Field studies: Hands-on, real-science research. *The Science Teacher, 63,* 48-51.

Dagher, Z.R. (1994). Does the use of analogies contribute to conceptual change? *Science Education, 78*(6), 601-614.

Dagher, Z.R. (1995a). Analysis of analogies used by science teachers. *Journal of Research in Science Teaching, 32*(3), 259-270.

Dagher, Z.R. (1995b). Review of studies on the effectiveness of instructional analogies in science education. *Science Education, 79*(3), 295-312.

Dagher, Z.R. & BouJaoude, S. (1997). Scientific views and religious beliefs of college students: The case of biological evolution. *Journal of Research in Science Teaching, 34,* 429-445.

Dagher, Z.R. & BouJaoude, S. (2005). Students' perceptions of the nature of evolutionary theory. *Science Education, 89*(3), 378-391.

Dagher, Z., Brickhouse, N., Shipman, H. & Letts, W. (2004). How some college students represent their understanding of scientific theories. *International Journal of Science Education, 26*(6), 735-755.

Daiute, C. (1997). Youth genres in the classroom: Can childrens' and teachers' cultures meet? In J. Flood, S.B. Heath & D. Lapp (Eds.), *Handbook of research on teaching literacy through the communicative and visual arts* (pp. 323-333). New York: Macmillan.

Dakers, J.R. (Ed.)(2006). *Defining technological literacy: towards an epistemological framework.* New York: PalgraveMacmillan.

Dana, T., Lorsbach, A., Hook, K. & Briscoe, C. (1992). Students showing what they know: A look at alternative assessments. In G. Kulm & S. Malcolm (Eds.), *Science assessment in the service of reform* (pp. 331-337). Washington, DC: American Association for the Advancement of Science.

Danserau, D.F. (1985). Learning strategy research. In J. Segal, S. Chipman & R. Glaser (Eds.), *Thinking and learning skills: Relating instruction to research,* Vol. 1 (pp. 209-239). Hillsdale, NJ: Erlbaum.

Darden, L. & Craver, C.F. (2002). Strategies in the interfield discovery of the mechanism of protein synthesis. *Studies in History and Philsophy of Biological and Biomedical Sciences, 33,* 1-28.

Dass, P.M. (2005). Understanding the nature of scientific enterprise (NOSE) through a discourse with its history: The influence of an undergraduate "history of science" course. *International Journal of Science and Mathematics Education, 3,* 87-115.

Daugs, D.R. (1970). Scientific literacy – Re-examined. *The Science Teacher, 37*(8), 10-11.

Davies, F. & Greene, T. (1984). *Reading for learning in the sciences.* Edinburgh: Oliver & Boyd.

Davies, I. (2004). Science and citizenship education. *International Journal of Science Education, 26*(14), 1751-1763.

Davies, P.C.W. (1984). *God and the new physics.* Harmondsworth: Penguin.

366

Davies, P.C.W. (1992). *The mind of God: The scientific basis for a rational world. Science and the search for ultimate meaning.* New York: Simon & Schuster.

Davies, T. & Gilbert, J. (2003). Modelling: Promoting creativity while forging links between science education and design and technology education. *Canadian Journal of Science, Mathematics and Technology Education, 3*(1), 67-82.

Davis, E.A. & Linn, M.C. (2000). Scaffolding students' knowledge integration: Prompts for reflection in KIE. *International Journal of Science Education, 22*, 819-837.

Davison, D. & Miller, K. (1998). An ethnoscience approach to curriculum issues for American Indian students. *School Science & Mathematics, 98*(5), 260-265.

Davson-Galle, P. (2008). Why compulsory science education should not include philosophy of science. *Science & Education, 17*, 677-716.

Dawes, L. (2004). Talk and learning in classroom science. *International Journal of Science Education, 26*(6), 677-695.

Dawkins, R. (1976). *The selfish gene.* Oxford: Oxford University Press.

Dawkins, R. (1981). *The extended phenotype: The gene as the unit of selection.* San Francisco, CA: Freeman.

Dawkins, R. (1986). *The blind watchmaker: Why the evidence of evolution reveals a universe without design.* New York: W.W. Norton.

Dawkins, R. (1995). *River out of Eden.* New York: Basic Books.

Dawkins, R. (1998). *Unweaving the rainbow: Science, delusion, and the appetite for wonder.* New York: Houghton Mifflin.

Dawkins, R. (2006). *The God delusion.* London: Bantam Books.

Dawson, V. & Schibeci, R. (2003). Western Australian school students' understanding of biotechnology. *International Journal of Science Education, 25*, 57-69.

DeBoer, G. (2001). Scientific literacy: Another look at its historical and contemporary meanings and its relationship to science education reform. *Journal of Research in Science Teaching, 37*(6), 582-601.

DeCoito, I. (2008). *Writing in science: Exploring teachers' and students' views of the nature of science in language enriched environments.* Toronto: Unpublished PhD thesis, University of Toronto.

de Hosson, C. & Kaminski, W. (2007). Historical controversy as an educational tool: Evaluating elements of a teaching-learning sequence conducted with the text "Dialogue on the ways that vision operates". *International Journal of Science Education, 29*(5), 617-642.

Delpit, L. (1988). The silenced dialogue: Power and pedagogy in educating other people's children. *Harvard Educational Review, 58*(3), 280-298.

Delpit, L. (1995). *Other people's children: Cultural conflict in the classroom.* New York: The New Press.

Demastes, S.S., Good, R.G. & Peebles, P. (1995). Students' conceptual ecologies and the process of conceptual change in evolution. *Science Education, 79*(6), 637-666.

Dembski, W. (1998a). Introduction. In W. Dembski (Ed.), *Mere creation: Science, faith and intelligent design.* Downer's Grove, IL: Intervarsity Press.

Dembski, W. (Ed.) (1998b). *Mere creation: Science, faith and intelligent design.* Downer's Grove, IL: Intervarsity Press.

Department of Education (Republic of South Africa) (2002). Revised national curriculum statement for grades R-9 (schools) – Natural sciences. Pretoria: Department of Education, *Government Gazette*, Vol. 443, No. 23406.

Department of Education & Science (1975). *A language for life* (The Bullock Report). London: HMSO.

Department of Education & Science/Welsh Office (1989). *Science in the national curriculum.* London: HMSO.

Department for Education & Skills (2002). *ImpaCT2 – Final Report.* London: DfES/ British Educational Communications and Technology Agency.

Désautels, J., Fleury, S. & Garrision, J. (2002). The enactment of epistemological practice as subversive social action, the provocation of power, and anti-modernism. In W-M. Roth & J. Désautels (Eds.), *Science education as/for sociopolitical action* (pp. 237-269). New York: Peter Lang.

De Solla Price, D.J. (1965). Is technology historically independent of science? *Technology and Culture, 6*, 553-567.

REFERENCES

De Solla Price, D.J. (1969). The structures of publication in science and technology. In W. Gruber & G. Marquis (Eds.), *Factors in the transfer of technology* (pp. 91-104). Cambridge, MA: MIT Press.

Dhingra, K. (2003). Thinking about television science: How students understand the nature of science from different program genres. *Journal of Research in Science Teaching, 40*, 234-256.

Dhingra, K. (2006). Science on television: Storytelling, learning and citizenship. *Studies in Science Education, 42*, 89-124.

Dibbs, D.R. (1982). *An investigation into the nature and consequences of teachers' implicit philosophies of science.* Birmingham: Unpublished PhD thesis, University of Aston.

Dillashaw, F.G. & Okey, J.R. (1980). Test of integrated process skills for secondary school science students. *Science Education, 64*(5), 601-608.

Dillon, J.T. (1994). *Using discussions in the classroom.* Buckingham: Open University Press.

Dimopoulos, K. & Koulaidis, V. (2003). Science and technology education for citizenship: The potential role of the press. *Science Education, 87*(2), 241-256.

Dirac, P.A.M. (1963). The evolution of the physicist's picture of nature. *Scientific American, 208*(5), 45-63.

Dirac, P.A.M. (1980). The excellence of Einstein's theory of gravitation. In M. Goldsmith, A. Mackay & J. Woudhuysen (Eds.), *Einstein: The first hundred years* (pp. 41-46). Oxford: Pergamon Press.

Dixon, B. (1973). *What is science for?* London: Collins.

Dobzhansky, T. (1973). Nothing in biology makes sense except in the light of evolution. *American Biology Teacher, 35*, 125-129.

Dogan, N. & Abd-El-Khalick, F. (2008). Turkish grade 10 students' and science teachers' conceptions of nature of science: A national study. *Journal of Research in Science Teaching, 45*(10), 1083-1112.

Donahue, D.J., Olin, J.S. & Harbottle, G. (2002). Determination of the radiocarbon age of parchment of the Vinland Map. *Radiocarbon, 44*, 45-52.

Donnelly, J. (2005). Reforming science in the school curriculum: A critical analysis. *Oxford Review of Education, 31*(2), 293-309.

Donnelly, J., Buchan, A., Jenkins, E., Laws, P. & Welford, G. (1996). *Investigations by order: Policy, curriculum and science teachers' work under the Education Reform Act.* Nafferton: Studies in Science Education.

Donnelly, L.A. & Boone, W.J. (2007). Biology teachers' attitudes toward and use of Indiana's evolution standards. *Journal of Research in Science Teaching, 44*(2), 236-257.

Dorfman, D.D. (1978). The Cyril Burt question: New findings. *Science, 201*(4362), 1177-1186.

Driver, R. (1989). Students' conceptions and the learning of science. *International Journal of Science Education, 11*, 481-490.

Driver, R., Leach, J., Millar, R. & Scott, P. (1996). *Young people's images of science.* Buckingham: Open University Press.

Driver, R., Newton, P. & Osborne, J. (2000). Establishing the norms of scientific argmentation in classrooms. *Science Education, 84*(3), 287-312.

Duit, R. (1991). On the role of analogies and metaphors in learning science. *Science Education, 75*(6), 649-672.

Duit, R. (2006). Bibliography STCSE (students' and teachers' conceptions and science education). Kiel: IPN Available at http://www.ipn.uni-kiel.de/aktuell/stcse/stcse.html

Duit, R., Roth, W-M., Komorek, M. & Wilbers, J. (2001). Fostering conceptual change by analogies – Between Scylla and Charybdis. *Learning and Instruction, 11*, 283-303.

Dunbar, K. (1993). Concept discovery in a scientific domain. *Cognitive Science, 17*, 397-434.

Dunbar, K. (1995). How scientists really reason: Scientific reasoning in real-world laboratories. In R.J. Sternberg & J.E. Davidson (Eds.), *The nature of insight* (pp. 365-395). Cambridge, MA: MIT Press.

Dunbar, K. & Blanchette, I. (2001). The in vivo/in vitro approach to cognition: The case of analogy. *Trends in Cognitive Science, 5*, 334-339.

Dunwoody, S. (1993). *Reconstructing science for public consumption: Journalism as science education.* Geelong: Deakin University Press.

Duschl, R.A. (1983). Science teachers' beliefs about the nature of science and the selection, implementation and development of instructional tasks: A case study. *Dissertation Abstracts International, 45*(2), 422-A.

Duschl, R.A. (1990). *Restructuring science education: The importance of theories and their development*. New York: Teachers College Press.

Duschl, R. (2001). Making the nature of science explicit. In R. Millar, J. Leach & J. Osborne (Eds.), *Improving science education: The contribution of research* (pp. 187-206). Buckingham: Open University Press.

Duschl, R.A. (2004). Relating history of science to learning and teaching science: Using and abusing. In L.B. Flick & N.G. Lederman (Eds.), *Scientific inquiry and nature of science: Implications for teaching, learning, and teacher education* (pp. 319-330). Dordrecht: Kluwer.

Duschl, R. (2008). Science education in three part harmony: Balancing conceptual, epistemic, and social learning goals. *Review of Research in Education, 32*(1), 268-291.

Duschl, R.A. (2008). Quality argumentation and epistemic criteria. In S. Erduran & M.P. Jiménez-Aleixandre (Eds.), *Argumentation in science education: Perspectives from classroom-based research* (pp. 159-178). Dordrecht: Kluwer.

Duschl, R. & Ellenbogen, K. (2002). Argumentation processes in learning science. Paper presented at the International Conference on Ontological, Epistemological, Linguistic and Pedagogical Considerations of Language and Science Literacy: Empowering Research and Informing Instruction, Victoria, BC, September.

Duschl, R.A. & Grandy, R.E. (2008). Reconsidering the character and role of inquiry in school science: Framing the debates. In R.A. Duschl & R.E. Grandy (Eds.), *Teaching scientific inquiry: Recommendations for research and implementation* (pp. 1-37). Rotterdam/Taipei: Sense.

Duschl, R.A. & Osborne, J. (2002). Supporting and promoting argumentation discourse in science education. *Studies in Science Education, 38*, 39-72.

Duschl, R.A., Schweingruber, H.A. & Shouse, A.W. (Eds.) (2007). *Taking science to school: Learning and teaching science in grades K-8*. Washington, DC: The National Academies Press.

Duster, T. (1990). *Backdoor to eugenics*. New York: Routledge. [2nd edition: 2003].

Duveen, J., Scott, L. & Solomon, J. (1993). Pupils' understanding of science: Description of experiments or "a passion to explain"? *School Science Review, 75*(271), 19-27.

Duveen, J. & Solomon, J. (1994). The great evolution trial: Use of role-play in the classroom. *Journal of Research in Science Teaching, 31*, 575-582.

Dyche, S., McClurg, P., Stepans, J. & Veath, M.L. (1993). Questions and conjectures concerning models, misconceptions, and spatial ability. *School Science & Mathematics, 93*(4), 191-197.

Dyson, S. (2005). *Ethnicity and screening for sickle cell/thalassaemia: Lessons for practices from the voices of experience*. New York: Elsevier Churchill Livingstone.

Earland, S. (2004). Researchers in residence put the fun back into science: Learning from the fossil record. *School Science Review, 85*(312), 63-70.

Ebbers, M. (2002). Science text sets: Using various genres to promote literacy and inquiry. *Language Arts, 80*(1), 40-50.

Eddington, A. (1928). *The nature of the physical world*. Cambridge: Cambridge University Press.

Edelson, D.C. (2001). Learning-for-use: A framework for the design of technology-supported inquiry activities. *Journal of Research in Science Teaching, 38*(3), 355-385.

Edelson, D., Pea, R. & Gomez, L. (1996). Constructivism in the collaboratory. In B.G. Wilson (Ed.), *Constructivist learning environments: Case studies in instructional design* (pp. 151-164). Englewood Cliffs, NJ: Educational Technolgy Publications.

Edwards, A.D. & Westgate, D.P.G. (1994). *Investigating classroom talk*. London: Falmer.

Egan, K. (1986). *Teaching as story telling*. Chicago, IL: University of Chicago Press.

Egan, K. (1988a). *Primary understanding*. New York: Routledge & Kegan Paul.

Egan, K. (1988b). Metaphors in collision: Objectives, assembly lines, and stories. *Curriculum Inquiry, 18*(1), 63-86.

Egan, K. (1989a). The shape of the science text: A function of stories. In S. de Castell, A. Like & C. Luke (Eds.), *Language, authority and criticism: Readings on the school textbook* (pp. 96-108). New York: Falmer Press.

Egan, K. (1989b). Memory, imagination, and learning: Connected by the story. *Phi Delta Kappan, 70*(6), 455-459.

Eilam, B. (2004). Drops of water and of soap solution: Students' constraining mental models of the nature of matter. *Journal of Research in Science Teaching, 41*, 970-993.

Einstein, E. (1940). Science and religion. *Nature, 146*(3706), 605-607.

Elbow, P. (1973). *Writing without teachers*. Oxford: Oxford University Press.

Elbow, P. (1981). *Writing with power*. Oxford: Oxford University Press.

Elby, A. & Hammer, D. (2001). On the substance of a sophisticated epistemology. *Science Education, 85*(5), 554-567.

El-Hani, C.N. & Bandeira, F.P.S. (2008). Valuing indigenous knowledge: to call it "science" will not help. *Cultural Studies of Science Education, 3*, 751-779.

El-Hani, C.N. & Mortimer, E.F. (2007). Multicultural education, pragmatism, and the goals of science teaching. *Cultural Studies of Science Education, 2*(3), 657-702.

El-Hindi, A.E. (2003). Integrating literacy and science in the classroom: From ecomysteries to readers theatre. *Reading Teacher, 56*(6), 536-538.

Elliott, P. (2006). Reviewing newspaper articles as a technique for enhancing the scientific literacy of student-teachers. *International Journal of Science Education, 28*(11), 1245-1265.

Ellis, B.F. (2000). The cottonwood: How I learned the importance of storytelling in science education. *Science and Children, 38*(4), 43-46.

Ellis, R.A. (2004). University student approaches to learning science through writing. *International Journal of Science Education, 26*(15), 1835-1853.

Enfield, M., Smith, E.L. & Grueber, D.J. (2008). "A sketch is like a sentence": Curriculum structures that support teaching epistemic practices of science. *Science Education, 92*(4), 608-630.

Engle, R.A. & Conant, F.R. (2002). Guiding principles for fostering productive disciplinary engagement: Explaining an emergent argument in a community of learners classroom. *Cognition and Instruction, 20*(4), 399-483.

Epstein, S. (1995). The construction of lay expertise: AIDS activism and the forging of credibility in the reform of clinical trials. *Science, Technology & Human Values, 20*, 408-437.

Epstein, S. (1997). Activism, drug regulation, and the politics of therapeutic evaluation in the AIDS era: A case study of ddC and the "surrogate markers" debate. *Social Studies of Science, 27*, 691-726.

Erduran, S. (2006). Promoting ideas, evidence and argument in initial teacher training. *School Science Review, 87*(321), 45-50.

Erduran, S. (2008). Methodological foundations in the study of argumentation in science classroom. In S. Erduran & M.P. Jiménez-Aleixandre (Eds.), *Argumentation in science education: Perspectives from classroom-based research* (pp. 47-69). Dordrecht: Kluwer.

Erduran, S., Bravo, A.A. & Naaman, R.M. (2007). Developing epistemologically empowered teachers: Examining the role of philosophy of chemistry in teacher education. *Science & Education, 16*(9&10), 975-989.

Erduran, S. & Duschl, R.A. (2004). Interdisciplinary characterizations of models and the nature of chemical knowledge in the classroom. *Studies in Science Education, 40*, 105-138.

Erduran, S., Simon, S. & Osborne, J. (2004). TAPping into argumentation: Developments in the application of Toulmin's argument pattern for studying science discourse. *Science Education, 88*(6), 915-933.

Ericsson, K.A. & Smith, J. (1991). *Toward a general theory of expertise*. Cambridge: Cambridge University Press.

Ertmer, P.A. & Newby, T.J. (1996). The expert learner: Strategic, self-regulated, and reflective. *Instructional Science, 26*, 1-26.

Etzkowitz, H., Kemelgor, C. & Uzzi, B. (Eds.) (2000). *Athena unbound: The advancement of women in science and technology*. Cambridge: Cambridge University Press.

Evans, C.A., Abrams, E.D., Rock, B.N. & Spencer, S.L. (2001). Student/scientist partnerships: A teachers' guide to evaluating the critical components. *American Biology Teacher, 63*(5), 318-323.

Falk, H., Brill, G. & Yarden, A. (2008). Teaching a biotechnology curriculum based on adapted primary literature. *International Journal of Science Education, 30*(14), 1841-1866.

Falk, J.H. & Dierking, L.D. (2000). *Learning from museums: Visitor experiences and the making of meaning*. Walnut Creek, CA: Altamira Press.

Fang, Z. (2005). Scientific literacy: A systematic functional linguistics perspective. *Science Education, 89*(2), 335-347.

Farrell, M.P. & Ventura, F. (1998). Words and understanding in physics. *Language and Education, 12*(4), 243-253.

Fausto-Sterling, A. (1985). *Myths of gender*. New York: Basic Books.

Feasey, R. (1998). *Primary science and literacy*. Hatfield: Association for Science Education.

Feasey, R. (Ed.) (2001). *Science is like a tub of ice cream – Cool and fun*. Hatfield (UK): Association for Science Education.

Fedigan, L. (1986). The changing role of women in models of human evolution. *Annual Review of Anthropology, 15*, 25-66.

Fee, E. (1979). Nineteenth Century craniology: The study of the female skull. *Bulletin of the History of Medicine, 53*, 415-433.

Felton, M. (2004). The development of discourse strategies in adolescent argumentation. *Cognitive Development, 19*, 35-52.

Feltovich, P.J., Spiro, R.J. & Coulson, R.L. (1997). Issues of expert flexibility in contexts characterized by complexity and change. In P.J. Feltovich, K.M. Ford & R.R. Hoffman (Eds.), *Expertise in context* (pp. 125-146). Menlo Park, CA: AAAI Press.

Fensham, P.J. (1990). Practical work and the laboratory in "Science for All". In E. Hegarty-Hazel (Ed.), *The student laboratory and the science curriculum* (pp. 291-311). London: Routledge.

Ferrari, M., Bouffard, T. & Rainville, L. (1998). What makes a good writer? Differences in good and poor writers' self-regulation of writing. *Instructional Science, 26*, 473-488.

Feyerabend, P.K. (1975). *Against method: Outline of an anarchistic theory of knowledge*. London: New Left Books.

Feynman, R. (1965). *The character of physical law*. Cambridge, MA: MIT Press.

Fields, D.A. (2009). What do students gain from a wek at science camp? Youth perceptions and the design of an immersive, research-oriented astronomy camp. *International Journal of Science Education, 31*(2), 151-171.

Fine, A. (1986). Unnatural attitudes: Realist and instrumentalist attachments to science. *Mind, 95*(378), 149-179.

Fine, A. (1991). The natural ontological attitude. In R. Boyd, P. Gasper & J.D. Trout (Eds.), *The philsophy of science* (pp. 261-277). Cambridge, MA: MIT Press.

Finson, K.D. (2002). Drawing a scientist: What we do and do not know after fifty years of drawing. *School Science & Mathematics, 102*(7), 335-345.

Finson, K.D., Beaver, J.B. & Cramond, B.L. (1995). Development and field test of a checklist for the draw-a-scientist test. *School Science & Mathematics, 95*(4), 195-205.

Finson, K.D., Pederson, J. & Thomas, J. (2006). Comparing science teaching styles to students' perceptions of scientists. *School Science & Mathematics, 106*(1), 8-15.

Fisher, E. (1992). Distinctive features of pupil-pupil classroom talk and their relationship to learning: How discursive exploration might be encouraged. *Language and Education, 7*(4), 239-257.

Flanders, N.A. (1970). *Analysing teaching behavior*. Reading, MA: Addison-Wesley.

Fleming, A. (1922). On a remarkable bacteriolytic element found in tissues and secretions. *Proceedings of the Royal Society, Series B, 93*, 306-317.

Fleming, A. (1929). On the antibacterial action of cultures of a Penicillium, with special reference to their use in the isolation of B. influenzae. *British Journal of Experimental Pathology, 10*, 226-236.

Fleischmann, M. & Pons, S. (1989). Electrochemically induced nuclear fusion of deuterium. *Journal of Electroanalytical Chemistry, 261*(2), 301-308.

Fletcher, R. (1991). *Science, ideology and the media: The Cyril Burt scandal*. New Brunswick, NJ: Transaction Publishers.

Flick, L. (1990). Scientists in residence program: Improving children's image of science and scientists. *School Science & Mathematics, 90*(3), 204-214.

Flick, L. (1991). Analogy and metaphor: Tools for understanding inquiry science methods. *Journal of Science Teacher Education, 2*(3), 61-66.

Flick, L.B. (1995). Navigating a sea of ideas: Teacher and students negotiate a course toward mutual relevance. *Journal of Research in Science Teaching, 32*, 1065-1082.

Florence, M.K. & Yore, L.D. (2004). Learning to write like a scientist: Coauthoring as an enculturation task. *Journal of Research in Science Teaching, 41*(6), 637-668.

Ford, B. (1998). Critically evaluating scientific claims in the popular press. *American Biology Teacher, 60*, 174-180.

Ford, D.J. (2004). Scaffolding preservice teachers' evaluation of children's science literature: Attention to science-focused genres and use. *Journal of Science Teacher Education, 15*(2), 133-153.

Ford, D.J., Brickhouse, N.W., Lottero-Perdue, P. & Kittleson, J. (2006). Elementary girls' science reading at home and school. *Science Education, 90*(2), 270-288.

Ford, M. (2008a). Disciplinary authority and accountability in scientific practice and learning. *Science Education, 92*(3), 404-423.

Ford, M. (2008b). "Grasp of practice" as a reasoning resource for inquiry and nature of science understanding. *Science & Education, 17*(2&3), 147-177.

Ford, M.J. & Forman, E.A. (2006). Redefining disciplinary learning in classroom contexts. *Review of Research in Education, 30*(1), 1-32.

Ford, M.J. & Wargo, B.M. (2007). Routines, roles, and responsibilities for aligning scientific and classroom practices. *Science Education, 91*(1), 133-157.

Fort, D.C. & Varney, H.L. (1989). How students see scientists: Mostly male, mostly white, and mostly benevolent. *Science and Children, 26*, 8-13.

Fourez, G. (1997). Scientific and technological literacy as a social practice. *Social Studies of Science, 27*(6), 903-936.

Fowler, M. (2003). Galileo and Einstein: Using history to teach basic physics to nonscientists. *Science & Education, 12*, 229-231.

Fox, R. (1971). *The caloric theory of gases.* Oxford: Oxford University Press.

France, B. (2000). The role of models in biotechnology education. In J. Gilbert & C. Boulter (Eds.), *Developing models in science education* (pp. 271-288). Dordrecht: Kluwer.

Francis, L.J. & Greer, J.E. (2001). Shaping adolescents' attitudes towards science and religion in Northern Ireland: The role of scientism, creationism and denominational schools. *Research in Science & Technological Education, 19*(1), 39-54.

Franco, C., Barros, H.L., Colinvaux, D., Krapas, S., Queiroz, G. & Alves, F. (1999). From scientists' and inventors' minds to some scientific and technological products: Relationships between theories, models, mental models and conceptions. *International Journal of Science Education, 21*(3), 277-291.

Frankel, H. (1979). The career of continental drift theory. *Studies in History and Philosophy of Science, 10*, 21-66.

Franks, F. (1981). *Polywater.* Cambridge, MA: MIT Press.

Fransella, F. & Bannister, D. (1977). *A manual for repertory grid technique.* New York: Academic Press.

Freire, P. (1972). *Pedagogy of the oppressed.* London: Penguin.

Fretz, E.B., Wu, H.K., Zhang, B., Davis, E.A., Krajcik, J.S. & Soloway, E. (2002). An investigation of software scaffolds supporting modeling practices. *Research in Science Education, 32*, 567-589.

Freundenrich, C.C. (2000). Sci-fi science. *The Science Teacher, 67*(8), 42-45.

Fryer, P. (1984). *Staying power: The history of black people in Britain.* London: Pluto Press.

Fung, Y.Y.H. (2002). A comparative study of primary and secondary school students' images of scientists. *Research in Science & Technological Education, 20*(2), 199-213.

Fysh, R. & Lucas, K.B. (1998). Religious beliefs in science classrooms. *Research in Science Education, 28*(4), 399-427.

Galbraith, D. (1999). Writing as a knowledge-constituting process. In D.G.M. Torrance (Ed.), *Knowing what to write: Conceptual processes in text production* (pp. 139-159). Amsterdam: Amsterdam University Press.

Galili, I. & Hazan, A. (2001). The effect of a history-based course in optics on students' views about science. *Science & Education, 10*, 7-32.

Galison, P. (1987). *How experiments end.* Chicago, IL: University of Chicago Press.

Gallagher, J. (1991). Prospective and practicing secondary science teachers' knowledge and beliefs about the philosophy of science. *Science Education, 75*, 121-134.

Gallagher, M., Knapp, P. & Noble, G. (1993). Genre in practice. In B. Cope & M. Kalatzis (Eds.), *The power of literacy: A genre approach to teaching writing* (pp. 179-202). Pittsburgh, PA: University of Pittsburgh Press.

Gallas, K. (1994). *The languages of learning: How children talk, write, dance, draw, and sing their understanding of the world.* New York: Teachers College Press.

Gallas, K. (1995). *Talking their way into science: Hearing children's questions and theories, responding with curricula.* New York: Teachers College Press.

Gaon, S. & Norris, S.P. (2001). The undecidable grounds of scientific expertise: Science education and the limits of intellectual independence. *Journal of Philosophy of Education, 35*(2), 187-201.

Garcia-Mila, M. & Andersen, C. (2008). Cognitive foundations of learning argumentation. In S. Erduran & M.P. Jiménez-Aleixandre (Eds.), *Argumentation in science education: Perspectives from classroom-based research* (pp. 29-46). Dordrecht: Kluwer.

Gardner, H. (1984). *Frames of mind: The theory of multiple intelligences.* London: Heinemann.

Gardner, P.L. (1975). Logical connectives in science: A summary of the findings. *Research in Science Education, 7,* 9-24.

Garfinkel, H. (1967). *Studies in ethnomethodology.* Englewood Cliffs, NJ: Prentice Hall.

Gaskell, J. & Willinsky, J. (Eds.) (1995). *Gender in/forms curriculum.* New York: Teachers College Press.

Gatto, J.T. (2002). *Dumbing us down: The hidden curriculum of compulsory schooling,* 2nd ed. Gabriola island (BC): New Society Publishers.

Gauld, C.F. (1982). The scientific attitude and science education: A critical reapopraisal. *Science Education, 66,* 109-121.

Gauld, C. (1991). History of science, individual development and science teaching. *Research in Science Education, 21,* 133-140.

Gauld, C.F. (2005). Habits of mind, scholarship and decision making in science and religion. *Science & Education, 14,* 291-308.

Gay, G. (2002). Preparing for culturally responsive teaching. *Journal of Teacher Education, 53*(2), 106-116.

Gayford, C.G. (2002). Environmental literacy: Towards a shared understanding. *Research in Science & Technological Education,* 20(1), 1470-1138.

Gee, J.P. (1996). *Social linguistics and literacies: Ideology in discourses,* 2nd ed. London: Taylor & Francis.

Gee, J. (1999). What is literacy? In C. Mitchell & K. Weiler (Eds.), *Rewriting literacy: Culture and the discourse of the other* (pp. 3-11). Westport, CT: Bergin & Garvin.

Gee, J. (2004). Language in the science classroom: Academic social languages at the heart of school-based literacy. In E.W. Saul (Ed.), *Crossing borders in literacy and science education: Perspectives in theory and practice* (pp. 13-32). Newark, DE: International Reading association/National Science Teachers Association.

Gentner, D., Holyoak, K. & Kokinov, B. (Eds.) (2000). *Analogy: Perspectives from cognitive science.* Cambridge, MA: MIT Press.

Gentner, D. & Stevens, A. (1983). *Mental models.* Hillsdale, NJ: Lawrence Erlbaum.

George, J. (1999a). Worldview analysis of knowledge in a rural village: Implications for science education. *Science Education, 83*(1), 77-95.

George, J.M. (1999b). Indigenous knowledge as a component of the school curriculum. In L.M. Semali & J.L. Kincheloe (Eds.), *What is indigenous knowledge? Voices from the academy* (pp. 79-94). New York: Falmer Press.

Gerber, B.L., Cavallo, A.M.L. & Marek, E.A. (2001). Relationships among informal learning environments, teaching procedures and scientific reasoning ability. *International Journal of Science Education, 23*(5), 535-549.

Gertz, S., Portman, D.J. & Sarquis, M. (1996). *Teaching physical science through children's literature.* New York: Learning Triangle Press/McGraw Hill.

Gess-Newsome, J. (2002). The use and impact of explicit instruction about the nature of science and science inquiry in an elementary science methods course. *Science & Education, 11,* 55-67.

Giddens, A. (1991). *Modernity and self-identity: Self and society in the late modern age.* Cambridge: Polity Press.

Giere, R.N. (1988). *Explaining science: A cognitive approach.* Chicago, IL: University of Chicago Press.

Giere, R.N. (1997). *Understanding scientific reason,* 4th ed. Orlando, FL: Harcourt Brace Jovanovich.

Giere, R.N. (1999). *Science without laws.* Chicago, IL: University of Chicago Press.

Gieryn, T.F. (1992). The ballad of Pons and Fleischmann: Experiment and narrative in the (un)making of cold fusion. In E. McMullin (Ed.), *The social dimensions of science* (pp. 217-243). Notre Dame, IN: University of Notrre Dame Press.

Gieryn, T.F. (1995). Boundaries of science. In S. Jasanoff, G. Markle, J. Peterson & T. Pinch (Eds.), *Handbook of science and technology studies* (pp. 393-443). Thousand Oaks, CA: Sage.

Gieryn, T.F. & Figert, A.E. (1986). Scientists protect their cognitive authority: The status degradation ceremony of Cyril Burt. In G. Boehme & N. Stehr (Eds.), *The knowledge society* (pp. 67-86). Dordrecht: Reidel.

Gilbert, A. & Yerrick, R. (2001). Same school, separate worlds: A sociocultural study of identity, resistance, and negotiation in a rural, lower track science classroom. *Journal of Research in Science Teaching, 38*, 574-598.

Gilbert, J. (2001). Science and its "other": Looking underneath "woman" and "science" for new directions in research on gender and science. *Gender and Education, 13*(3), 291-305.

Gilbert, J. & Calvert, S. (2003). Challenging accepted wisdom: Looking at the gender and science education question through a different lens. *International Journal of Science Education, 25*, 861-878.

Gilbert, J.K. (2004). Models and modelling: Routes to more authentic science education. *International Journal of Science and Mathematics Education, 2*(1), 115-130.

Gilbert, J.K. & Boulter, C.J. (1998). Learning science through models and modelling. In B.J. Fraser & K.G. Tobin (Eds.), *International handbook of science education* (pp. 53-66). Dordrecht: Kluwer.

Gilbert, J.K., Boulter, C. & Rutherford, M. (1998a). Models in explanations, Part 1: Horses for courses? *International Journal of Science Education, 20*(1), 83-97.

Gilbert, J.K., Boulter, C. & Rutherford, M. (1998b). Models in explanations, Part 2: Whose voice? Whose ears? *International Journal of Science Education, 20*(2), 187-203.

Gilbert, J. & Boulter, C. (2000). *Developing models in science education.* Dordrecht: Kluwer.

Gilbert, J.K. & Reiner, M. (2000). Thought experiments in science education: Potential and current realization. *International Journal of Science Education, 22*(3), 265-283.

Gilbert, N. & Mulkay, N. (1984). *Opening Pandora's box.* Cambridge: Cambridge University Press.

Gingerich, O. (1994). Dare a scientist believe in design? In J.M. Templeton (Ed.), *Evidence of purpose: Scientists discover the creator* (pp. 21-32). New York: Continuum.

Gingerich, O. (2006). *God's universe.* Cambridge, MA: Harvard University Press.

Girod, M. (2007). A conceptual overview of the role of beauty and aesthetics in science and science education. *Studies in Science Education, 43*, 38-61.

Giroux, H. (1992). *Border crossings: Cultural workers and the politics of education.* New York: Routledge.

Gish, D.T. (1978). *Evolution: The fossils say no!* San Diego, CA: Creation-Life.

Gish, D.T. (1985). *Evolution: The challenge of the fossil record.* El Cajon, CA: Master Books.

Gitari, W. (2003). An inquiry into the integration of indigenous knowledges and skills in the Kenyan secondary science curriculum: A case study of human health knowledge. *Canadian Journal of Science, Mathematics and Technology Education, 3*(2), 195-212.

Gitari, W. (2006). Everyday objects of learning about health and healing and implications for science education. *Journal of Research in Science Teaching, 43*(2), 172-193.

Glaser, R.E. & Carson, K.M. (2005). Chemistry is in the news: Taxonomy of authentic news media-based learning activities. *International Journal of Science Education, 27*(4), 1083-1098.

Glaser, R. & Chi, M.T.H. (1988). Overview. In M.T.H. Chi, R. Glaser & M.J. Farr (Eds.), *The nature of expertise* (pp. xv-xxviii). Hillsdale, NJ: Erlbaum.

Glasson, G.E. & Bentley, M.L. (2000). Epistemological undercurrents in scientists' reporting of research to teachers. *Science Education, 84*(4), 469-485.

Glynn, S.M. (1991). Explaining science concepts: A teaching with analogy model. In S.M. Glynn, H.R. Yeany & K.B. Britton (Eds.), *The psychology of learning* (pp. 219-240). Hillsdale, NJ: Lawrence Erlbaum.

Glynn, S.M. & Duit, R. (1995). Learning science meaningfully: Constructing conceptual models. In S.M. Glynn and R. Duit (Eds.), *Learning science in the schools: Research reforming practice* (pp. 3-33). Mahwah, NJ: Erlbaum

Glynn, S.M., Duit, R., & Thiele, R.B. (1995). Teaching science with analogies: A strategy for constructing knowledge. In S.M. Glynn, & R. Duit (Eds.), *Learning science in the schools: Research reforming practice* (pp. 247-273). Mahwah, NJ:Erlbaum.

Glynn, S.M. & Muth, K.D. (1994). Reading and writing to learn science: Achieving scientific literacy. *Journal of Research in Science Teaching, 31*(9), 1057-1073.

Glynn, S. M., & Takahashi, T. (1998). Learning from analogy-enhanced science text. *Journal of Research in Science Teaching, 35*(10), 1129-1149.

Gobert, J.D. & Pallant, A. (2004). Foster students' epistemologies of models via authentic model-based texts. *Journal of Science Education and Technology, 13*(1), 7-21.

Goldman, S.R. & Bisanz, G.L. (2002). Towqard a functional understanding of scientific genres: Implications for understanding and learning processes. In J. Otero, J.A. Leon & A.C. Graesser (Eds.), *The psychology of science text comprehension* (pp. 19-50). Mahwah, NJ: Erlbaum.

Goldstein, A., Sheehan, P. & Goldstein, P. (1973). Unsuccessful attempts to transfer morphine tolerance and passive avoidance by brain extracts. *Nature, 233,* 126-129.

Goodrum, D., Hackling, M. & Rennie, L. (2000). *The status and quality of teaching and learning science in Australian schools.* Canberra: Department of Education, Training and Youth Affairs.

Goodwin, A. (2001). Wonder in science teaching and learning: An update. *School Science Review, 83*(302), 69-74.

Gott, R. & Duggan, S. (1996). Practical work: its role in the understanding of evidence in science. *International Journal of Science Education, 18*(7), 791-806.

Gott, R. & Duggan, S. (2007). A framework for practical work in science and scientific literacy through argumentation. *Research in Science & Technological Education, 25*(3), 271-291.

Gott, R., Duggan, S. & Roberts, R. (2003). Concepts of evidence. Available online at: www.dur.ac.uk/rosalyn.roberts?Evidence/cofev.htm

Gough, N. (1993). *Laboratories in fiction: Science education and popular media.* Geelong: Deakin University Press.

Gould, S.J. (1981a). *The mismeasure of man.* Harmondsworth: Penguin.

Gould, S.J. (1981b). Evolution as fact and theory. *Discover, 2,* 33-37.

Gould, S.J. (1995). *Dinosaur in a haystack: Reflections in natural history.* New York: Crown Trade Paperbacks.

Gould, S.J. (1999). *Rock of ages: Science and religion in the fullness of life.* New York: Ballantin.

Government of Canada (1991). *Prosperity through competitiveness.* Ottawa: Ministry of Supply and Services Canada.

Gräber, W. & Bolte, C. (Eds.) (1997). *Scientific literacy: An international symposium.* Kiel: Institut fur die Padagogik der Naturwiseenchaften (IPN) an der Universitat Kiel.

Gräber, W., Nentwig, P., Becker, H-J., Sumfleth, E., Pitton, A., Wollweber, K. & Jorde, D. (2002) Scientific literacy: from theory to practice. In H. Behrendt, H. Dahncke, R. Duit, W. Gräber, M. Komorek, A. Kross & P. Reiska (Eds.), *Research in science education – Past, present and future* (pp. 61-70). Dordrecht: Kluwer.

Grandy, R.E. & Duschl, R.A. (2008). Consensus: Expanding the scientific method and school science. In R.A. Duschl & R.E. Grandy (Eds.), *Teaching scientific inquiry: Recommendations for research and implementation* (pp. 304-325). Rotterdam: Sense Publishers.

Graves, D.L. (1983). *Writing: Students and teachers at work.* Exeter, NH: Heinemann.

Greca, I.M. & Moreira, M.A. (2000). Mental models, conceptual models and modelling. *International Journal of Science Education, 22*(1), 1-11.

Greca, I.M. & Moreira, M.A. (2002). Mental, physical and mathematical models in the teaching and learning of physics. *Science Education, 86*(1), 106-121.

Greeno, J.G. & Hall, R.P. (1997). Practicing representation: Learning with and about representational forms. *Phi Delta Kappan, 78*(5), 361-367.

Griffin, J. (1998). Learning science through practical experiences in museums. *International Journal of Science Education, 20*(6), 643-653.

Griffith, J.A. & Brem, S.K. (2004). Teaching evolutionary biology: Pressures stress, and coping. *Journal of Research in Science Teaching, 41*(8), 791-809.

Griffiths, A.K. & Barman, C.R. (1995). High school students' views about the nature of science: Results from three countries. *School Science & Mathematics, 95*(2), 248-255.

Grimberg, B.I. & Hand, B. (2009). Cognitive pathways: Analysis of students' written texts for science understanding. *International Journal of Science Education, 31*(4), 503-521.

Groisman, B., Shapiro, B. & Willinsky, J. (1991). The potential of semiotics to inform understanding of events in science education. *International Journal of Science Education, 13,* 217-226.

Grosslight, L., Unger, C. & Jay, E. (1991). Understanding models and their use in science: Conceptions of middle and high school students and experts. *Journal of Research in Science Teaching, 28*(9), 799-822.

Gutierrez, K., Baquedano-Lopez, P. Alvarez, H. & Chiu, M. (1999). Building a culture of collaboration through hybrid lanaguage practices. *Theory into Practice, 8,* 87-93.

Guzzetti, B.J., Snyder, T.E. & Glass, G.V. (1992). Promoting conceptual change in science: Can texts be used effectively? *Journal of Reading, 35*(8), 642-649.

Guzzetti, B.J., Snyder, T.E., Glass, G.V. & Gamas, W.W. (1993). Promoting conceptual change in science: A comparative meta-analysis of interventions from reading education and science education. *Reading Research Quarterly, 28,* 116-159.

Guzzetti, B.J., Williams, W.O., Skeels, S.A. & Wu, S.M. (1997). Influence of text structure on learning counterintuitive physics concepts. *Journal of Research in Science Teaching, 34*(7), 701-719.

Gwimbi, E. & Monk, M. (2003). A study of the association of attitudes to the philosophy of science with classroom contexts, academic qualification and professional training, amongst A-level biology teachers in Harare, Zimbabwe. *International Journal of Science Education, 25*(4), 469-488.

Hacking, I. (1983). *Representing and intervening.* Cambridge: Cambridge University Press.

Hacking, I. (1991). Experimentation and realism. In In R. Boyd, P. Gasper & J.D. Trout (Eds.), *The philosophy of science* (pp. 247-260). Cambridge, MA: MIT Press.

Hacking, I. (1996). The disunities of the sciences. In P. Galison & D. Stump (Eds.), *The disunity of science* (pp. 37-74). Stanford, CA: Stanford University Press.

Hacking, I. (1999). *The social construction of what?* Cambridge, MA: MIT Press.

Hagen, J.B. & Kugler, C. (1992). Using history to teach principles of biology to college students: The case of cell theory. In G.L.C. Hills (Ed.), *Proceedings of the Second International Conference on History and Philosophy of Science and Science Teaching* (pp. 471- 481). Kingston. ON: Queen's University.

Haidar, A.H. (1999). Emirates pres-service and in-service teachers' views about the nature of science. *International Journal of Science Education, 21*(8), 807-822.

Haigh, M. (2007). Can investigative practical work in high school biology foster creativity? *Research in Science Education, 37*(2), 123-140.

Haigh, M., France, B. & Foret, M. (2005). Is "doing science" in New Zealand classrooms an expression of scientific inquiry? *International Journal of Science Education, 27*(2), 215-226.

Halkia, K. (2003). Teachers? views and attitudes towards the communication code and the rhetoric used in press science articles. In D. Psillos, P. Kariotoglou, V. Tselfes, E. Hatzikraniotis, G. Fasssoulopoulos & M. Kallery (Eds.), *Science education research in the knowledge-based society* (pp. 415-423). Dordrecht: Kluwer.

Halliday, M.A.K. (1993). Some grammatical problems in scientific English. In M. Halliday & J. Martin (Eds.), *Writing science: Literacy and discursive power* (pp. 69-85). London: Falmer.

Halliday, M.A.K. (1996). *Writing science: Literacy and discursive power.* London: Falmer.

Halloun, I. (2004). *Modeling theory in science education.* Dordrecht: Kluwer.

Halloun, I.A. (2007). Mediated modeling in science education. *Science & Education, 16*(7&8), 653-697.

Halloun, I. & Hestenes, D. (1998). Interpreting VASS dimensions and profiles for physics students. *Science & Education, 7,* 553-577.

Hallowell, C. & Holland, M.J. (1998). Journalism as a path to scientific literacy. *Journal of College Science Teaching, 28*(1), 29-32.

Halstead, J.H. (1996). Values and values education in schools. In J.M. Halstead & M.J. Taylor (Eds.), *Values in education and education in values* (pp. 3-14). Lewes: Falmer Press.

Hamilton-Ekeke, J-T. (2007). Relative effectiveness of expository and field trip methods of teaching on students' achievement in ecology. *International Journal of Science Education, 29*(15), 1847-1868.

Hammond, L. & Brandt, C. (2004). Science and cultural process: Defining an anthropological approach to science education. *Studies in Science Education, 40,* 1-47.

Hand, B.M. (Ed.) (2008). *Science inquiry, argument and language: A case for the science writing heuristic.* Rotterdam/Taipei: Sense.

Hand, B., Gunel, M. & Ulu, C. (2009). Sequencing embedded multimodal respresentations in a writing to learn approach to the teaching of electricity. *Journal of Research in Science Teaching, 46*(3), 225-247.

Hand, B., Hohenshell, L. & Prain, V. (2004a). Exploring students' responses to conceptual questions when engaged with planned writing experiences: A study with year 10 science students. *Journal of Research in Science Teaching, 41*(2), 186-210.

Hand, B.M. & Keys, C.W. (1999). Inquiry investigation: A new approach to laboratory reports. *The Science Teacher, 66*, 27-29.

Hand, B. & Prain, V. (1995). Using writing to help improve students' understanding of science knowledge. *School Science Review, 77*(278), 112-117.

Hand, B. & Prain, V. (2002). Teachers implementing writing-to-learn strategiesin junior secondary science: A case study. *Science Education, 86*(6), 737-755.

Hand, B. & Prain, V. (2006). Moving from border crossing to convergence of perspectives in language and science literacy research and practice. *International Journal of Science Education, 28*(2-3), 101-107.

Hand, B., Prain, V., Lawrence, C. & Yore, L. (1999). A writing in science framework designed to enhance science literacy. *International Journal of Science Education, 21*(10), 1021-1035.

Hand, B., Wallace, C. & Yang, G. (2004b). Using the science writing heuristic to enhance learning outcomes from laboratory activities in seventh grade science: Quantitative and qualitative aspects. *International Journal of Science Education, 26*, 131-149.

Hanrahan, M. (1999). Rethinking science literacy: Enhancing communication and participation in school science through affirmational dialogue journal writing. *Journal of Research in Science Teaching, 36*(6), 699-718.

Hansen, J.A., Barnett, M., MaKinster, J.G. & Keating, T. (2004). The impact of three-dimensional computational modelling on students' understanding of astronomy concepts: A qualitative analysis. *International Journal of Science Education, 26*, 1555-1575.

Hanson, N.R. (1958). *Patterns of discovery.* Cambridge: Cambridge University Press.

Hanson, N.R. (1972). *Observation and explanation.* London: Allen & Unwin.

Hansson, L. & Redfors, A. (2007). Physics and the possibility of a religious view of the universe: Swedish upper secondary students' views. *Science & Education, 16*(3-5), 461-478.

Hanuscin, D.L., Akerson, V.L. & Phillipson-Mower, T. (2006). Integrating nature of science instruction into a physical science content course for preservice elementary teachers: NOS views of teaching assistants. *Science Education, 90*(5), 912-935.

Hapgood, S., Magnusson, S.J. & Palincsar, A.S. (2004). Teacher, text, and experience: A case of young children's scientific inquiry. *Journal of the Learning Sciences, 13*(4), 455-505.

Haraway, D.J. (1989). *Primate visions: Gender, race, and nature in the world of modern science.* New York: Routledge.

Harding, P. & Hare, W. (2000). Portraying science accurately in classrooms: Emphasizing open-mindedness rather than relativism. *Journal of Research in Science Teaching, 37*(3), 225-236.

Harding, S. (1986). *The science question in feminism.* Ithaca, NY: Cornell University Press.

Harding, S. (1989). Is there a feminist method? In N. Tuana (Ed.), *Feminism and science* (pp. 119-131). Bloomington, IN: Indiana University Press.

Harding, S. (1991). *Whose science? Whose knowledge? Thinking from women's lives.* Ithaca, NY: Cornell University Press.

Harding, S. (1998). *Is science multicultural?* Bloomington, IN: Indiana University Press.

Harding, S.G. (Ed.) (2004). *The feminist standpoint theory reader.* New York: Routledge.

Harding, S.G. & Hintikka, M. (Eds.)(1983). *Discovering reality: Feminist perspectives on epistemology, metaphysics, methodology, and philosophy of science.* Dordrecht: Reidel.

Hardwig, J. (1991). The role of trust in knowledge. *Journal of Philosophy, 88*, 693-708.

Harms, J.M. & Lettow, L.J. (2000). Poetry and the environment. *Science & Children, 37*(6), 30-34.

Harris, D. & Taylor, M. (1983). Discovery learning in school science: The myth and the reality. *Journal of Curriculum Studies, 15*(3), 277-289.

Harrison, A.G. (2006). The affective dimension of analogy. In P.J. Aubusson, A.G. Harrison & S.M. Ritchie (Eds.), *Metaphor and analogy in science education* (pp. 51-63). Dordrecht: Kluwer.

Harrison, A.G. & de Jong, O. (2005). Exploring the use of multiple analogical models when teaching and learning chemical equilibrium. *Journal of Research in Science Teaching, 42*(10), 1135-1159.

Harrison, A.G. & Treagust, D.F. (1993). Teaching with analogies: A case study in Grade-10 optics. *Journal of Research in Science Teaching, 30*(10), p. 1291-1307.

Harrison, A.G. & Treagust, D.F. (1996). Secondary students' mental models of atoms and molecules: Implications for teaching chemistry. *Science Education, 80*(5), 509-534.

Harrison, A.G. & Treagust, D.F. (1998). Modelling in science lessons: Are there better ways to learn with models? *School Science & Mathematics, 98*(8), 420-429.

Harrison, A.G. & Treagust, D.F. (2000). A typology of school science models. *International Journal of Science Education, 22*(9), 1011-1026.

Harrison, A.G. & Treagust, D.F. (2006). Teaching and learning with analogies. In P.J. Aubusson, A.G. Harrison & S.M. Ritchie (Eds.), *Metaphor and analogy in science education* (pp. 11-24). Dordrecht: Kluwer.

Hart, C. (2008). Models in physics, models for physics learning, and why the distinction may matter in the case of electric circuits. *Research in Science Education, 38*(5), 529-544.

Hart, C., Mulhall, P., Berry, A., Loughran, J. & Gunstone, R. (2000). What is the purpose of this experiment? Or can students learn something from doing experiments? *Journal of Research in Science Teaching, 37*(7), 655-675.

Hart, P. (2007). Environmental education. In S.K. Abell & N.G. Lederman (Eds.), *Handbook on research in science education* (pp. 689-726). Mahwah, NJ: Lawrence Erlbaum.

Hartry, A.L, Keith-Lee, P. & Morton, W.D. (1964). Planaria: Memory transfer through cannibalism re-examined. *Science, 164*(3641), 274-275.

Hashweh, M.Z. (1996). Effects of science teachers' epistemological beliefs in teaching. *Journal of Research in Science Teaching, 33*, 47-63.

Hatano, G. & Inagaki, K. (1991). Sharing cognition through a collective comprehension activity. In R.L. Resnick, J. Levine & S. Teasley (Eds.), *Perspectives on socially shared cognition* (pp. 331-348). Washington, DC: American Psychological Association.

Haught, J.F. (1995). *Science and religion: From conflict to conversation.* New York: Paulist Press.

Havdala, R. & Ashkenazi, G. (2007). Coordination of theory and evidence: Effect of epistemological theories on students' laboratory practice. *Journal of Research in Science Teaching, 44*(8), 1134-1159.

Hawking, S.W. (1988). *A brief history of time: From the big bang to black holes.* London: Bantam Books.

Haynes, R.D. (1994). *From Faust to Strangelove.* Baltimore, MD: The Johns Hopkins University Press.

Hazari, Z. (2006). *Gender differences in introductory university physics: The influence of high school physics preparation and affect.* Toronto: Unpublished PhD thesis, University of Toronto.

Heap, R. (2006). *Myth busting and tenet building: Primary and early childhood teachers' understanding of the nature of science.* Auckland: Unpublished MEd thesis, University of Auckland.

Hearnshaw, L.S. (1979). *Cyril Burt, psychologist.* New York: Cornell University Press.

Heering, P. (2000). Getting shocks: Teaching secondary school physics through history. *Science & Education, 9*, 363-373.

Heering, P. (2003). History-science-epistemology: On the use of historical experiments in physics teachetr training. In W.F. McComas (Ed.), *Proceedings of the 6th International History, Philosophy and Science Teaching Group Meeting*, Denver, CO (available as File 58 on CD-ROM from IHPST.org).

Heering, P. & Muller, F. (2002). Cultures of experimental practice – An approach in a museum. *Science & Education, 11*, 203-214.

Heidelberger, M. & Stadler, F. (Eds.) (2002). *History of philosophy of science: New trends and perspectives.* Dordrecht: Kluwer.

Hein, H.S. (2000). *The museum in transition.* Washington, DC: Smithsonian Insttituion Press.

Heisenberg, W. (1971). *Physics and beyond: Encounters and conversations.* New York: Harper & Row.

Hekman, S.J. (1997). Truth and method: Feminist standpoint theory revisited. *Signs, 22*(21), 341-365.

Helms, J.V. (1998a). Learning about the dimensions of science through authentic tasks. In J. Wellington (Ed.), *Practical work in school science: Which way now?* (pp. 126-151). London: Routledge.

Helms, J.V. (1998b). Science and/in the community: Context and goals in practical work. *International Journal of Science Education, 20*(6), 643-653.

Hendrick, R.M. (1992). The role of history in teaching science – A case study. *Science & Education, 1*, 145-162.

Hennessey, M.G. (2003). Metacognitive aspects of students' reflective discourse: Implications for intentional conceptual change teaching and learning. In G.M. Sinatra & P.R. Pintrich (Eds.), *Intentional conceptual change* (pp. 103-132). Mahwah, NJ: Lawrence Erlbaum.

Hennessy, S. (1993). Situated cognition and cognitive apprenticeship: Implications for classroom learning. *Studies in Science Education, 22*, 1-41.

Henze, I., van Driel, J. & Verloop, N. (2007a). The change of science teachers' personal knowledge about teaching models and modelling in the context of science education reform. *International Journal of Science Education, 29*(15), 1819-1846.

Henze, I., van Driel, J.H. & Verloop, N. (2007b). Science teachers' knowledge about teaching models and modelling in the context of a new syllabus on public understanding of science. *Research in Science Education, 37*(2), 99-122.

Hermann, R.S. (2008). Evolution as a controversial issue: A review of instructional approaches. *Science & Education, 17*(8&9), 1011-1032.

Herrenkohl, L. & Guerra, M. (1998). Participant structures, scientific discourse, and student engagement in fourth grade. *Cognition and Instruction, 16*(4), 431-473.

Herrenkohl, L.P., Palinscar, A.S., De Water, L.S. & Kawasaki, K. (1999). Developing scientific communities in classrooms: A sociocognitive approach. *Journal of the Learning Sciences, 8*(3&4), 451-493.

Herschberger, R. (1948). *Adam's rib.* New York: Pelligrini & Cudaby.

Heselden, R. & Staples, R. (2002). Science teaching and literacy, Part 2: Reading. *School Science Review, 83*(304), 51-62.

Hesse, M. (1966). *Models and analogies in science.* Bloomington, IN: University of Notre Dame Press.

Hewitt, J., Pedretti, E., Bencze, L., Vaillancourt, B.D. & Yoon, S. (2003). New applications for multimedia cases: Promoting reflective practice in preservice teacher education. *Journal of Technology and Teacher Education, 11*(4), 483-500.

Heywood, D. (2002). The place of analogies in science education. *Cambridge Journal of Education, 32*(2), 233-247.

Heywood, D. & Parker, J. (1997). Confronting the analogy: primary teachers exploring the usefulness of analogies in the teaching and learning of electricity. *International Journal of Science Education, 19*, 869-885.

Hildebrand, G. (1998). Disrupting hegemonic writing practices in school science: Contesting the right way to write. *Journal of Research in Science Teaching, 35*(4), 345-362.

Hill, D.R. (1993). *Islamic science and engineering.* Edinburgh: Edinburgh University Press.

Hill, J.R. & Hannafin, M.J. (2001). Teaching and learning in digital environments: The resurgence of resource-based learning. *Educational Technology Research and Development, 49*(3), 37-52.

Hinton, M.E. & Nakkhleh, M.B. (1999). Students' microscopic, macroscopic, and symbolic representations of chemical reactions. *The Chemical Educator, 4*(4), 1-29.

Hipkins, R., Barker, M. & Bolstad, R. (2005). Teaching the "nature of science": Modest adaptations or radical reconceptions? *International Journal of Science Education, 27*(2), 243-254.

Hirst, P.H. (1974). Realms of meaning and forms of knowledge. In P.H. Hirst (Ed.), *Knowledge and the curriculum: A collection of philosophical papers* (pp. 54-68). London: Routledge & Kegan Paul.

Hmelo-Silver, C.E., Nagarajan, A. & Day, R.S. (2002). "It's harder than we thought it would be": A comparative case study of expert-novice experimental strategies. *Science Education, 86*(2), 219-243.

Ho, M-W. (1997). The unholy alliance. *The Ecologist, 27*(4), 152-158.

Hoadley, C.M. & Linn, M.C. (2000). Teaching science through online, peer discussions: SpeakEasy in the knowledge integration environment. *International Journal of Science Education, 22*(8), 839-857.

Hodson, D. (1986). Rethinking the role and status of observation in science education. *Journal of Curriculum Studies, 18*(4), 381-396.

Hodson, D. (1990a). Making the implicit explicit: A curriculum planning model for enhancing children's understanding of science. In D.E. Herget (Ed.), *More history and philosophy of science in science teaching* (pp. 292-310). Tallahassee, FL: Florida State University Press.

Hodson, D. (1990b). A critical look at practical work in school science. *School Science Review, 71*, 33-40.

Hodson, D. (1992a). In search of a meaningful relationship: An exploration of some issues relating to integration in science and science education. *International Journal of Science Education, 14*, 541-562.

Hodson, D. (1992b). Redefining and reorienting practical work in school science. *School Science Review, 73*, 65-78.

Hodson, D. (1992c). Assessment of practical work: Some considerations in philosophy of science. *Science & Education, 1*, 115-144.

Hodson, D. (1993a). Philosophic stance of secondary school science teachers, curriculum experiences, and children's understanding of science: Some preliminary findings. *Interchange, 24*(1&2), 41-52.

Hodson, D. (1993b). Re-thinking old ways: Towards a more critical approach to practical work in school science. *Studies in Science Education, 22*, 85-142.

Hodson, D. (1993c). Against skills-based testing in science. *Curriculum Studies, 1*(1), 127-148.

Hodson, D. (1994). Seeking directions for change: The personalisation and politicisation of science education. *Curriculum Studies, 2*, 71-98.

Hodson, D. (1996). Laboratory work as scientific method: Three decades of confusion and distortion. *Journal of Curriculum Studies, 28*(2), 115-135.

Hodson, D. (1998a). *Teaching and learning science: Towards a personalized approach.* Buckingham: Open University Press.

Hodson, D. (1998b). Science fiction: The continuing misrepresentation of science in the school curriculum. *Curriculum Studies, 6*(2), 191-216.

Hodson, D. (2001). Inclusion without assimilation: Science education from an anthropological and metacognitive perspective. *Canadian Journal of Science, Mathematics and Technology Education, 1*(2), 161-182.

Hodson, D. (2003). Time for action: Science education for an alternative future. *International Journal of Science Education, 25*(6), 645-670.

Hodson, D. (2005). Teaching and learning chemistry in the laboratory: A critical look at the research. *Educacion Quimica, 16*(1), 60-68.

Hodson, D. (2008). *Towards scientific literacy: A teachers' guide to the history, philosophy and sociology of science.* Rotterdam/Taipei: Sense.

Hodson, D. & Bencze, L. (1998). Becoming critical about practical work: Changing views and changing practice through action research. *International Journal of Science Education, 20*(6), 683-694.

Hodson, D. with Bencze, L., Nyhof-Young, J., Pedretti, E. & Elshof, L. (2002). *Changing science education through action research: Some experiences from the field.* Toronto: Imperial Oil Centre for Studies in Science, Mathematics & Technology Education in association with University of Toronto Press.

Hodson, D. & Hodson, J. (1998a). From constructivisn to social constructivism: A Vygotskian perspective on teaching and learning science. *School Science Review, 78*(289), 33-41.

Hodson, D. & Hodson, J. (1998b). Science education as enculturation: Some implications for practice. *School Science Review, 80*(290), 17-24.

Hodson, D. & Prophet, R.B. (1986). A bumpy start to science education. *New Scientist, 1521*, 25-28

Hofer, B.K. (2000). Dimensionality and disciplinary differences in personal epistemology. *Contemporary Educational Psychology, 25*, 378-405.

Hoffmann, R. (1995). *The same and not the same.* New York: Columbia University Press.

Hofstein, A., Navon, O., Kipnis, M. & Mamlok-Naaman, R. (2005). Developing students' ability to ask more and better questions resulting from inquiry-type chemistry laboratories. *Journal of Research in Science Teaching, 42*(7), 791-806.

Hofstein, A., Shore, R. & Kipnis, M. (2004). Providing high school chemistry students with opportunities to develop learning skills in an inquiry-type laboratory: A case study. *International Journal of Science Education, 26*(1), 47-62.

Hogan, K. (1998). Sociocognitive roles in science group discourse. Paper presented at the Annual Meeting of the National Association for Research in Science Teaching, San Diego, CA.

Hogan, K. (1999). Thinking aloud together: A test of an intervention to foster students' collaborative scientific reasoning. *Journal of Research in Science Teaching, 36*(10), 1085-1109.

Hogan, K. (2000). Exploring a process view of students' knowledge about the nature of science. *Science Education, 84*(1), 51-70.

Hogan, K. (2002). Small groups' ecological reasoniung while making an environmental management decision. *Journal of Research in Science Teaching, 39*, 341-368.

Hogan, K. & Maglienti, M. (2001). Comparing the epistemological underpinnings of students' and scientists' reasoning about conclusions. *Journal of Research in Science Teaching, 38*(6), 663-687.

Hogan, K., Nastasi, B.K., and Pressley, M. (2002). Discourse patterns and collaborative scientific reasoning in peer and teacher-guided discussions. *Cognition and Instruction, 17*, 379-432.

Hogan, K. & Pressley, M. (1997). *Scaffolding student learning: Instructional approaches and issues.* Cambridge, MA: Brookline Books.

Hohenshell, L.M. & Hand, B. (2006). Writing-to-learn strategies in secondary school cell biology: A mixed method study. *International Journal of Science Education, 28*(2&3), 261-289.

Holliday, W., Yore, L. & Alvermann, D. (1994). The reading-science learning-writing connection: Breakthroughs, barriers and promises. *Journal of Research in Science Teaching, 31*, 877-893.

Holton, G. (1975). On the role of themata in scientific thought. *Science, 188*(4186), 328-334..

Holton, G. (1978). *The scientific imagination: Case studies.* Cambridge: Cambridge University Press.

Holton, G. (1981). Thematic presuppositions and the direction of science advance. In A.F. Heath (Ed.), *Scientific explanation: Papers based on the Herbert Spencer lectures given in the University of Oxford* (pp. 1-27). Oxford: Oxford University Press.

Holton, G. (1986). *The advancement of science and its burdens: The Jefferson lectures and other essays.* Cambridge: Cambridge University Press.

Holton, G. (1988). *Thematic origins of scientific thought: Kepler to Einstein.* Cambridge, MA: Harvard University Press.

Holton, G. (1993). *Science and anti-science.* Cambridge, MA: Harvard University Press.

Holton, G. & Brush, S. (1973). *Introduction to concepts and theories in physical science.* Menlo Park, CA: Addision-Wesley.

Honda, M. (1994). *Linguistic inquiry in the science classroom: "It is science, but it's not like a science problem in a book".* Cambridge, MA: MIT working papers in linguistics. Cited by Smith et al. (2000).

Horton, R. (1967). African traditional thought and Western science. *Africa, 37*, 50-71. Reprinted in M.F.D. Young (Ed.), *Knowledge and control* (pp. 208-266). London: CollierMacmillan.

Hottecke, D. (2000). How and what can we learn from replicating historical experiments? A case study. *Science & Education, 9*, 343-362.

Hounsell, D. (1997). Contrasting conceptions of essay writing. In F. Marton, D. Hounsell & N.J. Entwistle (Eds.), *The experience of learning: Implications for teaching and studying in higher education*, 2nd ed. (pp.106-125). Edinburgh: Scottish Academic Press.

Howe, E.M. (2007). Untangling sickle-cell anemia and the teaching of heterozygote protection. *Science & Education, 16*(1), 1-19.

Howe, E.M. & Rudge, D.W. (2005). Recapitulating the history of sickle-cell anemia research. *Science & Education, 14*, 423-441.

Howes, E.V., Lim, M.. & Campos, J. (2009). Journeys into inquiry-based elementary science: Literacy practices, questioing, and empirical study. *Science Education, 93*(2), 189-217.

Hrdy, S.B. (1986). Empathy, polyandry, and the myth of the coy female. In R. Bleier (Ed.), *Feminist approaches to science* (pp. 119-146). New York: Pergamon.

Hubbard, R. (1988). Some thoughts about the masculinity of natural science. In M. McCanney Gergen (Ed.), *Feminist thought and the structure of knowledge* (pp. 1-15). New York: New York University Press.

Huber, R.A. & Burton, G.M. (1995). What do students think scientists look like? *School Science & Mathematics, 95*(7), 371-376.

Hull, D.L. (1973). *Darwin and his critics: The reception of Darwin's theory of evolution by the scientific community.* Cambridge, MA: Harvard University Press.

Hume, A. & Coll, R. (2008). Student experiences of carrying out a practical science investigation under direction. *International Journal of Science Education, 30*(9), 1201-1228.

Hymes, D. (1972). On communicative competence. In J.B. Pride & J. Holmes (Eds.), *Sociolinguistics: Selected readings* (pp. 269-293). Harmondsworth: Penguin.

Ibrahim, B., Buffler, A. & Lubben, F. (2009). Profiles of freshman physics students' views on the nature of science. *Journal of Research in Science Teaching, 46*(3), 248-264.

Iding, M.K. (1997). How analogies foster learning from science texts. *Instructional Science, 25*, 233-253.

Intemann, K. (2008). Increasing the number of feminist scientists: Why feminist aims are not srved by the underdetermination thesis. *Science & Education, 17*(10), 1065-1079.

Irwin, A.R. (2000). Historical case studies: Teaching the nature of science in context. *Science Education, 84*(1), 5-26.

Irwin, A., Dale, A. & Smith, D. (1996). Science and hell's kitchen: The local understanding of hazard issues. In A. Irwin & B. Wynne, B. (Eds.), *Misunderstanding science? The public reconstruction of science and technology* (pp. 47-64) Cambridge: Cambridge University Press.

Irwin, A. & Wynne, B. (Eds.)(1996). *Misunderstanding science? The public reconstruction of science and technology.* Cambridge: Cambridge University Press.

Irzik, G. & Irzik, S. (2002). Which multiculturalism? *Science & Education, 11*, 393-403.

Jackson, P.W. (1968). *Life in classrooms.* New York: Holt, Rinehart & Winston.

Jackson, S., Krajcik, J. & Soloway, E. (2000). Model-IT: A design retrospective. In M.J. Jacobson & R.B. Kozma (Eds.), *Innovations in science and mathematcis education: Advanced design for technologies of learning* (pp. 77-116). Mahwah, NJ: Erlbaum.

Jackson, T. (1992). Perceptions of scientists among elementary school children. *Australian Science Teachers Journal, 38*, 57-61.

Jacobs, F. (1985). *Breakthrough: The true story of penicillin.* New York: Dodd, Mead & Company.

Jakobson, B. & Wickman, P-O. (2007). Transformation through language use: Children's spontaneous metaphors in elementary school science. *Science & Education, 16*(3-5), 267-289.

Jakobson, B. & Wickman, P-O. (2008). The roles of aesthetic experiences in elementary school science. *Research in Science Education, 38*(1), 45-65.

James, M.C. & Scharmann, L.C. (2007). Using analogies to improve teaching performance of preservice teachers. *Journal of Research in Science Teaching, 44*(4), 565-585.

Janick-Buckner, D. (1997). Getting undergraduates to critically read and discuss primary literature. *Journal of College Science Teaching, 27*(1), 29-32.

Janson-Smith, D. (1980). Sociobiology. So what? In Brighton Women and Science Forum (Eds.), *Alice through the microscope: The power of science over women's lives* (pp. 62-86). London: Virago.

Jarman, R. (1996). Student teachers' use of analogies in science instruction. *International Journal of Science Education, 18*(7), 869-880.

Jarman, R. & McClune, B. (2002). A survey of the use of newspapers in science instruction by secondary teachers in Northern Ireland. *International Journal of Science Education, 24*(10), 997-1020.

Jarman, R. & McClune, B. (2004). Learning with newspapers. In M. Braund & M. Reiss (Eds.), *Learning science outside the classroom* (pp. 185-205). London: RoutledgeFalmer.

Jeanpierre, B., Oberhauser, K. & Freeman, C. (2005). Characteristics of professional development that effect change in secondary science teachers' classroom practices. *Journal of Research in Science Teaching, 42*(6), 668-690.

Jegede, O. (1998). Worldview presuppositions and science and technology education. In D. Hodson (Ed.), *Science and technology education and ethnicity: An Aotearoa/New Zealand perspective* (pp. 76-88). Wellington: The Royal Society of New Zealand.

Jegede, O.J. (1995). Collateral learning and the eco-cultural paradigm in science and mathematics education in Africa. *Studies in Science Education, 25*, 97-137.

Jegede, O. & Okebukola, P.A. (1991). The relationship between African traditional cosmology and students' acquisition of a science process skill. *International Journal of Science Education, 13*(1), 37-47.

Jenkins, E. (1989). Why the history of science? In M. Shortland & A. Warwick (Eds.), *Teaching the history of science* (pp. 19-29). Oxford: Basil Blackwell.

Jenkins, E. (2000). "Science for all": Time for a paradigm shift? In R. Millar, J. Leach & J. Osborne (Eds.), *Improving science education: The contribution of research* (pp. 207-236). Buckingham: Open University Press.

Jenkins, E.W. (2006). School science and citizenship: Whose science and whose citizenship? *The Curriculum Journal, 17*(3), 197-211.

Jenkins, E. (2007). School science: A questionable construct? *Journal of Curriculum Studies, 39*(3), 265-282.

Jeong, H., Songer, N.B. & Lee, S-Y. (2007). Evidentiary competence: Sixth graders' understanding for gathering and interpreting evidence in scientific investigations. *Research in Science Education, 37*, 75-97.

Jiménez-Aleixandre, M.P. (2008). Designing argumentation learning environments. In S. Erduran & M.P. Jiménez-Aleixandre (Eds.), *Argumentation in science education: Perspectives from classroom-based research* (pp. 91-116). Dordrecht: Kluwer.

Jiménez-Aleixandre, M.P. & Erduran, S. (2008). Argumentation in science education; An overview. In S. Erduran & M.P. Jiménez-Aleixandre (Eds.), *Argumentation in science education: Perspectives from classroom-based research* (pp. 3-27). Dordrecht: Kluwer.

Jiménez-Aleixandre, M.P. & Reigosa, C. (2006). Contextualizing practices across epistemic levels in the chemistry laboratory. *Science Education, 90*(4), 707-733.

Jiménez-Aleixandre, M.P., Rodriguez, A.B. & Duschl, R.A. (2000). "Doing the lesson" or "doing science": Argument in high school genetics. *Science Education, 84*(6), 757-792.

Johanson, D. & Blake, E. (2001). *From Lucy to language*. London: Cassell.

Johnson, J.R. (1989). *Technology: Report of the Project 2061 phase I technology panel*. Washington, DC: American Association for the Advancement of Science.

Johnson-Laird, P.N. (1983). *Mental models*. Cambridge: Cambridge University Press.

Johnston, A., Southerland, S.A. & Sowell, S. (2006). Dissatisfied with the fruitfulness of "learning ecologies". *Science Education, 90*(5), 907-911.

Johnstone, A.H. (1991). Why is science difficult to learn? Things are seldom what they seem. *Journal of Computer Assisted Learning, 7*, 75-83.

Jonassen, D.H. & Carr, C.S. (2000). Mindtools: Affording multiple knowledge representations for learning. In S. LaJoie (Ed), *Computers as cognitive tools* (pp. 165-196). Mahwah, NJ: Lawrence Erlbaum Associates.

Joravsky, D. (1970). *The Lysenko affair*. Chicago, IL: University of Chicago Press.

Joynson, R.B. (1989). *The Burt affair*. New York: Routledge.

Judson, H.F. (1980). *The search for solutions*. New York: Holt, Rinehart & Winston.

Judson, H.F. (2004). *The great betrayal: Fraud in science*. Orlando, FL: Harcourt.

Juell, P. (1985). The course journal. In A. Ruggles Gere (Ed.), *Roots in the sawdust: Writing to learn across the disciplines* (pp. 187-201). Urbana, IL: National Council of Teachers of English.

Justi, R. (2000). Teaching with historical models. In J.K. Gilbert & C.J. Boulter (Eds.), *Developing models in science education* (pp. 209-226). Dordrecht: Kluwer.

Justi, R. & Gilbert, J.K. (1999). History and philosophy of science through models: The case of kinetics. *Science & Education, 8*(3), 287-307.

Justi, R.S. & Gilbert, J.K. (2002a). Modelling, teachers' views on the nature of modelling and implications for the education of modellers. *International Journal of Science Education, 24*(4), 369-388.

Justi, R.S. & Gilbert, J.K. (2002b). Science teachers' knowledge about and attitudes towards the use of models in learning science. *International Journal of Science Education, 24*(12), 1273-1292.

Justi, R.S. & Gilbert, J.K. (2002c). Models and Modelling in Chemical Education. In J.K. Gilbert, O. de Jong, R.Justi, D.F. Treagust & J.H. van Driel (Eds.), *Chemical education: Towards research-based practice* (pp. 47-68). Dordrecht: Kluwer.

Justi, R.S. & Gilbert, J.K. (2003). Teachers' views on the nature of models. *International Journal of Science Education, 25*(11), 1369-1386.

Justi, R. & Gilbert, J. (2006). The role of analog models in the understanding of the nature of models in chemistry. In P.J. Aubusson, A.G. Harrison & S.M. Ritchie (Eds.), *Metaphor and analogy in science education* (pp. 119-130). Dordrecht: Kluwer.

Justi, R. & van Driel, J. (2005). The development of science teachers' knowledge on models and modelling: Promoting, characterizing, and understanding the process. *International Journal of Science Education, 27*(5), 549-573.

Kachan, M.R., Guilbert, S.M. & Bisanz, G.L. (2006). Do teachers ask students to read news in secondary science? Evidence from the canadian context. *Science Education, 90*(3), 496-521.

Kafai, Y.B. & Gilliland-Swetland, A.J. (2001). The use of historical materials in elementary science classrooms. *Science Education, 85*(4), 349-367.

Kahle, J.B. & Meece, J. (1994). Research on gender issues in the classroom. In D. Gabel (Ed.), *Handbook of research on science teaching and learning* (pp. 542-557). New York: Macmillan.

Kalman, C. (2009). The need to emphasize epistemology in teaching and research. *Science & Education, 18*(3&4), 325-347.

Kanari, Z. & Millar, R. (2004). Reasoning from data: How students collect and interpret data in scientific investigations. *Journal of Research in Science Teaching, 41*(7), 748-769.

Kang, N-H. (2007). Elementary teachers' epistemological and ontological understanding of teaching for conceptual learning. *Journal of Research in Science Teaching, 44*(9), 1292-1317.

Kang, N-H. & Wallace, C.S. (2005). Secondary science teachers' use of laboratory activities: Linking epistemological beliefs, goals, and practices. *Science Education, 89*(1), 140-165.

Kang, S., Scharmann, L.C. & Noh, T. (2005). Examining students' views on the nature of science: Results from Korean 6th, 8th, and 10th graders. *Science Education, 89*(2), 314-334.

Kawagley, A.O. & Barnhardt , R.(1999). Education indigenous to place: Western science meets Native reality. In G.A. Smith & D.R. Williams (Eds.), *Ecological education in action: On weaving education, culture, and the environment* (pp. 117-140). Albany, NY: SUNY Press.

Kawagley, A.O., Norris-Tull, D. & Norris-Tull, R.A. (1995). Incorporation of the world views of indigenous cultures: A dilemma in the practice and teaching of Western science In F. Finley, D. Allchin, D. Rhees & S.Fifield (Eds.), *Proceedings of the Third International History, Philosophy and Science Teaching Conference, Vol. 1* (pp. 583-588). Minneapolis, MN: University of Minnesota Press.

Kawagley, A.O., Norris-Tull, D. & Norris-Tull, R.A. (1998). The indigenous worldview of Yupiaq culture: Its scientific nature and relevance to the practice and teaching of science. *Journal of Research in Science Teaching, 35*, 133-144.

Kawasaki, K. (1996). The concepts of science in Japanese and Western education. *Science & Education, 5*, 1-20.

Kawasaki, K., Herrenkohl, L.P. & Yeary, S.A. (2004). Theory building and modelling in a sinking and floating unit: A case study of third and fourth grade students' developing epistemologies of science. *International Journal of Science Education, 26*(11), 1299-1324.

Keane, M. (2008). Science education and worldview. *Cultural Studies of Science Education, 3*(3), 587-613.

Kearney, M. (1984). *World view.* Novato, CA: Chandler & Sharp

Keller, E.F. (1985). *Reflections on gender and science.* New Haven, CT: Yale University Press.

Keller, E.F. (1992). How gender matters, or, why it's so hard for us to count past two. In G. Kirkup & L.S. Keller (Eds.), *Inventing women: Science, technology and gender* (pp. 42-56). Cambridge, MA: Polity Press.

Keller, E.F. & Longino, H.E. (Eds.) (1996). *Feminism and science.* Oxford: Oxford University Press.

Kelly, G.J. (2007). Discourse in science classrooms. In S.K. Abell & N.G. Lederman (Eds.), *Handbook on research in science education* (pp. 443-469). Mahwah, NJ: Lawrence Erlbaum.

Kelly, G.J. & Bazerman, C. (2003). How students argue scientific claims: A rhetorical-semantic analysis. *Applied Linguistics, 24*(1), 28-55.

Kelly, G.J., Brown, C. & Crawford, T. (2000). Experiments, contingencies, and curriculum: Providing opportunities for learning through improvisation in science teaching. *Science Education, 84*(5), 624-657.

Kelly, G. & Chen, C. (1999). The sound of music: Constructing science as sociocultural practices through oral and written discourse. *Journal of Research in Science Teaching, 36*, 883-915.

Kelly, G.J., Chen, C. & Prothero, W. (2000). The epistemological framing of a discipline: Writing science in university oceanography. *Journal of Research in Science Teaching, 17*, 691-718.

Kelly, G.J. & Green, J. (1998). The social nature of knowing: Toward a sociocultural perspective on conceptual change and knowledge construction. In B.J. Guzzetti & C.R. Hynd (Eds.), *Perspectives on conceptual change: Multiple ways to understand knowing and learning in a complex world* (pp. 145-181). Mahweh, NJ: Lawrence Erlbaum Associates.

Kelly, G.J., Regev, J. & Prothero, W. (2008). Analysis of lines of reasoning in written argumentation. In S. Erduran & M.P. Jiménez-Aleixandre (Eds.), *Argumentation in science education: Perspectives from classroom-based research* (pp. 137-158). Dordrecht: Kluwer.

Kelly, G.J. & Takao, A. (2002). Epistemic levels in argument: An anlysis of university oceanography students' use of evidence in writing. *Science Education, 86*(3), 314-342.

Kempa, R.F. (1986). *Assessment in science.* Cambridge: Cambridge University Press.

Kempa, R.F. & Ayob, A. (1995). Learning from group work in science. *International Journal of Science Education, 17*, 743-754.

Kenyon, L., Kuhn, L. & Reiser, B.J. (2006). Using students' epistemologies of science to guide the practice of argumentation. In S.A. Barab, K.E. Hay & T.D. Hickey (Eds.), *Proceedings of the 7th international conference of the learning sciences* (pp. 321-327). Mahwah, NJ: Lawrence Erlbaum.

Keranto, T. (2001). The perceived credibility of scientific claims, paranormal phenomena and miracles among primary teacher students: A comparative study. *Science & Education, 10*, 493-511.

Keselman, A., Kaufman, D.R. & Patel, V.L. (2004). "You can exercise your way out of HIV" and other stories: The role of biological knowledge in adolescents' evaluation of myths. *Science Education, 88*(4), 548-573.

Key, S.G. (2003). Enhancing the science interest of African American students using cultural inclusion. In S.M. Hines (Eds.), *Multicultural science education: theory, practice, and promise* (pp. 87-101). New York: Peter Lang.

Keys, C. (1995). An interpretative study of students' use of scientific reasoning during a collaborative report writing intervention in ninth grade general science. *Science Education, 79*(4), 415-435.

Keys, C.W. (1998). A study of grade 6 students generating questions and plans for open-ended science investigations. *Research in Science Education, 28*(3), 301-316.

Keys, C.W. (1999). Revitalizing instruction in scientific genres: Connecting knowledge production with writing to learn in science. *Science Education, 83*, 115-130.

Keys, C.W. (2000). Investigating the thinking processes of eighth grde writers during the composition of a scientific laboratory report. *Journal of Research in Science Teaching, 37*(7), 676-690.

Keys, C.W., Hand, B., Prain, V. & Collins, S. (1999). Using the science writing heuristic as a tool for learning from laboratory investigations in science. *Journal of Research in Science Teaching, 36*, 1065-1084.

Khan, S. (2007). Model-based inquiries in chemistry. *Science Education, 91*(6), 877-905.

Khishfe, R. (2008). The development of seventh graders' views of nature of science. *Journal of Research in Science Teaching, 45*(4), 470-496.

Khishfe, R. & Abd-El-Khalick, F. (2002). Influences of explicit and reflective versus implicit inquiry-oriented instruction on sixth graders' views of nature of science. *Journal of Research in Science Teaching, 39*(7), 551-578.

Khishfe, R. & Lederman, N. (2006). Teaching nature of science within a controversial topic: Integrated versus non-integrated. *Journal of Research in Science Teaching, 43*(4), 395-418.

Khishfe, R. & Lederman, N. (2007). Relationship between instructional context and views of the nature of science. *International Journal of Science Education, 29*(8), 939-961.

Kichawen, P., Swain, J. & Monk, M. (2004). Views on the philosophy of science among undergraduate science students and their tutors at the University of Papua New Guinea: Origins, progression, enculturation and destinations. *Research in Science & Technological Education, 22*(1), 81-98.

Killerman, W. (1996). Biology education in Germany: Research into the effectiveness of different teaching methods. *International Journal of Science Education, 18*, 333-346.

Kim, B.S., Ko, E.K., Lederman, N.G. & Lederman, J.S. (2005). A developmental continuum of pedagogical content knowledge for nature of science instruction. Paper presented at the Annual Meeting of the National Association for Research in Science Teaching, Dallas, TX, April.

Kim, H. & Song, J. (2006). The features of peer argumentation in middle school students' scientific inquiry. *Research in Science Education, 36*(3), 211-233.

Kim, J.J-H.(2009). *Teaching and learning grade 7 science concepts by elaborate analogies: Mainstream and East and South Asian ESL students' experiences*. Toronto: Unpublished PhD thesis, University of Toronto.

Kim, M.C., Hannafin, M.J. & Bryan, L.A. (2007). Technology-enhanced inquiry tools in science education: An emerging pedaagogical framework for classrpom practice. *Science Education, 91*(6), 1010-1030.

Kimball, M.E. (1967). Understanding the nature of science: A comparison of scientists and teachers. *Journal of Research in Science Teaching, 5*(2), 110-120.

King, A.R. & Brownell, J.A. (1966). *The curriculum and the disciplines of knowledge: A theory of curriculum practice*. New York: John Wiley.

King, B.B. (1991). Beginning teachers' knowledge of and attitudes towards history and philosophy of science. *Science Education, 75*(1), 135-141.

King, M.D. & Bruce, M.C. (2003). Inspired by real science. *Science and Children, 40*(5), 30-34.

Kintgen, E.R. (1988). Literacy, literacy. *Visible Language, 1*(2/3), 149-168.

Kipnis, N. (1996). The "historical-investigative" approach to teaching science. *Science & Education, 5*, 277-292.

Kipnis, N. (2007). Discovery in science and in teaching science. *Science & Education, 16*(9&10), 883-920.

Kisiel, J. (2006). An examination of fieldtrip strategies and their implementation within a natural history museum. *Science Education, 90*(3), 434-452.

Kitcher, P. (1982). *Abusing science: The case against creationism.* Cambridge, MA: MIT Press.

Kitcher, P. (1993). *The advancement of science: Science without legend, objectivity without illusions.* Oxford: Oxford University Press.

Kithinji, W. (2000). *An inquiry into the integration of Indigenous knowledges and skills in the Kenyan secondary science curriculum.* Toronto: Unpublished PhD thesis, University of Toronto.

Klasson, S. (2006a). A theoretical framework for contextual science teaching. *Interchange, 37*(1&2), 31-62.

Klasson, S. (2006b). The scientific thought experiment: How might it be used profitably in the classroom? *Interchange, 37*(1&2), 77-96.

Klassen, S. (2009). The construction and analysis of a science story: A proposed methodology. *Science & Education, 18*(3&4), 401-423.

Klein, M.J. (1972). The use and abuse of historical teaching in physics. In S.G. Brush & A.L. King (Eds.), *History in the teaching of physics* (pp. 12-18). Hanover, NH: University of New England Press.

Klein, P.D. (1999). Reopening inquiry into cognitive processes in writing-to-learn. *Educational Psychology Review, 11*(3), 203-270.

Klein, P.D. (2000). Elementary students' strategies for writing-to-learn science. *Cognition and Instruction, 18*, 317-348.

Klein, P.D. (2006). The challenges of scientific literacy: from the viewpoint of second-generation cognitive science. *International Journal of Science Education, 28*(2&3), 143-178.

Klein, S., Richardson, B., Grayson, D.A., Fox, L.H., Kramarae, C., Pollard, D.S. & Dwyer, C.A. (2007). *Handbook for achieving gender equity through education*, 2nd ed. Mahwah, NJ: Lawrence Erlbaum Asociates.

Kleinman, S.S. (1998). Overview of feminist perspectives on the ideology of science. *Journal of Research in Science Teaching, 35*(8), 837-844.

Klopfer, L.E. (1969). Science education in 1991. *School Review, 77*(3-4), 199-217.

Klopfer, L.E. (1976). Scientific literacy reexamined. *Science Education, 60*(1), 95.

Klotz, I.M. (1980). The N-ray affair. *Scientific American, 242*(5), 122-131.

Knain, E. (2001). Ideologies in school science textbooks. *International Journal of Science Education, 23*(3), 319-329.

Knorr-Cetina, K. (1992). The couch, the cathedral, and the laboratory: On the relationship between experiment and laboratory in science. In A. Pickering (Ed.), *Science as practice and culture* (pp. 113-137). Chicago, IL: University of Chicago Press.

Knorr-Cetina, K. (1999). *Epistemic cultures: How the sciences make knowledge.* Cambridge, MA: Harvard University Press.

Knox, K.L., Moynihan, J.A. & Markowitz, D.G. (2003). Evaluation of short-term impact of a high-school summer science program on students' perceived knowledge and skills. *Journal of Science Education & Technology, 12*, 471-478.

Knudtson, P. & Suzuki, D. (1992). *Wisdom of the elders.* Toronto: Stoddart.

Koertge, N. (2004). How might we put gender politics into science? *Philosophy of Science, 71*, 868-879.

Kohn, A. (1986). *False prophets.* Oxford: Bail Blackwell.

Kolstø, S.D. (2006). Patterns in students' argumentation confronted with a risk-focused socio-scientific issue. *International Journal of Science Education, 28*(14), 1689-1716.

Kolstø, S.D. (2008). Science education for democratic citizenship through the use of history of science. *Science & Education, 17*(8&9), 977-997.

Koponen, I.T. (2007). Models and modelling in physics education: A critical re-analysis of philosophical underpinnings and suggestions for revisions. *Science & Education, 16*(7&8), 751-773.

Korpan, C.A., Bisanz, G.L., Bisanz, J. & Henderson, J.M. (1997). Assessing literacy in science: Evaluation of scientific news briefs. *Science Education, 81*, 515-532.

Koslowski, B. (1996). *Theory and evidence: The development of scientific reasoning.* Cambridge, MA: MIT Press.

Kosma, R., Chin, E., Russell, J. & Marx, N. (2000). The role of representations and tools in the chemistry laboratory and their implications for chemistry learning. *Journal of the Learning Sciences, 9*(2), 105-143.

Kosma, R.B. & Russell, J. (1997). Multimedia and understanding: Expert and novice responses to different representations of chemical phenomena. *Journal of Research in Science Teaching, 34*(9), 949-968.

Kosso, P. (2009). The large-scale structure of scientific method. *Science & Education, 18*(1), 33-42.

Koul, R. (2003a). The relevance of public image of science in science education policy and practice. *Science & Education, 12*, 115-124.

Koul, R. (2003b). Revivalist thinking and student conceptualizations of science/religion. *Studies in Science Education, 39*, 103-124.

Koulaidis, V. & Ogborn, J. (1988). Use of systematic networks in the development of a questionnaire. *International Journal of Science Education, 10*(3), 497-509.

Koulaidis, V. & Ogborn, J. (1989). Philosophy of science: An emoirical study of teachers' views. *International Journal of Science Education, 11*(2), 173-184.

Koulaidis, V. & Ogborn, J. (1995). Science teachers' philosophical assumptions: How well do we understand them? *International Journal of Science Education, 17*(3), 273-283.

Kourany, J. (2003). A philosophy of science for the twenty-first century. *Philosophy of Science, 70*(1), 1-14.

Kozma, R. (2003). The material features of multiple representations and their cognitive and social affordances for science understanding. *Learning and Instruction, 13*, 205-226.

Krapfel, P. (1999). Deepening children's participation through local ecological investigations. In G.A.Smith & D.R.Williams (Eds.), *Ecological education in action: On weaving education, culture, and the environment* (pp. 47-64). Albany, NY: SUNY Press.

Krajcik, J. (2008). Commentary on Chinn's and Samarapungavan's paper. In R.A. Duschl & R.E. Grandy (Eds.), *Teaching scientific inquiry: Recommendations for research and implementation* (pp. 226-232). Rotterdam/Taipei: Sense.

Krajcik, J., Blumenfeld, P., Marx, R. & Soloway, E. (2000). Instructional, curricular, and technological supports for inquiry in science classrooms. In J. Minstrell & E. van Zee (Eds.), *Inquiring into inquiry learning and teaching in science* (pp. 283-315). Washington, DC: American Association for the Advancement of Science.

Kress, G., Jewitt, C., Ogborn, J. & Tsatsarelis, C. (2001). *Multimodal teaching and learning: The rhetorics of the science classroom.* London: Continuum.

Kreuzman, H. (1995). Stepping into science: Historical experiments and scientific method. In F. Finley, D. Allchin, D. Rhees & S.Fifield (Eds.), *Proceedings of the Third International History, Philosophy and Science Teaching Conference*, Vol. 1 (pp. 636-647). Minneapolis, MN: University of Minnesota Press.

Kruglanski, A.W. (1989). *Lay epistemics and human knowledge: Cognitive and motivational bases.* New York: Plenum.

Kuhn, D. (1991). *The skills of argument.* Cambridge: Cambridge University Press.

Kuhn, D. (1992). Thinking as argument. *Harvard Educational Review, 62*(2), 155-178.

Kuhn, D. (1993). Science as argument: Implications for teaching and learning scientific thinking. *Science Education, 77*, 319-337.

Kuhn, D. (1997). Developmental psychology and science education. *Review of Educational Research, 67*, 141-150.

Kuhn, D. (2005). *Education for thinking.* Cambridge, MA: Harvard University Press.

Kuhn, D. (2007). Reasoning about multiple variables: Control of variables is not the only challenge. *Science Education, 91*(5), 710-726.

Kuhn, D., Amsel, E. & O'Loughlin, M. (1988). *The development of scientific thinking skills.* London: Academic Press.

Kuhn, D., Black, J., Keselman, A. & Kaplan, D. (2000). The development of cognitive skills to support inquiry. *Cognition and Instruction, 18*(4), 495-523.

Kuhn, D. & Franklin, S. (2006). The second decade: What develops (and how)? In D. Kuhn & R. Siegler (Eds.), *Handbook of child psychology. Vol. 2: Cognition, perception, and language* (pp. 953-993). Hoboken, NJ: Wiley.

Kuhn, D., Garcia-Mila, M., Zohar, A. & Andersen, C. (1995). *Strategies of knowledge acquisition.* Monographs of the Society for Research in Child Development, Serial No. 245, 60(4).

Kuhn, D. & Pearsall, S. (2000). Developmental origins of scientific thinking. *Journal of Cognition and Devlopment, 1,* 113-129.

Kuhn, T.S. (1963). The essential tension: Tradition and innovation in scientific research. In C.W. Taylor & F. Barron (Eds.), *Scientific creativity: Its recognition and development* (pp. 341-354). New York: John Wiley.

Kuhn, T.S. (1970). *The structure of scientific revolutions,* 2nd ed. Chicago, IL: University of Chicago Press.

Kuhn, T.S. (1977). *The essential tension: Selected studies in scientific traditions and change.* Chicago, IL: University of Chicago Press.

Kuldell, N. (2003). Read like a scientist to write like a scientist. *Journal of College Science Teaching, 33*(2), 32-35.

Kurtz, K., Miao, C-H. & Gentner, D. (2001). Learning by analogical bootstrapping. *Journal of the Learning Sciences, 10,* 417-446.

Kutnick, P. & Rogers, C. (Eds.) (1994). *Groups in schools.* London: Cassell.

Labinger, J.A. (1995). Science as culture: A view from the Petri dish. *Social Studies of Science, 25,* 285-306.

Lacey, H. (1996). On relations between science and religion. *Science & Education, 5*(2), 143-153.

Ladson-Billings, G. (1995). But that's just good teaching! The case for culturally relevant pedagogy. *Theory into Practice, 34*(3), 159-165.

Ladyman, J. (2002). *Understanding philosophy of science.* London: Routledge.

Lagoke, B.A., Jegede, O.J. & Oyebanji, P.K. (1997). Towards an elimination of the gender gulf in science concept attainment through the use of environmental analogs. *International Journal of Science Education, 19*(4), 365-380.

Lakatos, I. (1978a). Introduction: science and pseudoscience. In J. Worrall & G. Currie (Eds.), *Imre Lakatos: Philosophical papers volume 1.* Cambridge: Cambridge University Press.

Lakatos, I. (1978b). *The methodology of scientific research programmes* (Eds. J. Worrall & G. Currie). Cambridge: Cambridge University Press.

Lakin, S. & Wellington, J. (1994). Who will teach the "nature of science"? Teachers' views of science and their implications for science education. *International Journal of Science Education, 16*(2), 175-190.

Lambert, H. & Rose, H. (1996). Disembodied knowledge? Making sense of medical science. In A. Irwin & B. Wynne (Eds.), *Misunderstanding science? The public reconstruction of science and technology* (pp. 65-83). Cambridge: Cambridge University Press.

Landau, I. (2008). Problems with feminist standpoint theory in science education. *Science & Education, 17*(10), 1081-1088.

Lankshear, C., Gee, J.P. & Hull, G. (1996). *The new work order: Behind the language of the new capitalism.* Boulder, CO: Westview Press.

Lantz, O. & Kass, H. (1987). Chemistry teachers' functional paradigms. *Science Education, 71*(1), 117-134.

Larkin, J.H. (1983). The role of problem representation in physics. In D. Gentner & A.L. Stevens (Eds.), *Mental models* (pp. 75-98). Hillsdale, NJ: Lawrence Erlbaum.

Larson, J.O. (1995). Fatima's rules and other elements of an unintended chemistry curriculum. Paper presented at the American Educational Research Association annual meeting, San Francisco, CA, April.

Latour, B. (1987). *Science in action: How to follow scientists and engineers through society.* Cambridge, MA: Harvard University Press.

Latour, B. (1990). Drawing things together. In M. Lynch & S. Woolgar (Eds.), *Representation in scientific practice* (pp. 19-68). Cambridge, MA: MIT Press.

Laudan, L. (1977). *Progress and its problems: Toward a theory of scientific growth.* Berkeley, CA: University of California Press.

Laudan, L. (1981). A confutation of convergent realism. *Philosophy of Science, 48*(1), 19-49.

Laudan, L. (1983). The demise of the demarcation problem. In R. Cohen & L. Laudan (Eds.), *Physics, philosophy and psychoanalysis* (pp. 111-128). Dordrecht: Reidel.

Laudan, L. (1984a). *Science and values: The aims of science and their role in scientific debate.* Berkeley, CA: University of California Press.

Laudan, L. (1984b). Explaining the success of science: Beyond epistemic realism and relativism. In J.T. Cushing, C.F. Delaney & G.M. Gutting (Eds.), *Science and reality: Recent work in the philosophy of science* (pp. 83-105). Notre Dame, IN: Notre Dame University Press.

Laugksch, R.C. (2000). Scientific literacy: A conceptual overview. *Science Education, 84*(1), 71-94.

Lave, J. (1988). *Cognition in practice: Mind, mathematics and culture in everyday life.* New York: Cambridge University Press.

Lave, J. (1991). Situated learning in communities of practice. In L.B. Resnick, J.M. Devine & S.D. Teasley (Eds.), *Perspectives on socially shared cognition* (pp. 63-82). Washington, DC: American Psychological Association.

Lave, J. & Wenger, E. (1991). *Situated learning: Legitimate peripheral participation.* Cambridge: Cambridge University Press.

Lawless, J.G. & Rock, B.N. (1998). Student scientist partnerships and data quality. *Journal of Science Education & Technology, 7*(1), 5-13.

Lawson, A.E. (2002). Sound and faulty arguments generated by preservice biology teachers when testing hypotheses involving unobservable entities. *Journal of Research in Science Teaching, 39*(3), 237-252.

Layton, D. (1988). Revaluing the T in STS. *International Journal of Science Education, 10,* 367-378.

Layton, D. (1991). Science education and praxis: The relationship of school science to practical action. *Studies in Science Education, 19,* 43-79.

Layton, D. (1993). *Technology's challenge to science education: Cathedral, quarry or company store?* Buckingham: Open University Press.

Layton, D, Jenkins, E., MacGill, S. & Davey, A. (1993). *Inarticulate science?* Driffield: Studies in Education.

Lazarowitz, R. & Hertz-Lazarowitz, R.P. (1998). Cooperative learning in the science curriculum. In B. Fraser & K. Tobin (Eds.), *International handbook of science education* (pp. 449-470). Dordrecht: Kluwer.

Leach, J., Driver, R., Millar, R. & Scott, P. (1996). Progression in learning about "the nature of science": Issues of conceptualisation and methodology. In M. Hughes (Ed.), *Progress in learning* (pp. 109-139). Clevedon (UK): Multilingual Matters.

Leach, J., Hind, A. & Ryder, J. (2003). Designing and evaluating short teaching interventions about the epistemology of science in high school classrooms. *Science Education, 87*(3), 831-848.

Leach, J., Millar, R., Ryder, J. & Séré, M-G. (2000). Epistemological understanding in science learning: The consistency of representations across contexts. *Learning and Instruction, 10*(6), 497-527.

Leatherdale, W.H. (1974). *The role of analogy, model and metaphor in science.* Amsterdam: North-Holland.

Lecourt, D. (1977). *Proletarian science? The case of Lysenko.* Manchester: Manchester University Press.

Lederman, L., with Teresi, D. (1993). *The God particle: If the universe is the answer, what is the question?* Boston, MA: Houghton Mifflin.

Lederman, M. & Bartsch, I. (Eds.) (2001). *The gender and science reader.* New York: Routledge.

Lederman, N.G. (1986). Relating teahing behaviour and classroom climate too changes in students' conceptions of the nature of science. *Science Education, 70*(1), 3-19.

Lederman, N.G. (1992). Students' and teachers' conceptions of the nature of science: A review of the research. *Journal of Research in Science Teaching, 29*(4), 331-359.

Lederman, N.G. (1999). Teachers' understanding of the nature of science and classroom practice: Factors that facilitate or impede the relationship. *Journal of Research in Science Teaching, 36*(8), 916-929.

Lederman, N. G. (2006). Research on nature of science: Reflections on the past, anticipations of the future. Asia-Pacific Forum on Science Learning and Teaching. Available on-line at: http://www.ied.edu.hk/apfslt/v7_issue1/foreword/index.html

REFERENCES

Lederman, N.G. (2007). Nature of science: Past, present, and future. In S.K. Abell & N.G. Lederman (Eds.), *Handbook of research on science education* (pp. 831-879). Mahwah, NJ: Lawrence Erlbaum Associates.

Lederman, N.G. & Abd-El-Khalick, F. (1998). Avoiding de-natured science: Activities that promote understandings about the nature of science. In W.F. McComas (Ed.), *The nature of science in science education: Rationales and strategies* (pp. 83-126). Dordrecht: Kluwer.

Lederman, N.G., Abd-El-Khalick, F., Bell, R.L. & Schwartz, R.S. (2002). Views of nature of science questionnaire: Toward valid and meaningful assessment of learners' conceptions of nature of science. *Journal of Research in Science Teaching, 39*(6), 497-521.

Lederman, N.G. & O'Malley, M. (1990). Students' perceptions of tentativeness in science: Development, use and sources of change. *Science Education, 74*(2), 225-239.

Lederman, N.G., Schwartz, R.S., Abd-El-Khalick, F. & Bell, R.L. (2001). Pre-service teachers' understanding and teaching of nature of science: An intervention study. *Canadian Journal of Science, Mathematics and Technology Education, 1*(2), 135-160.

Lederman, N., Wade, P. & Bell, R.L. (1998). Assessing the nature of science: What is the nature of our assessments? *Science & Education, 7*(6), 595-615.

Lederman, N., Wade, P. & Bell, R.L. (2000). Assessing understanding of the nature of science: A historical perspective. In W.F. McComas (Ed.), *The nature of science in science education: Rationales and strategies* (pp. 331-350). Dordrecht: Kluwer Academic.

Lee, H-S. & Songer, N.B. (2003). Making authentic science accessible to students. *International Journal of Science Education, 25*(8), 923-948.

Lee, O. (1997). Scientific literacy for all? What is it, and how can we achieve it? *Journal of Research in Science Teaching, 34*, 219-222.

Lee, O. (2002). Science inquiry for elementary students from diverse backgrounds. *Review of Research in Education, 26*, 23-69.

Lee, O. (2004). Teacher change in beliefs and practices in science and literacy instruction with English language learners. *Journal of Research in Science Teaching, 41*, 65-93.

Lee, O. & Fradd, S.H. (1996). Interactional patterns of linguistically diverse students and teachers: Insights for promoting science learning. *Linguistics and Education, 8*, 269-297.

Lee, O. & Fradd, S.H. (1998). Science for all, including students from non-English-language backgrounds. *Educational Researcher, 17*, 12-21.

Lee, Y-C. (2008). Exploring the roles and nature of science: A case study of severe acute respiratory syndrome. *International Journal of Science Education, 30*(4), 515-541.

LeGrand, H.E. (1988). *Drifting continents and shifting theories*. Cambridge: Cambridge University Press.

Lehrer, R. & Schauble, L. (2000). Modeling in mathematics and science. In R. Glaser (Ed.), *Advances in instructional psychology: Vol. 5 Educational design and cognitive science* (pp. 101-159). Mahwah, NJ: Erlbaum.

Lehrer, R. & Schauble, L. (2003). Origins and evolution of model-based reasoning in mathematics and science. In R. Lesh & H.M. Doerr (Eds.), *Beyond constructivism: A models and modeling petrspective on mathematics problem-solving* (pp. 59-70). Mahwah, NJ: Erlbaum.

Lehrer, R. & Schauble, L. (2005). Developing modeling and argument in the elementary grades. In T. Romberg & T.P. Carpenter (Eds.), *Understanding mathematics and science matters* (pp. 29-53). Mahwah, NJ: Erlbaum.

Lehrer, R. & Schauble, L. (2006). Scientific thinking and science literacy. In K.A. Renninger & I.E. Sigel (Eds.), *Handbook of child psychology: Vol. 4 Child psychology and practice* (pp. 153-196). Hoboken, NJ: John Wiley.

Leite, L. (2002). History of science in science education: Development and validation of a checklist for analysing the historical content of science textbooks. *Science & Education, 11*, 333-359.

Lemke, J.L. (1987). Social semiotics and science education. *American Journal of Semiotics, 5*, 217-232.

Lemke, J.L. (1990). *Talking science: Language, learning and values*. Norwood, NJ: Ablex.

Lemke, J.L. (1992). Intertextuality and educational research. *Linguistics and Education, 4*, 257-267.

Lemke, J.L. (1998). Teaching all the languages of science: Words symbols, images, and actions. Available at http://academic.brooklyn.cuny.edu/education/jlemke/papers/barcelon.htm

Lemke, J.L. (2001). Articulating communities: Sociocultural perspectives on science education. *Journal of Research in Science Teaching, 38*(3), 296-316.

Lemke, J.L. (2004). The literacies of science. In E.W. Saul (Ed.), *Crossing borders in literacy and science instruction: Perspectives on theory and practice* (pp. 33-47). Newark, DE: International Reading Association/National Science Teachers Association.

Lemons, J. (1995). The role of the university, scientists, and educators in promotion of environmental literacy. In B. Jickling (Ed.), *A colloquium on environment, ethics, and education* (pp. 90-109). Whitehorse, Yukon: Yukon College, July 14–16.

Lennon, K. (2004). Feminist epistemology. In I. Niiniluoto, I., Sintonen, M. & Wolenski, J. (Eds.), *Handbook of epistemology* (pp. 1013-1026). Dodrecht: Kluwer.

Levere, T.H. (2006). What can history teach us about science: Theory and experiment, data and evidence. *Interchange, 37*(1&2), 115-128.

Levine, T. & Geldman-Caspar, Z. (1997). What can be learned from informal student writings in a science context. *School Science & Mathematics, 97*(7), 359-367.

Levinson, R. & Turner, S. (2001). *Valuable lessons: The teaching of social and ethical issues in the school curriculum arising from developments in biomedical research – A research study of teachers.* London: The Wellcome Trust.

Levi-Strauss, C. (1968). *The savage mind.* New York: Weidenfeld & Nicholson.

Lewenstein, B.V. (1995). Science and the media In S. Jasanoff, G.E. Markle, J.C. Petersen & T. Pinch (Eds.), *Handbook of science and technology studies* (pp. 343-360). London: Sage.

Lewenstein, B.V. (2001). Who produces science information for the public? In J.H. Falk, E. Donovan & R. Woods (Eds.), *Free-choice science education: How we learn science outside of school?* (pp. 21-43). New York: Teachers College Press.

Lewis, J. & Leach, J. (2006). Discussion of socio-scientific issues: The role of science knowledge. *International Journal of Science Education, 28*(11), 1267-1287.

Lewis, T. & Gagel, C. (1992). Technological literacy: A critical analysis. *Journal of Curriculum Studies, 24*(2), 117-138.

Lewontin, R.C. (2002). The politics of science. *The New York Review of Books, 49*(8), 28-32.

Lewontin, R. & Levins, R. (1976). The problem of Lysenkoism. In H. Rose & S. Rose (Eds.), *The radicalisation of science* (pp. 32-64). London: Macmillan.

Liao, S.J., Lee, M.W.H. & Ng, L.K.Y. (1994). *Principles and practice of contemporary acupuncture.* New York: Marcel Dekker.

Liben, L.S. & Downs, R.M. (1993). Understanding person-space-map relations: Cartographic and developmental perspectives. *Developmental Psychology, 29*, 739-752.

Lidar, M., Lundqvist, E. & Östman, L. (2006). Teaching and learning on the science classroom: The interplay between teachers' epistemological moves and students' practical epistemology. *Science Education, 90*(1), 148-163.

Lin, H. (1998). The effectiveness of teaching chemistry through the history of science. *Journal of Chemical Education, 75*, 1326-1330.

Lin, H.S. & Chen, C.C. (2002). Promoting preservice chemistry teachers' understanding about the nature of science through history. *Journal of Research in Science Teaching, 39*, 773-792.

Lin, H.S., Chiu, H-L & Chou, C-Y. (2004). Student understanding of the nature of science and problem-solving strategies. *International Journal of Science Education, 26*(1), 101-112.

Lin, J-W. & Chiu, M-H. (2007). Exploring the characteristics and diverse sources of students' mental models of acids and bases. *International Journal of Science Education, 29*(6), 771-803.

Lin, J-Y. (2007). Responses to anomalous data obtained from repeatable experiments in the laboratory. *Journal of Research in Science Teaching, 44*(3), 506-528.

Lindley, D. (1990). The embarrassment of cold fusion. *Nature, 344*, 375-376.

Linn, M.C., Clark, D. & Slotta, J.D. (2003). WISE design for knowledge integration. *Science Education, 87*(4), 517-538.

Linn, M.C., Shear, L., Bell, R. & Slotta, J.D. (1999). Organizing principles for science education partnerships: Case studies of Students' learning about "rats in space" and "deformed frogs". *Educational Technology Research and Development, 47*, 61-85.

Linnet, J.W. (1970). Structure of polywater. *Science, 167*(3926), 1719-1720.

Lippincott, E.R., Stromberg, R.R., Grant, W.H. & Cessac, G.L. (1969). Polywater: Vibrational spectra indicate unique stable polymeric structure. *Science, 164*(3887), 1482-1487.

Liu, S-Y. & Lederman, N.G. (2002). Taiwanese gifted students' views of nature of science. *School Science & Mathematics, 102*(3), 114-123.

Liu, S-Y. & Lederman, N.G. (2007). Exploring prospective teachers' worldviews and conceptions of nature of science. *International Journal of Science Education, 19*(10), 1281-1307.

Liu, S-Y. & Tsai, C-C. (2008). Differences in the scientific epistemological views of undergraduate students. *International Journal of Science Education, 30*(8), 1055-1073.

Lochhead, J. & Dufresne, R. (1989). Helping students understand difficult science concepts through the use of dialogues with history. In D.E. Herget (Ed.), *History and philosophy of science in science teaching* (pp. 221-229). Tallahassee, FA: Florida State University.

Lock, R. (1987). Practical work. In R. Lock & D. Foster (Eds.), *Teaching science 11–13* (pp. 45-66). London: Croom Helm.

Lock, R. (1998). Fieldwork in the life sciences. *International Journal of Science Education, 20*(6), 633-642.

Locust, C. (1988). Wounding the spirit: Discrimination and traditional American Indian belief systems. *Harvard Educational Review, 58*(3), 315-330.

Loh, B., Reiser, B.J., Radinsky, J., Edelson, D.C., Gomez, L.M. & Marshall, S. (2001). Developing reflective inquiry practices: A case study of software, the teacher, and students. In K. Crawley, C.D. Schunn & T. Okada (Eds.), *Designing for science: Implications from everyday, classroom, and professional settings* (pp. 279-323). Mahwah, NJ: Lawrence Erlbaum.

Long, M. & Steinke, J. (1996). The thrill of everyday science: Images of science and scientist on children's educational science programs in the United States. *Public Understanding of Science, 5*, 101-119.

Longino, H.E. (1990). *Science as social knowledge: Values and objectivity in scientific inquiry.* Princeton, NJ: Princeton University Press.

Longino, H. (1994). The fate of knowledge in social theories of science. In F.T. Schmitt (Ed.), *Socializing epistemology: The social dimensions of knowledge* (pp. 135-157). Lanham, MD: Rowman and Littlefield.

Longino, H. (1995). Gender, politics, and the theoretical virtues. *Synthese, 104*, 383-397.

Longino, H.E. (2002). *The fate of knowledge.* Princeton, NJ: Princeton University Press.

Loo, S.P. (1996). The four horsemen of Islamic science: A critical analysis. *International Journal of Science Education, 18*(3), 285-294.

Loo, S.P. (1999). Scientific understanding: Control of the environment and science education. *Science & Education, 8*, 79-87.

Loo, S.P. (2001). Islam, science and science education: Conflict or concord? *Studies in Science Education, 36*, 45-78.

Lopes, J.B. & Costa, N. (2007). The evaluation of modelling competencies: Difficulties and potentials for the learning of the sciences. *International Journal of Science Education, 29*(7), 811-851.

Losee, J. (1993). *A historical introduction to the philosophy of science*, 3rd ed. Oxford: Oxford University Press.

Losh, S.C., Wilke, R. & Pop, M. (2008). Some methodological issues with "draw a scientist tests" among young children. *International Journal of Science Education, 30*(6), 773-792.

Lotter, C., Harwood, W.S. & Bonner, J.J. (2007). The influence of core teaching conceptions on teachers' use of inquiry teaching practices. *Journal of Research in Science Teaching, 44*(9), 1318-1347.

Lottero-Perdue, P.S. & Brickhouse, N.W. (2002). Learning on the job: The acquisition of scientific competence. *Science Education, 86*, 756-782.

Loughran, J. & Derry, N. (1997). Researching teaching for understanding: The students' perspective. *International Journal of Science Education, 19*, 925-938.

Lovejoy, C.O. (1981). The origin of man. *Science, 211*, 341-350.

Loving, C.C. (1991). The scientific theory profile: A philosophy of science model for teachers. *Journal of Research in Science Teaching, 28*(9), 823-838.

Loving, C.C. (1997). From the summit of truth to its slippery slopes: Science education's journey through positivist-postmodern territory. *American Educational Research Journal, 34*(3), 421-452.

Loving, C.C. (1998). Nature of science activities using the scientific profile: From the Hawking-Gould dichotomy to a philosophical checklist. In W.F. McComas (Ed.), *The nature of science in science education: Rationales and strategies* (pp. 137-150). Dordrecht: KluwerAcademic Publishers.

Loving, C.C. & de Montellano, B.R.O. (2003). Good versus bad culturally relevant science: Avoiding the pitfalls. In S.M. Hines (Ed.), *Multicultural science education: Theory, practice, and promise* (pp. 147-166). New York: Peter Lang.

Loving, C.C. & Foster, A. (2000). The religion-in-the-science-classroom issue: Seeking graduate student conceptual change. *Science Education, 84*, 445-468.

Lubben, F. & Millar, R. (1996). Children's ideas about the reliability of experimental data. *International Journal of Science Education, 18*, 955-968.

Lubben, F., Netshisaulu, T. & Campbell, B. (1999). Students' use of cultural metaphors and their scientific understandings related to heating. *Science Education*, 761-774.

Lucas, A. (1975). Hidden assumptions in measures of "knowledge about science and scientists". *Science Education, 59*(4), 481-485.

Lucas, B. & Roth, W-M. (1996). The nature of scientific knowledge and student learning: Two longitudinal case studies. *Research in Science Education, 26*(1), 103-127.

Lunzer, E. & Gardner, K. (1979). *Learning from the written word.* Edinburgh: Oliver & Boyd.

Luykx, A. & Lee, O. (2007). Measuring instructional congruence in elemntary science classrooms: Pedagogical and methodological components of a theoretical framework. *Journal of Research in Science Teaching, 44*(3), 424-447.

Lynch, M.C. (2002). *The online educator: A guide to creating the virtual classroom.* London: Routledge.

MacDonald, L. & Sherman, A. (2007). Student perspectives on mentoring in a science outreach project. *Canadian Journal of Science, Mathematics and Technology Education, 7*(2/3), 133-147.

Mach, E. (1960). *The science of mechanics: A critical and historical account of its development.* Trans. T.J. McCormack [originally published 1893]. La Salle, IL: Open Court.

MacKintosh, N.J. (1995). *Cyril Burt: Fraud or framed?* Oxford: Oxford University Press.

Maddock, M.N. (1981). Science education: An anthropological viewpoint. *Studies in Science Education, 8*, 1-26.

Mahner, M. & Bunge, M. (1996). Is religious education compatible with science education? *Science & Education, 5*, 101-123.

Mahoney, M.J. (1979). Psychology of the scientist: An evaluative review. *Social Studies of Science, 9*(3), 349-375.

Maia, P.F. & Justi, R. (2009). Learning of chemical equilibrium through modelling-based teaching. *International Journal of Science Education, 31*(5), 603-630.

Maksić, Z.B. (1990). Modelling – A search for simplicity. Prologue. In Z.B. Maksić (Ed.), *Atomic hypothesis and the concept of molecular structure* (pp. xiii- xxviii). Berlin: Springer Verlag.

Malcolm, C. (2007). The value of science in African countries. In D. Corrigan, J. Dillon & R. Gunstone (Eds.), *The re-emergence of values in science education* (pp. 61-76). Rotterdam/Taipei: Sense.

Malinowski, B. (1971). *Myth in primitive society.* Westport, CN: Negro Universities Press.

Mallove, E.F. (1991). *Fire from ice: Searching for the truth behind the cold fusion furore.* New York: John Wiley.

Mallow, J.V. (1991). Reading science. *Journal of Reading, 34*, 324-328.

Mamlok-Naaman, R., Ben-Zvi, R., Hofstein, A., Menis, J. & Erduran, S. (2005). Learning science through a historical approach: Does it affect the attitudes of non-science-oriented students towards science? *International Journal of Science & Mathematics Education, 3*, 485-507.

Mansour, N. (2008). The experiences and personal religious beliefs of Egyptian science teachers as a framework for understanding the shaping and reshaping of their beliefs and practices about science-technology-society (STS). *International Journal of Science Education, 30*(12), 1605-1634.

Martin, A.M. & Hand, B. (2009). Factors affecting the implementation of argument in the elementary science classroom: A longitudinal case study. *Research in Science Education, 39*(1), 17-38.

Martin, B.E. & Brouwer, W. (1991). The sharing of personal science and the narrative element in science education. *Science Education, 75*, 707-722.

Martin, E. (1991). The egg and the sperm: How science has constructed a romance based on stereotypical male-female roles. *Signs, 16*(3), 485-501.

Martin, M. (1972). *Concepts of science education.* Glenview, IL: Scott Foresman.

Martin, M. (1994). Pseudoscience, the paranormal, and science education. *Science & Education, 3*, 357-371.

Martín-Díaz, M.J. (2006). Educational background, teaching experience and teachers' views on the inclusion of nature of science in the science curriculum. *International Journal of Science Education, 28*(10), 1161-1180.

Martin-Hanson, L.M. (2008). First-year college students' conflict with religion and science. *Science & Education, 17*(4), 317-357.

Martins, I., Mortimer, E., Osborne, J., Tsatsarelis, C. & Jiménez-Aleixandre, M.P. (2001). Rhetoric and science education. In H Behrendt, H. Dahncke, R. Duit, W. Graber, M. Komorek, A. Kross & P. Reiska (Eds.), *Research in science education – Past, present and future* (pp. 189-198). Dordrecht: Springer.

Masnick, A.M. & Klahr, D. (2003). Error matters: An initial exploration of elementary school children's understanding of experimental error. *Journal of Cognition and Development, 4*, 67-98.

Mason, C.L., Kahle, J.B. & Gardner, A.L. (1991). Draw-a-scientist test: Future implications. *School Science & Mathematics, 91*(5), 193-198.

Masters, R. & Nott, M. (1998). Implicit knowledge and science practical work in schools. In J. Wellington (Ed.), *Practical work in school science: Which way now?* (pp. 206-219). London: Routledge.

Matthews, B. (1994). What does a chemist look like? *Education in Chemistry*, 127-129.

Matthews, B. (1996). Drawing scientists. *Gender and Education, 8*(2), 231-243.

Matthews, M.R. (1992). History, philosophy and science teaching: The present rapprochement. *Science & Education, 1*(1), 11-48.

Matthews, M.R. (1994). *Science teaching: The role of history and philosophy of science.* New York: Routledge.

Matthews, M.R. (1998). In defence of modest goals when teaching about the nature of science. *Journal of Research in Science Teaching, 35*(2), 161-174.

Matthews, R.A. J. (2001). Testing Murphy's Law: Urban myths as a source of school science projects. *School Science Review, 83*(302), 23-28.

May, D.B., Hammer, D. & Roy, P. (2006). Children's analogical reasoning in a third grade science discussion. *Science Education, 90*(2), 318-330.

Mayberry, M. (1998). Reproductive and resistant pedagogies: The comparative roles of collaborative learning and feminist pedagogy in science education. *Journal of Research in Science Teaching, 35*, 443-459.

Mayberry, M. & Rose, E.C. (Eds.) (1999). *Meeting the challenge: Innovative feminist pedagogies in action.* New York: Routledge.

Mayer, R. (1993). Illustrations that instruct. In R. Glaser (Ed.), *Advances in instructional psychology, Vol. 4* (pp. 253-284). Hillsdale, NJ: Lawrence Erlbaum.

Mayr, E. (1988). *Towards a new philosophy of biology: Observations of an evolutionist.* Cambridge, MA: The Belknap Press of Harvard University Press.

Mayr, E. (1997). *This is biology: The science of the living world.* Cambridge, MA: Harvard University Press.

Mayr, E. (2004). *What makes biology unique? Considerations on the autonomy of a scientific discipline.* Canbridge: Cambridge University Press.

Mbajiorgu, N.M. & Iloputaife, E.C. (2001). Combating stereotypes of the scientist among pre-service science teachers in Nigeria. *Research in Science &Technological Education, 19*(1), 55-67.

McAllister, J.W. (1996). *Beauty and revolution in science.* Ithaca, NY: Cornell University Press.

McClune, B. & Jarman, R. (2000). Have I got news for you: Using newspapers in the secondary science classroom. *Media Education Journal, 28*, 10-16.

McComas, W.F. (1998). The principal elements of the nature of science: Dispelling the myths. In W.F. McComas (Ed.), *The nature of science in science education: Rationales and strategies* (pp. 41-52). Dordrecht: Kluwer Academic Publishers.

McComas, W.F. (2008). Seeking historical examples to illustrate key aspects of the nature of science. *Science & Education, 17*(2&3), 249-263.

McConnell, J.V. (1962). Memory transfer through cannibalism in planarians. *Journal of Neurophysiology, 3*, 42-48.

McConnell, J.V. (1965). Failure to interpret planarian data correctly: A reply to Bennett and Calvin. Unpublished manuscript. Ann Arbor, MI: University of Michigan. Cited in Collins, H.M. &

Pinch, T.J. (1993) *The golem: What everyone should know about science* . Cambridge: Cambridge University Press, p. 11.

McConnell, J.V. & Jacobson, A.L. (1974). Learning in invertebrates. In D.A. Dewsbury & D.A. Rethlingschafer (Eds.), *Comparative psychology: A modern survey* (pp. 429-470). Tokyo: McGraw Hill Kogashuka.

McCook, A. (2006). Is peer review broken? *The Scientist, 20*(2), 26-31. Available at: www.the_scientist.com

McCrone, W.C. (1976). Authenticity in medieval document tested by small particle analysis. *Analytical Chemistry, 48*, 676A-679A.

McFarlane, G. (1984). *Alexander Fleming: The man and the myth.* Oxford: Oxford University Press.

McFarlane, T.J. (Ed.) (2002). *Einstein and Buddha: The parallel sayings.* Berkeley, CA: Ulysses Press.

McGee, R.J. & Warms, R.L. (2004). *Anthropological theory: An introductory history.* New York: McGrawHill.

McGinley, W. & Tierney, R.J. (1989). Traversing the topical landscape: Reading and writing as ways of knowing. *Written Communication, 6*, 246-269.

McGinn, M.K. & Roth, W-M. (1999). Preparing students for competent scientific practice: Implications of recent research in science and technology studies. *Educational Researcher, 28*(3), 14-24.

McGinnis, J.R., Parker, A. & Graeber, A.O. (2004). A cultural perspective of the induction of five reform-minded new specialist teachers of matehamatics and science. *Journal of Research in Science Teaching, 41*, 720-747.

McKeon, F. (2000). Literacy and secondary science – Building on primary experience. *School Science Review, 81*(297), 45-50.

McKinley, E. (2000). Māori science education: The urgent need for research. In *Ministry of Education, Exploring issue in science education* (pp. 33-40). Wellington: Learning Media.

McKinley, E. (2005). Locating the global: Culture, language and science education for indigenous students. *International Journal of Science Education, 27*(2), 227-241.

McKinley, E. & Keegan, P.J. (2008). Curriculum and language in Aotearoa New Zealand: From science to putaiao. *L1-Educational Studies in Language and Literature, 8*(1), 135-147.

McLellan, H. (Ed.) (1996). *Situated learning perspectives.* Englewood Cliffs, NJ: Educational Technology Publications.

McNeill, K.L. (2009). Teachers' use of curriculum to support students in writing scientific arguments to explain phenomena. *Science Education, 93*(2), 233-268.

McNeill, K.L. & Krajcik, J. (2007). Middle school students' use of appropriate and inappropriate evidence in writing scientific explanations. In M. Lovett & P. Shah (Eds.), *Thinking with data* (pp. 233-265). New York: Taylor & Francis.

McNeill, K.L. & Krajcik, J. (2008). Scientific explanations: Characterizing and evaluating the effects of teachers' instructional practices on student learning. *Journal of Research in Science Teaching, 45*(1), 53-78.

McNeill, K.L, Lizotte, D.J., Krajcik, J. & Marx, R.W. (2006). Supporting students' construction of scientific explanations by fading scaffolds in instructional materials. *Journal of the Learning Sciences, 15*(2), 153-191.

McPeck, J.E. (1981). *Critical thinking and education.* Oxford: Martin Robertson.

McSharry, G. (2002). Television programming and advertisements: Help or hindrance to effective science education? *International Journal of Science Education, 24*(5), 487-497.

Medawar, P.B. (1969). *Induction and intuition in scientific thought.* London: Methuen.

Medawar, P.B. (1982). *Plato's republic.* Oxford: Oxford University Press.

Mehan, H. (1979). *Learning lessons: Social organization in the classroom.* Cambridge, MA: Harvard University Press.

Meichtry, Y.J. (1992). Influencing student understanding of the nature of science: Data from a case of curriculum development. *Journal of Research in Science Teaching, 29*, 389-407.

Mellado, V. (1998). The classroom practice of preservice teachers and their conceptions of teaching and learning science. *Science Education, 82*(2), 197-212.

Menzies, C.R. (2003). *Forests for the future.* Vancouver: Department of Anthropology, University of British Columbia.

Mercer, N. (1995). *The guided construction of knowledge: Talk amongst teachers and learners.* Clevedon (UK): Multilingual Matters.

Mercer, N. (1996). The quality of talk in children's collaborative activity in the classroom. *Learning and Instruction, 6*(4), 359-375.

Mercer, N. (2000). *Words and minds: How we use language to think together.* London: Routledge.

Mercer, N., Wegerif, R. & Dawes, L. (1999) Children's talk and the development of reasoning in the classroom. *British Educational Research Journal, 25*(1), 95-110.

Merton, R.K. (1973). *The sociology of science: Theoretical and empirical investigations.* Chicago, IL: University of Chicago Press.

Metz, D., Klassen, S., McMillan, B., Clough, M. & Olson, J. (2007). Building a foundation for the use of historical narratives. *Science & Education, 16*(3&5), 313-334.

Metz, K.E. (1995). Reassessment of developmental constraints on children's science instruction. *Review of Educational Research, 65,* 93-127.

Metz, K.E. (2000). Young children's inquiry in biology: Building the knowledge base to empower independent inquiry. In J. Minstrell & E.H. van Zee (Eds.), *Inquiry into inquiry learning and teaching in science* (pp. 371-404). Washington, DC: American Association for the Advancement of Science.

Metz, K.E. (2004). Children's understanding of scientific inquiry: Their conceptualization of uncertainty in investigations of their own design. *Cognition and Instruction, 22*(2), 219-290.

Meyling, H. (1997). How to change students' conceptions of the epistemology of science. *Science & Education, 6,* 397-416.

Michaelson, M.G. (1987). Sickle cell anaemia: An "interesting pathology". In D. Gill & L. Levidow (Eds.), *Anti-racist science teaching* (pp. 59-75). London: Free Association Press.

Millar, R. (1993). Science education and public understanding of science. In R. Hill (Ed.), *ASE secondary science teachers' handbook* (pp. 357-374). Hemel Hempstead: Simon & Schuster.

Millar, R. (1996). Towards a science curriculum for public understanding. *School Science Review, 77*(280), 7-18.

Millar, R. (1997). Science education for democracy: What can the school curriculum achieve? In R. Levinson & J. Thomas (Eds.), *Science today: Problem or crisis?* (pp. 87-101). London: Routledge

Millar, R. (1998). Rhetoric and reality: What practical work in science education is really for. In J. Wellington (Ed.), *Practical work in school science: Which way now?* (pp. 16-31). London: Routledge.

Millar, R., Gott, R., Lubben, F. & Duggan, S. (1996). Children's performance of investigative tasks in science: A framework for considering progression. In M. Hughes (Ed.), *Progress in learning* (pp. 82-108). Clevedon (UK): Multilingual Matters.

Millar, R. & Lubben, F. (1996). Investigative work in science: The role of prior expectations and evidence in shaping conclusions. *Education 3-13, 24*(1), 28-34.

Millar, R. & Osborne, J. (Eds.)(1998). *Beyond 2000: Science education for the future.* London: King's College London School of Education.

Millard, E. & Marsh, J. (2001). Sending Minnie the Minx home: Comics and reading choices. *Cambridge Journal of Education, 31*(1), 25-38.

Miller, A.I. (2006). A thing of beauty. *New Scientist, 189*(2557), 50-52 (Feb 4-10).

Miller, J.D. (1992). Toward a scientific understanding of the public understanding of science and technology. *Public Understanding of Science, 1,* 23-26.

Miller, J.D. (1993). Theory and measurement in the public understanding of science: A rejoinder to Bauer and Schoon. *Public Understanding of Science, 2,* 235-243.

Miller, J.D. (2000). The development of civic scientific literacy in the United States. In D.D. Kumar & D. Chubin (Eds.), *Science, technology, and society: A sourcebook on research and practice* (pp. 21-47). New York: Plenum Press.

Miller, J.D. (2004). Public understanding of, and attitudes toward, scientific research: What we know and what we need to know. *Public Understanding of Science, 13,* 273-294.

Miller, J.D. & Kimmel, L.G. (2001). *Biomedical communications: Purposes, audiences, and strategies.* New York: Academic Press.

Miller, P.E. (1963). A comparison of the abilities of secondary teachers and students of biology to understand science. *Iowa Academy of Science, 70,* 510-513.

Miller, P.H., Swalinski Blessing, J. & Schwartz, S. (2006). Gender differences in high-school students' views about science. *International Journal of Science Education, 28*(4), 363-381.

Milne, C. (1998). Philosophically correct science stories? Examining the implications of heroic science stories for school science. *Journal of Research in Science Teaching, 35*(2), 175-187.

Milne, C.E. & Taylor, P.C. (1998). Between a myth and a hard place: Situating school science in a climate of critical cultural reform. In W.W. Cobern (Ed.), *Socio-cultural perspectives on science education* (pp. 25-48). Dordrecht: Kluwer.

Ministry of Education, New Zealand (1993). *Science in the New Zealand curriculum.* Wellington: Learning Media.

Ministry of Education, New Zealand (1996). *Pūtaiao: i roto i te marautanga o Aotearoa.* Te Whangauia Tar: Te Karauna.

Misgeld, W., Ohly, K.P. & Strobl, G. (2000). The historical-genetical approach to science teaching at the Oberstufen-Kolleg, Bielefeld. *Science & Education, 9*, 333-341.

Mitcham, C. (1994). *Thinking through technology: The path between engineering and philosophy.* Chicago, IL: University of Chicago Press.

Mitchell, S.D. (2003). *Biological complexity and integrative pluralism.* Cambridge: Cambridge University Press.

Mitroff, I.I. & Mason, R.O. (1974). On evaluating the scientific contribution of the Apollo missions via information theory: A study of the scientist-scientist relationship. *Management Science: Applications, 20*, 1501-1513.

Moje, E.B. (1995). Talking about science: An interpretation of the effects of teacher talk in a high school classroom. *Journal of Research in Science Teaching, 32*, 349-371.

Moje, E.B., Ciechanowski, K.M., Kramer, K., Ellis, L., Carrillo, R. & Collazo, T. (2004). Working toward third space in content area literacy: An examination of everyday funds of knowledge and discourse. *Reading Research Quarterly, 39*, 38-70.

Moje, B.E., Collazo, T., Carrillo, R. & Marx, W.R. (2001). "Maestro, what is quality?": Language, literacy, and discourse in project-based science. *Journal of Research in Science Teaching, 38*, 469-498.

Moje, E.B. & Shepardson, D.P. (1998). Social interactions and children's changing understanding of electric circuits. In B. Guzzetti & C. Hynd (Eds.), *Perspectives on conceptual change* (pp. 17-26). Mahwah, NJ: Erlbaum.

Monk, M. & Osborne, J. (1997). Placing the history and philosophy of science on the curriculum: A model for the development of pedagogy. *Science Education, 81*, 405-424.

Monhardt, M. & Monhardt, L. (2000). Children's literature and environmental issues: Heart over mind? *Reading Horizons, 40*(3), 175-178.

Monk, M. & Osborne, J. (1997). Placing history and philosophy of science on the curriculum: A model for the development of pedagogy. *Science Education, 81*, 405-424.

Moore, F.M. (2007). Language in science education as a gatekeeper to learning, teaching, and professional development. *Journal of Science Teacher Education, 18*, 319-343.

Morrison, M. & Morgan, M.S. (1999). Models as mediating instruments. In M.S. Morgan & M. Morrison (Eds.), *Models as mediators: Perspectives on natural and social sciences* (pp. 10-37). Cambridge: Cambridge University Press.

Morvillo, N. & Brooks, J.G. (1995). Headline science: Popular news stories spark student interest in biology. *The Science Teacher, 62*, 20-23.

Moss, D.M., Abrams, E.D. & Robb, J. (2001). Examining student conceptions of the nature of science. *International Journal of Science Education, 23*(8), 771-790.

Mortimer, E.F. & Scott, P.H. (2003). *Meaning making in secondary science classrooms.* Maidenhead: Open University Press.

Muench, S.B. (2000). Choosing primary literature in biology to achieve specific educational goals. *Journal of College Science Teaching, 29*(4), 255-260.

Mulkay, M.J. (2005). Knowledge and utility: Implications for the sociology of knowledge. In N. Stehr & V. Meja (Eds.), *Society and knowledge: Contemporary perspectives in the sociology of knowledge,* 2nd ed. (pp. 93-112). New Brunswick, NJ: Transaction Publications.

Munby, H. (1980). Analyzing teaching for intellectual independence. In H. Munby, G. Orpwood & T. Russell (Eds.), *Seeing curriculum in a new light: Essays from science education* (pp. 11-33). Toronto: OISE Press.

Munby, H. & Roberts, D. (1998). Intellectual independence: A potential link between science teaching and responsible citizenship. In D. Roberts & L. Östman (Eds.), *Problems of meaning in science curriculum* (pp. 101-114). New York: Teachers College Press.

Murcia, K. (2009). Re-thinking the development of scientific literacy through a rope metaphor. *Research in Science Education, 39*(2), 215-229.

Murcia, K. & Schibeci, R. (1999). Primary student teachers' conceptions of the nature of science. *International Journal of Science Education, 21*(11), 1123-1140.

Myers, G. (1990a). Making a discovery: Narratives of split genes. In C. Nash (Ed.), *Narrative in culture: The uses of storytelling in the sciences, philosophy, and literature* (pp. 102-126). London: Routledge.

Myers, G. (1990b). Every picture tells a story: Illustrations in E.O. Wilson's Sociobiology. In M. Lynch & S. Woolgar (Eds.), *Representation in scientific practice* (pp. 231-265). Cambridge, MA: MIT Press.

Myers, G. (1992). Textbooks and the sociology of scientific knowledge. *English for Specific Purposes, 11*, 3-17.

Myers, G. (1997). Words and pictures in a biology textbook. In T. Miller (Ed.), *Functional approaches to written text* (pp. 93-104). Paris: USIA.

Nadeau, R. & Désautels, J. (1984). *Epistemology and the teaching of science.* Ottawa: Science Council of Canada.

Nagel, E. (1987). *The structure of science.* Indianapolis, IN: Hackett.

Nakhleh, M.B., Polles, J. & Malina, E. (2002). Learning chemistry in a laboratory environment. In J. Gilbert, O. de Jong, R. Justi, D.F. Treagust & J.H. van Driel (Eds.), *Chemical education: Towards research-based practice* (pp. 69-94). Dordrecht: Kluwer.

Nanda, M. (1997). The science wars in India. *Dissent, 44*(1), 79-80.

Nandy, A. (Ed.) (1988). *Science, hegemony and violence: A requiem for modernity.* Tokyo: The United Nations University.

Nashon, S. (2000). Teaching physics through analogy. In D. Hodson (Ed.), *OISE Papers in STSE Education*, Vol. 1 (pp. 209-223). Toronto: Imperial Oil Centre for Studies in Science, Mathematics and Technology Education, OISE-UT, in collaboration with University of Toronto Press.

Nashon, S.M. (2003). Teaching and learning high school physics in Kenyan classrooms using analogies. *Canadian Journal of Science, Mathematics and Technology Education, 3*(3), 333-345.

Nashon, S.M. (2004). The nature of analogical explorations: High school physics teachers' use in Kenya. *Research in Science Education, 34*(4), 475-502.

National Research Council (1996). *National science education standards.* Washington, DC: National Academy Press.

National Science Foundation (2007). NSF graduate teaching fellows in K-12 education. Available from www.nsf.gov/funding/education

Naylor, S., Keogh, B. & Downing, B. (2007). Argumentation and primary science. *Research in Science Education, 37*(1), 17-39.

Neate, B. (1992). *Finding out about finding out.* Sevenoaks: Hodder & Stoughton.

Negrete, A. & Lartigue, C. (2004). Learning from education to communicate science as a good story. *Endeavour, 28*(3), 120-124.

Nelkin, D. (1982). *The creation controversy: Science or scripture in schools.* Boston, MA: Beacon Press.

Nelkin, D. (1995). *Selling science: How the press covers science and technology* (Revised ed.). New York: Freeman & Co.

Nersessian, N. (1992). How do scientists think? Capturing the dynamics of conceptual change in science. In R.N. Giere (Ed.), *Cognitive models of science* (pp. 3-44). Minneapolis, MN: University of Minnesota Press.

Nersessian, N.J. (1995). Should physicists preach what they practice? *Science & Education, 4*(3), 203-226.

Nersessian, N.J. (2008). Model-based reasoning in scientific practice. In R.A. Duschl & R.E. Grandy (Eds.), *Teaching scientific inquiry: Recommendations for research and implementation* (pp. 57-79). Rotterdam/Taipei: Sense.

Newport, F. & Strausberg, M. (2001). Americans' beliefs in psychic and paranormal phenomena is up over the last decade. Gallup News Service (Poll Analyses, June 8). Available from www.gallup.com

Newton, D.P. (2002). *Talking sense in science: Helping children understand through talk.* London: RoutledgeFalmer.

Newton, D.P. & Newton, L.D. (1992). Young children's perceptions of science and the scientist. *International Journal of Science Education, 14*, 331-348.

Newton, D.P. & Newton, L.D. (1995). Using analogy to help young children understand. *Educational Studies, 21*(3), 379-391.

Newton, L.D. & Newton, D.P. (1998). Primary children's conceptions of science and the scientist: Is the impact of a National Curriculum breaking down the stereotype? *International Journal of Science Education, 20*(9), 1137-1149.

Newton, P., Driver, R. & Osborne, J. (1999). The place of argumentation in the pedagogy of school science. *International Journal of Science Education, 21*(5), 553-576.

Newton, R. (1997). *The truth of science: Physical theories and reality.* Cambridge, MA: Harvard University Press.

Newton-Smith, W.H. (1981). *The rationality of science.* London: Routledge & Kegan Paul.

Niaz, M. (1998). From cathode rays to alpha particles to quantum of action: A rational reconstruction of the atom and its implications for chemistry textbooks. *Science Education, 82*, 527-552.

Niaz, M. (2000). A rational reconstruction of the kinetic molecular theory of gases based on history and philosophy of science and its implications for chemistry textbooks. *Instructional Science, 28*, 23-50.

Nielsen, W.S., Nashon, S. & Anderson, D. (2009). Metacognitive engagement during field-trip experiences: A case study of students in an amusement park physics program. *Journal of Research in Science Teaching, 46*(3), 265-288.

Ninnes, P. (2003). Rethinking multicultural science education: Representations, identities, and texts. In S.M. Hines (Ed.), *Multicultural science education: Theory, practice, and promise* (pp. 167-186). New York: Peter Lang.

Nisbett, R.E. (2003). *The geography of thought: How Asians think differently and why.* New York: The Free Press.

Noddings, N. & Witherell, C. (1991). Epilogue: Themes remembered and foreseen. In C. Witherell & N. Noddings (Eds.), *Stories lives tell* (pp. 279-280). New York: Teachers College Press.

Norris, S.P. (1985). The philosophical basis of observation in science and science education. *Journal of Research in Science Teaching, 22*(9), 817-833.

Norris, S.P. (1995). Learning to live with scientific expertise: Toward a theory of intellectual communalism for guiding science teaching. *Science Education, 79*(2), 201-217.

Norris, S.P., Guilbert, S.M., Smith, M.L., Hakimelahi, S. & Phillips, L.M. (2005). A theoretical framework for narrative explanation in science. *Science Education, 89*(4), 535-563.

Norris, S.P. & Phillips, L.M. (1994). Interpreting pragmatic meaning when reading popular reports of science. *Journal of Research in Science Teaching, 31*, 947-967.

Norris, S.P. & Phillips, L.M. (2003). How literacy in its fundamental sense is central to scientific literacy. *Science Education, 87*(2), 224-240.

Norris, S.P. & Phillips, L.M. (2008). Reading as inquiry. In R.A. Duschl & R.E. Grandy (Eds.), *Teaching scientific inquiry: Recommendations for research and implementation* (pp. 233-262). Rotterdam/Taipei: Sense.

Norris, S.P., Phillips, L.M. & Korpan, C.A. (2003). University students' interpretation of media reports of science and its relationship to background knowledge, interest and reading difficulty. *Public Understanding of Science, 12*, 123-145.

Norris, S.P., Phillips, L.M., Smith, M.L., Guilbert, S.M., Stange, D.M., Baker, J.J. & Weber, A.C. (2008). Learning to read scientific text: Do elementary school commercial reading programs help? *Science Education, 92*(5), 765-798.

Norton, J. (1996). Are thought experiments just what you thought? *Canadian Journal of Philosophy, 26*(3), 333-366.

Nott, M. & Smith, R. (1995). "Talking your way out of it", "rigging it", and "conjuring": What science teachers do when practicals go wrong. *International Journal of Science Education, 17*, 399-410.

Nott, M. & Wellington, J. (1993). Your nature of science profile: An activity for science teachers. *School Science Review, 75*(270), 109-112.

Nott, M. & Wellington, J. (1996). Probing teachers' views of the nature of science: How should we do it and where should we be looking? In G. Welford, J. Osborne & P. Scott (Eds.), *Science education research in Europe* (pp. 283-294). London: Falmer Press.

Nott, M. & Wellington, J. (1998). Eliciting, interpreting and developing teachers' understandings of the nature of science. *Science & Education, 7*, 579-594.

Nott, M. & Wellington, J. (2000). A programme for developing understanding of the nature of science in teacher education. In W.F. McComas (Ed.), *The nature of science in science education: Rationales and strategies* (pp. 293-313). Dordrecht: Kluwer Academic.

Novak, A.M. & Krajcik, J.S. (2004). Using technology to support inquiry in middle school science. In L.B. Flick & N.G. Lederman (Eds.), *Scientific inquiry and nature of science: Implications for teaching, learning, and teacher education* (pp. 75-101). Dordrecht: Kluwer.

Numbers, R.L. & Stenhouse, J. (2000). Antievolutionism in the Antipodes: From protesting evolution to promoting creationism in New Zealand. *British Journal for the History of Science, 33*(3), 335-350.

Nussbaum, E.M. & Bendixen, L.M. (2003). Approaching and avoiding arguments: The role of epistemological beliefs, need for cognition, and extraverted personality traits. *Contemporary Educational Psychology, 28*, 573-595.

Nussbaum, E.M., Hartley, K., Sinatra, G.M., Reynolds, R.E. & Bendixen, L.M. (2004). Personality interactions and scaffolding in on-line discussions. *Journal of Educational Computing Research, 30*, 113-137.

Nussbaum, E.M. & Kardash, C.M. (2005). The effect of goal instructions and text on the generation of cunterarguments during writing. *Journal of Educational Psychology, 97*, 157-169.

Nussbaum, E.M. & Sinatra, G.M. (2003). Argument and conceptual engagement. *Contemporary Educational Psychology, 28*, 384-395.

Nussbaum, E.M., Sinatra, G.M. & Poliquin, A. (2008). Arole of epistemic beliefs and scientific rgumentation in science learning. *International Journal of Science Education, 30*(15), 1977-1999.

Nye, D.E. (1990). *Electrifying America: Social meanings of a new technology, 1880–1940.* Cambridge, MA: MIT Press.

Nye, M.J. (1980). N-rays: An episode in the history and psychology of science. *Historical Studies in the Physical Sciences, II*(1), 125-156.

Oakeshott, M. (1962). Rational conduct. In M. Oakeshott (Ed.), *Rationalism in politics and other essays* (pp. 80-110). London: Methuen.

O'Byrne, B. (2009). Knowing more than words can say: using multimodal assessment tools to excavate and construct knowledge about wolves. *International Journal of Science Education, 31*(4), 523-539.

Ogawa, M. (1995). Science education in a multi-science perspective. *Science Education, 79*, 583-593.

Ogawa, M. (1998). A cultural history of science education in Japan: An epic description. In W.W. Cobern (Ed.), *Socio-cultural perspectives on science education* (pp. 139-161). Dordrecht: Kluwer.

Ogborn, J. & Martins, I. (1996). Metaphorical understandings and scientific ideas. *International Journal of Science Education, 18*(6), 631-652.

Ogunniyi, M.B. (2007a). Teachers' stances and practical arguments regarding a science-indigenous knowledge curriculum: Part 1. *International Journal of Science Education, 29*(8), 963-986.

Ogunniyi, M.B. (2007b). Teachers' stances and practical arguments regarding a science-indigenous knowledge curriculum: Part 2. *International Journal of Science Education, 29*(11), 1189-1207.

Ogunniyi, M.B., Jegede, O.J., Ogawa, M., Yandila, C.D. & Oladele, F.K. (1995). Nature of worldview presuppositions among science teachers in Botswana, Indonesia, Japan, Nigeria and the Philippines. *Journal of Research in Science Teaching, 32*, 817-831.

Ohlsson, S. (1996). Learning to do and learning to understand? A lesson and a challenge for cognitive modelling. In P. Reimann & H. Spada (Eds.), *Learning in humans and machines* (pp. 37-62). Oxford: Elsevier.

Olin. J.S. (2003). Evidence that the Vinland Map is medieval. *Analytical Chemistry, 75*, 6745-6747.

Oliva, J.M., Azcarate, P. & Navarrete, A. (2007). Teaching models in the use of analogies as a resource in the science classroom. *International Journal of Science Education, 29*(1), 45-66.

Oliveira, A.W., Sadler, T.D. & Suslak, D.F. (2007). The linguistic construction of expert identity in professor-student discussions of science. *Cultural Studies in Science Education, 2*(1), 119-150.

O'Loughlin, M. (1992). Rethinking science education: Beyond Piagetian constructivism toward a sociocultural model of teaching and learning. *Journal of Research in Science Teaching, 29*, 791-820.

O'Neill, D.K. (2001). Knowing when you've brought them in: Scientific genre knowledge and communities of practice. *Journal of the Learning Sciences, 10*, 223-264.

O'Neill, K.D. & Polman, J.L. (2004). Why educate "little scientists"? Examining the potential of practice-based scientific literacy. *Journal of Research in Science Teaching, 41*(3), 234-266.

O'Rafferty, M.H. (1995). Developing sociological insights on scientific norms using case studies on misconduct in science. In F. Finley, D. Allchin, D. Rhees & S. Fifield (Eds.), *Proceedings of the Third International History, Philosophy and Science Teaching Conference, Vol. 2* (pp. 905-912). Minneapolis, MN: University of Minnesota Press.

Organization for Economic Cooperation and Development (OECD) (1999). Scientific literacy. In *OECD, Measuring student knowledge and skills* (pp. 59-75). Paris: OECD.

Organization for Economic Cooperation and Development (OECD) (2006). *Assessing scientific, reading and mathematical literacy: A framework for PISA 2006.* Paris: OECD.

Orgill, M.K. & Bodner, G.M. (2006). An analysis of the effectiveness of analogy use in college-level biochemistry textbooks. *Journal of Research in Science Teaching, 43*(10), 1040-1060.

Orr, D.W. (1992). *Ecological literacy: Education and the transition to a postmodern world.* Albany, NY: SUNY Press.

Osborne, J.F. (1996). Beyond constructivism. *Science Education, 80*(1), 53-82.

Osborne, J. (1997). Practical alternatives. *School Science Review, 78*(285), 61-66.

Osborne, J. (1998). Science education without a laboratory? In J. Wellington (Ed.), *Practical work in school science: Which way now?* (pp. 156-173). London: Routledge.

Osborne, J. (2001). Promoting argument in the science classroom: A rhetorical perspective. *Canadian Journal of Science, Mathematics and Technology Education, 1*(3), 271-290.

Osborne, J. (2002). Science without literacy: A ship without a sail? *Cambridge Journal of Education, 32*(2), 203-218.

Osborne, J. & Collins, S. (2000). *Pupils' and parents' views of the school science curriculum.* London: King's College.

Osborne, J., Collins, S., Ratcliffe, M., Millar, R. & Duschl, R. (2003). What "ideas-about-science" should be taught in school science? A Delphi study of the expert community. *Journal of Research in Science Teaching, 40*(7), 692-720.

Osborne, J., Erduran, S. & Simon, S. (2004). Enhancing the quality of argumentation in school science. *Journal of Research in Science Teaching, 41*(10), 994-1010.

Osmo, R. & Landau, R. (2001). The need for explicit argumentation in ethical decision-making in social work. *Social Work Education, 20*(4), 483-492.

Żstman, L. (1998). How companion meanings are expressed by science education discourse. In D. Roberts & L. Östman (Eds.), *Problems of meaning in science curriculum* (pp. 54-70). New York: Teachers College Press.

Pacey, A. (1983). *The culture of technology.* Oxford: Basil Blackwell.

Painter, J., Tretter, T.R., Jones, G.M. & Kubasko, D. (2006). Pulling back the curtain: Uncovering and changing students' perceptions of scientists. *School Science & Mathematics, 106*(4), 181-190.

Pais, A. (1982). *Subtle is the Lord? The science and the life of Albert Einstein.* Oxford: Oxford University Press.

Palefau, T.H. (2005). *Perspectives on scientific and technological literacy in Tonga: Moving forward in the 21st century.* Toronto: Unpublished PhD thesis, University of Toronto.

Palincsar, A.S. & Brown, L.A. (1984). Reciprocal teaching of comprehension-fostering and comprehension-monitoring activities. *Cognition and Instruction, 1*, 117-175.

Palincsar, A.S. & Magnusson, S.J. (2001). The interplay of first-hand and second-hand investigations to model and support the development of scientific knowledge and reasoning. In S. Carver & D. Klahr (Eds.), *Cognition and instruction: Twenty-five years of progress* (pp. 151-193). Mahwah, NJ: Lawrence Erlbaum.

Palmer, B. & Marra, R.M. (2004). College student epistemological perspectives across knowledge domains: A proposed grounded theory. *Higher Education, 47*, 311-335.

Palmquist, B.C. & Finley, F.N. (1997). Preservice teachers' views of the nature of science during a postbaccalauraeate science teaching program. *Journal of Research in Science Teaching, 34*(6), 595-615.

Pappas, C.C. (2006). The information book genre: Its role in integrated science literacy research and practice. *Reading Research Quarterly, 41*(2), 226-250.

Pappas, C.C. & Zecker, L.B. (2001a). Transforming curriculum genres in urban schools: The political significance of collaborative classroom discourse. In C.C. Pappas & L.B. Zecker (Eds.), *Transforming literacy curriculum genres: Working with teacher researchers in urban classrooms* (pp. 325-333). Mahwah, NJ: Lawrence Erlbaum.

Pappas, C.C. & Zecker, L.B. (2001b). Urban teacher-researchers' struggles in sharing power with their students: Exploring changes in literacy curriculum genres. In C.C. Pappas & L.B. Zecker (Eds.), *Transforming literacy curriculum genres: Working with teacher researchers in urban classrooms* (pp. 1-31). Mahwah, NJ: Lawrence Erlbaum.

Paris, N.A., & Glynn, S.M. (2004). Elaborate analogies in science text: Tolls for enhancing preservice teachers' knowledge and attitudes. *Contemporary Educational Psychology, 29*, 230-247.

Paris, S.G., Yambor, R.M. & Packard, B.W. (1998). Hands-on biology: A museum-school-university partnership for enhancing students' interest and learning in science. *The Elementary School Journal, 98*, 267-288.

Park, J., Jang, K-A. & Kim, I. (2009). An anlysis of the actual processes of physicists' research and the implications for teaching scientific inquiry in schools. *Research in Science Education, 39*(1), 111-129.

Park, J., Kim, I., Kwon, S. & Song, J. (2001). An analysis of thought experiments in the history of physics and implications for physics teaching. In R. Pinto & S. Sirinach (Eds.), *Physics teacher education beyond 2000* (pp. 347-351). Paris: Elsevier.

Parker, L.H., Rennie, L.J. & Fraser, B.J. (Eds.) (1995). *Gender, science and mathematics: Shortening the shadow*. Dordrecht: Kluwer.

Parker, W.C. & McDaniel, J.E. (1992). Bricolage: Teachers do it daily. In E.W. Ross, J.W. Cornett & G. McCutcheon (Eds.), *Teacher personal theorising: Connecting curriculum practice, theory, and research* (pp. 97-114). Albany, NY: State University of New York Press.

Parkin, G. (1992). Do bond-stretch isomers really exist? *Accounts of Chemical Research, 25*, 455-460.

Park Rogers, M.A. & Abell, S.K. (2008). The design, enactment, and experience of inquiry-based instruction in undergraduate science education: A case study. *Science Education, 92*(4), 591-607.

Parsons, E.C. (1997). Black high school females' images of the scientist: Expression of culture. *Journal of Research in Science Teaching, 34*(7), 745-768.

Paton, R. (2002). Metaphors in scientific thinking. Manuscript available from: http://www.csc.liv.ac.uk/ rcp/metaphor.html

Patronis, T., Potari, D. & Spiliotopoulou, V. (1999). Students' argumentation in decision-making on a socio-scientific issue: Implications for teaching. *International Journal of Science Education, 21*, 745-754.

Paulsen, M.B. & Wells, C.T. (1998). Domain differences in the epistemological beliefs of college students. *Research in Higher Education, 39*(4), 365-384.

Peacock, A. & Weedon, H. (2002). Children working with text in science: Disparaties with "literacy hour" practice. *Research in Science & Technological Education, 20*(2), 185-197.

Peacocke, A. (2001). *Paths from science towards God: The end of all our exploring*. Oxford: Oneworld.

Pearson, P.D., Rochler, L.R., Dole, J.A. & Duffy, G.G. (1992). Developing expertise in reading comprehension. In S.J. Samuels & A.E. Farstrup (Eds.), *What research has to say about reading instruction*, 2nd ed. (pp. 145-199). Newark, DE: International Reading Association.

Peat, F.D. (1995). *Blackfoot physics: A journey into the Native American universe*. London: Fourth Estate.

Pedretti, E. (2002). T.Kuhn meets T.Rex: Critical conversations and new directions in science centres and science museums. *Studies in Science Education, 37*, 1-42.

Pedretti, E. (2004). Perspectives on learning through critical issues-based science center exhibits. *Science Education, 88*(suppl.1), S34-S47.

Pedretti, E., Mayer-Smith, J. & Woodrow, J. (1998). Technology, text, and talk: Students' perspectives on teaching and learning in a technology-enhanced secondary science classroom. *Science Education, 82*(5), 569-589.

Penner, D.E., Giles, N.D., Lehrer, R. & Schauble, L. (1997). Building functional models: designing an elbow. *Journal of Research in Science Teaching, 34*(2), 125-143.

Penney, K., Norris, S.P., Phillips, L.M. & Clark, G. (2003). The anatomy of junior high school science textbooks: An analysis of textual characteristics and a comparison to media reports of science. *Canadian Journal of Science, Mathematics and Technology Education, 3*, 415-436.

Pennock, R.T. (2002). Should creationism be taught in the public schools? *Science & Education, 11*, 111-133.

Pepper, S.C. (1942). *World hypotheses: A study in evidence*. Berkeley, CA: University of California Press.

Perkins, D.N. & Grotzer, T.A. (2005). Dimensions of causal understanding: The role of complex causal models in students' understanding of science. *Studies in Science Education, 41*, 117-166.

Perkins, S. (2004). What's wrong with this picture? *Science News, 166*(16), 250 (online publication).

Pessoa de Carvalho, A.M. & Vannucchi, A.I. (2000). History, philosophy and science teaching: Some ansers to "how"? *Science & Education, 9*(5), 427-448.

Peterson, G.R. (2002). The intelligent-design movement: Science or ideology? *Zygon, 37*(1), 7-23.

Phillips, L.M. & Norris, S.P. (1999). Interpreting popular reports of science: What happens when the reader's world meets the world on paper? *International Journal of Science Education, 21*(3), 317-327.

Pickersgill, S. & Lock, R. (1991). Students' understanding of selected non-technical words in science. *Research in Science & Technological Education, 9*(1), 71-79.

Pilkington, R. & Walker, A. (2003). Facilitating debate in networked learning: Reflecting on online synchronous discussion in higher eduaction. *Instructional Science, 31*, 41-63.

Pinch. T.J. & Collins, H.M. (1984). Private science and public knowledge: The committee for the scientific investigation of the paranormal and its use of the literature. *Social Studies of Science, 14*, 521-546.

Pinnick, C. (1994). Feminist epistemology: Implications for philosophy of science. *Philosophy of Science, 61*, 646-657.

Pinnick, C.L. (2008). Science education for women: Situated cognition, feminist standpoint theory, and the status of women in science. *Science & Education, 17*(10), 1055-1063.

Pintrich, P.R., Marx, R.W. & Boyle, R.A. (1993). Beyond cold conceptual change: The role of Motivational beliefs and classroom contextual factors in the process of conceptual change. *Review of Educational Research, 63*, 167-199.

Pittman, K.M. (1999). Student-generated analogies: Another way of knowing? *Journal of Research in Science Teaching, 36*(1), 1-22.

Plimer, I. (1994). *Telling lies for God: Reason vs creationism*. Milsons Point, NSW: Random House.

Poincaré, H. (1952). *Science and method*. New York: Dover (originally published in 1908).

Polanyi. M. (1958). *Personal knowledge: Towards a post-critical philosophy*. London: Routledge & Kegan Paul.

Polkinghorne, J.C. (1994). *The faith of a physicist*. Princeton, NJ: Princeton University Press.

Polkinghorne, J.C. (1996). *Beyond science*. Cambridge: Cambridge University Press.

Polkinghorne, J.C. (1998). *Belief in God in an age of science*. New Haven, CT: Yale University Press..

Polkinghorne, J.C. (2000). *Faith, science and understanding*. New Haven, CT: Yale University Press.

Polkinghorne, J. (2005). The continuing interaction of science and religion. *Zygon, 40*, 43-50.

Polman, J. (2004). Dialogic activity structures for project-based learning environments. *Cognition and Instruction, 22*, 431-466.

Pomeranz, B. (1987). Scientific basis of acupuncture. In G. Stuz & B. Pomeranz (Eds.), *Acupuncture: Textbook and Atlas* (pp. 1-34). Berlin: Springer-Verlag.

Pomeroy, D. (1993). Implications of teachers' beliefs about the nature of science: Comparison of the beliefs of scientists, secondary science teachers, and elementary teachers. *Science Education, 77*, 261-278.

Pomeroy, D. (1994). Science education and cultural diversity: Mapping the field. *Studies in Science Education, 24*, 49-73.

Poock, J.A., Burke, K.A., Greenbowe, T.J. & Hand, B.M. (2007). Using the science writing heuristic to improve students' academic performance. *Journal of Chemical Education, 84*, 1371-1379.

Poole, M. (1996). ... for more and better religious education. *Science & Education, 5*(2), 165-174.

Poole, M.W. (1998). Science and science education: A Judeo-Christian perspective. In W.W. Cobern (Ed.), *Socio-cultural perspectives on science education* (pp. 181-201). Dordrecht: Kluwer.

Poole, M. (2007). *User's guide to science and belief*. Oxford: Lion/Hudson.

Pope, M. & Denicolo, P. (1993). The art and science of constructivist research in teacher thinking. *Teaching & Teacher Education, 9*(5&6), 529-544.

Popper, K.R. (1959). *The logic of scientific discovery.* London: Hutchinson.

Popper, K.R. (1980). Evolution. *New Scientist, 87*(1215), 611.

Popper, K.R. (1988). Natural selection and the emergence of mind. *Dialectica, 32*(3&4), 339-355.

Posch, P. (1993). Research issues in environmental education. *Studies in Science Education, 21*, 21-48.

Posner, G.J., Strike, K.A., Hewson, P.J. & Gertzog, W.A. (1982). Accommodation of a scientific conception: Toward a theory of conceptual change. *Science Education, 66*, 211-227.

Postman, N. & Weingartner, C. (1971). *Teaching as a subversive activity.* London: Penguin/Pitman.

Prain, V. (2006). Learning from writing in secondary science: Some theoretical and practical implications. *International Journal of Science Education, 28*(2&3), 179-201.

Prain, V. & Hand, B. (1996). Writing and learning in secondary science: Rethinking practices. *Teaching and Teacher Education, 12*, 609-626.

Prawat, R.S. (1993). The value of ideas: Problems versus possibilities in learning. *Educational Researcher, 22*, 5-16.

Preece, P.F.W. & Baxter, J.H. (2000). Scepticism and gullibility: The superstitious and pseudo-scientific beliefs of secondary school students. *International Journal of Science Education, 22*(11), 1147-1156.

Price, D. (1984). Creationist and fundamentalist apologetics: Two branches of the same tree. *Creation/Evolution, 14*, 19-31. Available online from National Center for Science Education (http://ncseweb.org/media/creation-evolution_journal).

Pringle, S. (1997). Sharing science. In R. Levinson & J. Thomas (Eds.), *Science today; problem or crisis?* (pp. 206-223). London: Routledge.

Prins, G.T., Bulte, A.M.W., van Driel, J.H. & Pilot, A. (2008). Selection of authentic modelling practices as contexts for chemistry education. *International Journal of Science Education, 30*(14), 1867-1890.

Prophet, B. & Towse, P. (1999). Pupils' understanding of some non-technical words in science. *School Science Review, 81*(295), 79-86.

Putnam, H. (1975). *Mathematics, matter and method: Philosophical papers Volume 1.* Cambridge: Cambridge University Press.

Putnam, H. (1987). *The many faces of realism.* LaSalle, IL: Open Court Press.

Qualifications and Curriculum Authority (QCA) (1998). *Education for citizenship and the teaching of democracy in schools.* London, QCA.

Radner, D. & Radner, M. (1982). *Science and unreason.* Belmont, CA: Wadsworth.

Raghavan, K. & Glaser, R. (1995). Model-based analysis and reasoning in science: The MARS curriculum. *Science Education, 79*(1), 37-62.

Raghavan, K., Sartoris, M.L. & Glaser, R. (1998a). Impact of the MARS curriculum: The mass unit. *Science Education, 82*, 53-91.

Raghavan, K., Sartoris, M.L. & Glaser, R. (1998b). Why does it go up? The impact of the MARS curriculum as revealed through changes in student explanations of a helium balloon. *Journal of Research in Science Teaching, 35*, 547-567.

Raham, R.G. (2004). *Teaching science fact with science fiction.* Portsmouth, NH: Heinemann.

Rahm, J. (2007). Youths' and scientists' authoring of and positioning within science and scientists' work. *Cultural Studies of Science Education, 1*(3), 517-544.

Rahm, J. (2008). Urban youths' hybrid positioning in science practices at the margin: A look inside a school-museum-scientist partnership project and an after-school science prohgram. *Cultural Studies of Science Education, 3*(1), 97-121.

Rahm, J. & Charbanneau, P. (1997). Probing stereotypes through students' drawings of scientists. *American Journal of Physics, 65*(8), 774-778.

Rahm, J., Miller, H.C., Hartley, L. & Moore, J.C. (2003). The value of an emergent notion of authenticity: Examples from two student/teacher- scientist partnership programs. *Journal of Research in Science Teaching, 40*, 737-756.

Ratcliffe, M. (1999). Evaluation of abilities in interpreting media reports of scientific research. *International Journal of Science Education, 21*(10), 1085-1099.

Ratcliffe, M. (2007). Values in the science classroom – The "enacted" curriculum. In D. Corrigan, J. Dillon & R. Gunstone (Eds.), *The re-emergence of values in science education* (pp. 119-132). Rotterdam/Taipei: Sense.

Ravetz, J.R. (1971). *Scientific knowledge and its social problems.* Oxford: Clarendon Press.

Reid, D. (1990). The role of pictures in learning. *Journal of Biological Education, 24*(3), 161-172.

Reid, D.J. & Beveridge, M. (1986). Effects of text illustration on children's learning of a school science topic. *British Journal of Educational Psychology, 56,* 294-303.

Reigosa, C. & Jiménez-Aleixandre, M-P. (2007). Scaffolded problem-solving in the physics and chemistry laboratory: Difficulties hindering students' assumption of responsibility. *International Journal of Science Education, 29*(3), 307-329.

Reiner, M. (1998). Thought experiments and collaborative learning in physics. *International Journal of Science Education, 20*(9), 1043-1059.

Reiner, M. (2000). Thought experiments and embodied cognition. In J.K. Gilbert & C.J. Boulter (Eds.), *Developing models in science education* (pp. 157-176). Dordrecht: Kluwer.

Reiner, M. (2006). The context of thought experiments in physics learning. *Interchange, 37*(1&2), 97-113.

Reiner, M. & Burko, L.M. (2003). On the limitations of thought experiments in physics and the consequences for physics education. *Science & Education, 12,* 365-385.

Reiner, M. & Gilbert, J. (2000). Epistemological resources for thought experimentation in science learning. *International Journal of Science Education, 22*(5), 489-506.

Reis, P. & Galvão, C. (2004). The impact of socio-scientific controversies in Portuguese natural science teachers' conceptions and practices. *Research in Science Education, 34*(2), 153-171.

Reis, P. & Galvão, C. (2007). Reflecting on scientists' activity based on science fiction stories written by secondary students. *International Journal of Science Education, 29*(10), 1245-1260.

Reiser, B. (2002). Why scaffolding should sometimes make tasks more difficult for learners. In G. Stahl (Ed.), Computer support for collaborative learning: Foundations for a computer supported collaborative learning community (pp.255-264). Hillsdale, NJ: Erlbaum.

Reiser, B., Tabak, I., Sandoval, W., Smith, B., Steinmuller, F. & Leone, A. (2001). BGuILE: Strategic and conceptual scaffolds for scientific inquiry in biology classrooms. In S.M. carver & D. Klahr (Eds.), *Cognition and instruction: Twenty-five years of progress* (pp. 263-305). Mahwah, NJ: Lawrence Erlbaum.

Reiss, M. (1999). Teaching ethics in science. *Studies in Science Education, 34,* 115-140.

Reiss, M.J. (2007). Teaching about origins in science: Where now? In L. Jones & M.J. Reiss (Eds.), *Teaching about scientific origins: Taking account of creationism* (pp. 197-208). New York: Peter Lang.

Reiss, M.J. (2008). Should science educators deal with the science/religion issue? *Studies in Science Education, 44*(2), 157-186.

Rennie, L.J. (1998). Gender equity: Toward clarification and a research direction for science teacher education. *Journal of Research in Science Teaching, 35*(8), 951-961.

Rennie, L.J. (2003). "Pirates can be male or female": Investigating gender-inclusivity in a years 2/3 classroom. *Research in Science Education, 33*(4), 515-528.

Rennie, L.J. (2007). Learning science outside of school. In S.K. Abell & N.G. Lederman (Eds.), *Handbook of research on science education* (pp. 125-167). Mahwah, NJ: Lawrence Erlbaum Associates.

Rennie, L.J. & Jarvis, T. (1995a). Children's choice of drawings to communicate their ideas about technology. *Journal of Research in Science Teaching, 37,* 784-806.

Rennie, L.J. & Jarvis, T. (1995b). English and Australian children's perceptions about technology. *Research in Science & Technological Education, 13*(1), 37-52.

Rennie, L.J. & Williams, G.F. (2006). Communication about science in a traditional museum: Visitors' and staff's perceptions. *Cultural Studies of Science Education, 1*(4), 791-820.

Rennie, L.J. & Williams, G. (2006). Adults' learning about science in free-choice settings. *International Journal of Science Education, 28*(8), 871-893.

Reveles, J.M. & Brown, B.A. (2008). Contextual shifting: Teachers emphasizing students' academic identity to promote scientific literacy. *Science Education, 92*(6), 1015-1041.

Reveles, J., Cordova, R. & Kelly, G. (2002). Science literacy and academic identity formulation. *Journal of Research in Science Teaching, 41,* 1111-1144.

REFERENCES

Rice, R.E. (1998). Scientific writing: A course to improve the writing of science studemnts. *Journal of College Science Teaching, 27*(4), 267-272.

Richardson, R.C. (1984). Biology and ideology: The interpenetration of science and values. *Philosophy of Science, 51*(2), 396-420.

Richert, A.E. (1992). The content of student teachers' reflection within different structures for facilitating the reflective process. In T. Russell & H. Munby (Eds.), *Teachers and teaching: From classroom to reflection* (pp. 171-191). London: Falmer Press.

Richmond, G., Howes, E., Kurth, L. & Hazelwood, C. (1998). Connections and critique: Feminist oedagogy and science teacher education. *Journal of Research in Science Teaching, 35*, 897-918.

Richmond, G. & Kurth, L.A. (1999). Moving from outside to inside: High school students' use of apprenticeships as vehicles for entering the culture and practice of science. *Journal of Research in Science Teaching, 36*(6), 677-697.

Richmond, G. & Striley, J. (1996). Making meaning in classrooms: Social processes in small group discourse and scientific knowledge building. *Journal of Research in Science Teaching, 33*(8), 839-858.

Rigano, D.L. & Ritchie, S.M. (1995). Student disclosures of fraudulent practice in school laboratories. *Research in Science Education, 25*(4), 353-363.

Rijlaarsdam, G., Couzijn, M., Janssen, T., Braaksma, M. & Keift, M. (2006). Writing experiment manuals in science education: The impact of writing, genre, and audience. *International Journal of Science Education, 28*(2&3), 203-233.

Ritchie, S.M. & Rigano, D.L. (1996). Laboratory apprenticeship through a student research project. *Journal of Research in Science Teaching, 33*(7), 799-815.

Ritchie, S. Rigano, D. & Duane, A. (2008). Writing an ecological mystery in class: Merging genres and learning science. *International Journal of Science Education, 30*(2), 143-166.

Ritchie, S. & Tobin, K. (2001). Actions and discourses for transformative understanding in a middle school science class. *International Journal of Science Education, 23*(3), 283-299.

Rivard, L. (1994). A review of writing to learn in science: Implications for practice and research. *Journal of Research in Science Teaching, 31*(9), 969-983.

Rivard, L.P. (2004). Are language-based activities in science effective for all students, including low achievers? *Science Education, 88*(3), 420-442.

Rivard, L.P. & Straw, S.B. (2000). The effect of talk and writing on learning in science: An exploratory study. *Science Education, 84*, 566-593.

Roach, L.E. & Wandersee, J.H. (1993). Short story science. *The Science Teacher, 60*(6), 18-21.

Roach, L.E. & Wandersee, J.H. (1995). Putting people back into science: Using historical vignettes. *School Science & Mathematics, 95*(7), 365-370.

Roberts, D.A. (2007). Scientific literacy/science literacy. In S.K. Abell & N.G. Lederman (Eds.), *Handbook of research on science education* (pp. 729-780). Mahwah, NJ: Lawrence Erlbaum Associates.

Roberts, L.F. & Wassersug, R.J. (2009). Does doing scientific research in high school correlate with students staying in science? A half-century retrospective study. *Research in Science Education, 39*(2), 251-256.

Roberts, M. (1998). Indigenous knowledge and Western science: Perspectives from the Pacific. In D. Hodson (Ed.), *Science and technology education and ethnicity: An Aotearoa/New Zealand perspective* (pp. 59-75). Wellington: The Royal Society of New Zealand.

Robertson, I. (1999). Key evidence in testing hypotheses. In M. Bandiera, S. Caravita, E. Torracca & M. Vicenti (Eds.), *Research in science education in Europe* (pp. 193-200). Dordrecht: Kluwer.

Robinson, G. (2004). Developing the talents of teacher/scientists. *Journal of Secondary Gifted Eduucation, 15*(4), 155-161.

Robinson, S.L. (1990). Differing levels of superstitious beliefs among three groups: Psychiatric inpatients, churchgoers, and students. Paper presented at the Annual Meeting of the Southeastern Psychological Association, Atlanta, GA, April.

Robinson, W. (2003). Chemistry problem-solving: Symbol, macro, micro, and process aspoects. *Journal of Chemical Education, 80*, 978-982.

Rodriguez, A.J. (1998). Strategies for counterresistance: Towards sociotransformative constructivism and learning to teach science for diversity and understanding. *Journal of Research in Science Teaching, 35*(6), 589-622.

Rodriguez, M.A. & Niaz, M. (2002). How in spite of the rhetoric, history of chemistry has been ignored in presenting atomic structure in textbooks. *Science & Education, 11*, 423-441.

Roe, A. (1961). The psychology of the scientist. *Science, 134*(3477), 456-459.

Roehrig, G.H. & Luft, J.A. (2004). Constraints experienced by beginning secondary science teachers in implementing scientific inquiry lessons. *International Journal of Science Education, 26*, 3-24.

Rogoff, B. (1990). *Apprenticeship in thinking: Cognitive development in context.* Oxford: Oxford University Press.

Rohlf, J.W. (1994). *Modern physics from alpha to z.* New York: John Wiley.

Rolin, K. (2004). Why gender is a relevant factor in the social epistemology of scientific inquiry. *Philosophy of Science, 71*, 880-891.

Rolin, K. (2008). Gender and physics: Feminist philosophy and science education. *Science & Education, 17*(10), 1111-1125.

Roller, D.E. (1950). *The early development of the concepts of temperature and heat: The rise and decline of the caloric theory.* Cambridge, MA: Harvard University Press.

Rose, H. (1997). Science wars: My enemy's enemy is – only perhaps – my friend. In R. Levinson & J. Thomas (Eds.), *Science today: Problem or crisis?* (pp. 51-64). London: Routledge

Rose, S. (2001). What sort of science broadcasting do we want for the 21st century? *Science as Culture, 10*(1), 113-119.

Rosebery, A., Warren, B. & Conant, F. (1992). Appropriating scientific discourse: Findings from language minority classrooms. *Journal of the Learning Sciences, 2*, 61-94.

Rosenthal, D.B. (1993). Images of scientist: A comparison of biology and liberal studies majors. *School Science & Mathematics, 93*(4), 212-216.

Rosser, S.V. (1997). *Re-engineering female-friendly science.* New York: Teachers College Press.

Roth, K. & Anderson, C. (1988). Promoting conceptual change learning from science textbooks. In P. Ramsden (Ed.), *Improving learning: New perspectives* (pp. 109-141). London: Kogan Page.

Roth, W-M. (1994). Experimenting in a constructivist high school physics laboratory. *Journal of Research in Science Teaching, 31*(2), 197-223.

Roth, W-M. (1995). *Authentic school science: Knowing and learning in open-inquiry science laboratories.* Dordrecht: Kluwer.

Roth, W-M. & Alexander, T. (1997). The interaction of students' scientific and religious discourses: Two case studies. *International Journal of Science Education, 19*, 125-146.

Roth, W-M. & Barton, A.C. (2004). *Rethinking scientific literacy.* New York: RoutledgeFalmer.

Roth, W-M. & Lee, S. (2002). Scientific literacy as collective praxis. *Public Understanding of Science, 11*, 33-56.

Roth, W-M. & Lee, S. (2004). Science education as/for participation in the community. *Science Education, 88*, 263-291.

Roth, W-M. & Roychoudhury, A. (1993). The development of science process skills in authentic contexts. *Journal of Research in Science Teaching, 30*(2), 127-152.

Roth, W-M. & Roychoudhury, A. (1994). Physics students' epistemologies and views about knowing and learning. *Journal of Research in Science Teaching, 31*, 5-30.

Roughgarden, J. (2006). *Evolution and Christian faith: Reflections of an evolutionary biologist.* Washington, DC: Island Press.

Roup, R., Gal, S., Drayton, B. & Pfister, M. (Eds.)(1993). *LabNet: Towards a community of practice.* Hillsdale, NJ: Lawrence Erlbaum.

Rowell, J.A. & Cawthron, E.R. (1982). Images of sciences: An empirical study. *European Journal of Science Education, 4*(1), 79-94.

Rowell, P. (1997). Learning in school science: The promises and practices of writing. *Studies in Science Education, 30*, 19-56.

Royal Society, The (1985). *The public understanding of science.* London: Royal Society.

Rubba, P. (1976). *Nature of scientific knowledge scale.* Bloomington, IN: Indiana University School of Education.

Rubba, P.A. & Anderson, H.O. (1978). Development of an instrument to assess secondary school students' understanding of the nature of scientific knowledge. *Science Education, 62*(4), 449-458.

Rubin, E., Bar, V. & Cohen, A. (2003). The images of scientists and science among Hebrew- and Arabic-speaking pre-service teachers in Israel. *International Journal of Science Education, 25*(7), 821-846.

Rudd, J.A., Greenbowe, T.J., Hand, B.M. & Legg, M.L. (2001). Using the science writing heuristic to move toward an inquiry-based laboratory curriculum. *Journal of Chemical Education, 78*, 1680-1686.

Rudolph, J.L. (2000). Reconsidering the "nature of science" as a curriculum component. *Journal of Curriculum Studies, 32*(3), 403-419.

Rudolph, J.L. (2008). Commentary on "inquiry, activity, and epistemic practice". In R.A. Duschl & R.E. Grandy (Eds.), *Teaching scientific inquiry: Recommendations for research and implementation* (pp. 118-122). Rotterdam/Taipei: Sense.

Rudolph, J.L. & Stewart, J. (1998). Evolution and the nature of science: On the historical discord and its implications for education. *Journal of Research in Science Teaching, 35*(10), 1069-1089.

Ruse, M. (1979a). *Sociobiology: Sense or nonsense?* Dordrecht: Reidel.

Ruse, M. (1979b). *The Darwinian revolution.* Chicago, IL: University of Chicago Press.

Ruse, M. (1988b). *Philosophy of biology today.* Albany, NY: SUNY Press.

Ruse, M. (Ed.) (1988a). *But is it science? The philosophical question in the creation/evolution controversy.* Buffalo, NY: Prometheus Books.

Ruse, M. (1989). Making use of creationism: A case study for the philosophy of science classroom. *Studies in Philosophy and Education, 10*(1), 81-92.

Ruse, M. (Ed.) (1998). *Philosophy of biology.* Amherst, NY: Prometheus Books.

Ruse, M. (2000). *The evolution wars: A guide to the debates.* Santa Barbara, CA: ABC-CLIO.

Ruse, M. (2005). *The evolution-creation struggle.* Cambridge, MA: Harvard University Press.

Ruse, M. & Hull, D.L. (Eds.) (1998). *The philosophy of biology.* Oxford: Oxford University Press.

Rushton, J.P. (1997). Racial research and the final solution: Review essay. *Society, 34*(3), 78-82.

Rushton, J.P. (2000). Race, evolution and behaviour: A life history perspective. Port Huron, MI: Charles Darwin Research Institute.

Rushton, J.P. & Jensen, A.R. (2005). Thirty years of research on race differences in cognitive ability. *Psychology, Public Policy and Law, 11*, 235-294.

Russ, R.S., Scherr, R.E., Hammer, D. & Mikeska, J. (2008). Recognizing mechanistic reasoning in student scientific inquiry: A framwork for discourse analysis developed from philosophy of science. *Science Education, 92*(3), 499-525.,

Russell, C.A. (2002). The conflict of science and religion. In G.B. Ferngren (Ed.), *Science & religion: An historical introduction* (pp. 3-12). Baltimore, MD: The Johns Hopkins University Press.

Ryan, A (2008). Indigenous knowledge in the science curriculum: Avoiding neo-colonialism. *Cultural Studies of Science Education, 3*(3), 663-702.

Ryan, A.G. (1987). High school graduates' beliefs about science-technology-society. IV. The characteristics of scientists. *Science Education, 71*, 489-510.

Ryan, A.G. & Aikenhead, G.S. (1992). Students' preconceptions about the epistemology of science. *Science Education, 76*, 559-580.

Ryder, J. (2009). Enhancing engagement with science/technology-related issues. In A.T. Jones & M.J. deVries (Eds.), *International handbook of research and development in technology education* (pp. 349-360). Rotterdam/Taipei: Sense.

Ryder, J. & Leach, J. (2000). Interpreting experimental data: The views of upper secondary school and university science students. *International Journal of Science Education, 22*(10), 1069-1084.

Ryder, J. & Leach, J. (2008). Teaching about the epistemology of science in upper secondary schools: An anlysis of teachers' classroom talk. *Science & Education, 17*, 289-315.

Ryder, J., Leach, J. & Driver, R. (1999). Undergraduate science students' images of science. *Journal of Research in Science Teaching, 36*(2), 201-219.

Saari H. & Viiri, J. (2003). A research-based teaching sequence for teaching the concept of modelling to seventh-grade students. *International Journal of Science Education, 25*(11), 1333-1352.

Sadler, P.M., Gould, R.R., Leiker, P.S., Antonucci, P.R.A., Kimbeck, R., Deutsch, F.S., Hoffman, B., Dussault, M., Contos, A. Brecher, K. & French, L. (2001). Micro observatry net: A network of automated remote telescopes dedicated to educational use. *Journal of Science Education and Technology, 10*(1), 39-55.

Sadler, T.D. (2004). Informal reasoning regarding socioscientific issues: A critical review of research. *Journal of Research in Science Teaching, 41*(5), 513-536.

Sadler, T.D. (2006). Promoting discourse and argumentation in science teacher education. *Journal of Science Teacher Education, 17*, 323-346.

Sadler, T.D. & Donnelly, L.A. (2006). Socioscientific argumentation: The effects of content knowledge and morality. *International Journal of Science Education, 28*(12), 1463-1488.

Sadler, T.D. & Fowler, S.R. (2006). A threshold model of content knowledge transfer for socioscientific argumentation. *Science Education, 90*(6), 986-1004.

Sadler, T.D. & Zeidler, D.L. (2004). Students' conceptualizations of the nature of science in response to a socioscientific issue. *International Journal of Science Education, 26*(4), 387-409.

Sadler, T.D. & Zeidler, D.L. (2005). The significance of content knowledge for informal reasoning regarding socioscientific issues: Applying genetics knowledge to genetic engineering issues. *Science Education, 89*(1), 71-93.

Sagan, C. (1988). Introduction. In S.W. Hawking (1988). *A brief history of time: From the big bang to black holes* (pp. xiii-xiv). London: Bantam Books.

Sagan, C. (1995). *The demon-haunted world: Science as a candle in the dark.* New York: Random House.

Samarpungavan, A. (1992). Children's judgements in theory choice tasks: Scientific rationality in childhood. *Cognition, 45*(1), 1-32.

Samarpungavan, A. (1997). Children's beliefs about the boundaries of knowledge: Examples from biology learning. Paper presented at the Annual Meeting of the American Educational Research Association. Chicago, IL: March.

Samarpungavan, A., Mantzicopoulos, P. & Patrick, H. (2008). Learning science through inquiry in kindergarten. *Science Education, 92*(5), 868-908.

Samarpungavan, A., Westby, E.L. & Bodner, G.M. (2006). Contextual epistemic development in science: A comparison of chemistry students and research chemists. *Science Education, 90,* 468-495.

Sampson, V. & Clark, D. (2006). Assessment of argument in science education: A critical review of the literature. In *Proceedings of the 7th International Conference of the Learning Sciences* (pp. 655-661). Bloomington, IN, June. International Society of the Learning Sciences (www.isls.org).

Sampson, V. & Clark, D.B. (2008). Assessment of the ways students generate arguments in science education: Current perspectives and recommendations for future directions. *Science Education, 92*(3), 447-472.

Sandoval, W.A. (2005). Understanding students' practical epistemologies and their influence on learning through inquiry. *Science Education, 89*(4), 634-656.

Sandoval, W.A. & Millwood, K.A. (2005). The quality of students' use of evidence in written scientific explanations. *Cognition and Instruction, 23*(1), 23-55.

Sandoval, W.A. & Millwood, K.A. (2008). What can argumentation tell us about epistemology? In S. Erduran & M.P. Jiménez-Aleixandre (Eds.), *Argumentation in science education: Perspectives from classroom-based research* (pp. 71-90). Dordrecht: Kluwer.

Sandoval, W. & Morrison, K. (2003). High school students' ideas about theories and theory change after a biological inquiry unit. *Journal of Research in Science Teaching, 40,* 369-392.

Sandoval, W.A. & Reiser, B.J. (2004). Explanation-driven inquiry: Integrating conceptual and epistemic scaffolds for scientific inquiry. *Science Education, 88*(3), 345-372.

Santayana, G. (1955b). *The sense of beauty: Being the outline of aesthetic theory.* New York: Dover (originally published 1896).

Santayana, G. (1955a). Reason in common sense (Chapter 10: Flux and constancy in human nature). In G. Santayana, *The life of reason,* One volume edition. New York: Charles Scribner.

Sardar, Z. (1988). *The touch of Midas: Science, values and environment in Islam and the West.* Petaling Jaya (Malaysia): Pelanduk Publications.

Sardar, Z. (1989). *Explorations in Islamic science.* London:Mansell.

Scantlebury, K. (1998). An untold story: Gender, constructivism & science education. In W.W. Cobern (Ed.), *Socio-cultural perspectives on science education* (pp. 99-120). Dordrecht: Kluwer.

Scantlebury, K. & Baker, D. (2007). Gender issues in science education research: Remembering where the difference lies. In S.K. Abell & N.G. Lederman (Eds.), *Handbook of research on science education* (pp. 257-285). Mahwah, NJ: Lawrence Erlbaum Associates.

Scanlon, E., Morris, E., Di Paolo, T. & Copper, M. (2002). Contemporary approaches to learning science: technologically-mediated practical work. *Studies in Science Education, 38,* 73-114.

Schailles, M. & Lembens, A. (2002). Student learning by research. *Journal of Biological Education, 37*(1), 13-17.

Schank, R. (1990). *Tell me a story: A new look at real and artificial memory.* New York: Macmillan.

Scharmann, L.C., Smith, M.U., James, M.C. & Jensen, M. (2005). Explicit reflective nature of science instruction: Evolution, intelligent design, and umbrellaology. *Journal of Science Teacher Education, 16,* 27-41.

Schauble, L. (1996). The development of scientific reasoning in knowledge-rich contexts. *Developmental Psychology, 32*(1), 102-119.

Schauble, L. (2008). Three questions about development. In R.A. Duschl & R.E. Grandy (Eds.), *Teaching scientific inquiry: Recommendations for research and implementation* (pp. 50-56). Rotterdam/Taipei: Sense.

Schauble, L., Glaser, R., Duschl, R., Schulze, S. & John, J. (1995). Students' understanding of the objectives and procedures of experimentation in the science classroom. *Journal of the Learning Sciences, 4*(2), 131-166.

Schauble, L., Glaser, R., Raghavan, K. & Reiner, M. (1991a). Causal models and experimentation strategies in scientific reasoning. *Journal of the Learning Sciences, 1*(2), 201-238.

Schauble, L., Klopfer, L.E. & Raghavan, K. (1991b). Students' transition from an engineering model to a scientific model of experimentation. *Journal of Research in Science Teaching, 28*(9), 859-882.

Schenkel, L.A. (2002). Hands on and feet first: Linking high-ability students to marine scientists. *Journal of Secondary Gifted Education, 13*(4), 173-191.

Scherz, Z. & Oren, M. (2006). How to change students' images of science and technology. *Science Education, 90*(6), 965-985.

Schibeci, R. & Lee, L. (2003). Portrayals of science and scientists, and "science for citizenship". *Research in Science & Technological Education, 21*(2), 177-192.

Schiebinger, I. (1999). *Has feminism changed science.* Cambridge, MA: Harvard University Press.

Schilpp, P.A. (1951). *Albert Einstein: Philosopher-scientist.* New York: Tudor.

Schnotz, W. & Lowe, R. (2003). External and internal representations in multimedia learning. *Learning and Instruction, 13,* 117-123.

Schommer, M. & Walker, K. (1997). Epistemological beliefs and valuing school: Considerations for college admissions and retentions. *Research in Higher Education, 38*(2), 173-186.

Schroeder, M., McKeough, A., Graham, S., Stock, H. & Bisanz, G. (2009). The contribution of trade books to early science literacy: In and out of school. *Research in Science Education, 39,* 231-250.

Schwab, J.J. (1962). The teaching of science as enquiry. In J.J. Schwab & P.F. Brandwein (Eds.), *The teaching of science* (pp. 3-103). Cambridge, MA: Harvard University Press.

Schwartz, A.T. (1995a). The paradoxical Dr Priestley. *Chemistry Review, 4*(4), 18-21.

Schwartz, A.T. (1995b). Three decades of humanizing the scientists and simonizing the humanist: Some retrospective reflections. In F. Finley, D. Allchin, D. Rhees & S.Fifield (Eds.), *Proceedings of the Third International History, Philosophy and Science Teaching Conference, Vol. 2* (pp. 1031-1041). Minneapolis, MN: University of Minnesota Press.

Schwartz, R.S. & Crawford, B.A. (2004). Authentic scientific inquiry as context for teaching nature of science. In L.B. Flick & N.G. Lederman (Eds.), *Scientific inquiry and nature of science: Implications for teaching, learning, and teacher education* (pp. 331-355). Dordrecht: Kluwer.

Schwartz, R.S. & Lederman, N.G. (2002). "It's the nature of the beast": The influence of knowledge and intentions on learning and teaching the nature of science. *Journal of Research in Science Teaching, 39*(3), 205-236.

Schwartz, R.S. & Lederman, N.G. (2006). Exploring contextually-based views of NOS and scientific inquiry: What scientists say [tentativeness, creativity, scientific method, and justification]. Paper presented at Annual Meeting of the National Association for Research in Science Teaching, San Francisco, CA, April.

Schwartz, R. & Lederman, N. (2008). What scientists say: Scientists' views of nature of science and relation to science context. *International Journal of Science Education, 30*(6), 721-771.

Schwartz, R.S., Lederman, N.G., & Crawford, B.A. (2004). Developing views of nature of science in an authentic context: An explicit approach to bridging the gap between nature of science and scientific inquiry. *Science Education, 88,* 610-645.

Shwarz, B.B., Neuman, Y., Gil, J. & Ilya, M. (2003). Construction of collective and individual knowledge in argumentative activity. *Journal of the Learning Sciences, 12*(2), 219-256.

Scott, E.C. (2005). *Evolution vs. creationism: An introduction.* Berkeley, CA: University of California Press.

Scott, P. (1998). Teacher talk and meaning making in science classrooms: A Vygotskian analysis and review. *Studies in Science Education, 32*, 45-80.

Scott, P.H., Mortimer, E.F. & Aguiar, O.G. (2006). The tension between authoritative and dialogi discourse: A fundamental characteristic of meaning making interactions in high school science lessons. *Science Education, 90*(4), 605-631.

Secondary School Curriculum Review (SSCR) (1987). *Better science: Working for a multicultural society*. London: Heinemann/Association for Science Education.

Seiler, G. (2001). Reversing the "standard" direction: Science emerging from the lives of African American students. *Journal of Research in Science Teaching, 38*, 1000-1014.

Seldon, S. (2000). Eugenics and the social construction of merit, race and disability. *Journal of Curriculum Studies, 32*(2), 235-252.

Select Committee on Science and Technology, House of Lords (2000). Science and society. 3rd Report, Session 1999-2000. London: HMSO.

Sensevy, G., Tiberghien, A., Santini, J., Laubé, S. & Griggs, P. (2008). An epistemological approach to modeling: Cases studies and implications for science teaching. *Science Education, 92*(3), 424-446.

Séré, M-G. (2002). Towards renewed research questions from the outcomes of the European project labwork in science education. *Science Education, 86*(5), 624-644.

Service, R.F. (2002). Bell labs fires star physicist found guilty of forging data. *Science, 298*, 30-31.

Service, R.F. (2003). More of Bell labs physicist papers retracted. *Science, 299*, 31.

Shahn, E. (1988). On science literacy. *Educational Philosophy and Theory, 20*(2), 42-52.

Shamos, M.H. (1993). STS: A time for caution. In R.E. Yager (Ed.), *The science, technology, society movement* (pp. 65-72). Washington, DC: National Science Teachers Association.

Shanks, N. (2004). *God, the Devil and Darwin: A critique of intelligent design theory*. New York: Oxford University Press.

Shapiro, B.L. (1994). *What children bring to light: A constructivist perspective on children's learning in science*. New York: Teachers College Press.

Shapiro, B.L. (1996). A case study of change in elementary student teacher thinking during an independent investigation in science: Learning about the "face of science that does not yet know". *Science Education, 80*, 535-560.

Shapiro, B. (1998). An approach to consider semiotic messages of school science learning culture. *Journal of Science Teacher Education, 9*, 221-240.

Sharkawy, A. (2006). *An inquiry into the use of stories about scientists from diverse sociocultural backgrounds in broadening grade one students' images of science and scientists*. Toronto: Unpublished PhD thesis, University of Toronto.

Sharp, R. & Green, A. (1975). *Education and social control: A study of progressive primary education*. London: Routledge.

She, H-C. (1995). Elementary and middle school students' image of science and scientists related to current science textbooks in Taiwan. *Journal of Science Education and Technology, 4*(4), 283-294.

She, H-C. (1998). Gender and grade level differences in Taiwan students' stereotypes of science and scientists. *Research in Science & Technological Education, 16*(2), 125-135.

Sheffield, L.J. (1997). From Doogie Howser to dweebs' Or how we went in search of Bobby Fischer and found that we are dumb and dumber. *Mathematics Teaching in the Middle School, 2*(6), 376-379.

Shen, B.S.P. (1975). Scientific literacy and the public understanding of science. In S.B. Day (Ed.), *The communication of scientific information* (pp. 44-52). Basel: Karger.

Shen J. & Confrey, J. (2007). From conceptual change to transformative modelling: A case study of an elementary teacher in learning astronomy. *Science Education, 91*(6), 948-956.

Shepardson, D.P. & Britsch, S.J. (2001). The role of children's journals in elementary school science activities. *Journal of Research in Science Teaching, 38*(1), 43-69.

Shepherd, L. (1993). *Lifting the veil: The feminine face of science*. Boston, MA: Shambala Publications.

Sherborne, T. (2004). Immediate inspiration: Ready-made resources for teaching ethics. *School Science Review, 86*(315), 67-72.

Sherin, B.L. (2001). How students understand physics equations. *Cognition and Instruction, 19*(4), 479-541.

Shibley, I.A. (2003). Using newpapers to examine nature of science. *Science & Education, 12*(7), 691-702.

Shipman, H.L., Brickhouse, N.W., Dagher, Z. & Letts, W.J. (2002). Changes in student views of religion and science in a college astronomy course. *Science Education, 86*, 526-547.

Shulman, L.S. (1986). Those who understand: Knowledge growth in teaching. *Educational Researcher, 15*, 4-14.

Shulman, L.S. (1987). Knowledge and teaching: Foundations of the new reform. *Harvard Educational Review, 57*, 1-22.

Siegel, H. (1991). The rationality of science, critical thinking, and science education. In M.R. Matthews (Ed.), *History, philosophy and science teaching: Selected readings* (pp. 45-62). Toronto: OISE Press.

Simon, S., Erduran, S. & Osborne, J. (2006). Learning to teach argumentation: Research and development in the science classroom. *International Journal of Science Education, 28*(2-3), 235-260.

Simon, S., Naylor, S., Keogh, B., Maloney, J. & Downing, B. (2008). Puppets promoting engagement and talk in science. *International Journal of Science Education, 30*(9), 1229-1248.

Simmonneaux, L. (2001). Roleplay or debate to promote students' argumentation and justification on an issue in animal tramsgenesis. *International Journal of Science Education, 23*(9), 903-927.

Simpson, J.S. & Parsons, E.C. (2009). African American perspectives and informal science educational experiences. *Science Education, 93*(2), 293-321.

Sinatra, G. (2005). The "warming trend" in conceptual change research: The legacy of Paul R. Pintrich. *Educational Psychologist, 40*(2), 107-115.

Sinatra, G. & Pintrich, P.R. (2003). *Intentional conceptual change*. Mahwah, NJ: Lawrence Erlbaum.

Sinatra, G., Southerland, S.A., McConaughy, F. & Demastes, J.W. (2003). Intentions and beliefs in students' understanding and acceptance of biological evolution. *Journal of Research in Science Teaching, 40*, 510-528.

Sinclair, J. & Coulthard, M. (1975). *Towards an analysis of discourse*. Oxford: Oxford University Press.

Slater, M.H. (2008). How to justify teaching false science. *Science Education, 92*(3), 526-542.

Slezak, P. (1994). Sociology of scientific knowledge and scientific education: Part 1. *Science & Education, 3*, 265-294.

Sloan, C.P. (1999). Feathers for T Rex? *National Geographic, 196*(5), 98-107.

Smit, J.J.A. & Finegold, M. (1995). Models in physics: Perceptions held by final-year prospective physical science teachers studying at South African universities. *International Journal of Science Education, 17*(5), 621-634.

Smith, C.L., Maclin, D., Houghton, C. & Hennessey, M.G. (2000). Sixth grade students' epistemologies of science: The impact of school science experiences on epistemological development. *Cognition and Instruction, 18*(3), 349-422.

Smith, C.L. & Wenk, L. (2006). Relations among three aspects of first-year college students' epistemologies of science. *Journal of Research in Science Teaching, 43*(8), 747-785.

Smith, G.R. (2001). Guided literature explorations: Introducing students to the primary literature. *Journal of College Science Teaching, 30*(7), 465-469.

Smith, M.U. & Scharmann, L.C. (1999). Defining versus describing the nature of science: A pragmatic analysis for classroom teachers and science educators. *Science Education, 83*(4), 493-509.

Smith, M.U. & Scharmann, L.C. (2008). A multi-year program developing an explicit reflective pedagogy for teaching pre-service teachers the nature of science by ostention. *Science & Education, 17*(2&3), 219-248.

Smolicz, J.J. & Nunan, E.E. (1975). The philosophical and sociological foundations of science education: The demythologizing of school science. *Studies in Science Education, 2*, 101-143.

Snir, J., Smith, C.L. & Raz, G. (2003). Linking phenomena with competing underlying models: A software tool for introducing students to the particulate nature of matter. *Science Education, 87*(6), 794-830.

Snively, G. & Corsiglia, J. (2001). Discovering indigenous science: Implications for science education. *Science Education, 85*, 6-34.

Snow, C.P. (1962). *The two cultures and the scientific revolution*. Cambridge: Cambridge University Press.

Snyder, B.R. (1973). *The hidden curriculum*. Cambridge, MA: MIT Press.

Sodian, B., Zaitchik, D. & Carey, S. (1991). Young children's differentiation of hypothetical beliefs from evidence. *Child Development, 62*, 753-766.

Solomon, J. (1995). Higher level understanding of the nature of science. *School Science Review, 76*(276), 15-22.

Solomon, J. (1997). Girls' science education: choice, solidarity and culture. *International Journal of Science Education, 19*, 407-417.

Solomon, J. (1998). About argument and discussion. *School Science Review, 80*(291), 57-62.

Solomon, J. (2002). Science stories and science texts: What can they do for our students? *Studies in Science Education, 37*, 85-106.

Solomon, J., Duveen, J., Scott, L. & McCarthy, S. (1992). Teaching about the nature of science through history: Action research in the classroom. *Journal of Research in Science Teaching, 29*, 409-421.

Solomon, J., Duveen, J. & Scott, L. (1994a). Pupils' images of scientific epistemology. *International Journal of Science Education, 16*, 361-373.

Solomon, J. (with Duveen, J. & Scott, L.) (1994b). *Exploring the nature of science*. Hatfield: Association for Science Education.

Solomon, J., Scott, L. & Duveen, J. (1996). Large-scale exploration of pupils' understanding of the nature of science. *Science Education, 80*(5), 493-508.

Solomon, M. (2001). *Social empiricism*. Cambridge, MA: MIT Press.

Somerville, J. (1941). Umbrellaology, or, methodology in social science. *Philosophy of Science, 8*(4), 557-566.

Song, J. & Kim, K.-S. (1999). How Korean students see scientists: The images of the scientist. *International Journal of Science Education, 21*(9), 957-977.

Songer, N.B. (2007). Digital resources versus cognitive tools: A discussion of learning with technology. In S.K. Abell & N.G. Lederman (Eds.), *Handbook of research on science education* (pp. 471-491). Mahwah, NJ: Lawrence Erlbaum Associates.

Songer, N.B., Devaul, H., Hester, P., Crouch, S., Kam, R., Lee, H.S. & Lee, S.Y. (1999). *Kids as global scientists: Weather! An eight-week inquiry curriculum for middle school atmospheric science*. Ann Arbor, MI: University of Michigan.

Songer, N.B., Lee, H-S. & McDonald, S. (2003). Research towards an expanded understanding of inquiry science beyond one idealized standard. *Science Education, 87*(4), 490-516.

Songer, N.B. & Linn, M.C. (1991). How do students' views of science influence knowledge integration? *Journal of Research in Science Teaching, 28*, 761-784.

Sorensen, R.A. (1992). *Thought experiments*. Oxford: Oxford University Press.

Southerland, S.A. (2000). Epistemic universalism and the shortcomings of curricular multicultural science education. *Science & Education, 9*, 289-307.

Southerland, S.A., Gess-Newsome, J. & Johnston, A. (2003). Portraying science in the classroom: The manifestation of scientists' beliefs in classroom practice. *Journal of Research in Science Teaching, 40*(7), 669-691.

Southerland, S.A., Johnston, A. & Sowelll, S. (2006). Describing teachers' conceptual ecologies for the nature of science. *Science Education, 90*(5), 874-906.

Spier-Dance, L., Mayer-Smith, J., Dance, N. & Khan, S. (2005). The role of student-generated analoies in promoting conceptual understanding for undergraduate chemistry students. *Research in Science & Technological Education, 23*(2), 163-178.

Spiro, R.J., Feltovich, P.J., Coulson, R.L. & Anderson, D.K. (1989). Multiple analogies for complex concepts: Antidotes for analogy-induced misconception in advanced knowledge acquisition. In S. Vosniadou & A. Ortony (Eds.), *Similarity and anaological reasoning* (pp. 498- 531). Cambridge: Cambridge University Press.

Stanley, A. (1983). Women hold up two-thirds of the sky: Notes for a revised history of technology. In J. Rothschild (Ed.), *Machina ex Dea: Feminist perspectives on technology* (pp. 5-22). Oxford: Pergamon Press.

Stanley, W.B. & Brickhouse, N.W. (1994). Multiculturalism, universalism, and science education. *Science Education, 78*(4), 387-398.

Stanley, W.B. & Brickhouse, N.W. (2001). Teaching sciences: The multicultural question revisited. *Science Education, 85*(1), 35-49.

Stannard, R. (2001). Communicating physics through story. *Physics Education, 36*(1), 30-34.

413

Staples, R. & Heselden, R. (2001). Science teaching and literacy, Part 1: Writing. *School Science Review, 83*(303), 35-46.

Staples, R. & Heselden, R. (2002). Science teaching and literacy, Part 3: Speaking and listening, spelling and vocabulary. *School Science Review, 84*(306), 83-95.

Stathopoulou, C. & Vosnidou, S. (2007). Conceptual change in physics and physics-related epistemological beliefs: A relationship under scrutiny. In S. Vosnidou, A. Baltas & X. Vamvaloussi (Eds.), *Re-framing the problem of conceptual change in learning and instruction* (pp. 145-163). Amsterdam: Elsevier.

Stavy, R. (1991). Using analogy to overcome misconceptions about conservation of matter. *Journal of Research in Science Teaching, 28*(4), 305-313.

Stein, N.L. & Albro, E.R. (2001). The origin and nature of arguments: Studies in conflict understanding, emotion, and negotiation. *Discourse Processes, 32*(2&3), 113-133.

Steinberg, M.S., Brown, D.E. & Clement, J. (1990). Genius is not immune to persistent misconceptions: Conceptual difficulties impeding Isaac Newton and contemporary physics students. *International Journal of Science Education, 12*(3), 265-273.

Stenhouse, L. (1975). *An introduction to curriculum research and development.* London: Heinemann.

Stewart, G. (2005). Māori in the science curriculum: Developments and possibilities. *Educational Philosophy & Theory, 37*(6), 851-870.

Stewart, G. (2007). Narrative pedagogy for teaching and learning about the nature of pūtaiao (Māori-medium science). *New Zealand Journal of Educational Studies, 42*(1&2), 129-142.

Stewart, J. & Rudolph, J.L. (2001). Considering the nature of scientific problems when designing science curricula. *Science Education, 85*(3), 207-222.

Stinner, A. (1989). The teaching of physics and the contexts of inquiry: From Aristotle to Einstein. *Science Education, 73*(5), 591-605.

Stinner, A. (1992). Contextual teaching in physics: From science stories to large-context problems. *Alberta Journal of Science Education, 26*(1), 20-29.

Stinner, A. (1994a). Providing a contextual base and a theoretical structure to guide the teaching of high school physics. *Physics Education, 29*, 375-383.

Stinner, A. (1994b). The story of force: From Aristotle to Einstein. *Physics Education, 29*(2), 77-85.

Stinner, A. (1995). Contextual settings, science stories, and large context problems: Toward a more humanistic science education. *Science Education, 79*(5), 555-581.

Stinner, A. (1996). Providing a contextual base and a theoretical structure to guide the teaching of science from early years to senior years. *Science & Education, 5*, 247-266.

Stinner, A. (2006). The large context problem (LCP) approach. *Interchange, 37*(1&2), 19-30.

Stinner, A., McMillan, B.A., Metz, D., Jilek, J.M. & Klassen, S. (2003). The renewal of case studies in science education. *Science & Education, 12*(7), 617-643.

Stinner, A. & Teichmann, J. (2003). Lord Kelvin and the-age-of-the-earth debate: A dramatization. *Science & Education, 12*, 213-228.

Stinner, A. & Williams, H. (1993). Conceptual change, history and science stories. *Interchange, 24*(1&2), 87-103.

Stinner, A. & Williams, H. (1998). History and philosophy of science in the science curriculum. In B.J. Fraser & K.G. Tobin (Eds.), *The international handbook of science education* (pp. 1027-1045). Dordrecht: Kluwer.

Stolberg, T. (2007). The religio-scientific frameworks of pre-service primary teachers: An analysis of their influence on their teaching of science. *International Journal of Science Education, 29*(7), 909-930.

Stone, C.A. (1993). What is missing in the metaphor of scaffolding? In E.A. Forman, N. Minick & C.A. Stone (Eds.), *Contexts for learning: Sociocultural dynamics in children's development* (pp. 169-183). New York: Oxford University Press.

Stone, C.A. (1998). The metaphor of scaffolding: Its utility for the field of learning disabilities. *Journal of Learning Disabilities, 31*, 344-364.

Strube, P. (1990). Narrative in science education. *English in Education, 24*(1), 53-60.

Suchting, W.A. (1995). The nature of scientific thought. *Science & Education, 4*, 1-22.

Sumrall, W.J. (1995). Reasons for the perceived images of scientists by race and gender of students in grades 1-7. *School Science & Mathematics, 95*(2), 83-90.

Suppe, F. (1998). The structure of a scientific paper. *Philosophy of Science, 65*, 381-405.

Sutherland, D. (2005). Resiliency and collateral learning in science in some students of Cree ancestry. *Science Education, 89*(4), 595-613.

Sutherland, D. & Dennick, R. (2002). Exploring culture, language and the perception of the nature of science. *International Journal of Science Education, 24*(1), 1-25.

Sutton, C. (1989). Writing and reading in science: The hidden messages. In R. Millar (Ed.), *Doing science: Images of science in science education* (pp. 137-159). Lewes: Flamer.

Sutton, C. (1992). *Words, science and learning.* Buckingham: Open University Press.

Sutton, C. (1996). Beliefs about science and beliefs about language. *International Journal of Science Education, 18*, 1-18.

Sutton, C. (1998). New persectives on language in science. In B.J. Fraser & K.G. Tobin (Eds.), *International handbook of science education* (pp. 27-38). Dordrecht: Kluwer.

Symington, D. & Spurling, H. (1990). The "draw a scientist test": Interpreting the data. *Research in Science & Technological Education, 8*(1), 75-77.

Taber, K.S. (2003). Mediating mental models of metals: Acknowledging the priority of the learner's prior learning. *Science Education, 87*(5), 732-756.

Taber, K.S. (2008). Towards a curricular model of the nature of science. *Science & Education, 17*(2&3), 179-218.

Tairab, H.H. (2001). How do pre-service and in-service science teachers view the nature of science and technology. *Research in Science & Technological Education, 19*(2), 235-250.

Takao, A.Y. & Kelly, G.J. (2003). Assessment of evidence in university students' scientific writing. *Science & Education, 12*(4), 341-363.

Tal, T. & Morag, O. (2007). School visits to natural history museums: Teaching or enriching? *Journal of Research in Science Teaching, 44*(5), 747-769.

Tal, T. & Steiner, L. (2006). Patterns of teacher-museum staff relationships: School visits to the educatiuonal centre of a science museum. *Canadian Journal of Science, Mathematics and Technology Education, 6*(1), 25-46.

Tala, S. (2009). Unified view of science and technology for education: technoscience and technoscoence education. *Science & Education, 18*(3&4), 275-298.

Talbot, C. (2000). Ideas and evidence in science. *School Science Review, 82*(298), 13-22.

Tao, P-K. (2002). A study of students' focal awareness when studying science stories designed for fostering understanding of the nature of science. *Research in Science Education, 32*(1), 97-120.

Tao, P-K. (2003). Eliciting and developing junior secondary students' understanding of the nature of science through a peer collaboration instruction in science stories. *International Journal of Science Education, 25*(2), 147-171.

Taylor, A.R., Jones, G.M., Broadwell, B. & Oppewal, T. (2008). Creativity, inquiry, or accountability? Scientists' and teachers' perceptions of science education. *Science Education, 92*(6), 1058-1075.

Taylor, C. (1994). The politics of recognition. In A. Gutmann (Ed.), *Multiculturalism* (pp. 25-73). Princeton, NJ: Princeton University Press.

Taylor, C.A. (1996). *Defining science: A rhetoric of demarcation.* Madison, WI: University of Wisconsin Press.

Taylor, I., Barker, M. & Jones, A. (2003). Promoting mental model building in astronomy education. *International Journal of Science Education, 25*(10), 1205-1225.

Taylor, J.A. & Dana, T.M. (2003). Secondary school physics teachers' conceptions of scientific evidence: An exploratory case study. *Journal of Research in Science Teaching, 40*(8), 721-736.

Tenopir, C. & King, D.W. (2004). *Communication patterns of engineers.* Hoboken, NJ: John Wiley.

Thagard, P.R. (1980). Why astrology is a pseudoscience. In E.D. Klemke, R. Hollinger & A.D. Kline (Eds.), *Introductory readings in the philosophy of science* (pp. 66-73). Buffalo, NY: Prometheus.

Thagard, P. (1992). Analogy, explanation, and education. *Journal of Research in Science Teaching, 29*(6), 537-544.

Thiele, R.B. & Treagust, D.F. (1991). Using analogies in secondary chemistry teaching. *Australian Science Teachers Journal, 37*(2), 10-14.

Thiele, R.B. & Treagust, D.F. (1994a). An interpretive examination of high school chemistry teachers' analogical explanations. *Journal of Research in Science Teaching, 31*(3), 227-242.

Thiele, R.B., & Treagust, D.F. (1994b). The nature and extent of analogies in secondary chemistry textbooks. *Instructional Science, 22*, 61-74.

Thier, M. & Daviss, B. (2002). *The new science literacy: Using language skills to help students learn science*. Portsmouth, NH: Heinemann.

Thoermer, C. & Sodian, B. (2002). Science undergraduates' and graduates' epistemologies of science: The notion of interpretive frameworks. *New Ideas in Psychology, 20*, 263-283.

Thomas, E.L. & Robinson, H.A. (1972). *Improving reading in every class: A sourcebook for teachers*. Boston, MA: Allyn & Bacon.

Thomas, G. & Durant, J. (1987). Why should we promote the public understanding of science? In M. Shortland (Ed.), *Scientific literacy papers* (pp. 1-14). Oxford: Oxford University Department for External Studies.

Thomas, G. & McRobbie, C.J. (1999). Using metaphors to probe students' conceptions of learning. *International Journal of Science Education, 21*(6), 667-685.

Thomas, G. & McRobbie, C.J. (2001). Using a metaphor for learning to improve students' metacognition in the chemistry classroom. *Journal of Research in Science Teaching, 35*(2), 222-259.

Thomas, J. (1997). Informed ambivalence: Changing attitudes to the public understanding of science. In R. Levinson & J. Thomas (Eds.), *Science today: Problem or crisis?* (pp. 137-150). London: Routledge

Thomas, J.A., Pederson, J.E. & Finson, K. (2001). Validation of the draw-a-science-teacher-test checklist (DASTT-C): Exploring mental models and teacher beliefs. *Journal of Science Teacher Education, 12*(4), 295-310.

Thompson, D., Praia, J. & Marques, L. (2000). The importance of history and epistemology in the designing of earth science curriculum materials for general science education. *Research in Science & technological Education, 18*(1), 45-62.

Tipler, F. (1995). *The physics of immortality: Modern cosmology, God and the resurrection of the dead*. London: Macmillan.

Tobin, K. (1990). Changing metaphors and beliefs: A master switch for teaching. *Theory into Practice, 29*(2), 122-127.

Tobin, K. (1993). Referents for making sense of science teaching. *International Journal of Science Education, 15*(3), 241-254.

Tobin, K. & McRobbie, C.J. (1997). Beliefs about the nature of science and the enacted curriculum. *Science & Education, 6*(4), 335-371.

Tobin, K. & Tippens, D. (1996). Metaphors as seeds of conceptual change and the improvement of science teaching. *Science Education, 80*(6), 711-730.

Todd, P. (1984). Teaching evolutionary theory as general education. *Journal of General Education, 35*(4), 212-228.

Toh, K-A., Boo, H-K. & Yeo, K-H. (1997). Open-ended investigations: Performances and effects of pre-training. *Research in Science & Technological Education, 27*(1), 131-140.

Toplis, R. (2007). Evaluating science investigations at ages 14–16: Dealing with anomalous results. *International Journal of Science Education, 29*(2), 127-150.

Toth, E.E. & Klahr, D. (2001). "We are doomed!" Children's scientific reasoning during experimentation. Paper presented at the Annual Meeting of the American Educarional Research Association, Seattle, WA.

Toth, E.E., Klahr, D. & Chen, Z. (2000). Bridging research and practice: A research-based classroom intervention for teaching experimentation skills to elementary school children. *Cognition and Instruction, 18*(4), 423-459.

Toth, E.E., Suthers, D.D. & Lesgold, A.M. (2002). "Mapping to know": The effects of representational guidance and reflective assessment on scientific inquiry. *Science Education, 86*(2), 264-286.

Towe, K.M. (2004). The Vinland Map ink is not medieval. *Analytical Chemistry, 76*, 863-865.

Townes, C. (1995). *Making waves*. Woodbury, NY: American Institute of Physics.

Toulmin, S.E. (1958). *The uses of argument*. Cambridge: Cambridge University Press.

Toulmin, S.E. (1963). *The philosophy of science*. London: Hutchinson.

Tran, L.U. (2007). Teaching science in museums: The pedagogy and goals of museum educators. *Science Education, 91*(2), 278-297.

Treagust, D.F. (1993). The evolution of an approach for using analogies in teaching and learning science. *Research in Science Education, 23*, 293-301.

Treagust, D. (2007). General instructional methods and strategies. In S.K. Abell & N.G. Lederman (Eds.), *Handbook on research in science education* (pp. 373-391). Mahwah, NJ: Lawrence Erlbaum.

Treagust, D.F., Chittleborough, G. & Mamiala, T.L. (2002). Students' understanding of the role of scientific models in learning science. *International Journal of Science Education, 24*(4), 357-368.

Treagust, D.F., Chittleborough, G. & Mamiala, T.L. (2004). Students' understanding of the descriptive and predictive nature of teaching models in organic chemistry. *Research in Science Education, 34*(1), 1-20.

Treagust, D., Duit, R., Joslin, P. & Lindauer, I. (1992). Science teachers' use of analogies: Observations from classroom practice. *International Journal of Science Education, 14*(4), 413-422.

Treagust, D.F., Harrison, A.G. & Venville, G.J. (1998). Teaching science effectively with analogies: An approach for preservice and inservice teacher education. *Journal of Science Teacher Education, 9*(2), 85-101.

Trumbull, D.J., Bonney, R., Bascom, D. & Cabral, A. (2000). Thinking scientifically during participation in a citizen-science project. *Science Education, 84*(2), 265-275.

Trumbull, D.J., Scarano, G. & Bonney, R. (2006). Relations among two teachers' practices and beliefs, conceptualizations of the nature of science, and their implementation of student independent inquiry projects. *International Journal of Science Education, 28*(14), 1717-1750.

Tsai, C-C. (1998). An analysis of scientific epistemological beliefs and learning orientations of Taiwanese eighth graders. *Science Education, 82*(4), 473-489.

Tsai, C-C. (1999). "Laboratory exercises help me to memorize the scientific truths": A study of eight graders' scientific epistemtological views and learning in laboratory activities. *Science Education, 83*, 654-674.

Tsai, C-C. (2000). Relationships between student scientific elistemological beliefs and perceptions of constructivist learning environments. *Educational Research, 42*(2), 193-205.

Tsai, C-C. (2002). Nested epistemologies: Science teachers' beliefs of teaching, learning and science. *International Journal of Science Education, 24*(8), 771-783.

Tsai, C-C. (2003). The interplay between philosophy of science and the practice of science education. *Curriculum and Teaching, 18*(1), 27-43.

Tsai, C-C. (2007). Teachers' scientific epistemological views: The coherence with instruction and students' views. *Science Education, 91*(2), 222-243.

Tsai, C-C. & Liu, S-Y. (2005). Developing a multi-dimensional instrument for assessing students' epistemological views toward science. *International Journal of Science Education, 27*(13), 1621-1638.

Tsaparlis, G. (1997). Atomic and molecular structure in chemical education. *Journal of Chemical Education, 74*(8), 922-925.

Tucker-Raymond, E., Varelas, M., Pappas, C.C., Korzah, H.A. & Wentland, A. (2007). "They probably aren't named Rachel": Young children's scientist identities as emergent multimodal narratives. *Cultural Studies of Science Education, 1*(3), 559-592.

Tyler, R.W. (1949). *Basic principles of curriculum and instruction.* Chicago, IL: University of Chicago Press.

Tytler, R. & Peterson, S. (2004). From "try it and see" to strategic exploration: Characterizing young children's scientific reasoning. *Journal of Research in Science Teaching, 41*(10, 94-118.

Tytler, R. & Peterson, S. (2005). A longitudinal study of children's developing knowledge and reasoning in science. *Research in Science Education, 35*(1), 63-98.

UNESCO (1993). International forum on scientific and technological literacy for all. Final Report. Paris: UNESCO.

Ungar, G., Desiderio, D.M. & Parr, W. (1972). Isolation, identification and synthesis of a specific-behaviour-inducing brain peptide. *Nature, 238*(5361), 198-202.

van der Valk, T., van Driel, J.H. & de Vos, W. (2007). Common characteristics of models in present-day scientific practice. *Research in Science Education, 37*(4), 469-488.

van Driel, J.H. & Verloop, N. (1999). Teachers' knowledge of models and modelling in science. *International Journal of Science Education, 21*(11), 1141-1153.

van Driel, J.H. & Verloop, N. (2002). Experienced teachers' knowledge of teaching and learning models and modelling in science education. *International Journal of Science Education, 24*(12), 1255-1272.

van Eijck. M. (2007). Towards authentic forms of knowledge. *Cultural Studies of Science Education, 2*(3), 606-613.

417

van Eijck, M. & Roth, W-M. (2008). Representations of scientists in Canadian high school and college textbooks. *Journal of Research in Science Teaching, 45*(9), 1059-1082.

van Waterschoot van der Gracht, W.A.J.M. (Ed.) (1928). *Theory of continental drift: A symposium on the orign and movement of land masses both inter-continental and intra-continental, as proposed by Alfred Wegener.* Proceedings of Conference, November 15, 1926. Tulsa, OK: American Association of Petroleum Geologists.

van Zee, E.H., Iwasyk, M.,. Kurose, A., Simpson, D. & Wild, J. (2001). Stiudent and teacher questioning during conversations about science. *Journal of Research in Science Teaching, 38*(2), 159-190.

van Zee, E.H. & Minstrell, J. (1997). Reflective discourse: Developing shared understandings in a physics classroom. *International Journal of Science Education, 19*(2), 209-228.

Varelas, M., Becker, J., Luster, B. & Wenzel, S. (2002). When genres meet: Inquiry into a sixth-grade urban science class. *Journal of Research in Science Teaching, 39*(7), 579-605.

Varelas, M., Pappas, C.C. & Rife, A. (2006). Exploring the role of intertextuality in concept construction: Urban second graders make sense of evaporation, bioling, and condensation. *Journal of Research in Science Teaching, 43*(7), 637-666.

Varelas, M. Pappas, C.C., Kane, J.M., Arsenault, A., Hankes, J. & Cowan, B.M. (2008). Urban primary-grade children think and talk science: Curricular and instructional practices that nurture prticipation and argmentation. *Science Education, 92*(1), 65-95.

Veel, R. (1997). Learning how to mean – Scientifically speaking: Apprenticeship into scientific discourse in the secondary school. In F. Christie & J.R. Martin (Eds.), *Genre and institutions: Social processes in the workplace and school* (pp. 161-195). London: Cassell.

Veerman, A. (2003). Constructive discussions through electronic dialogue. In J. Andriessen, M. Baker & D. Suthers (Eds.), *Arguing to learn: Confronting cognitions in computer-supported collaborative learning environments* (pp. 117-143). Dordrecht: Kluwer.

Venville, G., Bryer, L. & Treagust, D. (1994). Training students in the use of analogies to enhance understanding in science. *Australian Science Teachers Journal, 40*(2), 60-68.

Venville, G.J. & Treagust, D.F. (1996). The role of analogies in promoting conceptual change in biology. *Instructional Science, 24*, 295-320.

Vhurumuku, E., Holtman, L., Mikalsen, O. & Kolsto, S.D. (2006). An investigation of Zimbabwe high school chemistry students' laboratory work-based images of the nature of science. *Journal of Research in Science Teaching, 43*(2), 127-149.

Vílchez-González, J.M. & Perales Palacios, F.J. (2006). Image of science in cartoons and its relationship with the image in comics. *Physics Education, 41*, 240-249.

von Aufschnaiter, C., Erduran, S., Osborne, J. & Simon, S. (2008). Arguing to learn and learning to argue: Case studies of how students' argumentation relates to their scientific knowledge. *Journal of Research in Science Teaching, 45*(1), 101-131.

Vygotsky, L.S. (1962). *Thought and language.* Cambridge, MA: MIT Press.

Vygotsky, L.S. (1978). *Mind in society: The development of higher psychological processes.* Cambridge, MA: Harvard University Press.

Wade, N. (1976). IQ and heredity: Suspicion of fraud beclouds classic experiment. *Science, 194*(4268), 916-919.

Wajcman, J. (1995). Feminist theories of technology. In S. Jasanoff, G.E. Markle, J.C. Petersen & T. Pinch (Eds.), *Handbook of science and technology studies* (pp. 189-204). London: Sage.

Walker, K.A. & Zeidler, D.L. (2003). Students' understanding of the nature of science and their reasoning on socioscientific issues: A web-based learning inquiry. Paper presented at the annual meeting of the National Association for Research in Science Teaching, Philadelphia, PA, March.

Wallace, C.S. (2004). Framing new research in science literacy and language use: Authenticity, multiple discourses, and the "third space". *Science Education, 88*(6), 901-914.

Wallace, C.S., Tsoi, M.Y., Calkin, J. & Darley, W.M. (2003). Learning from inquiry-based laboratories in nonmajor biology: An interpretive study of the relationships among inquiry experience, epistemologies, and conceptual growth. *Journal of Research in Science Teaching, 40*, 986-1024.

Wallace, R.M., Kupperman, J., Krajcik, J. & Soloway, E. (2000). Science on the web: Students on-line in a sixth-grade classroom. *Journal of the Learning Sciences, 9*, 75-104.

Wallsgrove, R. (1980). The masculine face of science. In L. Birke, W. Faulkner, S. Best, D. Janson-Smith & K. Overfield (Eds.), *Alice through the microscope* (pp. 228-240). London: Virago.

Walsh, D. (1992). A mythology of reason: The persistence of pseudo-science in the modern world. In S.A. McKnight (Ed.), *Science, pseudo-science, and utopianism in early modern thought* (pp. 141-166). Columbia. MO: University of Missouri Press.

Walton, D.N. (1996). *Argumentation schemes for presumptive reasoning*. Mahwah, NJ: Lawrence Erlbaum.

Wandersee, J.H. (1985). Can the history of science help science educators anticipate students' misconceptions? *Journal of Research in Science Teaching, 23*(7), 581-597.

Wandersee, J.H. (1990). On the value and use of the history of science in teaching today's science: Constructing historical vignettes. In D.E. Herget (Ed.), *More history and philosophy of science in science teaching* (pp. 278-283). Tallahassee, FA: Florida State University.

Wandersee, J.H. (1999). Designing an image-based biology test. In J.J. Mintzes, J.H. Wandersee & J.D. Novack (Eds.), *Assessing science understanding: A human constructivist view* (pp. 129-140). San Diego, CA: Academic Press.

Wang, H.A. & Marsh, D.D. (2002). Science instruction with a humanistic twist: Teachers' perception and practice in using the history of science in their classrooms. *Science & Education, 11*(2), 169-189.

Ward, A. (1986). Magician in a white coat. *School Science Review, 68*(243), 348-350.

Warren, B., Ballenger, C., Ogonowski, M., Rosebery, A.S. & Hudicourt-Barnes, J. (2001). Rethinking diversity in learning science: The logic of everyday sense-making. *Journal of Research in Science Teaching, 38*(5), 529-552.

Watanabe, M. (1974). The conception of nature in Japanese culture. *Science, 183*(4122), 279-282.

Waters-Adams, S. (2006). The relationship between understanding of the nature of science and practice: The influence of teachers' beliefs about education, teaching and learning. *International Journal of Science Education, 28*(8), 919-944.

Watson, J.D. (1980). *The double helix: A personal account of the discovery of the structure of DNA*, edited by G.S. Stent. New York: Norton.

Watson, J.D. & Crick, F.H.C. (1953). Molecular structure of nucleic acids: A structure for deoxyribose nucleic acid. *Nature, 171*, 737-738.

Watts, M. (Ed.) (2000). *Creative trespass*. Hatfield (UK): Association for Science Education

Watts, M. (2001). Science and poetry: passion vs. prescription in school science. *International Journal of Science Education, 23*(2), 197-208.

Webb, M.E. (1994). Beginning computer-based modeling in primary schools. *Computers in Education, 22*(1), 129-144.

Webb, P. & Treagust, D.F. (2006). Using exploratory talk to enhance problem-solving and reasoning skills in grade-7 science classrooms. *Research in Science Education, 36*(4), 381-401.

Wegener, A. (1915). *The origin of continents and oceans*. Trans. J. Biram. Republished in 1966 by Dover: New York.

Wegner, G.P. (1991). Schooling for a new mythos: race, anti-semitism and the curriculum materials of a Nazi race educator. *Paedogogica Historica, 27*(2), 189-213.

Weinberg, S. (1994). Response to Steve Fuller. *Social Studies of Science, 24*, 748-750.

Weiner, J.S. (1955). *The Piltdown forgery*. Oxford: Oxford University Press.

Weinstein, M. (2006). Slash writers and guinea pigs as models for scientific multiliteracy. *Educational Theory and Philosophy, 38*(5), 583-599.

Weitkamp, E. & Burnet, F. (2007). The chemedian brings laughter to the chemistry classroom. *International Journal of Science Education, 29*(15), 1911-1929.

Welch, W.W. (1969a). *Science process inventory*. Minneapolis, MN: University of Minnesota.

Welch, W.W. (1969b). *Wisconsin inventory of science processes*. Madison, WI: University of Wisconsin Scientific Literacy Research Center.

Wellington, J. (1991). Newspaper science, school science: Friends or enemies? *International Journal of Science Education, 13*(4), 363-372.

Wellington, J. (1993). Using newspapers in science education. *School Science Review, 74*(268), 47-52.

Wellington, J. (2001). What is science education for? *Canadian Journal of Science, Mathematics and Technology Education, 1*(1), 23-38.

Wellington, J. & Britto, J. (2004). Learning science through ICT at home. In M. Braund & M. Reiss (Eds.), *Learning science outside the classroom* (pp. 207-223). London: RoutledgeFalmer.

REFERENCES

Wellington, J. & Osborne, J. (2001). *Language and literacy in science education*. Buckingham: Open University Press.

Wells, G. (1986). *The meaning makers: Children learning language and using language to learn*. London: Hodder & Stoughton.

Wells, G. (1990). Talk about text: Where literacy is learned and taught. *Curriculum Inquiry, 20*, 369-405.

Wells, G. (1993). Text, talk and inquiry: Schooling as semiotic apprenticeship. Paper presented at the Language in Education Conference, Hong Kong.

Wells, G. (Ed.)(1999a). *Dialogic inquiry: Towards a sociocultural practice and theory of education*. Cambridge: Cambridge University Press.

Wells, G. (1999b). Putting a tool to different uses: A reevaluation of the IRF sequence. In G. Wells (Ed.), *Dialogic inquiry: Towards a sociocultural practice and theory of education* (pp. 167-208). Cambridge: Cambridge University Press.

Wells, G. & Chang-Wells, G.L. (1992). *Constructing knowledge together: Classrooms as centers of inquiry and literacy*. Portsmouth, NH: Heinemann.

Welzel, M. & Roth, W-M. (1998). Do interviews really assess students' knowledge. *International Journal of Science Education, 20*, 25-44.

Wenger, E. (1998). *Communities of practice: Learning, meaning and identity*. Cambridge: Cambridge University Press.

Wertheim, M. (1995). *Pythagoras' trousers: God, physics, and the gender wars*. New York: Random House.

West, L.H.T. & Pines, A.L. (1983). How "rational" is rationality? *Science Education, 67*, 37-39.

White, B. & Fredericksen, J. (1998). Inquiry, modeling, and metacognition: Making science accessible to all students. *Cognition and Instruction, 16*, 3-118.

White, B. & Fredericksen, J. (2000). Metacognitive facilitation: An approach to making scientific inquiry accessible to all. In J. Minstrell & E.H. van Zee (Eds.), *Inquiry into inquiry learning and teaching in science* (pp. 331-370). Washington, DC: American association for the Advancement of Science.

Whitfield, J. & Dalton, R. (2002). Ink analysis raises storm over Viking map. *Nature, 418*, 574.

Wible, J.R. (1992). Fraud in science: An economic approach. *Philosophy of the Social Sciences, 22*(1), 5-27.

Wilber, K. (1995). *Sex, ecology, spirituality, the spirit of evolution*. Boston, MA: Shambhala.

Wilkinson, J. & Ward, M. (1997). A comparative study of students' and their teachers' perceptions of laboratory work in secondary schools. *Research in Science Education, 27*(4), 599-610.

Williams, J. (1993). Fakes, frauds and fluorine. *School Science Review, 74*(268), 41-46.

Williams, J.D. (2002). Ideas and evidence in science: The portrayal of scientists in GCSE textbooks. *School Science Review, 84*(307), 89-101.

Williams, P.A. (2001). *Doing without Adam and Eve: Sociobiology and original sin*. Minneapolis, MN: Fortress Press.

Wilson, E.O. (1975). *Sociobiology: The new synthesis*. Cambridge, MA: Harvard University Press.

Wilson, E.O. (1981). *Genes, mind and culture: The coevolutionary process*. Cambridge, MA: Harvard University Press.

Wilson, E.O. (2004). *On human nature*. Cambridge, MA: Harvard University Pres.. •

Wilson, E.O. (2006). *The creation: An appeal to save life on earth*. New York: Norton.

Winberg, T.M. & Berg, C.A. (2007). Students' cognitive focus during a chemistry laboratory exercise: Effects of a computer-simulated prelab. *Journal of Research in Science Teaching, 44*, 1108-1133.

Winchester, I. (1990). Thought experiments and conceptual revision. *Studies in Philosophy and Education, 10*, 73-80.

Winchester, I. (1993). "Science is dead. We have killed it you and I': how attacking the presuppositional structures or our scientific age can doom the interrogation of nature. *Interchange, 24*, 191-198.

Winchester, I. (2006). Large context problems and their applications to education: Some contemporary examples. *Interchange, 37*(1&2), 7-17.

Windschitl, M. M. (2003). Inquiry projects in science teacher education: What can investigative experiences reveal about teacher thinking and eventual classroom practice? *Science Education, 87*(1), 112-143.

Windschitl, M. (2004). Caught in the cycle of reproducing folk theories of "inquiry": How pre-service teachers reproduce the discourse and practices of an atheoretical scientific method. *Journal of Research in Science Teaching, 41*(5), 481-512.

Windschitl, M. (2008). Our challenges in disrupting popular folk theories of "doing science". In R.A. Duschl & R.E. Grandy (Eds.), *Teaching scientific inquiry: Recommendations for research and implementation* (pp. 292-303). Rotterdam/Taipei: Sense.

Windschitl, M. & Andre, &. (1998). Using computer simulations to enhance conceptual changes: The roles of contructivist instruction and student epistemological beliefs. *Journal of Research in Science Teaching, 35*(2), 145-160.

Windschitl, M. & Thompson, J. (2006). Transcending simple forms of school science investigation: The impact of pre-service instruction on teachers' understandings of model-based inquiry. *American Educational Research Journal, 43*(4), 783-835.

Windschitl, M., Thompson, J. & Braaten, M. (2008). Beyond the scientific method: Model-based inquiry as a new paradigm of preference for school science investigations. *Science Education, 92*(5), 941-967.

Winn, W., Stahr, F., Sarason, C., Fruland, R., Oppenheimer, P. & Lee, Y. (2006). Learning oceanography from a computer simulation compared with direct experience at sea. *Journal of Research in Science Teaching, 43*, 25-42.

Winner, L. (1977). *Autonomous technology: Technics-out-of-control as a theme in political thought.* Cambridge, MA: MIT Press.

Wolfe, L.F. (1989). Analysing science lessons: A case study with gifted children. *Science Education, 73*(1), 87-100.

Wolpert, L. (1992). *The unnatural nature of science.* London: Faber & Faber.

Wolpert, L. (1997). In praise of science. In R. Levinson & J. Thomas (Eds.), *Science today: Problem or crisis?* (pp. 9-21). London: Routledge.

Wong, E.D. (1993a). Self-generated analogies as a tool for constructing and evaluating explanations of scientific phenomena. *Journal of Research in Science Teaching, 30*, 367-380.

Wong, E.D. (1993b). Understanding the generative capacities of analogies as a tool for explanation. *Journal of Research in Science Teaching, 30*(10), 1259-1272.

Wong, S.L. & Hodson, D. (2009). From the horse's mouth: What scientists say about scientific investigation and scientific knowledge. *Science Education, 93*(1), 109-130.

Wong, S.L. & Hodson, D. (in press). More from the horse's mouth: What scientists say about science as a social practice. *International Journal of Science Education.*

Wong, S.L., Hodson, D., Kwan, J. & Yung, B.H.W. (2008). Turning crisis into opportunity: Enhancing student-teachers' understanding of nature of science and scientific inquiry through a case study of the scientific research in severe acute respiratory syndrome. *International Journal of Science Education, 30*(11), 1417-1439.

Wong, S.L., Kwan, J., Hodson, D. & Yung, B.H.W. (2009). Turning crisis into opportunity: Nature of science and scientific inquiry as illustrated in the scientific research on severe acute respiratory syndrome. *Science & Education, 18*(1), 95-118.

Wong, S.L., Yung, B.H.W., Cheng, M.W., Lam, K.L. & Hodson, D. (2006). Setting the stage for developing pre-service science teachers' conceptions of good science teaching: The role of classroom videos. *International Journal of Science Education, 28*(1), 1-24.

Wood, A. & Lewthwaite, B. (2008). Māori science education in Aotearoa New Zealand. He pūtea whakarawe: Aspirations and realities. *Cultural Studies of Science Education, 3*(3), 625-662.

Wood, D., Bruner, J.S. & Ross, G. (1976). The role of tutoring in problem solving. *Journal of Child Psychology and Psychiatry, 17*, 89-100.

Wood, K., Lapp, D. & Flood, J. (1992). *Guiding reading through text: A review of study guides.* Newark, DE: International Reading Association.

Wood, R.W. (1904). The N-rays. *Nature, 70*, 530-531.

Woodruff, E. & Meyer, K. (1997). Explanations from intra- and inter-group discourse: Students building knowledge in the science classroom. *Research in Science Education, 27*(1), 25-39.

Woody, A. (1995). The explanatoiry power of our models: A philosophical analysis with some implications for science education. In F. Finley, D. Allchin, D. Rhees & S. Fifield (Eds.), *Proceedings of the Third International History, Philosophy and Science Teaching Conference, Vol. 2* (pp. 1295-1304). Minneapolis, MN.

REFERENCES

Woolnough, B.E. (1996). On the fruitful compatibility of religious education and science. *Science & Education, 5*(2), 175-183.

Woolnough, B.E. (2000). Appropriate practical work for school science – Making it practical and making it science. In J. Minstrell & E.H. van Zee (Eds.), *Inquiry into inquiry learning and teaching in science* (pp. 434-446). Washington, DC: American Association for the Advancement of Science.

Wray, D. & Lewis, M. (1995). *Developing children's non-fiction writing.* Leamington Spa: Scholastic.

Wray, D. & Lewis, M. (1997). *Extending literacy: Children reading and writing non-fiction.* London: Routledge.

Wray, D. & Lewis, M. (2000). Extending literacy: Learning and teaching. In M. Lewis & D. Wray (Eds.), *Literacy in the secondary school* (pp. 17-24). London: David Fulton.

Wu, H-K. & Krajcik, J.S. (2006). Exploring middle school students' use of inscriptions in project-based science classrooms. *Science Education, 90*(5), 852-873.

Wynne, B. (1991). Knowledges in context. *Science, Technology and Human Values, 16*(1), 111-121.

Wynne, B. (1995). Public understanding of science. In S. Jasanoff, G.E. Markle, J.C. Petersen & T. Pinch (Eds.), *Handbook of science and technology studies* (pp. 361-388). London: Sage.

Wynne, B. (1996). Misunderstood misunderstandings: Social identities and public uptake of science. In A. Irwin & B. Wynne (Eds.), *Misunderstanding science? The public reconstruction of science and technology* (pp. 19-46). Cambridge: Cambridge University Press.

Xing, X. (2000). Response to feathers for T Rex? *National Geographic, 197*(3), unnumbered pages (March).

Xing, X., Zhou, Z. & Wang, X. (2000). The smallest known non-avian theropod dinosaur. *Nature, 408*(Dec. 7th), 705-708.

Yang, C.N. (1982). Beauty and theoretical physics. In D.W. Curtin (Ed.), *The aesthetic dimension of science. 1980 Nobel conference* (pp. 25-40). New York: Philosophical Library.

Yanowitz, K.L. (2001). Using analogies to improve elementary school students' inferential reasoning about scientific concepts. *School Science and Mathematics, 101*(3), 133-142.

Yarden, A., Brill, G. & Falk, H. (2001). Primary literature as a basis for a high-school biology curriculum. *Journal of Biological Education, 35*(4), 190-195.

Yates, G.C.R. & Chandler, M. (2000). Where have all the sceptics gone? Patterns of New Age beliefs and anti-scientific attitudes in preservice primary teachers. *Research in Science Education, 30*(4), 377-387.

Yerrick, R.K., Pedersen, J.E. & Arnason, J. (1998). "We're just spectators": A case study of science teaching, and classroom management. *Science Education, 82*(6), 619-648.

Yilmaz-Tuzun, O. & Topcu, M.S. (2008). Relationships among preservice science teachers' episte-mological beliefs, epistemological world views, and self-efficacy beliefs. *International Journal of Science Education, 30*(1), 65-85.

Yip, D. (2001). Promoting the development of a conceptual change model of science instruction in prospective secondary biology teachers. *International Journal of Science Education, 23*, 755-770.

Yip, D-N. (2006). Using history to promote understanding of nature of science in science teachers. *Teaching Education, 17*(2), 157-166.

Yore, L.D., Bisanz, G.L. & Hand, B.M. (2003). Examining the literacy component of science literacy: 25 years of language arts and science research. *International Journal of Science Education, 25*(6), 689-725.

Yore, L.D., Florence, M.K., Pearson, T.W. & Weaver, A.J. (2006). Written discourse in scientific communities: A conversation with two scientists about their views of science, use of language, role of writing in doing science, and compatibility between their epistemic views and language. *International Journal of Science Education, 28*(2&3), 109-141.

Yore, L.D., Hand, B.M. & Florence, M.K. (2004). Scientists' views of science, models of writing, and science writing practice. *Journal of Research in Science Teaching, 41*(4), 338-369.

Yore, L.D., Hand, B.M. & Prain, V. (2002). Scientists as writers. *Science Education, 86*, 672-692.

Yore, L.D. & Treagust, D.F. (2006). Current realities and future possibilities: Language and science literacy – Empowering research and informing instruction. *International Journal of Science Education, 28*(2-3), 291-314.

Young, M.F.D. (1976). The schooling of science. In G. Whitty & M.F.D. Young (Eds.), *Explorations in the politics of school knowledge* (pp. 47-61). Driffield: Nafferton Books.

422

Young, R.M. (1987). Racist society, racist science. In D. Gill & L. Levidow (Eds.), *Anti-racist science teaching* (pp. 16-42). London: Free Association Books.

Yousif, A.F. (2001). Islamic science: Controversies, influence and future possibilities for science education in Brunei Darussalam. Paper presented at the Energising Science, Mathematics and Technical Education for All Conference, University of Brunei, Brunei Darussalam, May.

Yoxen, E. (1989). Play up and play the game: A simulation of hard and soft fraud in science. In M. Shortland & A. Warwick (Eds.), *Teaching the history of science* (pp. 185-200). Oxford: Blackwell.

Yung, B.H.W. (2001). Three views of fairness in a school-based assessment scheme of practical work in biology. *International Journal of Science Education, 23*, 985-1005.

Yung, B.H.W., Wong, A.S.L., Cheng, M.W., Hui, C.S. & Hodson, D. (2007). Benefits of progressive video reflection on pre-service teachers' conceptions of good science teaching. *Research in Science Education, 37*(3), 239-259.

Yung, B.H.W., Wong, A.S.L., Cheng, M.W., Lo, F.Y. & Hodson, D. (2008). Preparing students for examinations: A divided view between teachers' and students' conceptions of good science teaching. In Y.J. Lee & A.l. Tan (Eds.), *Science education at the nexus of theory and practice* (pp. 181-202). Rotterdam/Taipei: Sense.

Zeidler, D.L. (1997). The central role of fallacious thinking in science education. *Science Education, 81*(4), 483-495.

Zeidler, D.L. & Lederman, N.G. (1987). The effect of teachers' language on students' conceptions of the nature of science. Paper presented at the Annual Meeting of the National Association for Research in Science Teaching, Washington, DC.

Zeidler, D.L., Osborne, J., Erduran, S., Simon, S. & Monk, M. (2003). The role of argument during discourse about sociosvcientific issues. Zeidler, D.L. (Ed.), *The role of moral reasoninbg on socioscientific issues and discourse in science education* (pp. 97-116). Dordrecht: Kluwer.

Zeitoun, H.H. (1984). Teaching scientific analogies: A proposed model. *Research in Science & Technological Education, 2*(2), 107-125.

Zembal-Saul, C., Munford, D., Crawford, B., Friedrichsen, P. & Land, S. (2002). Scaffolding pre-service sciencev teachers' evidence-based arguments during an investigation of natural selection. *Research in Science Education, 32*(4), 437-463.

Zhou, Z., Clarke, J.A. & Zhang, F. (2002). Archaeoraptor's better half. *Nature, 420*, 253-344.

Ziman, J. (1998). Essays on science and society: Why must scientists become more ethically sensitive than they used to be? *Science, 282*(5395), 1813-1814.

Ziman, J. (2000). *Real science: What it is, and what it means.* Cambridge: Cambridge University Press.

Zimmerman, C. (2000). The development of scientific reasoning skills. *Developmental Review, 20*, 99-149.

Zion, M., Slezak, M., Shapira, D., Link, E., Bashan, N., Brumer, M., Orian, T., Nussinowitz, R., Court, D., Agrest, B., Mendelovici, R. & Valanides, N. (2004). Dynamic, open inquiry in biology learning. *Science Education, 88*(5), 728-753.

Zohar, A. & Nemet, F. (2002). Fostering students' knowledge and argumentation skills through dilemmas in human genetics. *Journal of Research in Science Teaching, 39*(1), 35-62.

Zoller, U., Ben-Chaim, D., Pentimalli, R. & Borsese, A. (2000). The disposition towards critical thinking of high school and university science students: An inter-intra Israeli-Italian study. *International Journal of Science Education, 22*, 571-582.

INDEX

425